T0214822

Lecture Notes in Computer Science   **9470**

Commenced Publication in 1973
Founding and Former Series Editors:
Gerhard Goos, Juris Hartmanis, and Jan van Leeuwen

Advanced Research in Computing and Software Science

Subline of Lecture Notes in Computer Science

More information about this series at http://www.springer.com/series/7409

Evangelos Markakis · Guido Schäfer (Eds.)

Web and Internet Economics

11th International Conference, WINE 2015
Amsterdam, The Netherlands, December 9–12, 2015
Proceedings

 Springer

Editors
Evangelos Markakis
Athens University of Economics
   and Business
Athens
Greece

Guido Schäfer
CWI and VU Amsterdam
Amsterdam
The Netherlands

ISSN 0302-9743   ISSN 1611-3349 (electronic)
Lecture Notes in Computer Science
ISBN 978-3-662-48994-9   ISBN 978-3-662-48995-6 (eBook)
DOI 10.1007/978-3-662-48995-6

Library of Congress Control Number: 2015955360

LNCS Sublibrary: SL3 – Information Systems and Applications, incl. Internet/Web, and HCI

Printed on acid-free paper

Springer-Verlag GmbH Berlin Heidelberg is part of Springer Science+Business Media
(www.springer.com)

Preface

This volume contains the papers and extended abstracts presented at WINE 2015: The 11th Conference on Web and Internet Economics, held during December 9–12, 2015, at Centrum Wiskunde & Informatica (CWI), Amsterdam, The Netherlands.

Over the past decade, researchers in theoretical computer science, artificial intelligence, and microeconomics have joined forces to tackle problems involving incentives and computation. These problems are of particular importance in application areas like the Web and the Internet that involve large and diverse populations. The Conference on Web and Internet Economics (WINE) is an interdisciplinary forum for the exchange of ideas and results on incentives and computation arising from these various fields. WINE 2015 built on the success of the Conference on Web and Internet Economics series (named Workshop on Internet and Network Economics until 2013), which was held annually from 2005 to 2014.

WINE 2015 received 142 submissions, which were all rigorously peer-reviewed and evaluated on the basis of originality, soundness, significance, and exposition. The Program Committee decided to accept only 38 papers, reaching a competitive acceptance ratio of 27%. To accommodate the publishing traditions of different fields, authors of accepted papers could ask that only a one-page abstract of the paper appears in the proceedings. Among the 38 accepted papers, the authors of eight papers opted for the publication as a one-page abstract. The program also included four invited talks by Michal Feldman (Tel Aviv University, Israel), Paul Goldberg (University of Oxford, UK), Ramesh Johari (Stanford University, USA), and Paul Milgrom (Stanford University, USA). In addition, WINE 2015 featured three tutorials on December 9: "Some Game-Theoretic Aspects of Voting" by Vincent Conitzer (Duke University, USA), "Polymatroids in Congestion Games" by Tobias Harks (University of Augsburg, Germany), and "Polymatrix Games: Algorithms and Applications" by Rahul Savani (University of Liverpool, UK).

We would like to thank DIAMANT, EATCS, Facebook, Google, Microsoft, NWO, and Springer for their generous financial support and CWI for hosting the event. We thank Susanne van Dam for her excellent local arrangements work and Irving van Heuven van Staereling, Chris Wesseling, and Niels Nes for their help with the conference website and online registration site.

We also acknowledge the work of the 37 members of the Program Committee, Krzysztof Apt for organizing the tutorials, Anna Kramer and Alfred Hofmann at Springer for helping with the proceedings, and the EasyChair paper management system.

December 2015

Evangelos Markakis
Guido Schäfer

Organization

Program Committee Chairs

Evangelos Markakis   Athens University of Economics and Business, Greece
Guido Schäfer   CWI and VU Amsterdam, The Netherlands

Program Committee

Saeed Alaei	Cornell University, USA
Elliot Anshelevich	Rensselaer Polytechnic Institute, USA
Moshe Babaioff	Microsoft Research, Israel
Liad Blumrosen	Hebrew University of Jerusalem, Israel
Yang Cai	McGill University, Canada
Ioannis Caragiannis	University of Patras, Greece
Giorgos Christodoulou	University of Liverpool, UK
José Correa	Universidad de Chile, Chile
Ulle Endriss	University of Amsterdam, The Netherlands
Michal Feldman	Tel Aviv University, Israel
Dimitris Fotakis	National Technical University of Athens, Greece
Vasilis Gkatzelis	Stanford University, USA
Laurent Gourvès	CNRS and University of Paris Dauphine, France
Tobias Harks	University of Augsburg, Germany
Patrick Hummel	Google Inc., California, USA
Nicole Immorlica	Microsoft, New England, USA
Max Klimm	Technical University Berlin, Germany
Piotr Krysta	University of Liverpool, UK
Ron Lavi	Israel Institute of Technology, Israel
Stefano Leonardi	Sapienza University of Rome, Italy
Brendan Lucier	Microsoft, New England, USA
Aranyak Mehta	Google Research, California, USA
Hervé Moulin	University of Glasgow, UK
Ahuva Mu'alem	Holon Institute of Technology, Israel
Evdokia Nikolova	The University of Texas at Austin, USA
Hans Peters	Maastricht University, The Netherlands
Georgios Piliouras	Singapore University of Technology and Design, Singapore
Maria Polukarov	University of Southampton, UK
Qi Qi	Hong Kong University of Science and Technology, SAR China
Alexander Skopalik	Paderborn University, Germany

Nicolas Stier-Moses   Facebook Core Data Science, California, USA
Vasilis Syrgkanis   Microsoft, New York, USA
Troels Bjerre Sørensen   IT-University of Copenhagen, Denmark
Orestis Telelis   University of Piraeus, Greece
Carmine Ventre   Teesside University, UK
Adrian Vetta   McGill University, Canada
Onno Zoeter   Booking.com, Amsterdam, The Netherlands

Additional Reviewers

Abraham, Ittai
Adamczyk, Marek
Adamic, Lada
Andreoli, Jean-Marc
Aziz, Haris
Babichenko, Yakov
Bade, Sophie
Balkanski, Eric
Balseiro, Santiago
Barman, Siddharth
Basu, Soumya
Bienkowski, Marcin
Biro, Peter
Bjelde, Antje
Bonifaci, Vicenzo
Bousquet, Nicolas
Brânzei, Simina
Busch, Costas
Caskurlu, Bugra
Chen, Jing
Chen, Zhou
Cheung, Yun Kuen
Colini Baldeschi, Riccardo
Cominetti, Roberto
Cord-Landwehr, Andreas
De Jong, Jasper
De Keijzer, Bart
Deligkas, Argyrios
Devanur, Nikhil
Dhamal, Swapnil
Disser, Yann
Drees, Maximilian
Dütting, Paul
Eden, Alon
Ehlers, Lars

Elbassioni, Khaled
Fadaei, Salman
Faliszewski, Piotr
Fanelli, Angelo
Fatima, Shaheen
Fearnley, John
Feldotto, Matthias
Ferraioli, Diodato
Filos-Ratsikas, Aris
Friedler, Ophir
Fu, Hu
Gairing, Martin
Gerding, Enrico
Giannakopoulos, Yiannis
Gollapudi, Sreenivas
Golomb, Iddan
Gonczarowski, Yannai A.
Gravin, Nick
Guo, Mingyu
Guzman, Cristobal
Göbel, Oliver
Haghpanah, Nima
Hoefer, Martin
Holzman, Ron
Hou, Mark
Huang, Chien-Chung
Jain, Shweta
Jalaly Khalilabadi, Pooya
Kanellopoulos, Panagiotis
Kanoria, Yash
Karakostas, George
Kash, Ian
Kesselheim, Thomas
Khani, Reza
Koenemann, Jochen

Kollias, Konstantinos
Korula, Nitish
Krimpas, George
Kroer, Christian
Kutty, Sindhu
Kyropoulou, Maria
Lambert, Nicolas
Lang, Jérôme
Lattanzi, Silvio
Lenzner, Pascal
Lev, Omer
Lianeas, Thanasis
Lobel, Ilan
Lykouris, Thodoris
Malekian, Azarakhsh
McKenzie, Andy
Monnot, Barnabé
Monnot, Jérôme
Morgenstern, Jamie
Moscardelli, Luca
Munagala, Kamesh
Müller, Rudolf
Naldi, Maurizio
Nath, Swaprava
Nava, Francesco
Nazerzadeh, Hamid
Nehama, Ilan
Nguyen, Thanh
Nisan, Noam
Noulas, Athanasios
Obraztsova, Svetlana
Olver, Neil
Oren, Sigal
Paes Leme, Renato
Panageas, Ioannis
Post, Ian
Postl, John
Pountourakis, Emmanouil
Pradelski, Bary
Proietti, Guido
Rastegari, Baharak

Riechers, Sören
Ronquillo, Lorena
Rosén, Adi
Rubinstein, Aviad
Saure, Denis
Savani, Rahul
Schmand, Daniel
Sekar, Shreyas
Serafino, Paolo
Sethuraman, Jay
Sgouritsa, Alkmini
Shah, Nisarg
Silander, Tomi
Solomon, Shay
Sun, Xiaorui
Talgam-Cohen, Inbal
Tang, Bo
Thraves, Charles
Tönnis, Andreas
Tsipras, Dimitris
Turchetta, Stefano
Tzamos, Christos
Tzoumas, Vasileios
Vegh, Laszlo
Vidali, Angelina
Vohra, Rakesh
Voudouris, Alexandros
Wang, Zizhuo
Weinberg, S. Matthew
Wilkens, Christopher
Wu, Zhiwei Steven
Yan, Xiang
Yang, Ger
Yang, Liu
Yazdanbod, Sadra
Yukun, Cheng
Zampetakis, Emmanouil
Zhang, Qiang
Ziani, Juba
Zick, Yair

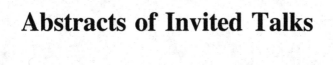

Abstracts of Invited Talks

Resolving Combinatorial Markets
via Posted Prices
(Invited Talk)

Michal Feldman

Blavatnic School of Computer Science,
Tel-Aviv University, Tel Aviv-Yafo, Israel
michal.feldman@cs.tau.ac.il

Abstract. In algorithmic mechanism design, we would like desired economic properties to cause no (or modest) additional loss in social welfare beyond the loss already incurred due to computational constraints. In this talk we review two recent results showing black-box reductions from welfare approximation algorithms to mechanisms that preserve desired economic properties. In particular: (1) we give a poly-time dominant strategy incentive compatible mechanism for Bayesian submodular (and more generally, fractionally subadditive) combinatorial auctions that approximates the social welfare within a constant factor. (2) we give a poly-time mechanism for arbitrary (known) valuation functions that, given a black-box access to a social welfare algorithm, provides a conflict free outcome that preserves at least half of its welfare. Both mechanisms are based on posted prices.

Approximate Nash Equilibrium Computation
(Invited Talk)

Paul W. Goldberg

University of Oxford, Oxford, UK
paul.goldberg@cs.ox.ac.uk

Abstract. Nash equilibrium computation is complete for the complexity class PPAD, even for two-player normal-form games. Should we understand this to mean that the computational challenge is genuinely hard? In this talk, I explain PPAD, what PPAD-completeness means for equilibrium computation, and possible ways to escape the worst-case hardness. Following the PPAD-completeness results, attention turned to the complexity of computing approximate Nash equilibria. In an approximate equilibrium, the usual "no incentive to deviate" requirement is replaced with "bounded incentive to deviate", where a parameter epsilon denotes a limit on any player's incentive to deviate. I review some of the progress that was made, and reasons to hope for a polynomial-time approximation scheme. I also discuss recent work suggesting that a quasi-polynomial time algorithm is the best thing we can hope to achieve.

Algorithms and Incentives in the Design of Online Platform Markets (Invited Talk)

Ramesh Johari

Stanford University, Stanford, USA
rjohari@stanford.edu

Abstract. Since the advent of the first online marketplaces nearly two decades ago, commerce in nearly every sector of the industry is being transformed: transportation (Lyft, Uber), lodging (Airbnb), delivery (Instacart, Postmates), labor markets (Amazon Mechanical Turk, LinkedIn, Taskrabbit, Upwork), etc. In this talk, we will survey challenges and opportunities that arise in the design of these markets, with an emphasis on how operational and algorithmic challenges interlace with incentives to dictate market outcomes.

Adverse Selection and Auction Design for Internet Display Advertising (Invited Talk)

Paul Milgrom

Department of Economics, Stanford University, Stanford, USA
milgrom@stanford.edu

Abstract. We model an online display advertising environment in which "performance" advertisers can measure the value of individual impressions, whereas "brand" advertisers cannot. If advertiser values for ad opportunities are positively correlated, second-price auctions for impressions can be very inefficient. Bayesian-optimal auctions are complex, introduce incentives for false-name bidding, and disproportionately allocate low-quality impressions to brand advertisers. We introduce "modified second bid" auctions as the unique auctions that overcome these disadvantages. When advertiser match values are drawn independently from heavy tailed distributions, a modified second bid auction captures at least 94.8% of the first-best expected value. In that setting and similar ones, the benefits of switching from an ordinary second-price auction to the modified second bid auction may be large, and the cost of defending against shill bidding and adverse selection may be low.

Contents

Abstracts

Sequential Posted Price Mechanisms with Correlated Valuations

Marek Adamczyk[1], Allan Borodin[2], Diodato Ferraioli[3](✉),
Bart de Keijzer[1], and Stefano Leonardi[1]

[1] Sapienza University of Rome, Rome, Italy
{adamczyk,dekeijzer,leonardi}@dis.uniroma1.it
[2] University of Toronto, Toronto, Canada
bor@cs.toronto.edu
[3] University of Salerno, Fisciano, SA, Italy
dferraioli@unisa.it

Abstract. We study the revenue performance of sequential posted price mechanisms and some natural extensions, for a general setting where the valuations of the buyers are drawn from a correlated distribution. Sequential posted price mechanisms are conceptually simple mechanisms that work by proposing a "take-it-or-leave-it" offer to each buyer. We apply sequential posted price mechanisms to single-parameter multi-unit settings in which each buyer demands only one item and the mechanism can assign the service to at most k of the buyers. For standard sequential posted price mechanisms, we prove that with the valuation distribution having finite support, no sequential posted price mechanism can extract a constant fraction of the optimal expected revenue, even with unlimited supply. We extend this result to the case of a continuous valuation distribution when various standard assumptions hold simultaneously. In fact, it turns out that the best fraction of the optimal revenue that is extractable by a sequential posted price mechanism is proportional to the ratio of the highest and lowest possible valuation. We prove that for two simple generalizations of these mechanisms, a better revenue performance can be achieved: if the sequential posted price mechanism has for each buyer the option of *either* proposing an offer *or* asking the buyer for its valuation, then a $\Omega(1/\max\{1, d\})$ fraction of the optimal revenue can be extracted, where d denotes the "degree of dependence" of the valuations, ranging from complete independence ($d = 0$) to arbitrary dependence ($d = n - 1$). When we generalize the sequential posted price mechanisms further, such that the mechanism has the ability to make a take-it-or-leave-it offer to the i-th buyer that depends on the valuations of all buyers except i, we prove that a constant fraction $(2 - \sqrt{e})/4 \approx 0.088$ of the optimal revenue can be always extracted.

This work is supported by the EU FET project MULTIPLEX no. 317532, the ERC StG Project PAAl 259515, the Google Research Award for Economics and Market Algorithms, and the Italian MIUR PRIN 2010-2011 project ARS TechnoMedia – Algorithmics for Social Technological Networks.

E. Markakis and G. Schäfer (Eds.): WINE 2015, LNCS 9470, pp. 1–15, 2015.
DOI: 10.1007/978-3-662-48995-6_1

1   Introduction

A large body of literature in the field of mechanism design focuses on the design of auctions that are optimal with respect to some given objective function, such as maximizing the social welfare or the auctioneer's revenue. This literature mainly considered direct revelation mechanisms, in which each buyer submits a bid that represents his valuation for getting the service, and the mechanism determines the winners and the payments. The reason for this is the *revelation principle* (see, e.g., [9]), which implies that one may study only direct revelation mechanisms for many purposes. Some of the most celebrated mechanisms follow this approach, such as the VCG mechanism [12,17,29] and the Myerson mechanism [23].

A natural assumption behind these mechanisms is that buyers will submit truthfully whenever the utility they take with the truthful bid is at least as high as the utility they may take with a different bid. However, it has often been acknowledged that such an assumption may be too strong in a real world setting. In particular, Sandholm and Gilpin [27] highlight that this assumption usually fails because of: (1) a buyer's unwillingness to fully specify their values, (2) a buyer's unwillingness to participate in ill understood, complex, unintuitive auction mechanisms, and (3) irrationality of a buyer, which leads him to underbid even when there is nothing to be gained from this behavior.

This has recently motivated the research about auction mechanisms that are conceptually simple. Among these, the class of *sequential posted price mechanisms* [11] is particularly attractive. First studied by Sandholm and Gilpin [27] (and called "take-it-or-leave-it mechanisms"), these mechanisms work by iteratively selecting a buyer that has not been selected previously, and offering him a price. The buyer may then accept or reject that price. When the buyer accepts, he is allocated the service. Otherwise, the mechanism does not allocate the service to the buyer. In the sequential posted-price mechanism we allow both the choice of buyer and the price offered to that buyer to depend on the decisions of the previously selected buyers (and the prior knowledge about the buyers' valuations). Also, randomization in the choice of the buyer and in the charged price is allowed. Sequential posted price mechanisms are thus conceptually simple and buyers do not have to reveal their valuations. Moreover, they possess a trivial dominant strategy (i.e., buyers do not have to take strategic decisions) and are individually rational (i.e., participation is never harmful to the buyer).

Sequential posted price mechanisms have been mainly studied for the setting where the valuations of the buyers are each drawn independently from publicly known buyer-specific distributions, called the *independent values* setting. In this paper, we study a much more general setting, and assume that the entire vector of valuations is drawn from one publicly known distribution, which allows for arbitrarily complex dependencies among the valuations of the buyers. This setting is commonly known as the *correlated values* setting. Our goal is to investigate the revenue guarantees of sequential posted price mechanisms in the correlated value setting. We quantify the quality of a mechanism by comparing its expected revenue to that of the *optimal mechanism*, that achieves the highest expected revenue among all dominant strategy incentive compatible and ex-post individually rational mechanisms (see the definitions below).

We assume a standard Bayesian, transferable, quasi-linear utility model and we study the *unit demand, single parameter, multi-unit* setting: there is one service (or type of item) being provided by the auctioneer, there are n buyers each interested in receiving the service once, and the *valuation* of each buyer consists of a single number that reflects to what extent a buyer would profit from receiving the service provided by the auctioneer. The auctioneer can charge a price to a bidder, so that the utility of a bidder is his valuation (in case he gets the service), minus the charged price. In this paper, our focus is on the k-limited supply setting, where service can be provided to at most k of the buyers. This is an important setting because it is a natural constraint in many realistic scenarios, and it contains two fundamental special cases: the *unit supply* setting (where $k = 1$), and the *unlimited supply* setting where $k = n$.

Related Work. There has been substantial work [5,14,19,20,26] on *simple* mechanisms. Babaioff et al. [5] highlight the importance of understanding the strength of simple versus complex mechanisms for revenue maximization.

As described above, sequential posted price mechanisms are an example of such a simple class of mechanisms. Sandholm and Gilpin [27] have been the first ones to study sequential posted price mechanisms. They give experimental results for the case in which values are independently drawn from the uniform distribution in [0, 1]. Moreover, they consider the case where multiple offers can be made to a bidder, and study the equilibria that arise from this. Blumrosen and Holenstein [8] compare fixed price (called symmetric auctions), sequential posted price (called discriminatory auctions) and the optimal mechanism for valuations drawn from a wide class of i.i.d. distributions. Babaioff et al. [3] consider *prior-independent* posted price mechanisms with k-limited supply for the setting where the only information known is that all valuations are independently drawn from the same distribution with support [0, 1]. Posted-price mechanisms have also been previously studied in [6,7,21], albeit for a non-Bayesian, on-line setting. In a recent work Feldman et al. [16] study on-line posted price mechanisms for combinatorial auctions when valuations are independently drawn.

The works of Chawla et al. [11] and Gupta and Nagarajan [18] are closest to our present work, although they only consider sequential posted price mechanisms in the independent values setting. In particular, Chawla et al. [11] prove that such mechanisms can extract a constant factor of the optimal revenue for single and multiple parameter settings under various constraints on the allocations. They also consider on-line (called *order-oblivious* in [11]) sequential posted price mechanisms in which the order of the buyers is fixed and adversarially determined. They use on-line mechanisms to establish results for the more general multi-parameter case. Yan [30], and Kleinberg and Weinberg [22] build on this work and strengthen some of the results of Chawla et al. [11].

Gupta and Nagarajan [18] introduce a more abstract stochastic probing problem that includes Bayesian sequential posted price mechanisms. Their approximation bounds were later improved by Adamczyk et al. [1] who in particular matched the approximation of Chawla et al. [11] for single matroid settings.

All previous work only consider the independent setting. In this work we instead focus on the correlated setting. The lookahead mechanism of Ronen [24] is a fundamental reference for the correlated setting. It also resembles some of the mechanisms considered in this work. However, as we will indicate, it turns out to be different in substantial ways. Cremer and McLean [13] made a fundamental contribution to auction theory in the correlated value setting, by exactly characterizing for which valuation distributions it is possible to extract the full optimal social welfare as revenue. Segal [28] gives a characterization of optimal ex-post incentive compatible and ex-post individually rational optimal mechanisms. Roughgarden and Talgam-Cohen [25] study the even more general *interdependent* setting. They show how to extend the Myerson mechanism to this setting for various assumptions on the valuation distribution. There is now a substantial literature [10,15,25] that develops mechanisms with good approximation guarantees for revenue maximization in the correlated setting. These mechanisms build on the lookahead mechanism of Ronen [24] and thus they also differ from the mechanisms proposed in this work.

Contributions and Outline. We first define some preliminaries and notation. In Sect. 2 we give a simple sequence of instances which demonstrate that for (unrestricted) correlated distributions, sequential posted price (SPP) mechanisms cannot obtain a constant approximation with respect to the revenue obtained by the optimal dominant strategy incentive compatible and ex-post individually rational mechanism. This holds for any value of k (i.e., the size of the supply). We extend this impossibility result by proving that a constant approximation is impossible to achieve even when we assume that the valuation distribution is continuous and satisfies all of the following conditions simultaneously: the valuation distribution is supported everywhere, is entirely symmetric, satisfies *regularity*, satisfies the *monotone hazard rate* condition, satisfies *affiliation*, all the induced marginal distributions have finite expectation, and all the conditional marginal distributions are non-zero everywhere.

Given these negative results, we consider a generalization of sequential posted price mechanisms that are more suitable for settings with limited dependence among the buyers' valuations: *enhanced sequential posted price (ESPP) mechanisms*. An ESPP mechanism works by iteratively selecting a buyer that has not been selected previously. The auctioneer can either offer the selected buyer a price or ask him to report his valuation. As in sequential posted price mechanisms, if the buyer is offered a price, then he may accept or reject that price. When the buyer accepts, he is allocated the service. Otherwise, the mechanism does not allocate the service to the buyer. If instead, the buyer is asked to report his valuation, then the mechanism does not allocate him the service. Note that the ESPP mechanism requires that some fraction of buyers reveal their valuation truthfully. Thus, the property that the bidders not have to reveal their preferences is *partially* sacrificed, for a more powerful class of mechanisms and (as we will see) a better revenue performance. For the ESPP mechanisms, again there are instances in which the revenue is not within a constant fraction of the optimal revenue. However, these mechanisms can extract a fraction $\Theta(1/n)$ of the optimal revenue, regardless of the valuation distribution.

This result seems to suggest that to achieve a constant approximation of the optimal revenue it is *necessary* to collect all the bids truthfully. Consistent with this hypothesis, we prove that a constant fraction of the optimal revenue can be extracted by dominant strategy IC *blind offer mechanisms*: these mechanisms inherit all the limitations of sequential posted price mechanisms (i.e., buyers are considered sequentially in an order independent of any bids; buyers are only offered a price when selected; and the buyer gets the service only if he accepts the offered price), except that the price offered to a bidder i may now depend on the bids submitted by all players other than i. This generalization sacrifices entirely the property that buyers valuations need not be revealed. Blind offer mechanisms are thus necessarily direct revelation mechanisms. However, this comes with the reward of a revenue that is only a constant factor away from optimal. In conclusion, blind offer mechanisms achieve a constant approximation of the optimal revenue, largely preserve the conceptual simplicity of sequential posted price mechanisms, and are easy to grasp for the buyers participating in the auction. In particular, buyers have a conceptually simple and practical strategy: to accept the price if and only if it is not above their valuation, regardless of how the prices are computed. We stress that, even if blind offer mechanisms sacrifice some simplicity (and practicality), we still find it theoretically interesting that a mechanism that allocates items to buyers *in any order* and thus not necessarily in an order that maximizes profit, say as in [24], is able to achieve a constant approximation of the optimal revenue even with correlated valuations. Moreover, blind offer mechanisms provide the intermediate step en-route to establishing revenue approximation bounds for other mechanisms. We will show how blind offer mechanisms serve this purpose in Sect. 3.

We highlight that our positive results do not make any assumptions on the marginal valuation distributions of the buyers nor the type of correlation among the buyers. However, in Sect. 3 we consider the case in which the degree of dependence among the buyers is limited. In particular, we introduce the notion of *d-dimensionally dependent distributions*. This notion informally requires that for each buyer i there is a set S_i of d other buyers such that the distribution of i's valuation when conditioning on the vector of other buyers' valuations can likewise be obtained by only conditioning on the valuations of S_i. Thus, this notion induces a hierarchy of n classes of valuation distributions with increasing degrees of dependence among the buyers: for $d = 0$ the buyers have independent valuations, while the other extreme $d = n - 1$ implies that the valuations may be dependent in arbitrarily complex ways. Note that d-dimensional dependence does not require that the marginal valuation distributions of the buyers themselves satisfy any particular property, and neither does it require anything from the type of correlation that may exist among the buyers. This stands in contrast with commonly made assumptions such as *symmetry, affiliation*, the *monotone-hazard rate assumption,* and *regularity*, that are often encountered in the auction theory and mechanism design literature.

Our main positive result for ESPP mechanisms then states that if the valuation distribution is d-dimensionally dependent, there exists an ESPP mechanism that extracts an $\Omega(1/d)$ fraction of the optimal revenue. The proof of this result consists of three key ingredients: (i) An upper bound on the optimal ex-post IC, ex-post IR revenue in terms of the solution of a linear program. This part of the proof generalizes a linear programming characterization introduced by Gupta and Nagarajan [18] for the independent distribution setting. (ii) A proof that incentive compatible blind offer mechanisms are powerful enough to extract a constant fraction of the optimal revenue of any instance. This makes crucial use of the linear program mentioned above. (iii) A conversion lemma showing that blind offer mechanisms can be turned into ESPP mechanisms while maintaining a fraction $\Omega(1/d)$ of the revenue of the blind offer mechanism.

Many proofs and various important parts of the discussion have been omitted from this version of our paper, due to space constraints. We refer the reader to [2] for full proofs and a complete discussion of our work and results.

Preliminaries. For $a \in \mathbb{N}$, $[a]$ denotes the set $\{1, \ldots, a\}$. For a vector $\vec{v}$ and an arbitrary element a, let $(a, \vec{v}_{-i})$ be the vector obtained by replacing v_i with a.

We face a setting where an auctioneer provides a service to n buyers, and is able to serve at most k of the buyers. The buyers have valuations for the service offered, which are drawn from a *valuation distribution* π, i.e., a probability distribution on $\mathbb{R}_{\geq 0}^n$. We will assume throughout this paper that π is discrete, except where otherwise stated.

We will use the following notation for conditional and marginal probability distributions. Let π be a discrete finite probability distribution on $\mathbb{R}^n$, let $i \in [n]$, $S \subset [n]$ and $\vec{v} \in \mathbb{R}^n$. For an arbitrary probability distribution π, denote by supp(π) the support of π, by $\vec{v}_S$ the vector obtained by removing from $\vec{v}$ the coordinates in $[n] \setminus S$, by π_S the distribution induced by drawing a vector from π and removing the coordinates corresponding to index set $[n] \setminus S$, by $\pi_{\vec{v}_S}$ the distribution of π conditioned on the event that $\vec{v}_S$ is the vector of values on the coordinates corresponding to index set S, and by $\pi_{i,\vec{v}_S}$ the marginal distribution of the coordinate of $\pi_{\vec{v}_S}$ that corresponds to buyer i. In the subscripts we sometimes write i instead of $\{i\}$ and $-i$ instead of $[n] \setminus \{i\}$.

An *instance* is a triple (n, π, k), where n is the number of participating buyers, π is the valuation distribution, and $k \in \mathbb{N}_{\geq 1}$ is the supply, i.e., the number of services that the auctioneer may allocate to the buyers. A *deterministic mechanism* f is a function from $\times_{i \in [n]} \Sigma_i$ to $\{0, 1\}^n \times \mathbb{R}_{\geq 0}^n$, for any choice of *strategy sets* $\Sigma_i, i \in [n]$. When $\Sigma_i = \text{supp}(\pi_i)$ for all $i \in [n]$, mechanism f is called a deterministic *direct revelation mechanism*. A *randomized mechanism* M is a probability distribution over deterministic mechanisms. For $i \in [n]$ and $\vec{s} \in \times_{j \in [n]} \Sigma_j$, we will denote i's *expected allocation* $\mathbf{E}_{f \sim M}[f(\vec{s})_i]$ by $x_i(\vec{s})$ and i's *expected payment* $\mathbf{E}_{f \sim M}[f(\vec{s})_{n+i}]$ by $p_i(\vec{s})$. For $i \in [n]$ and $\vec{s} \in \times_{j \in [n]} \Sigma_j$, the *expected utility* of buyer i is $x_i(\vec{s})v_i - p_i(\vec{s})$. The auctioneer is interested in maximizing the *revenue*

$\sum_{i \in [n]} p_i(\vec{s})$, and is assumed to have full knowledge of the valuation distribution, but not of the actual valuations of the buyers.

Mechanism M is *dominant strategy incentive compatible (dominant strategy IC)* iff for all $i \in [n]$ and $\vec{v} \in \times_{j \in [n]} \mathrm{supp}(\pi_j)$ and $\vec{v} \in \mathrm{supp}(\pi)$, $x_i(v_i, \vec{v}_{-i})v_i - p_i(v_i, \vec{v}_{-i}) \geq x_i(\vec{v})v_i - p_i(\vec{v})$. Mechanism M is *ex-post individually rational (ex-post IR)* iff for all $i \in [n]$ and $\vec{v} \in \mathrm{supp}(\pi)$, $x_i(v)v_i - p_i(v) \geq 0$. For convenience we usually will not treat a mechanism as a probability distribution over outcomes, but rather as the result of a randomized procedure that interacts with the buyers. In this case we say that a mechanism is *implemented by* that procedure.

A *sequential posted price (SPP) mechanism* for an instance (n, π, k) is any mechanism that is implementable by iteratively selecting a buyer $i \in [n]$ that has not been selected in a previous iteration, and proposing a price p_i for the service, which the buyer may accept or reject. If i accepts, he gets the service and pays p_i, resulting in a utility of $v_i - p_i$ for i. If i rejects, he pays nothing and does not get the service, resulting in a utility of 0 for i. Once the number of buyers that have accepted an offer equals k, the process terminates. Randomization in the selection of the buyers and prices is allowed. We will initially be concerned with only sequential posted price mechanisms. Later in the paper we define the two generalizations of SPP mechanisms that we mentioned in the introduction.

Our focus in this paper is on the maximum expected revenue of the SPP mechanisms, and some of its generalizations. Note that each buyer in a SPP mechanism has an obvious dominant strategy: he will accept whenever the price offered to him does not exceed his valuation, and he will reject otherwise. Also, a buyer always ends up with a non-negative utility when participating in a SPP mechanism. Thus, by the revelation principle (see, e.g., [9]), a SPP mechanism can be converted into a dominant strategy IC and ex-post IR direct revelation mechanism with the same expected revenue. Therefore, we compare the maximum expected revenue $REV(M)$ achieved by an SPP mechanism M to OPT, where OPT is defined as the maximum expected revenue that can be obtained by a mechanism that is dominant strategy IC and ex-post IR.

A more general solution concept is formed by the *ex-post incentive compatible*, ex-post individually rational mechanisms. Specifically, let (n, π, k) be an instance and M be a randomized direct revelation mechanism for that instance. Mechanism M is *ex-post incentive compatible (ex-post IC)* iff for all $i \in [n]$, $s_i \in \mathrm{supp}(\pi_i)$ and $\vec{v} \in \mathrm{supp}(\pi)$, $x_i(\vec{v})v_i - p_i(\vec{v}) \geq x_i(s_i, \vec{v}_{-i})v_i - p_i(s_i, \vec{v}_{-i})$. In other words, a mechanism is *ex-post IC* if it is a pure equilibrium for the buyers to always report their valuation. In this work we sometimes compare the expected revenue of our (dominant strategy IC and ex-post IR) mechanisms to the maximum expected revenue of the more general class of ex-post IC, ex-post IR mechanisms. This strengthens our positive results. We refer the reader to [25] for a further discussion of and comparison between various solution concepts.

2 Sequential Posted Price Mechanisms

We are interested in designing a posted price mechanism that, for any given n and valuation distribution π, achieves an expected revenue that is a constant approximation of the optimal expected revenue achievable by a dominant strategy IC, ex-post IR mechanism. Theorem 1 shows that this is impossible.

Theorem 1. *For all* $n \in \mathbb{N}_{\geq 2}$, *there exists a valuation distribution* π *such that for all* $k \in [n]$ *there does not exist a sequential posted price mechanism for instance* (n, π, k) *that extracts a constant fraction of the expected revenue of the optimal dominant strategy IC, ex-post IR mechanism.*

Proof sketch. Fix $m \in \mathbb{N}_{\geq 1}$ arbitrarily, and consider the case where $n = 1$ and the valuation v_1 of the single buyer is taken from $\{1/a : a \in [m]\}$ distributed so that $\pi_1(1/a) = 1/m$ for all $a \in [m]$. In this setting, an SPP mechanism will offer the buyer a price p, which the buyer accepts iff $v_1 \geq p$. After that, the mechanism terminates. We show that this mechanism achieve only a fraction $\frac{1}{H(m)}$ of the social welfare. We then extend this example to a setting where the expected revenue of the optimal dominant strategy IC, ex-post IR mechanism is equal to the expected optimal social welfare.   □

The above impossibility result holds also in the continuous case, even if a large set of popular assumptions hold simultaneously, namely, the valuation distribution π has support $[0, 1]^n$; the expectation $\mathbf{E}_{\vec{v} \sim \pi}[v_i]$ is finite for any $i \in [n]$; π is symmetric in all its arguments; π is continuous and nowhere zero on $[0, 1]^n$; the conditional marginal densities $\pi_{i|\vec{v}_{-i}}$ are nowhere zero for any $\vec{v}_{-i} \in [0, 1]^{n-1}$ and any $i \in [n]$; π has a monotone hazard rate and is regular; π satisfies affiliation.

Roughgarden and Talgam-Cohen [25] showed that when all these assumptions are simultaneously satisfied, the optimal ex-post IC and ex-post IR mechanism is the Myerson mechanism that is optimal also in the independent value setting. Thus, these conditions make the correlated setting in some sense similar to the independent one with respect to revenue maximization. Yet our result show that, whereas SPP mechanism can achieve a constant approximation revenue for independent distributions, this does not hold for correlated ones.

A Revenue Guarantee for Sequential Posted Price Mechanisms. More precisely, in our lower bound instances constructed in the proof of Theorem 1, it is the case that the expected revenue extracted by every posted price mechanism is a $\Theta(1/\log(r))$ fraction of the optimal expected revenue, where r is the ratio between the highest valuation and the lowest valuation in the support of the valuation distribution. A natural question that arises is whether this is the worst possible instance in terms of revenue extracted, as a function of r. It turns out that this is indeed the case, asymptotically. The proofs use a standard bucketing technique (see, e.g., [4]) and can be found in the full paper [2].

We start with the unit supply case. For a valuation distribution π on $\mathbb{R}^n$, let $v_\pi^{\max}$ and $v_\pi^{\min}$ be $\max\{v_i : v \in \mathrm{supp}(\pi), i \in [n]\}$ and $\min\{\max\{v_i : i \in [n]\} : v \in \mathrm{supp}(\pi)\}$ respectively. Let $r_\pi = v_\pi^{\max}/v_\pi^{\min}$ be the ratio between the highest and lowest coordinate-wise maximum valuation in the support of π.

Proposition 1. *Let* $n \in \mathbb{N}_{\geq 1}$, *and let* π *be a probability distribution on* $\mathbb{R}^n$. *For the unit supply case there exists a SPP mechanism that, when run on instance*

$(n, \pi, 1)$, *extracts in expectation at least an $\Omega(1/\log(r_\pi))$ fraction of the expected revenue of the optimal social welfare (and therefore also of the expected revenue of the optimal dominant strategy IC and ex-post IR auction).*

This result can be generalized to yield revenue bounds for the case of k-limited supply, where $k > 1$. The above result does not always guarantee a good revenue; for example in the extreme case where $v_\pi^{\min} = 0$. However, it is easy to strengthen the above theorem such that it becomes useful for a wide class of of distributions.

3 Enhanced Sequential Posted Price Mechanisms

We propose a generalization of sequential posted price mechanisms, in such a way that they possess the ability to retrieve valuations of some buyers.

Specifically, an *enhanced sequential posted price (ESPP) mechanism* for an instance (n, π, k) is a randomized mechanism that can be implemented by iteratively selecting a buyer $i \in [n]$ that has not been selected in a previous iteration, and performing exactly one of the following actions on buyer i:

- Propose service at price p_i to buyer i, which the buyer may accept or reject. If i accepts, he gets the service and pays p_i, resulting in a utility of $v_i - p_i$ for i. If i rejects, he pays nothing and does not get the service, resulting in a utility of 0 for i.
- Ask i for his valuation. (Buyer i pays nothing and does not get service.)

This generalization is still dominant strategy IC and ex-post IR.

Next we analyze the revenue performance of ESPP mechanisms. For this class of mechanisms we prove that, it is unfortunately still the case that no constant fraction of the optimal revenue can be extracted. Specifically, the next theorem establishes an $O(1/n)$ bound for ESPP mechanisms.

Theorem 2. *For all $n \in \mathbb{N}_{\geq 2}$, there exists a valuation distribution π such that for all $k \in [n]$ there does not exist a ESPP mechanism for instance (n, π, k) that extracts more than a $O(1/n)$ fraction of the expected revenue of the optimal dominant strategy IC, ex-post IR mechanism.*

Proof sketch. Let $n \in \mathbb{N}$ and $m = 2^n$. We specify an instance I_n with n buyers, and prove that $\lim_{n \to 0} RM(I_n)/OR(I_n) = 0$, where $RM(I_n)$ is the largest expected revenue achievable by any ESPP mechanism on I_n, and $OR(I_n)$ is the largest expected revenue achievable by a dominant strategy IC, ex-post IR mechanism. I_n is defined as follows. Fix ϵ such that $0 < \epsilon < 1/nm^2$. The valuation distribution π is the one induced by the following process: (i) Draw a buyer $i^\star$ from the set $[n]$ uniformly at random; (ii) Draw numbers $\{c_j \colon j \in [n] \setminus \{i^\star\}\}$ independently from $[m]$ uniformly at random; (iii) For all $j \in [n] \setminus \{i^\star\}$, set $v_j = c_j \epsilon$; (iv) Set $v_{i^\star} = ((\sum_{j \in [n] \setminus \{i^\star\}} c_j) \bmod m + 1)^{-1}$. □

However, ESPP mechanisms turn out to be more powerful than the standard sequential posted price. Indeed, contrary to SPP mechanisms, the ESPP mechanisms can be shown to extract a fraction of the optimal revenue that is independent of the valuation distribution. More precisely, the $O(1/n)$ bound turns out to be asymptotically tight. Our main positive result for ESPP mechanisms is that when dependence of the valuation among the buyers is limited, then a constant fraction of the optimal revenue can be extracted. Specifically, we will define the concept of d-dimensional dependence and prove that for a d-dimensionally dependent instance, there is an ESPP mechanism that extracts an $\Omega(1/d)$ fraction of the optimal revenue.

It is natural to identify the basic reason(s) why, in the case of general correlated distributions, standard and enhanced sequential posted price mechanisms may fail to achieve a constant approximation of the optimum revenue. There are two main limitations of these mechanisms: (i) such mechanisms do not solicit bids or values from all buyers, and (ii) such mechanisms award items in a sequential manner. Although it is crucial to retrieve the valuation of *all* (but one of the) buyers, we show that it is possible to achieve a constant fraction of the optimum revenue by a mechanism that allocates items sequentially in an on-line manner, in contrast to previously known approximation results.

Randomized mechanism M is a *blind offer mechanism* iff it can be implemented as follows. Let (n, π, k) be an instance and let $\vec{b}$ be the submitted bid vector. Then,

1. Terminate if $\vec{b} \notin \text{supp}(\pi)$.
2. Either terminate or select a buyer i from the set of buyers that have not yet been selected, such that the choice of i does not depend on $\vec{b}$.
3. Offer buyer i the service at price p_i, where p_i is drawn from a probability distribution that depends only on $\pi_{i, \vec{b}_{-i}}$ (hence the distribution of p_i is determined by $\vec{b}_{-i}$ and in particular does not depend on b_i).
4. Restart if the number of buyers who have accepted offers does not exceed k.

Note that the price offered to a buyer is entirely determined by the valuations of the remaining buyers, and is independent of what is reported by buyer i himself. Also the iteration in which a buyer is picked cannot be influenced by his bid. Nonetheless, blind offer mechanisms are in general not incentive compatible due to the fact that a bidder may be incentivized to misreport his bid in order to increase the probability of supply not running out before he is picked. However, blind offer mechanisms can easily be made incentive compatible as follows: let M be a non-IC blind offer mechanism, let $\vec{b}$ be a bid vector and let $z_i(\vec{b})$ be the probability that M picks bidder i before supply has run out. When a bidder is picked, we adapt M by *skipping* that bidder with a probability $p_i(\vec{b})$ that is chosen in a way such that $z_i(\vec{b})p_i(\vec{b}) = \min\{z_i(b_i\vec{b}_{-i}): b_i \in \text{supp}(\pi_i)\}$. This is a blind offer mechanism in which buyer i has no incentive to lie, because now the probability that i is made an offer is independent of his bid. Doing this iteratively for all buyers yields a dominant strategy IC mechanism M'. Note that the act of *skipping* a bidder can be implemented by offering a price that is so high that a bidder will never accept it, thus M' is still a blind offer mechanism. Moreover, if the probability that any bidder in M is made an offer is lower bounded by a constant c, then in M' the probability that any bidder is offered a price is at least c. We apply this principle in the proof of Theorem 3 below in order to obtain a dominant strategy IC mechanism with a constant factor revenue performance.

It is not hard to see that the classical Myerson mechanism for the *independent* single-item setting belongs to the class of blind offer mechanisms. Thus blind offer mechanisms are optimal when buyers' valuations are independent. We will prove next that when buyer valuations are *correlated*, blind offer mechanisms can always extract a constant fraction of the optimal revenue, even against the ex-post IC, ex-post IR solution concept. Other mechanisms that achieve a constant approximation to the optimal revenue have been defined by Ronen [24], and then by Chawla et al. [10] and Dobzinski et al. [15]. However, these mechanisms allocate the items to profit-maximizing buyers. Thus, they are different from blind offer mechanisms in which the allocation is on-line.

Theorem 3. *For every instance (n, π, k), there is a dominant strategy IC blind offer mechanism for which the expected revenue is at least a $(2 - \sqrt{e})/4 \approx 0.088$ fraction of the maximum expected revenue that can be extracted by an ex-post IC, ex-post IR mechanism.*

We need to establish some intermediate results in order to build up to a proof for the above theorem. First, we derive an upper bound on the revenue of the optimal ex-post IC, ex-post IR mechanism. For a given instance (n, π, k), consider the linear program with variables $(y_i(\vec{v}))_{i\in[n], \vec{v}\in\text{supp}(\pi)}$ where the objective is

$$\max \sum_{i\in[n]} \sum_{\vec{v}_{-i}\in\text{supp}(\pi_{-i})} \pi_{-i}(\vec{v}_{-i}) \sum_{v_i\in\text{supp}(\pi_{i,\vec{v}_{-i}})} \Pr_{v_i'\sim\pi_{i,\vec{v}_{-i}}} [v_i' \geq v_i] v_i y_i(v_i, \vec{v}_{-i}) \text{ sub-}$$

ject to the constraints $\forall i \in [n], \vec{v}_{-i} \in \text{supp}(\pi_{-i}): \sum_{v_i\in\text{supp}(\pi_{i,\vec{v}_{-i}})} y_i(\vec{v}) \leq 1; \forall \vec{v} \in \text{supp}(\pi): \sum_{i\in[n]} \sum_{v_i'\in\text{supp}(\pi_{i,\vec{v}_{-i}}): v_i'\leq v_i} y_i(v_i', \vec{v}_{-i}) \leq k; \vec{v} \in \text{supp}(\pi): y_i(\vec{v}) \geq 0$ $\forall i \in [n]$. The next lemma states that the solution to this linear program forms an upper bound on the revenue of the optimal mechanism.

Lemma 1. *For any instance (n, π, k), above linear program upper bounds the maximum expected revenue achievable by an ex-post IC, ex-post IR mechanism.*

Proof sketch. We first prove that a monotonicity constraint holds on the set of possible allocations that a ex-post IC, ex-post IR mechanism can output. Moreover, we show that the prices charged by the mechanism cannot exceed a certain upper bound given in terms of allocation probabilities. Then, we formulate a new linear program whose optimal value equals the revenue of the optimal ex-post IC, ex-post IR mechanism. We finally rewrite this new linear program into the one given above. This proof adapts the approach introduced in [18]. $\quad\square$

We can now proceed to prove our main result about blind offer mechanisms. Let (n, π, k) be an arbitrary instance. Let $(y_i^*(\vec{v}))_{i\in[n]}$ be the optimal solution to the linear program given above corresponding to this instance. Let M_π^k be the blind offer mechanism that does the following: let $\vec{v}$ be the vector of submitted valuations. Iterate over the set of buyers such that in iteration i, buyer i is picked. In iteration i, select one of the following options: offer service to buyer i at a price p for which it holds that $y_i^*(p, \vec{b}_{-i}) > 0$, or skip buyer i. The probabilities with which these options are chosen are as follows: Price p

is offered with probability $y_i^*(p, \vec{b}_{-i})/2$, and buyer i is skipped with probability $1 - \sum_{p' \in \text{supp}(\pi_{i, \vec{b}_{-i}})} y_i^*(p, \vec{b}_{-i})/2$. The mechanism terminates if k buyers have accepted an offer, or at iteration $n + 1$.

Proof sketch (of Theorem 3). We will show that the expected revenue of M_π^k is at least $\frac{2 - \sqrt{e}}{4} \cdot \sum_{i \in [n]} \sum_{\vec{v}_{-i} \in \text{supp}(\pi_{-i})} \pi_{-i}(\vec{v}_{-i}) \sum_{v_i \in \text{supp}(\pi_{i, \vec{v}_{-i}})} \Pr_{v_i' \sim \pi_{i, \vec{v}_{-i}}} [v_i' \geq v_i] v_i y_i^*(v_i, \vec{v}_{-i})$, which, by Lemma 1 and the LP above, is a $(2 - \sqrt{e})/4$ fraction of the expected revenue of the optimal ex-post IC, ex-post IR mechanism.

For a vector of valuations $\vec{v} \in \text{supp}(\pi)$ and a buyer $i \in [n]$, denote by $D_{i, \vec{v}_{-i}}$ the probability distribution from which mechanism $M_\pi^k(\vec{v})$ draws a price that is offered to buyer i, in case iteration $i \in [n]$ is reached. We let V be a number that exceeds $\max\{v_i : i \in [n], \vec{v} \in \text{supp}(\pi)\}$ and represent by V the option where $M_\pi^k(\vec{v})$ chooses to skip buyer i, so that $D_{i, \vec{v}_{-i}}$ is a probability distribution on the set $\{V\} \cup \{v_i : y_i^*(v_i, \vec{v}_{-i}) > 0\}$. Then, $\mathbf{E}_{\vec{v} \sim \pi}[\text{revenue of } M_\pi^k(\vec{v})] \geq$

$$\sum_{i \in [n]} \sum_{\vec{v} \in \text{supp}(\pi)} \pi(\vec{v}) \sum_{\substack{p_i \in \text{supp}(D_{i, \vec{v}_{-i}}) \\ : \, p_i \leq v_i}} \frac{p_i y_i^*(p_i, \vec{v}_{-i})}{2} \Pr_{p_i \sim D_{i, \vec{v}_{-i}}} [|\{j \in [n-1] : p_j \leq v_j\}| <$$

$k]$. Then, by applying a Chernoff bound, we can prove that $\Pr_{\forall i : p_i \sim D_{i, \vec{v}_{-i}}}[|\{j \in [n-1] : p_j \leq v_j\}| < k] \geq 1 - \left(\frac{e}{4}\right)^{k/2} \geq 1 - \left(\frac{e}{4}\right)^{1/2} = \frac{2 - \sqrt{e}}{2}$. Hence, we have a lower bound of $(2 - \sqrt{e})/2$ on the probability that all players get selected. The theorem follows by combining this with the principle explained above that allows us to transform M_π^k into a dominant strategy IC blind offer mechanism.   □

Revenue Guarantees for ESPP Mechanisms. Finally, in this section we evaluate the revenue guarantees of the ESPP mechanisms in the presence of a form of limited dependence that we will call *d-dimensional dependence*, for $d \in \mathbb{N}$. These are probability distributions for which it holds that the valuation distribution of a buyer conditioned on the valuations of the rest of the buyers can be retrieved by only looking at the valuations of a certain subset of d buyers. Formally, a probability distribution π on $\mathbb{R}^n$ is *d-dimensionally dependent* iff for all $i \in [n]$ there is a subset $S_i \subseteq [n] \setminus \{i\}, |S_i| = d$, such that for all $\vec{v}_{-i} \in \text{supp}(\pi_{-i})$ it holds that $\pi_{i, \vec{v}_{S_i}} = \pi_{i, \vec{v}_{-i}}$. Note that if $d = 0$, then π is a product of n independent probability distributions on $\mathbb{R}$. On the other hand, the set of $(n-1)$-dimensionally dependent probability distributions on $\mathbb{R}^n$ equals the set of all probability distributions on $\mathbb{R}^n$. This notion is useful in practice for settings where it is expected that a buyer's valuation distribution has a reasonably close relationship with the valuation of a few other buyers. As an example of one of these practical settings consider the case that there exists a true objective valuation v for the item or service, an expert buyer that knows this valuation precisely, and remaining buyers whose valuation is influenced by independent noise. It is then sufficient to know the valuation of a single buyer, namely the expert one, in order to retrieve the conditional distribution of any other buyer.

In general, d-dimensional dependence is relevant to many practical settings in which it is not necessary to have complete information about the valuations of all the other buyers in order to say something useful about the valuation of a particular buyer. This rules out the extreme kind of dependence defined in the proof of Theorem 2; there the distributions are not $(n - 2)$-dimensionally dependent, because for each buyer i it holds that the valuations of all buyers $[n] \setminus \{i\}$ are necessary in order to extract the valuation distribution of i conditioned on the others' valuations.

It is important to realize that the class of d-dimensionally dependent distributions is a strict superset of the class of *Markov random fields of degree d*. A Markov random field of degree d is a popular model to capture the notion of limited dependence. Anyway, d-dimensionally dependent distributions are more general: we show in [2] that there are distributions on $\mathbb{R}^n$ that are 1-dimensionally dependent, but are not a Markov random field of degree less than $n/2$.

Theorem 4. *For every instance (n, π, k) where π is d dimensionally dependent, there exists an ESPP mechanism of which the expected revenue is at least a $(2 - \sqrt{e})/(16d) \geq 1/(46d) \in \Omega(1/d)$ fraction of the maximum expected revenue that can be extracted by an ex-post IC, ex-post IR mechanism.*

As a corollary we have that the bound of Theorem 2 is asymptotically tight.

Theorem 4 follows by combining Theorem 3 with the following lemma.

Lemma 2. *Let $\alpha \in [0, 1]$ and let (n, π, k) be an instance such that π is d-dimensionally dependent. If there is a blind offer mechanism that extracts in expectation at least an α fraction of the expected revenue of the optimal dominant strategy IC, ex-post IR mechanism, then there is an ESPP mechanism that extracts in expectation at least a $\alpha/\max\{4d, 1\}$ fraction of the expected revenue of the optimal ex-post IC, ex-post IR mechanism.*

4   Open Problems

Besides improving approximation bounds established in the present paper, there are many other interesting further research directions. For example, it would be interesting to investigate revenue guarantees under the additional constraint that the sequential posted price mechanism be *on-line*, i.e., the mechanism has no control over which buyers to pick, and should perform well for any possible ordering. We are also interested in the role of randomization in our ESPP mechanism that extracts $O(1/d)$ of the optimal revenue: in the current proof buyers are picked uniformly at random. Does there exist a deterministic ESPP mechanism that attains the same revenue guarantee, or is randomness a necessity?

An obvious and interesting research direction is to investigate more general auction problems. In particular, to what extent can ESPP mechanisms be applied to auctions having non-identical items? Additionally, can such mechanisms be applied to more complex allocation constraints or specific valuation functions for the buyers? The agents may have, for example, a demand of more than one item, or there may be a matroid feasibility constraint.

Acknowledgments. We thank Joanna Drummond, Brendan Lucier, Tim Roughgarden and anonymous referees for their constructive comments.

References

1. Adamczyk, M., Sviridenko, M., Ward, J.: Submodular stochastic probing on matroids. In: STACS 2014 (2014)
2. Adamczyk, M., Borodin, A., Ferraioli, D., de Keijzer, B., Leonardi, S.: Sequential posted price mechanisms with correlated valuations. CoRR, abs/1503.02200 (2015)
3. Babaioff, M., Dughmi, S., Kleinberg, R., Slivkins, A.: Dynamic pricing with limited supply. In: EC 2012 (2012)
4. Babaioff, M., Immorlica, N., Kleinberg, R.: Matroids, secretary problems, and online mechanisms. In: SODA 2007 (2007)
5. Babaioff, M., Immorlica, N., Lucier, B., Weinberg, S.M.: A simple and approximately optimal mechanism for an additive buyer. In: FOCS 2014 (2014)
6. Blum, A., Hartline, J.D.: Near-optimal online auctions. In: SODA 2005 (2005)
7. Blum, A., Kumar, V., Rudra, A., Wu, F.: Online learning in online auctions. Theor. Comput. Sci. **324**, 137–146 (2004)
8. Blumrosen, L., Holenstein, T.: Posted prices vs. negotiations: an asymptotic analysis. In: EC 2008 (2008)
9. Börgers, T.: An Introduction to the Theory of Mechanism Design. Oxford University Press, Oxford (2015)
10. Chawla, S., Fu, H., Karlin, A.: Approximate revenue maximization in interdependent value settings. In: EC 2014, pp. 277–294 (2014)
11. Chawla, S., Hartline, J.D., Malec, D.L., Sivan, B.: Multi-parameter mechanism design and sequential posted pricing. In: STOC 2010 (2010)
12. Clarke, E.H.: Multipart pricing of public goods. Public Choice **11**(1), 17–33 (1971)
13. Cremer, J., McLean, R.P.: Full extraction of the surplus in bayesian and dominant strategy auctions. Econometrica **56**(6), 1247–1257 (1988)
14. Devanur, N.R., Morgenstern, J., Syrgkanis, V., Weinberg, S.M.: Simple auctions with simple strategies. In: EC 2015, pp. 305–322 (2015)
15. Dobzinski, S., Fu, H., Kleinberg, R.D.: Optimal auctions with correlated bidders are easy. In: STOC 2011 (2011)
16. Feldman, M., Gravin, N., Lucier, B.: Combinatorial auctions via posted prices. In: SODA 2015 (2015)
17. Groves, T.: Incentives in teams. Econometrica **41**, 617–631 (1973)
18. Gupta, A., Nagarajan, V.: A stochastic probing problem with applications. In: Goemans, M., Correa, J. (eds.) IPCO 2013. LNCS, vol. 7801, pp. 205–216. Springer, Heidelberg (2013)
19. Hart, S., Nisan, N.: Approximate revenue maximization with multiple items. In: EC 2012 (2012)
20. Hartline, J.D., Roughgarden, T.: Simple versus optimal mechanisms. SIGecom Exch. **8**(1), 5:1–5:3 (2009)
21. Kleinberg, R., Leighton, T.: The value of knowing a demand curve: bounds on regret for online posted-price auctions. In: FOCS 2003 (2003)
22. Kleinberg, R., Weinberg, S.M.: Matroid prophet inequalities. In: STOC 2012 (2012)
23. Myerson, R.B.: Optimal auction design. Math. Oper. Res. **6**(1), 58–73 (1981)
24. Ronen, A.: On approximating optimal auctions. In: EC 2001 (2001)
25. Roughgarden, T., Talgam-Cohen, I.: Optimal and near-optimal mechanism design with interdependent values. In: EC 2013 (2013)
26. Rubinstein, A., Weinberg, S.M.: Simple mechanisms for a subadditive buyer and applications to revenue monotonicity. In: EC 2015, pp. 377–394 (2015)

27. Sandholm, T.W., Gilpin, A.: Sequences of take-it-or-leave-it offers: near-optimal auctions without full valuation revelation. In: Faratin, P., Parkes, D.C., Rodríguez-Aguilar, J.-A., Walsh, W.E. (eds.) AMEC 2003. LNCS (LNAI), vol. 3048, pp. 73–91. Springer, Heidelberg (2004)
28. Segal, I.: Optimal pricing mechanisms with unknown demand. Am. Econ. Rev. **93**(3), 509–529 (2003)
29. Vickrey, W.: Counterspeculation, auctions, and competitive sealed tenders. J. Financ. **16**(1), 8–37 (1961)
30. Yan, Q.: Mechanism design via correlation gap. In: SODA 2011 (2011)

Price Competition in Networked Markets: How Do Monopolies Impact Social Welfare?

Elliot Anshelevich and Shreyas Sekar[✉]

Rensselaer Polytechnic Institute, Troy, NY, USA
eanshel@cs.rpi.edu, sekars@rpi.edu

Abstract. We study the efficiency of allocations in large markets with a network structure where every seller owns an edge in a graph and every buyer desires a path connecting some nodes. While it is known that stable allocations can be very inefficient, the exact properties of equilibria in markets with multiple sellers are not fully understood, even in single-source single-sink networks. In this work, we show that for a large class of buyer demand functions, equilibrium always exists and allocations can often be close to optimal. In the process, we characterize the structure and properties of equilibria using techniques from min-cost flows, and obtain tight bounds on efficiency in terms of the various parameters governing the market, especially the number of monopolies M.

Although monopolies can cause large inefficiencies in general, our main results for single-source single-sink networks indicate that for several natural demand functions the efficiency only drops linearly with M. For example, for concave demand we prove that the efficiency loss is at most a factor $1 + \frac{M}{2}$ from the optimum, for demand with monotone hazard rate it is at most $1 + M$, and for polynomial demand the efficiency decreases logarithmically with M. In contrast to previous work that showed that monopolies may adversely affect welfare, our main contribution is showing that monopolies may not be as 'evil' as they are made out to be. Finally, we consider more general, multiple-source networks and show that in the absence of monopolies, mild assumptions on the network topology guarantee an equilibrium that maximizes social welfare.

1 Introduction

The mechanism governing large decentralized markets is often straightforward: sellers post prices for their goods and buyers buy bundles that meet their requirements. Given this framework, the challenge faced by researchers has been to characterize the equilibrium states at which these markets operate. More concretely, consider a market with multiple sellers that can be represented by a directed graph G as follows:

- Every seller owns an item, which is a link in the network.
- Every infinitesimal buyer seeks to purchase a path in the network (set of items) connecting some pair of nodes.

This work was partially supported by NSF awards CCF-1101495 and CNS-1218374.

E. Markakis and G. Schäfer (Eds.): WINE 2015, LNCS 9470, pp. 16–30, 2015.
DOI: 10.1007/978-3-662-48995-6_2

In addition to actual bandwidth markets where users purchase capacity on links for routing traffic, networks are commonly used in the literature to model combinatorial markets where the items are a mix of substitutes and complements [3,14,18]. For instance, in a computer market, each link could represent some component (e.g., a processor or video card) and buyers require a set of parts to assemble a complete computer system. In ad-markets, the buyers (advertisers) may want to purchase ads from a satisfactory combination of websites to reach a target audience. Our goal in this paper is to analyze the effects of *price competition* in such networked markets, i.e., the pricing strategies employed by competing sellers and their effect on equilibrium welfare.

An extensive body of work has culminated in the design of pricing mechanisms for a variety of markets with a single central seller (for example, see [9,15,17] and the references therein). In contrast, there has been very little focus on even simple decentralized markets where multiple price-setting sellers operate, and buyers require bundles of goods. With the exception of a few specific but incomparable settings (homogeneous goods [7,8], single buyer [10]), our understanding of how different parameters affect equilibrium in markets with price competition is quite limited. With this in mind, we seek to answer the following questions:

1. What conditions on the market structure guarantee equilibrium existence?
2. How efficient are the equilibrium allocations and how do they depend on *buyer demand* and *network topology*?

Model and Equilibrium Concept. We model the interaction between buyers and sellers as a two-stage pricing game. Each seller e controls a single good or link in a network G; he can produce any quantity x of this good incurring a production cost of $C_e(x)$. Every buyer i in the market wants to purchase an infinitesimal amount of some path connecting a source and a sink node for which she receives a value v_i. For the majority of this work, we will focus on *single-source, single-sink networks*, i.e., markets where every buyer wants to purchase a path between the same source node s and sink node t. Such networks capture combinatorial markets where buyers are interested in a single type of good; e.g., all buyers desire a computer but may have different valuations (v_i) for the same.

We consider a full information game where sellers can estimate the aggregate demand. In the first stage of the game, sellers set prices on the edges and in the second stage, buyers buy edges along a path. For any seller e, if at a price of p_e per unit amount of the good, a population x_e of buyers purchase the good, then the profit is $p_e x_e - C_e(x_e)$. The buyer's utility is v_i minus the total price paid. A solution is said to be a Nash Equilibrium if (i) Every buyer receives a utility maximizing bundle, i.e., the cheapest s-t path with price at most v_i, (ii) No seller can unilaterally change his price and improve his profit at the new allocation.

Bertrand Competition with Monopolies. Our work is most closely related to the model of Bertrand Competition in networks with supply limited sellers studied in [14] and later in [13]. Our model is more general as the production costs (that we consider) are a substantial generalization of limited supply. The behavior of Bertrand networks with seller costs was posed as an open question in [14]. We address this question by applying techniques from the theory of min-cost flows. The above papers also considered the efficiency of supply-limited markets and showed that in the worst case the equilibrium solution can be arbitrarily worse than the social optimum, and in some special cases it decreases exponentially with the length of longest s-t path in the network. Our paper provides a nuanced understanding of efficiency in terms of the buyer demand and the network topology. One of our high-level contributions in this paper is breaking down the dependence of efficiency on topology into a single parameter M: the number of monopoly edges in the graph $G = (V, E)$.

(Set of monopolies in G) $\mathcal{M} := \{e \mid (s,t) \text{ are disconnected in } (V, E - \{e\})\}$.

Monopolies offer a natural market interpretation: these are the items which are not substitutable. It is not surprising, although also not obvious, that monopolies are the main cause of inefficiencies in markets where the items are a mix of substitutes and complements. What may be extremely surprising, and what we view as one of the main contributions of our paper, is that in many reasonable settings the effect of the monopolies on equilibrium efficiency is very limited. Our results show that having a few monopolies is still not so bad: high inefficiency only occurs when the number of monopolies is large. This is in contrast to conventional wisdom that monopolies are 'evil', and even a single monopoly can cause a significant loss in social welfare [22]. More concretely, our main result is that for a large class of natural demand functions, equilibrium not only exists, but the loss in efficiency is at most a factor $(1 + M)$ from the optimum solution. We interpret this as a positive result for the following reason:

- Given previous results [13,14] that in the worst-case, social welfare can drop exponentially as M increases, a linear loss in welfare for many natural market types establishes a crucial separation between theoretical worst-case analysis and settings that are more likely to arise.

The Inverse Demand Function. In this work, our primary focus will be on single-source single-sink networks where every buyer has a different value v_i, although we do look at more general models in Sect. 5. In markets with many buyers, it is common to consider a 'full information in the large' game where the sellers know exactly how many buyers value the s-t path at v or more. This can be estimated, for instance, using prior data. Formally, we define an inverse demand function $\lambda(x)$ such that for any v, $\lambda(x) = v$ implies that exactly x amount of buyers value the path at v or larger. For example, suppose that $\lambda(x) = 1 - x$. Then, $\lambda(0.25) = 0.75$, i.e., one-fourth of the buyers have a value of 0.75 or more for the s-t paths.

1.1 Our Contributions

Our objective in this paper is to characterize the quality of equilibrium in terms of the inverse demand function, and specifically to show the effect of monopolies on efficiency. Therefore, our efficiency bounds depend only on the number of monopolies $M = |\mathcal{M}|$. Note that we define efficiency to be the ratio of the optimum social welfare of the market to that at equilibrium.

Single-Source Single-Sink Networked Markets

Our first results concern existence and uniqueness. We show that:

1. There exists a Nash Equilibrium Pricing in every market under a very mild assumption on the demand function. Moreover, there exists a Nash Equilibrium Pricing satisfying several desirable properties, including individual rationality, Pareto-optimality, and robustness to small perturbations. We call such a solution a *focal equilibrium*.
2. We further prove the uniqueness of focal equilibria. Our result is constructive: we explicitly characterize the prices and allocations in the focal equilibrium and provide an algorithm to efficiently compute them.

Since the focal equilibrium solution is the unique one satisfying many properties that one would expect from a market equilibrium, we believe that this is the correct equilibrium to study, and the one which is likely to arise in a real system. Because of this we mainly focus on analyzing the efficiency of focal equilibria.

Efficiency. We consider the following hierarchy of inverse demand functions

$$\text{Uniform} \subset \text{Polynomial} \subset \text{Concave} \subset \text{Log-Concave} = \text{MHR}.$$

Our main result is that for every function in this hierarchy, the efficiency of the focal equilibrium drops only linearly as the number of monopolies increases. Specifically, we show the following,

(Informal Theorem). *If the inverse demand function has a monotone hazard rate (MHR), the loss in efficiency at equilibrium is bounded by a factor of $1 + M$.*

This result is quite general as the MHR class encapsulates all demand functions satisfying log-concavity. Moreover, some of the popular demand functions considered in the literature happen to be Concave or Polynomial (see Sect. 2 for examples). We show improved efficiency bounds for these classes, namely,

- (Uniform Demand) The Nash equilibrium maximizes welfare.
- (Polynomial Demand) Efficiency drops logarithmically as M increases.
- (Concave Demand) The efficiency loss is $1 + \frac{M}{2}$.

All of our efficiency bounds are tight. The main conclusion to draw from this is that monopolies do not completely destroy efficiency: it crucially depends on the nature of buyer demand and the *number* of these monopolies. We reiterate that since production costs strictly generalize limited supply, all of our efficiency bounds hold for the type of market considered in [13, 14] as well. We make absolutely no assumption on the production cost function other than convexity, which is standard in the literature.

Multiple-Source Networks. We provide a first step towards understanding efficiency in multiple-source networked markets by tackling a question of special interest: what conditions cause equilibrium to be fully efficient in such markets? Our main result is the following: *even when buyers desire different paths, as long as the network has a series-parallel topology, the absence of monopolies guarantees an efficient equilibrium.* In contrast, without the series-parallel structure, even simple networks with no monopolies may have inefficient equilibria. We also show conditions on the buyer demand that lead to optimal equilibrium. We briefly discuss our novel contributions and the techniques that enable our results:

1. Production costs are a non-trivial addition to the Bertrand model. In particular, the pricing strategies used in [13,14] do not extend to our model as we cannot price all non-monopoly edges at zero and choose equal prices for the monopolies. Instead, we extensively apply techniques from the theory of min-cost flows to compute equilibrium prices. Specifically, the property that the flow is 'balanced' across paths is utilized to set prices on the edges.
2. In order to show efficiency bounds for MHR demand, we establish a new connection between the sellers' profit and the 'lost welfare' at equilibrium. This approach may be useful in other settings involving MHR functions.

Relation to Other Concepts. For markets with multiple buyers and sellers, the standard solution concept used in the literature is the *Walrasian Equilibrium*: a set of prices such that when both buyers and sellers act as 'price-takers', the market clears. Walrasian Equilibria are indeed attractive: they always exist in large markets [6] and are often guaranteed to be optimal. However, the idea that prices are just 'handed out' so that the market clears may not be applicable in a decentralized market. In contrast, the body of work on *price-setting* sellers (e.g., [2,7,21]) takes the view that the sellers control their own prices in order to maximize profit. Therefore, our motivation is to analyze the two-stage game where sellers set prices and buyers purchase bundles. Our work also differs from the papers in Mechanism Design that study settings with strategic buyers and a single seller [17]. Instead, we consider a market with many strategic sellers and a continuum of buyers. In such a model, it is reasonable to assume that buyers behave as price-taking agents since their individual demand is infinitesimal.

1.2 Related Work

As mentioned earlier, the study of Bertrand competition in networks was initiated in [13,14], which gave worst case bounds on efficiency over all demand functions. Despite our model being more general, we show that for many important classes of demand, the efficiency is much better than the bound shown in the above papers. Price competition between sellers was also studied in [10], where it was shown that in markets with a single buyer, equilibrium allocations are efficient. The Uniform demand case that we study is similar in spirit to what they consider, but our main results are for more complex demand functions. Finally, our work bears broad similarities to recent papers that also study existence or efficiency in somewhat specific settings with multiple sellers [7,8,19]. However, their models are not comparable to ours. In [8], all sellers possess a single homogeneous good but buyers may not have access to all of them; in [7,19], there is a single buyer but sellers may own more than one good. In contrast, we consider a market with multiple buyers where every seller controls one good but the goods are not homogeneous.

Some researchers have also considered more sophisticated pricing mechanisms like *non-linear pricing*(see [16,18]). While complex mechanisms do sometimes lead to an improvement in efficiency, they are not commonly used as they impose a large overhead on buyers who have to anticipate the change in price due to others' demands. In this work, we study the more natural fixed pricing mechanism and attempt to provide additional insight on the quality of equilibrium.

Finally, one line of research that has gained traction in recent years [2,3] is pricing in networked markets with congestion, i.e., buyers pay the price on each edge, but also incur a delay due to congestion. In contrast, we share the view taken by Shenker et al. [21] that 'congestion costs are inherently inaccessible to the network'. Due to the underlying complexities of this model, most of the results are only known for simple networks such as parallel paths. One exception is [20], which considers a unique one-sided model where the routing decisions are taken locally by sellers and not buyers as in our paper. They show that in the absence of monopolies, local decisions by sellers can result in efficient solutions.

2   Definitions and Preliminaries

An instance of our two-stage game is specified by a directed graph $G = (V, E)$, a source and a sink (s, t), an inverse demand function $\lambda(x)$ and a cost function $C_e(x)$ on each edge. There is a population T of infinitesimal buyers; every buyer wants to purchase edges on some s-t path and x amount of buyers hold a value of $\lambda(x)$ or more for these paths. A buyer is satisfied if she purchases all the edges on some path connecting s and t and is indifferent among the different paths.

We define M to be the number of monopolies in the market: an edge e is a monopoly if removing it disconnects the source and sink. We make the following standard assumptions on the demand and cost functions.

1. The inverse demand function $\lambda(x)$ is continuous on $[0, T]$ and non-increasing, implying that demand decreases as price increases.
2. $C_e(x)$ is non-decreasing and convex $\forall e$, which is the standard way to model production costs. Moreover, $C_e(x)$ is continuous, differentiable, and its derivative $c_e(x) = \frac{d}{dx}C_e(x)$ satisfies $c_e(0) = 0$.

Nash Equilibrium Pricing. A solution of our two-stage game is a vector of prices on each item $\boldsymbol{p}$ and an allocation or flow $\boldsymbol{x}$ of the amount of each s-t path purchased, representing the strategies of the sellers and buyers respectively. The total flow or market demand is equal to the number of buyers with non-zero allocation $x = \sum_{P \in \mathbb{P}} x_P$, where $\mathbb{P}$ is the set of s-t paths. We can also decompose this flow x into the amount of each edge purchased by the buyers $(x_e)_{e \in E}$. Given this solution, the total utility of the sellers is $\sum_{e \in E}(p_e x_e - C_e(x_e))$ and the aggregate utility of the buyers is $\int_{t=0}^{x} \lambda(t)dt - \sum_{e \in E} p_e x_e$. The social welfare is simply $\int_{t=0}^{x} \lambda(t)dt - \sum_e C_e(x_e)$, i.e., prices are intrinsic to the system and do not appear in the welfare.

We use the standard definition of Nash equilibrium for two-stage games to model the stable states of our market. Formally, an allocation x is said to be a **best-response** by the buyers to prices p if buyers only buy the cheapest paths and for any cheapest path P, $\lambda(x) = \sum_{e \in P} p_e$. That is, buyers act as price-takers and any buyer whose value is at least the price of the cheapest path will purchase some such path. A solution (p, x) is a Nash equilibrium if x is a best-response allocation to the prices and, $\forall e$ if the seller unilaterally changes his price from p_e to p'_e, then for *every* feasible best-response flow (x'_e) for the new prices, seller e's profit cannot increase, i.e., $p_e x_e - C_e(x_e) \geq p'_e x'_e - C_e(x'_e)$. Our notion of equilibrium is quite strong as the seller does not have to anticipate the resulting flow: for every best-response by the buyers, the seller's profit should not increase.

Classes of inverse demand functions that we are interested in For ease of exposition, we assume that both the inverse demand and the production costs are continuously differentiable. However, **all our results** hold exactly even without this assumption. Note that $\lambda'(x)$ cannot be positive since $\lambda(x)$ is non-increasing. The reader is asked to refer to the full version of this paper [5] for additional discussion on these classes of demand.

Uniform Demand: $\lambda(x) = \lambda_0 > 0$ for $x \leq T$. In other words, a population of T buyers all have the same value λ_0 for the bundles.

Polynomial Demand: $\lambda(x) = \lambda_0(a - x^\alpha)$ for $\alpha \geq 1$. Polynomial demand functions are quite popular [11], especially linear inverse demand ($\lambda(x) = a - x$).

Concave Demand: $\lambda'(x)$ is a non-increasing function of x.

Monotone Hazard Rate (MHR) Demand: $\frac{\lambda'(x)}{\lambda(x)}$ is non-increasing or $h(x) = \frac{|\lambda'(x)|}{\lambda(x)}$ is non-decreasing in x. This is equivalent to the class of *log-concave* functions [4] where $\log(\lambda(x))$ is concave. Example function: $\lambda(x) = e^{-x}$.

It is not hard to see that Uniform[1] $\subset$ Polynomial $\subset$ Concave $\subset$ MHR. We remark that the MHR and Concave classes are quite general whereas Uniform or Polynomial demand are more common due their tractability.

Min-Cost Flows and the Social Optimum: Since an allocation vector x is equivalent to a s-t flow, we briefly dwell upon minimum cost flows. Formally, we define $R(x)$ to be the cost $\sum_e C_e(x_e)$ of the min-cost flow of magnitude $x \geq 0$ and $r(x)$, its derivative, i.e., $r(x) = \frac{d}{dx}R(x)$. Both the flow and its cost can be computed via a simple convex program given the graph. The min-cost function $R(x)$ obeys several desirable properties that we use later including:

---

[1] Uniform $= \lim_{\alpha \to \infty} \lambda_0(1 - x^\alpha)$.

Proposition 1. $R(x)$ *is continuous, non-decreasing, differentiable, and convex.*

From the KKT conditions, we have that for a min-cost flow $\boldsymbol{x}$, $r(x) = \sum_{e \in P} c_e(x_e)$ for any path P with non-zero flow. Using this property, we obtain the following characterization of the welfare maximizing solution in terms of $R(x)$.

Proposition 2. *The solution maximizing social welfare is a min-cost flow of magnitude x^* satisfying $\lambda(x^*) \geq r(x^*)$. Moreover, $\lambda(x^*) = r(x^*)$ unless $x^* = T$.*

3   Existence, Uniqueness, and Computation

In this section, we prove that a Nash equilibrium is guaranteed to exist under the very mild assumption that the demand function has a monotone *price elasticity*. Moreover, we show that there always exists a unique 'focal equilibrium' that satisfies several desirable properties. We also provide an algorithm to compute this important equilibrium.

Before proving our general existence result, it is important to understand the different types of equilibria that may exist in networked markets. In markets such as ours, an existence result by itself is meaningless because a large sub-class of instances admit trivial and unrealistic equilibria.

Trivial Equilibrium: In a networked market where all paths have a length of at least 2, it is easy to see that every seller setting an unreasonably high price (say larger than $\lambda(0)$) would result in a Nash equilibrium with zero flow. The existence of such unrealistic equilibria was also observed in [13], where they were referred to as trivial equilibria.

Our goal in this paper is to analyze the equilibrium operating states of actual markets. Given that our model admits such uninteresting equilibria, it is important that any existence result be characterized by properties that one might come to expect from equilibria that are likely to arise in practice; for example, one might expect that a meaningful equilibrium has non-zero flow, is not dominated by other equilibria and most importantly from the perspective of a large market, is robust to small perturbations (we define these formally below). Our main existence result is that under a very mild condition on the demand, there exists a 'nice' equilibrium that satisfies many such desiderata.

We first formally define what it means for the price elasticity of a demand function to be monotone. This condition is quite minimal: it is obeyed by almost all of the demand functions in the literature (for example: [1, 4, 11, 13]).

Definition 3. *Monotone Price Elasticity (MPE) An inverse demand function $\lambda(x)$ is said to have a monotone price elasticity if its price elasticity $\frac{x|\lambda'(x)|}{\lambda(x)}$ is a non-decreasing function of x which approaches zero as $x \to 0$.*

All the classes of demand functions listed in the previous section satisfy the MPE condition. At a high level, the MPE condition simply implies that a market's responsiveness at low prices cannot be too large compared to its responsiveness at a high price. Even more intuitively, MPE functions are concave if plotted on a log-log plot, and are essentially all functions which are "less convex" than x^{-r}.

Theorem 4. *For any given instance of a networked market where the inverse demand function $\lambda(x)$ has a monotone price elasticity, there exists a Nash Equilibrium $(p)_{e \in E}, (\tilde{x})_{e \in E}$ satisfying the following properties*

1. *Non-Trivial Pricing (Non-zero flow)*
2. *Recovery of Production Costs (Individual Rationality)*
3. *Pareto-Optimality*
4. *Local Dominance (Robustness to small perturbations)*

We now formally define these properties and argue why it is reasonable to expect an actual market equilibrium to satisfy them. For example, although they are not stable solutions for our price-setting sellers, it is not hard to see that Walrasian Equilibria satisfy all of these properties.

1. **(Non-Trivial Pricing):** Every edge that does not admit flow must be priced at 0. This guarantees that the equilibrium has non-zero flow.
2. **(Recovery of Production Costs):** Given an equilibrium (p, x), every item's price is at least $c_e(x_e)$. This property is similar in spirit to *individual rationality* and ensures that the prices are fair to the sellers. Suppose that $p_e < c_e(x_e)$, this means that the seller is selling at least some fraction of his items at price smaller than its cost of production, and therefore, would have no incentive to produce the given quantity of items.
3. **(Pareto-Optimality):** A Pareto-optimal solution over the space of equilibria is an equilibrium solution such that for any other equilibrium, at least one agent (buyer or seller) prefers the former solution to the latter. Pareto-Optimality is often an important criterion in games with multiple equilibria; research suggests that in Bertrand Markets, Pareto optimal equilibria are the solutions that arise in practice [12].
4. **(Local Dominance):** Given an equilibrium (p, x), consider a different flow assignment for the same prices (p, x'), differing only in which cheapest paths are taken by the buyers. Local Dominance means that the profit of each seller must be larger at the equilibrium solution than at any (p, x'). The essence of this property is that the solution is resilient against small buyer perturbations. In other words, if instead of changing his price (which we know no seller would do at equilibrium), a seller instead convinced some buyers to take different paths of the same total price, then this seller still could not benefit from the resulting new flow. If this were not the case, then a seller may be able to attract a small fraction of buyers towards his item and improve his profit, indicating that the original equilibrium is not robust.

(Proof Sketch of Theorem 4) The proof proceeds by analyzing the behavior of monopolies and non-monopolies at equilibrium: every monopoly behaves as if it is a part of a two-link serial network where the rest of the network can be composed into a single serial link. This allows us to derive a sufficient condition on a

monopoly's price at equilibrium that is independent of every other link (namely, $p_e = c_e(\tilde{x}) + \tilde{x}|\lambda'(\tilde{x})|$). In contrast, non-monopolies at equilibrium behave as if they are a part of a two-edge parallel network. A crucial ingredient of our result is the application of min-cost flows to link the behavior of monopolies and non-monopolies. Namely, the property that the (marginal) flow cost is balanced across all paths is used to choose the price of every edge. Once we have explicitly constructed the equilibrium prices, the rest of the theorem involves showing that these prices result in a non-trivial best-response flow. Note that standard techniques such as fixed point theorems cannot be used here since the solution space is not convex: small changes in price may result in large deviations. ■

The full proofs of all the theorems can be found in a full version of this paper [5]. The next corollary, which is the main ingredient in all of our efficiency bounds essentially characterizes the equilibrium structure by expressing the equilibrium flow $(\tilde{x})$ as a function of only the number of monopolies in the network.

Corollary 5. *For any demand λ satisfying the MPE condition, $\exists$ a Nash equilibrium with a min-cost flow $(\tilde{x}_e)$ of size $\tilde{x} \leq x^*$ such that,*

$$\text{Either} \quad \frac{\lambda(\tilde{x}) - r(\tilde{x})}{M} = \tilde{x}|\lambda'(\tilde{x})| \text{ or } \tilde{x} = x^*, \text{ the optimum solution.}$$

We now show that the equilibrium from Theorem 4 (which we will refer to as the *focal equilibrium*) is the unique solution that satisfies the useful desiderata defined above. In order to truly understand the equilibrium efficiency of our two-stage game, it does not make sense to show a blanket bound on all stable solutions since some of these are highly unrealistic (for example, Price of Anarchy is almost always unbounded due to the presence of trivial equilibria). However, since the focal equilibrium solution is the unique one satisfying many properties that one would expect from a market equilibrium, we focus on analyzing its efficiency in the rest of this paper.

Theorem 6. *For any given instance with strictly monotone MPE demand and non-zero costs, we are guaranteed that one of the following is always true:*

1. *There is a unique non-trivial equilibrium that satisfies Local Dominance. (or)*
2. *All non-trivial equilibria that satisfy Local Dominance maximize welfare.*

Moreover, we can compute this equilibrium efficiently.

For the purposes of studying efficiency, the above theorem provides a useful baseline: either all equilibria are fully efficient or it suffices to bound the efficiency of the unique equilibrium that satisfies Corollary 5 (which we do in Sect. 4). As always in the case of real-valued settings (e.g., convex programming, etc.), "computing" a solution means getting within arbitrary precision of the desired solution; the exact solution could be irrational.

4   Effect of Monopolies on the Efficiency of Equilibrium

In this paper, we are interested in settings where approximately efficient outcomes are reached despite the presence of self-interested sellers with monopolizing power. While for general functions $\lambda(x)$ obeying the MPE condition, the efficiency can be exponentially bad, we show that for many natural classes of functions it is much better, even in the presence of monopolies.

We begin with a more fundamental result that reinforces the fact that even in arbitrarily large networks (not necessarily parallel links), competition results in efficiency, i.e., when $M = 0$, the efficiency is 1. This result is only a starting point for us since it is the addition of monopolies that leads to interesting behavior.

Claim 7. *In any network with no monopolies (i.e., you cannot disconnect s, t by removing any one edge), there exists a focal Nash Equilibrium maximizing social welfare.*

We remark that our notion of a "no monopoly" graph is weaker than what has been considered in some other papers [16,20] and therefore, our result is stronger. We are now in a position to show our main theorem. The largest class of inverse demand functions that we consider are the MHR or Log-Concave functions. Note that all MHR functions satisfy the MPE condition and thus existence is guaranteed. Our main result is that for all demand functions in this class, the efficiency loss compared to the optimum solution is $1 + M$. We believe that this result has strong implications. First, log-concavity is a very natural assumption on the demand; these functions have received considerable attention in Economics literature(see [4] and follow-ups). Secondly, it is reasonable to assume that even in multi-item markets, the number of purely monopolizing goods is not too large: in such cases the equilibrium quality is high.

Theorem 8. *The social welfare of the Nash equilibrium from Sect. 3 is always within a factor of $1 + M$ of the optimum for MHR λ, and this bound is tight.*

(Proof Sketch). The proof relies crucially on our characterization of equilibria obtained in Corollary 5, and the following interesting claim for MHR functions linking the welfare loss at equilibrium to the profit made by all the sellers: 'the loss in welfare is at most a factor M times the total profit in the market at equilibrium'. In addition, it is also not hard to see that in any market, the profit cannot exceed the total social welfare of a solution.

Why is this claim useful? Using profit as an intermediary, we can now compare the welfare lost at equilibrium to the welfare retained. This implies that the welfare loss cannot be too high because that would mean that the profit and hence the welfare retained is also high. But then, the sum of welfare lost + retained is the optimum welfare and is bounded. Therefore, we can immediately bound the overall efficiency. Mathematically, our key claim is,

$$\text{Lost Welfare} = \int_{\tilde{x}}^{x^*} \lambda(x)dx - [R(x^*) - R(\tilde{x})] \leq M(p\tilde{x} - R(\tilde{x})),$$

where p is the payment made by every buyer, $\tilde{x}$ is the amount of buyers in the equilibrium solution and x^*, in the optimum. The integral in the LHS can be rewritten as $\int_{\tilde{x}}^{x^*} (\lambda(x) - r(x))dx$. Now, we apply some fundamental properties of MHR functions (λ) and show that for all $x \geq \tilde{x}$, the following is true, $\frac{\lambda(x)-r(x)}{|\lambda'(x)|} \leq \frac{\lambda(\tilde{x})-r(\tilde{x})}{|\lambda'(\tilde{x})|} = M\tilde{x}$. The final equality comes from our equilibrium characterization in Corollary 5. Therefore, we can prove the key claim as follows:

$$\int_{\tilde{x}}^{x^*} (\lambda(x) - r(x))dx \leq M\tilde{x} \int_{\tilde{x}}^{x^*} |\lambda'(x)|dx$$

$$\leq M\tilde{x}(\lambda(\tilde{x}) - \lambda(x^*)) \quad (\lambda(x) \text{ is non-increasing and } \tilde{x} \leq x^*)$$

$$\leq M(\lambda(\tilde{x})\tilde{x} - R(\tilde{x})) \quad (\lambda(x^*)\tilde{x} \geq r(x^*)\tilde{x} \geq r(\tilde{x})\tilde{x} \geq R(\tilde{x}))$$

The total payment p on any path must exactly equal $\lambda(\tilde{x})$ ∎

Tighter Bounds for Sub-classes. We now consider log-concave demand functions that satisfy additional requirements, namely Uniform, Polynomial, and Concave demand. For these classes, we show much stronger bounds on the efficiency loss at equilibrium.

Theorem 9. *The following bounds on the efficiency are tight*

- *Every instance with **Uniform** demand admits a fully efficient focal Nash equilibrium.*
- *For any instance with **Polynomial** demand, the inefficiency of focal equilibrium is at most $(1 + M\alpha)^{\frac{1}{\alpha}}$, where $\alpha \geq 1$ is the degree of the polynomial. When $\alpha \geq M$, this quantity is approximately $1 + \frac{\log(M\alpha)}{\alpha}$.*
- *When the demand is **Concave**, the inefficiency of focal equilibrium is $1 + \frac{M}{2}$.*

The efficiency bound for Polynomial demand extends to more general polynomials of the form $\lambda(x) = a_0 - \sum_{i=1}^{k} a_i x^{\alpha_i}$ with α now defined as $\min_i \alpha_i$.

Concave-Log Demand. In this paper, we considered MPE functions (concave on a log-log plot) and log-concave functions (concave on a semi-log plot). For the sake of completeness, we also consider functions that are not log-concave but still obey the MPE condition. One such important class consists of Concave-Log demand functions, which are in some sense the opposite of log-concave functions; in other words $\lambda(x)$ is concave against a logarithmically varying buyer demand (i.e., $\lambda(log(x))$ is concave). This class of functions was considered in [13], where an efficiency bound of e^D was shown: D being the length of the longest s-t path in the network which could potentially be much larger than M. We generalize their results to markets with cost functions, and further are able to improve upon the bound in [13].

Claim 10. *For any instance with concave-log demand, the inefficiency of the focal equilibrium is at most $\frac{M}{M-1}e^{M-1}$ for $M \geq 2$.*

We reiterate here that all of our results require no assumption on the graph structure and only the ones mentioned in Sect. 2 for the cost functions.

5   Generalizations: Multiple-Source Networks

We now move on to more general networks where different buyers have different s_i-t_i paths that they wish to connect and the demand function can be different for different sources. Unfortunately, our intuition from the previous sections does not carry over. Even when buyers have Uniform demand, Nash equilibrium may not even exist whereas in the single-source case, equilibrium was efficient. Perhaps more surprisingly, we give relatively simple examples in which perfect efficiency is no longer achieved in the absence of monopolies. Nevertheless, we prove that for some interesting special cases, fully efficient Nash equilibrium still exists even when buyers desire different types of bundles. In particular, we believe that our result on series-parallel networks is an important starting point for truly understanding multiple-source networks.

Claim 11. *There exist simple instances with two sources and one sink such that*

1. *Nash equilibrium may not exist even when the buyers at each source have Uniform demand.*
2. *All Nash equilibria are inefficient even when no edge is a monopoly.*

Series-Parallel Networks: In some sense, Claim 7 embodies the very essence of the Bertrand paradox, the fact that competition leads to efficiency. So it is surprising that this does not hold in general networks. However, we now show that for a large class of markets which have the series-parallel structure, the absence of monopolies still gives us efficient equilibria. Series-Parallel networks have been commonly used [16,18] to model the substitute and complementary relationship that exists between various products in combinatorial markets.

We define a multiple-source single-sink graph to be a series-parallel graph if the super graph of the given network obtained by adding a super-source and connecting it to all the sources has the series-parallel structure. The notion of "no-monopolies" for a complex network has the same idea as a single-source network: there is no edge in the graph such that its removal would disconnect any source from the sink. We are now in a position to show our result.

Theorem 12. *A multiple-source single-sink series-parallel network with no monopolies admits a welfare-maximizing Nash Equilibrium for any given demand.*

Finally, we show additional conditions on both the network topology and demand that lead to efficient equilibria, even in the presence of monopolies.

Claim 13. *There exists a fully efficient equilibrium in multiple-source multiple-sink networks with Uniform demand buyers at each source if one of the following is true: (i) Buyers have a large demand and production costs are strictly convex, (ii) Every source node is a leaf in the network.*

The second case commonly arises in several telecommunication networks, where the *last mile* between a central hub and the final user is often controlled by a local monopoly and thus the source is a leaf.

6   Conclusions

In this work, we initiate the study of Bertrand price competition in networked markets with *production costs*. Our results provide an improved understanding of how monopolies affect welfare in large, decentralized markets. Our main contribution is that as long as the inverse demand obeys a natural condition (monotone hazard rate), the efficiency loss is at most $1 + M$ for single-source single-sink networks, with stronger results for other important classes. Cast in the light of previous work [13,14], our result establishes that the inefficiency for commonly used demand is much better than the worst-case exponential inefficiency. Finally, for markets where buyers desire different paths, we identify series-parallel networks topology as a condition for efficiency. We believe this result is a useful first step in understanding the impact of monopolies on multiple-source networks. In a full version of this paper [5], we extend all our results to markets without a network structure where all buyers desire the same bundles.

References

1. Abolhassani, M., Bateni, M.H., Hajiaghayi, M.T., Mahini, H., Sawant, A.: Network cournot competition. In: Liu, T.-Y., Qi, Q., Ye, Y. (eds.) WINE 2014. LNCS, vol. 8877, pp. 15–29. Springer, Heidelberg (2014)
2. Acemoglu, D., Ozdaglar, A.: Competition and efficiency in congested markets. Math. Oper. Res. **32**(1), 1–31 (2007)
3. Acemoglu, D., Ozdaglar, A.: Competition in parallel-serial networks. IEEE J. Sel. Areas Commun. **25**(6), 1180–1192 (2007)
4. Amir, R.: Cournot oligopoly and the theory of supermodular games. Games Econ. Behav. **15**(2), 132–148 (1996)
5. Anshelevich, E., Sekar, S.: Price competition in networked markets: how do monopolies impact social welfare? arXiv preprint arXiv:1410.1113 (2015)
6. Azevedo, E.M., Weyl, E.G., White, A.: Walrasian equilibrium in large, quasilinear markets. Theor. Econ. **8**(2), 281–290 (2013)
7. Babaioff, M., Leme, R.P., Sivan, B.: Price competition, fluctuations and welfare guarantees. In: Proceedings of EC (2015)
8. Babaioff, M., Lucier, B., Nisan, N.: Bertrand networks. In: Proceedings of EC (2013)
9. Babaioff, M., Lucier, B., Nisan, N., Leme, R.P: On the efficiency of the walrasian mechanism. In: Proceedings of EC (2014)
10. Babaioff, M., Nisan, N., Leme, R.P: Price competition in online combinatorial markets. In: Proceedings of WWW (2014)
11. Bulow, J.I., Pfleiderer, P.: A note on the effect of cost changes on prices. J. Polit. Econ. **91**, 182–185 (1983)
12. Cabon-Dhersin, M.-L., Drouhin, N.: Tacit collusion in a one-shot game of price competition with soft capacity constraints. J. Econ. Manage. Strategy **23**(2), 427–442 (2014)
13. Chawla, S., Niu, F.: The price of anarchy in bertrand games. In: Proceedings of EC (2009)
14. Chawla, S., Roughgarden, T.: Bertrand competition in networks. In: Monien, B., Schroeder, U.-P. (eds.) SAGT 2008. LNCS, vol. 4997, pp. 70–82. Springer, Heidelberg (2008)

15. Chawla, S., Sivan, B.: Bayesian algorithmic mechanism design. SIGecom Exch. **13**(1), 5–49 (2014)

16. Correa, J.R., Figueroa, N., Lederman, R., Stier-Moses, N.E.: Pricing with markups in industries with increasing marginal costs. Math. Program., 1–42 (2008)

17. Hassidim, A., Kaplan, H., Mansour, Y., Nisan, N.: Non-price equilibria in markets of discrete goods. In: Proceedings of EC (2011)

18. Kuleshov, V., Wilfong, G.: On the efficiency of the simplest pricing mechanisms in two-sided markets. In: Goldberg, P.W. (ed.) WINE 2012. LNCS, vol. 7695, pp. 284–297. Springer, Heidelberg (2012)

19. Lev, O., Oren, J., Boutilier, C., Rosenschein, J.S.: The pricing war continues: on competitive multi-item pricing. In: Proceedings of AAAI (2015)

20. Papadimitriou, C.H., Valiant, G: A new look at selfish routing. In: Proceedings of ICS (2010)

21. Shenker, S., Clark, D., Estrin, D., Herzog, S.: Pricing in computer networks: reshaping the research agenda. ACM SIGCOMM Comput. Commun. Rev. **26**(2), 19–43 (1996)

22. Tullock, G.: The welfare costs of tariffs, monopolies, and theft. Econ. Inq. **5**(3), 224–232 (1967)

Computing Stable Coalitions: Approximation Algorithms for Reward Sharing

Elliot Anshelevich and Shreyas Sekar[✉]

Rensselaer Polytechnic Institute, Troy, NY, USA
eanshel@cs.rpi.edu, sekars@rpi.edu

Abstract. Consider a setting where selfish agents are to be assigned to coalitions or projects from a set $\mathcal{P}$. Each project $k \in \mathcal{P}$ is characterized by a valuation function; $v_k(S)$ is the value generated by a set S of agents working on project k. We study the following classic problem in this setting: "how should the agents divide the value that they collectively create?". One traditional approach in cooperative game theory is to study *core stability* with the implicit assumption that there are infinite copies of one project, and agents can partition themselves into any number of coalitions. In contrast, we consider a model with a finite number of non-identical projects; this makes computing both high-welfare solutions and core payments highly non-trivial.

The main contribution of this paper is a black-box mechanism that reduces the problem of computing a near-optimal core stable solution to the well-studied algorithmic problem of welfare maximization; we apply this to compute an approximately core stable solution that extracts one-fourth of the optimal social welfare for the class of subadditive valuations. We also show much stronger results for several popular sub-classes: anonymous, fractionally subadditive, and submodular valuations, as well as provide new approximation algorithms for welfare maximization with anonymous functions. Finally, we establish a connection between our setting and simultaneous auctions with item bidding; we adapt our results to compute approximate pure Nash equilibria for these auctions.

1 Introduction

"How should a central agency incentivize agents to create high value, and then distribute this value among them in a fair manner?" – this question forms the central theme of this paper. Formally, we model a combinatorial setting consisting of a set $\mathcal{P}$ of projects. Each project is characterized by a valuation function; $v_k(S)$ specifies the value generated when a set S of self-interested agents work on project k. The problem that we study is the following: compute an assignment of agents to projects to maximize social welfare, and provide *rewards* or *payments* to each agent so that no group of agents deviate from the prescribed solution.

For example, consider a firm dividing its employees into teams to tackle different projects. If these employees are not provided sufficient remuneration,

---

This work was partially supported by NSF awards CCF-1101495 and CNS-1218374.

E. Markakis and G. Schäfer (Eds.): WINE 2015, LNCS 9470, pp. 31–45, 2015.
DOI: 10.1007/978-3-662-48995-6_3

then some group could break off, and form their own startup to tackle a niche task. Alternatively, one could imagine a funding agency incentivizing researchers to tackle specific problems. More generally, a designer's goal in such a setting is to delicately balance the twin objectives of *optimality* and *stability*: forming a high-quality solution while making sure this solution is stable. A common requirement that binds the two objectives together is *budget-balancedness*: the payments provided to the agents must add up to the total value of the given solution.

Cooperative Coalition Formation. The question of how a group of agents should divide the value they generate has inspired an extensive body of research spanning many fields [4,8,16,20]. The notion of a 'fair division' is perhaps best captured by the **Core**: a set of payments so that no group of agents would be better off forming a coalition by themselves. Although the Core is well understood, implicit in the papers that study this notion is the underlying belief that there are infinite copies of one single project [2,7], which is often not realistic. For example, a tacit assumption is that if the payments provided are 'not enough', then every agent i can break off, and simultaneously generate a value of $v(i)$ by working alone; such a solution does not make sense when the number of projects or possible coalitions is limited. Indeed, models featuring selfish agents choosing from a finite set of *distinct* strategies are the norm while modeling real-life phenomena such as technological coordination, and opinion formation [12,14].

The fundamental premise of this paper is that many coalition formation settings feature multiple non-identical projects, each with its own (subadditive) valuation $v_k(S)$. Although our model allows for duplicate projects, the inherently combinatorial nature of our problem makes it significantly different from the classic problem with infinite copies of a single project. For example, in the classic setting with a single valuation $v(S)$, the welfare maximization problem is often trivial (complete partition when v is subadditive), and the stabilizing core payments are exactly the dual variables to the allocation LP [6]. This is not the case in our setting where even the welfare maximization problem is NP-Hard, and known approximation algorithms for this problem use LP-rounding mechanisms, which are hard to reconcile with stability. Given this, our main contribution is a poly-time approximation algorithm that achieves stability without sacrificing too much welfare.

Our Model. Given an instance $(\mathcal{N}, \mathcal{P}, (v_k)_{k \in \mathcal{P}})$ with N agents $(\mathcal{N})$ and m projects, a solution is an allocation $S = (S_1, \ldots, S_m)$ of agents to projects along with a vector of payments $(\bar{p})_{i \in \mathcal{N}}$. With unlimited copies of a project, *core stability* refers to the inability of any set of agents to form a group on their own and obtain more value than the payments they receive. The stability requirement that we consider is a natural extension of core stability to settings with a finite number of fixed projects. That is, when a set T of agents deviate to project k, they cannot displace the agents already working on that project (S_k). Therefore, the payments of the newly deviated agents (along with the payments of everyone else on that project) must come from the total value generated, $v_k(S_k \cup T)$. One could also take the Myersonian view [20] that 'communication is required for negotiation' and imagine that all the agents choosing project k $(S_k \cup T)$ together collaborate to improve their payments. Formally, we define a solution $(S, \bar{p})$ to be *core stable* if the following two conditions are satisfied,

(Stability) No set of agents can deviate to a project and obtain more total value for everyone in that project than their payments, i.e., for every set of agents T and project k, $\sum_{i \in T \cup S_k} \bar{p}_i \geq v_k(T \cup S_k)$.

(Budget-Balance) The total payments sum up to the social welfare (i.e., total value) of the solution: $\sum_{i \in \mathcal{N}} \bar{p}_i = \sum_{k \in \mathcal{P}} v_k(S_k)$.

It is not hard to see from the above properties that the value created from a project will go to the agents on that project only. Finally, we consider a full-information setting as it is reasonable to expect the central authority to be capable of predicting the value generated when agents work on a project.

(Example 1) We begin our work with an impossibility result: even for simple instances with two projects and four agents, a core stable solution need not exist. Consider $\mathcal{P} = \{1, 2\}$, and define $v_1(\mathcal{N}) = 4$ and $v_1(S) = 2$ otherwise; $v_2(S) = 1 + \epsilon$ for all $S \subseteq N$ $(v_1(\emptyset) = v_2(\emptyset) = 0)$. If all agents are assigned to project 1, then in a budget-balanced solution at least one agent has to have a payment of at most 1; such an agent would deviate to project 2. Instead, if some agents are assigned to project 2, then it is not hard to see that they can deviate to project 1 and the total utility goes from $3 + \epsilon$ to 4.

Approximating the Core. Our goal is to compute solutions that guarantee a high degree of stability. Motivated by this, we view core stability under the lens of approximation. Specifically, as is standard in cost-sharing literature [21], we consider relaxing one of the two requirements for core stability while retaining the other one. First, suppose that we generalize the Stability criterion as follows:

(α-Stability) For every set of agents T and every project k, $v_k(S_k \cup T) \leq \alpha \sum_{i \in S_k \cup T} \bar{p}_i$.

α-stability captures the notion of a 'switching cost' and is analogous to an *Approximate Equilibrium*; in our example, one can imagine that employees do not wish to quit the firm unless the rewards are at least a factor α larger. In the identical projects literature, the solution having the smallest value of α is known as the Multiplicative Least-Core [6]. Next, suppose that we only relax the budget-balance constraint,

(β-Budget Balance) The payments are at most a factor β larger than the welfare of the solution.

This generalization offers a natural interpretation: the central authority can subsidize the agents to ensure high welfare, as is often needed in other settings such as public projects or academic funding [4]. In the literature, this parameter β has been referred to as the *Cost of Stability* [1, 19].

We do not argue which of these two relaxations is the more natural one: clearly that depends on the setting. Fortunately, it is not difficult to see that these two notions of approximation are equivalent. In other words, every approximately core stable solution with α-stability can be transformed into a solution with α-budget balancedness by scaling the payments of every player by a factor α. Therefore, in the rest of this paper, we will use the term α-**core stable** without loss of generality to refer to either of the two relaxations. All our results can be interpreted either as forming fully budget-balanced payments which are α-stable, or equivalently as fully stable payments which are α-budget balanced. Finally, the problem that we tackle in this paper can be summarized as follows:

> (**Problem Statement**) Given an instance with subadditive valuation functions, compute an α-core stable solution $(S, (\bar{p})_{i \in \mathcal{N}})$ having as small a value of α as possible, that approximately maximizes social welfare.

1.1   Our Contributions

The problem that we face is one of bi-criteria approximation: to simultaneously optimize both social welfare and the stability factor α ($\alpha = 1$ refers to a core stable solution). For the rest of this paper, we will use the notation (α, c)-Core stable solution to denote an α-Core solution that is also a c-Approximation to the optimum welfare. In a purely algorithmic sense, our problem can be viewed as one of designing approximation algorithms that require the additional property of *stabilizability*.

Main Result. Our main result is the following black-box reduction that reduces the problem of finding an approximately core stable solution to the purely algorithmic problem of welfare maximization,

(Informal Theorem). For any instance where the projects have subadditive valuations, any LP-based α-approximation to the optimum social welfare can be transformed in poly-time to a $(2\alpha, 2\alpha)$-core stable solution.

The strength of this result lies in its versatility: our algorithm can stabilize *any input allocation* at the cost of half the welfare. The class of subadditive valuations is very general, and includes many well-studied special classes all of which use LP-based algorithms for welfare maximization; one can simply plug-in the value of α for the corresponding class to derive an approximately core stable solution. In particular, for general subadditive valuations, one can use the 2-approximation algorithm of Feige [11] and obtain a $(4, 4)$-Core stable solution. As is standard in the literature [10], we assume that our subadditive functions are specified in terms of a demand oracle (see Sect. 2 for more details).

For various sub-classes of subadditive valuations, we obtain stronger results by exploiting special structural properties of those functions. These results are summarized in Table 1. The classes that we study are extremely common and have been the subject of widespread interest in many different domains.

Lower Bounds. All of our results are 'almost tight' with respect to the theoretical lower bounds for both welfare maximization and stability. Even with anonymous functions, a $(2 - \epsilon)$ core may not exist; thus our $(2, 2)$-approximation

Table 1. Our results for different classes of complement-free valuations where Submodular $\subset$ XoS $\subset$ Subadditive, and Anonymous $\subset$ Subadditive. The results are mentioned in comparison to known computational barriers for welfare maximization for the same classes, i.e., lower bounds on c.

Valuation function class	Our results: (α, c)-Core	Lower bound for c
Subadditive	$(4, 4)$	2 [11]
Anonymous Subadditive	$(2, 2)$	2 [11]
Fractionally Subadditive (XoS)	$(1 + \epsilon, 1.58)$	1.58 [10]
Submodular	$(1 + \epsilon, 1.58)$ and $(1, 2)$	1.58 [22]

for this class is tight. For general subadditive functions, one cannot compute better than a 2-approximation to the optimum welfare efficiently [11], and so our $(4, 4)$ result has only a gap of 2 in both criteria. Finally, for XoS and Submodular functions, we get almost stable solutions ($(1 + \epsilon)$-Core) that match the lower bounds for welfare maximization.

Welfare Maximization for Anonymous Subadditive Functions. We devise a greedy 2-approximation algorithm for anonymous functions that may be of independent algorithmic interest. The only known 2-approximation algorithm even for this special class is the rather complex LP rounding mechanism for general subadditive functions [11]. In contrast, we provide an intuitive greedy algorithm that obtains the same approximation factor for welfare maximization.

Ties to Combinatorial Auctions with Item Bidding. We conclude by pointing out a close relationship between our setting and *simultaneous auctions* where buyers bid on each item separately [5,9]. Consider 'flipping' an instance of our problem to obtain the following combinatorial auction: every project $k \in \mathcal{P}$ is a buyer with valuation v_k, and every $i \in \mathcal{N}$ is an item in the market. We prove an equivalence between Core stable solutions in our setting and Pure Nash equilibrium for the corresponding flipped simultaneous second price auction. Adapting our lower bounds to the auction setting, we make a case for approximate Nash equilibrium by constructing instances where every exact Nash equilibrium requires buyers to *overbid* by a large factor ($O(\sqrt{N})$). Finally, we apply our earlier algorithms to efficiently compute approximate equilibria with small overbidding for two settings, namely, (*i*) a $\frac{1}{2}$-optimal, 2-*approximate equilibrium when buyers have anonymous subadditive valuations*, and (*ii*) a $(1 - \frac{1}{e})$-optimal, $1 + \epsilon$-*approximate equilibrium for submodular buyers.*

1.2   Related Work

Despite the staggering body of research on the *Core*, its non-existence in many important settings has prompted researchers to devise several natural relaxations: of these, the Cost of Stability [1,4,19], and the Multiplicative Least-Core [6] are the solution concepts that are directly analogous to our notion of an α-core stable solution. The overarching difference between our model, and almost all of the papers studying the core and its relatives is that while they study settings with **duplicate projects** having the same superadditive valuation, our model captures settings with multiple dissimilar projects where each project is a fixed resource with a subadditive valuation.

Although cooperative games traditionally do not involve any optimization, a number of papers have studied well-motivated games where the valuation or cost function is derived from an underlying combinatorial optimization problem [8, 16] such as vertex cover. Such settings are fundamentally different from ours because the hardness arises from the fact that the value of the cost function cannot be computed efficiently.

In the cooperative game theory literature, our setting is perhaps closest to the work studying coalitional structures where instead of forming the grand coalition, agents are allowed to arbitrarily partition themselves [1] or form overlapping coalitions [7]. This work has yielded some well-motivated extensions of the Core. Our work is similar in spirit to games where agents form coalitions to tackle specific tasks, e.g., threshold task games [7] or coalitional skill games [2]. In these games, there is still a single valuation function $v(S)$ which depends on the tasks that the agents in S can complete. Once again, the tacit assumption in these papers is that there are an infinite number of copies of each task.

Recently, there has been a lot of interest in designing cost-sharing mechanisms that satisfy strategy-proofness in settings where a single service is to be provided to a group of agents who hold private values for the same [15, 21]. In contrast, we look at a full information game with different projects where the central agency can exactly estimate the output due to a set of agents working on a project. Finally, [3] provides upper bounds on the efficiency of coalitional sink equilibria in transferable-utility settings; however, their results may not applicable here as it seems unlikely that our games are coalitionally smooth for general subadditive valuations. Moreover, their model of decentralized coalitional dynamics is somewhat orthogonal to our objective of computing stabilizable payments.

2   Model and Preliminaries

We consider a transferable-utility coalition formation game with a set $\mathcal{P}$ of m projects and a set $\mathcal{N}$ of N agents. Each project $k \in \mathcal{P}$ is specified by a monotone non-decreasing valuation function $v_k : 2^{\mathcal{N}} \to \mathbb{R}^+ \cup \{0\}$ with $v_k(\emptyset) = 0$. A solution consists of an allocation of agents to projects $S = (S_1, \ldots, S_m)$, and a payment scheme $(\bar{p})_{i \in \mathcal{N}}$ and is said to be (α, c)-core stable for $\alpha \geq 1$, $c \geq 1$ if

- The payments are fully budget-balanced, and for every project k, and set T of agents, $v_k(S_k \cup T) \leq \alpha \sum_{i \in S_k \cup T} \bar{p}_i$. An equivalent condition is that the payments are at most a factor α times the social welfare of the solution, and we have full stability, i.e., $v_k(S_k \cup T) \leq \sum_{i \in S_k \cup T} \bar{p}_i$.
- The allocation S is a c-approximation to the optimum allocation, i.e., the welfare of the solution S is at least $\frac{1}{c}$ times the optimum welfare.

Throughout this paper, we will use OPT to denote the welfare maximizing allocation as long as the instance is clear. Given an allocation $S = (S_1, \ldots, S_m)$, we use $SW(S) = \sum_{k=1}^{m} v_k(S_k)$ to denote the social welfare of this allocation.

Our solution concept is a reasonable extension of the Core for the setting at hand. Suppose that a set T of agents deviate to project k. Since, they cannot displace the agents already working on that project (S_k), the new welfare due to project k must come from both S_k and T ($v_k(S_k \cup T)$). No matter how this welfare is divided among these agents, our solution concept guarantees that not all agents will be happy after the deviation; implicit in this definition is the assumption that the agents in T cannot 'steal' the original utility (for e.g., funding) guaranteed to the agents in S_k by the central authority. Finally, in contrast to Strong Nash Equilibria [3], our solution concept does not impose a large communication overhead on the agents, for instance, when researchers (T) jointly choose projects, it is natural to prefer simple deviations (collaboration on a single project) instead of complex coordinations across many different projects.

Valuation Functions. Our main focus in this paper will be on monotone subadditive valuation functions. A valuation function v is said to be subadditive if for any two sets $S, T \subseteq \mathcal{N}$, $v(S \cup T) \leq v(S) + v(T)$, and monotone if $v(S) \leq v(S \cup T)$. The class of subadditive valuations encompasses a number of popular and well-studied classes of valuations, but at the same time is significantly more general than all of these classes. It is worth noting that when there are an unlimited number of allowed groups, subadditive functions are almost trivial to deal with: both the maximum welfare solution and the stabilizing payments are easily computable. For our setting, however, computing OPT becomes NP-Hard, and a fully core-stable solution need not exist. In addition, we are able to show stronger results for the following two commonly studied sub-classes of valuations.

Submodular Valuations For any two sets S, T with $T \subseteq S$, and any agent i,
$$v(S \cup \{i\}) - v(S) \leq v(T \cup \{i\}) - v(T).$$
Fractionally Subadditive (also called 'XoS') Valuations $\exists$ a set of additive functions $(a_1, \ldots, a_r)$ such that for any $T \subseteq \mathcal{N}$, $v(T) = \max_{j=1}^{r} a_j(T)$. These additive functions are referred to as *clauses*.

Recall that an additive function a_j has a single value $a_j(i)$ for each $i \in \mathcal{N}$ so that for a set T of agents, $a_j(T) = \sum_{i \in T} a_j(i)$. The reader is asked to refer to [10,11,18] for alternative definitions of the XoS class and an exposition on how both these classes arise naturally in many interesting applications.

Anonymous Subadditive Functions. In many project assignment settings in the literature [17,19], it is reasonable to assume that the value from a project depends only on the number of users working on that project. Mathematically, a valuation function is said to be anonymous or symmetric if for any two subsets S, T with $|S| = |T|$, we have $v(S) = v(T)$. One of our main contributions in this paper is a fast algorithm for the computation of Core stable solutions when the projects have anonymous subadditive functions. We remark here that anonymous subadditive functions form an interesting sub-class of subadditive functions that are quite different from submodular and XoS functions.

Demand Oracles. The standard approach in the literature while dealing with set functions (where the input representation is often exponential in size) is to assume the presence of an oracle that allows indirect access to the valuation by answering specific types of queries. In particular, when dealing with a subadditive function v, it is typical to assume that we are provided with a *demand oracle* that when queried with a vector of payments $\boldsymbol{p}$, returns a set $T \subseteq \mathcal{N}$ that maximizes the quantity $v(T) - \sum_{i \in T} p_i$ [10]. Demand oracles have natural economic interpretations, e.g., if $\boldsymbol{p}$ represents the vector of potential payments by a firm to its employees, then $v(T) - \sum_{i \in T} p_i$ denotes the assignment that maximizes the firm's revenue or surplus.

In this paper, we do not explicitly assume the presence of a demand oracle; our constructions are quite robust in that they do not make any demand queries. However, any application of our black-box mechanism requires as input an allocation which approximates OPT, and the optimum dual prices, both of which cannot be computed without demand oracles. For example, it is known [10] that one cannot obtain any reasonable approximation algorithm for subadditive functions in the absence of demand queries. That said, for several interesting valuations, these oracles can be constructed efficiently. For example, for XoS functions, this can be done in time polynomial in the number of input clauses.

2.1   Warm-Up Result: $(1, 2)$-Core for Submodular Valuations

We begin with an easy result: an algorithm that computes a core stable solution when all projects have submodular valuations, and also retains half the optimum welfare. Although this result is not particularly challenging, it serves as a useful baseline to highlight the challenges involved in computing stable solutions for more general valuations. Later, we show that by sacrificing an ϵ amount of stability, we can improve upon this result significantly.

Claim 1 *We can compute in poly-time a $(1, 2)$-Core stable solution for any instance with submodular project valuations.*

The above claim also implies that for every instance with submodular project valuations, there exists a Core stable solution. In contrast, for subadditive valuations, even simple instances (Example 1) do not admit a Core stable solution.

Proof: The proof uses the popular greedy half-approximation algorithm for submodular welfare maximization [18]. Initialize the allocation X to be empty. At every stage, add an agent i to project k so that the value $v_k(X_k \cup \{i\}) - v_k(X_k)$ is maximized. Set i's final payment $\bar{p}_i$ to be exactly the above marginal value. Let the final allocation once the algorithm terminates be S, so $\sum_{i \in S_k} \bar{p}_i = v_k(S_k)$. Consider any group of agents T, and some project k: by the definition of the greedy algorithm, and by submodularity, it is not hard to see that $\forall i \in T$, $\bar{p}_i \geq v_k(S_k \cup \{i\}) - v_k(S_k)$. Therefore, we have that $\sum_{i \in T} \bar{p}_i \geq v_k(S_k \cup T) - v_k(S_k)$, and since the payments are budget-balanced, the solution is core-stable. ∎

3   Computing Approximately Core Stable Solutions

In this section, we show our main algorithmic result, namely a black-box mechanism that reduces the problem of finding a core stable solution to the algorithmic problem of subadditive welfare maximization. We use this black-box in conjunction with the algorithm of Feige [11] to obtain a $(4, 4)$-Core stable solution, i.e., a 4-approximate core that extracts one-fourth of the optimum welfare. Using somewhat different techniques, we form stronger bounds $((2, 2)$-Core) for the class of anonymous subadditive functions. Our results for the class of anonymous functions are tight: there are instances where no $(2 - \epsilon, 2 - \epsilon)$-core stable solution exists. This indicates that our result for general subadditive valuations is close to tight (up to a factor of two).

We begin by stating the following standard linear program relaxation for the problem of computing the welfare maximizing allocation. Although the primal LP contains an exponential number of variables, the dual LP can be solved using the Ellipsoid method where the demand oracle serves as a separation oracle [10]. The best-known approximation algorithms for many classes of valuations use LP-based rounding techniques; of particular interest to us is the 2-approximation for Subadditive valuations [11], and $\frac{e}{e-1}$-approximation for XoS valuations [10].

$$\max \quad \sum_{k=1}^{M} \sum_{S \subseteq \mathcal{N}} x_k(S) v_k(S) \qquad (D) \quad \min \quad \sum_{i=1}^{N} p_i + \sum_{k=1}^{M} z_k$$

$$\text{s.t.} \quad \sum_{k=1}^{M} \sum_{S \ni i} x_k(S) \le 1 \quad \forall i \in \mathcal{N} \qquad \text{s.t.} \quad \sum_{i \in S} p_i + z_k \ge v_k(S) \quad \forall S, k$$

$$\sum_{S \subseteq \mathcal{N}} x_k(S) \le 1, \quad \forall k \in \mathcal{P} \qquad\qquad p_i \ge 0, \quad \forall i \in \mathcal{N}$$

$$x_k(S) \ge 0, \quad \forall S, \forall k \qquad\qquad z_k \ge 0, \quad \forall k \in \mathcal{P}$$

$$\tag{1}$$

As long as the instance is clear from the context, we will use $(\boldsymbol{p}^*, \boldsymbol{z}^*)$ to denote the optimum solution to the Dual LP, referring to $\boldsymbol{p}^*$ as the dual prices, and $\boldsymbol{z}^*$ as the slack.

Main Result. We are now in a position to show the central result of this paper. The following black-box mechanism assumes as input an *LP-based α-approximate allocation*, i.e., an allocation whose social welfare is at most a factor α smaller than the value of the LP optimum for that instance. LP-based approximation factors are a staple requirement for black-box mechanisms that explicitly make use of the optimum LP solution [15]. Along these lines, we make the assumption that the optimum dual variables (for the given instance) are available to the algorithm along with an input allocation.

Theorem 2. *Given any α-approximate solution to the LP optimum, we can construct a $(2\alpha, 2\alpha)$-Core Stable Solution in polynomial time as long as the projects have subadditive valuations.*

The above theorem in conjunction with the 2-approximation algorithm for general subadditive functions proposed in [11] yields the following corollary.

Corollary 3. *We can compute in poly-time a* $(4, 4)$*-Core stable solution for any instance with subadditive projects.*

Technical Challenges for Subadditive Valuations. At the heart of finding a Core allocation lies the problem of estimating 'how much is an agent worth to a coalition'. Unfortunately, the idea used in Claim 1 does not extend to more general valuations as the *marginal value* is no longer representative of an agent's worth. One alternative approach is to use the dual variables to tackle this problem: for example, in the classic setting with duplicate projects, **every** solution S along with the dual prices as payments yields an α-budget balanced core. Therefore, the challenge there is to bound the factor α using the integrality gap. However, this is no longer true in our combinatorial setting as there is no clear way of dividing the dual variables due to the presence of *slack*.

(Proof Sketch of Theorem 2): Since we cannot find every agent's exact worth, we will attempt to approximate this problem by identifying a 'small set of heavy users' who contribute significantly to the social welfare and providing large payments only to these users. The proof proceeds constructively via the following algorithm: initialize each user's payment to p_i^*, and then divide the slack z_k^* equally among all users on project k. Now, follow this up with a best-response (BR) phase, where agents are allowed to deviate to empty projects. Unfortunately, the BR phase may lead to a large welfare loss. However, we show that the projects which incur a significant loss in social welfare after the best-response phase lose at most half of the agents originally assigned to it, they form the set of heavy users. Therefore, we return these heavy agents to their original projects and provide them payments equal to their best available outside option. While this bounds the loss in welfare, the payments to the non-heavy users may no longer be enough to stabilize them. Therefore, in the final round, we once again do a best-response phase with the these agents to obtain core-stability. Note that in contrast to the traditional setting, we only use the dual variables as a 'guide' in the first BR step to ensure that $\forall k \in \mathcal{P}$, the payment given to users on that project is at least a good fraction of the value they generate.   □

3.1   Anonymous Functions

Our other main result in this paper is a $(2, 2)$-Core stable solution for the class of subadditive functions that are anonymous. Recall that for an anonymous valuation v, $v(T_1) = v(T_2)$ for any $|T_1| = |T_2|$. We begin with some existential lower bounds for approximating the core. From Example (1), we already know that the core may not exist even in simple instances. Extending this example, we show a much stronger set of results.

Claim 4 (*Lower Bounds*). *There exist instances having only two projects with anonymous subadditive functions such that*

1. *For any $\epsilon > 0$, no $(2 - \epsilon, c)$-core stable solution exists for any value c.*
2. *For any $\epsilon > 0$, no $(\alpha, 2 - \epsilon)$-core stable solution exists for any constant α.*

(Proof of Part 1) We show that no $(2 - \epsilon)$-budget-balanced core stable solution exists for a given $\epsilon > 0$. Consider an instance with N buyers. The valuations for the two projects are $v_1(S) = \frac{N}{2}$ $\forall S \subset \mathcal{N}$, and $v_1(\mathcal{N}) = N$; $v_2(S) = 2$ $\forall S \subseteq \mathcal{N}$. Assume by contradiction that there is a $(2 - \epsilon)$-core stable solution, then this cannot be achieved when all of the agents are assigned to project 1 because they would each require a payment of 2 to prevent them from deviating to project 2. On the other hand, suppose that some agents are assigned to project 2, then the social welfare of the solution is at most $\frac{N}{2} + 2$. If these agents cannot deviate to project 1, then, their payments would have to be at least $v_1(\mathcal{N}) = N$. For a sufficiently large N, we get that the budget-balance is $\frac{N}{N/2+2} > 2 - \epsilon$. ∎

We now describe an intuitive 2-approximation algorithm for maximizing welfare that may be of independent interest. To the best of our knowledge, the only previously known approach that achieves a 2-approximation for anonymous subadditive functions is the LP-based rounding algorithm for general subadditive functions [11]. Our result shows that for the special class of anonymous functions, the same factor can be achieved by a much faster, greedy algorithm.

> **Algorithm:** Starting with all agents unassigned, at each step we choose a set of agents T and project k maximizing $\frac{v_k(T \cup S_k) - v_k(S_k)}{|T|}$. Assign the agents to that project, i.e., $S_k = S_k \cup T$.

The next result uses the approximation algorithm to compute an approximately core stable solution. Although the Algorithm is straight-forward, the proof of the approximation factor is somewhat involved.

Theorem 5. *For any instance with anonymous subadditive projects, the allocation S returned by the greedy algorithm along with suitably chosen payments constitute a $(2, 2)$-core stable solution.*

Envy-Free Payments. One interpretation for projects having anonymous valuations is that all the agents possess the same level of skill, and therefore, the value generated from a project depends only on the number of agents assigned to it. In such scenarios, it may be desirable that the payments given to the different agents are 'fair' or envy-free, i.e., all agents assigned to a certain project must receive the same payment. We remark that the payments computed by Theorem 5 are also envy-free.

3.2 Submodular and Fractionally Subadditive (XoS) Valuations

Submodular and fractionally subadditive valuations are arguably the most popular classes of subadditive functions, and we show several interesting and improved results for these sub-classes. For instance, for XoS valuations, we can compute a $(1 + \epsilon)$-core using demand and XoS oracles (see [10] for a treatment of XoS oracles), whereas without these oracles, we can still compute a $(\frac{e}{e-1})$-core.

For submodular valuations, we provide an algorithm to compute a $(1+\epsilon)$-core even without a demand oracle. All of these solutions retain at least a fraction $(1-\frac{1}{e})$ of the optimum welfare, which matches the computational lower bound for both of these classes. We begin with a simple existence result for XoS valuations, that the optimum solution along with payments obtained using a XoS oracle form an exact core stable solution.

Proposition 6. *There exists a $(1,1)$-core stable solution for every instance where the projects have XoS valuations.*

Since *Submodular $\subset$ XoS*, this result extends to submodular valuations as well. Unfortunately, it is known that the optimum solution cannot be computed efficiently for either of these classes unless $P = NP$ [10]. However, we show that one can efficiently compute approximately optimal solutions that are almost-stable.

Theorem 7. *1. For any instance where the projects have XoS valuations, we can compute $(1+\epsilon, \frac{e}{e-1})$-core stable solution using Demand and XoS oracles, and a $(\frac{e}{e-1}, \frac{e}{e-1})$-core stable solution without these oracles.*
   2. For submodular valuations, we can compute a $(1+\epsilon, \frac{e}{e-1})$-core stable solution using only a Value oracle.

Note that for both the classes, a $(1+\epsilon)$-core can be computed in time polynomial in the input, and $\frac{1}{\epsilon}$. We conclude by pointing out that the results above are much better than what could have been obtained by plugging in $\alpha = \frac{e}{e-1}$ in Theorem 2 for Submodular or XoS valuations.

4   Relationship to Combinatorial Auctions

We now change gears and consider the seemingly unrelated problem of *Item Bidding Auctions*, and establish a surprising equivalence between Core stable solutions and pure Nash equilibrium in Simultaneous Second Price Auctions. Following this, we adapt some of our results specifically for the auction setting and show how to efficiently compute approximate Nash equilibrium when buyers have anonymous or submodular functions.

In recent years, the field of Auction Design has been marked by a paradigm shift towards 'simple auctions'; one of the best examples of this is the growing popularity of Simultaneous Combinatorial Auctions [5,9]. The auction mechanism is simple: every buyer submits one bid for each of the N items, the auctioneer then proceeds to run N-parallel single-item auctions (usually first-price or second-price). In the case of Second Price Auctions, each item is awarded to the highest bidder (for that item) who is then charged the bid of the second highest bidder. Each buyer's utility is her valuation for the bundle she receives minus her total payment.

We begin by establishing that for every instance of our utility sharing problem, there is a corresponding combinatorial auction, and vice-versa. Formally, given an instance $(\mathcal{N}, \mathcal{P}, (v)_{k \in \mathcal{P}})$, we define the following 'flipped auction': there is a set $\mathcal{N}$ of N items, and a set $\mathcal{P}$ of m buyers. Every buyer $k \in \mathcal{P}$ has a valuation function v_k for the items. In the simultaneous auction, the strategy of every buyer is a bid vector $\boldsymbol{b_k}$; $b_k(i)$ denotes buyer k's bid for item $i \in \mathcal{N}$. A profile of bid vectors along with an allocation is said to be a pure Nash equilibrium of the simultaneous auction if no buyer can unilaterally change her bids and improve her utility at the new allocation.

Over-Bidding. Nash equilibrium in simultaneous auctions is often accompanied by a rather strong *no-overbidding* condition that a player's aggregate bid for every set S of items is at most her valuation $v_k(S)$ for that set. In this paper, we study the slightly less stringent *weak no-overbidding* assumption considered in [13] which states that 'a player's total bid for her winning set is at most her valuation for that set'. Finally, to model buyers who overbid by small amounts, we focus on the following natural relaxation of no-overbidding known as γ-conservativeness that was defined by Bhawalkar and Roughgarden [5].

Definition 8. *(Conservative Bids) [5] For a given buyer $k \in \mathcal{P}$, a bid vector $\boldsymbol{b_k}$ is said to be γ-conservative if for all $T \subseteq \mathcal{N}$, we have $\sum_{i \in T} b_k(i) \leq \gamma \cdot v_k(T)$.*

We now state our main equivalence result that is based on a simple black-box transformation to convert a Core stable solution $(S, \bar{\boldsymbol{p}})$ to a profile of bids $(\boldsymbol{b_k})_{k \in \mathcal{P}}$ that form a Nash Equilibrium: $b_k(i) = \bar{p}_i$ if $i \in S_k$, and $b_k(i) = 0$ otherwise.

Theorem 9. *Every Core stable solution for a given instance of our game can be transformed into a Pure Nash Equilibrium (with weak no-overbidding) of the corresponding 'flipped' simultaneous second price auction, and vice-versa.*

A case for Approximate Equilibrium. The exciting connection between the two solution concepts unfortunately extends to negative results as well. One can extend our lower bound examples to show that even when all buyers have anonymous subadditive functions, there exist instances where every Nash equilibrium requires $O(\sqrt{N})$-conservative bids. The expectation that buyers will overbid by such a large amount appears to be unreasonable. In light of these impossibility results and the known barriers to actually compute a (no-overbidding) equilibrium [9], we argue that in many auctions, it seems reasonable to consider α-approximate Nash equilibrium that guarantee that buyers' utilities cannot improve by more than a factor α when they change their bids. In the following result, we adapt our previous algorithms to compute approximate equilibria with high social welfare for two useful settings. Moreover, these solutions require small over-bidding, and can be obtained via simple mechanisms, so it seems likely that they would actually arise in practice when pure equilibria either do not exist or require a large amount of overbidding.

Claim 10 *Given a Second Price Simultaneous Combinatorial Auction, we can compute in time polynomial in the input (and $\frac{1}{\epsilon}$ for a given $\epsilon > 0$)*

1. *A 2-approximate Nash equilibrium that extracts half the optimal social welfare as long as the buyers have anonymous subadditive valuations.*

2. A $(1 + \epsilon)$-*approximate Nash equilibrium that is a* $\frac{e}{e-1}$-*approximation to the optimum welfare when the buyers have submodular valuations.*

The first solution involves 4-*conservative bids, and the second solution involves* $(1 + \epsilon)$-*conservative bids.*

We conclude by remarking that despite the large body of work in Simultaneous Auctions, our theorems do not follow from any known results in that area, and we hope that our techniques lead to new insights for computing auction equilibria. Detailed proofs of all of our results can be found in the full version of this paper available at http://arxiv.org/abs/1508.06781.

References

1. Bachrach, Y., Elkind, E., Meir, R., Pasechnik, D., Zuckerman, M., Rothe, J., Rosenschein, J.S.: The cost of stability in coalitional games. In: Mavronicolas, M., Papadopoulou, V.G. (eds.) SAGT 2009. LNCS, vol. 5814, pp. 122–134. Springer, Heidelberg (2009)

2. Bachrach, Y., Parkes, D.C., Rosenschein, J.S.: Computing cooperative solution concepts in coalitional skill games. Artif. Intell. **204**, 1–21 (2013)

3. Bachrach, Y., Syrgkanis, V., Tardos, É., Vojnović, M.: Strong price of anarchy, utility games and coalitional dynamics. In: Lavi, R. (ed.) SAGT 2014. LNCS, vol. 8768, pp. 218–230. Springer, Heidelberg (2014)

4. Bejan, C., Gómez, J.C.: Theory Core extensions for non-balanced TU-games. Int. J. Game **38**(1), 3–16 (2009)

5. Bhawalkar, K., Roughgarden, T.: Welfare guarantees for combinatorial auctions with item bidding. In: Proceedings of SODA (2011)

6. Bousquet, N., Li, Z., Vetta, A.: Coalition games on interaction graphs: a horticultural perspective. In: Proceedings of EC (2015)

7. Chalkiadakis, G., Elkind, E., Markakis, E., Polukarov, M., Jennings, N.R.: Cooperative games with overlapping coalitions. J. Artif. Intell. Res. (JAIR) **39**, 179–216 (2010)

8. Deng, X., Ibaraki, T., Nagamochi, H.: Algorithmic aspects of the core of combinatorial optimization games. Math. Oper. Res. **24**(3), 751–766 (1999)

9. Dobzinski, S., Fu, H., Kleinberg, R.D.: On the complexity of computing an equilibrium in combinatorial auctions. In: Proceedings of SODA (2015)

10. Dobzinski, S., Nisan, N., Schapira, M.: Approximation algorithms for combinatorial auctions with complement-free bidders. Math. Oper. Res. **35**(1), 1–13 (2010)

11. Feige, U.: On maximizing welfare when utility functions are subadditive. SIAM J. Comput. **39**(1), 122–142 (2009)

12. Feldman, M., Friedler, O.: A unified framework for strong price of anarchy in clustering games. In: Halldórsson, M.M., Iwama, K., Kobayashi, N., Speckmann, B. (eds.) ICALP 2015. LNCS, vol. 9135, pp. 601–613. Springer, Heidelberg (2015)

13. Feldman, M., Fu, H., Gravin, N., Lucier, B.: Simultaneous auctions are (almost) efficient. In: Proceedings of STOC (2013)

14. Feldman, M., Lewin-Eytan, L., Naor, J.: Hedonic clustering games. In: Proceedings of SPAA (2012)

15. Georgiou, K., Swamy, C.: Black-box reductions for cost-sharing mechanism design. Games and Economic Behavior (2013)

16. Hoefer, M.: Strategic cooperation in cost sharing games. Int. J. Game Theory **42**(1), 29–53 (2013)
17. Kleinberg, J.M., Oren, S.: Mechanisms for (mis)allocating scientific credit. In: Proceedings of STOC (2011)
18. Lehmann, B., Lehmann, D., Nisan, N.: Combinatorial auctions with decreasing marginal utilities. Games Econ. Behav. **55**(2), 270–296 (2006)
19. Meir, R., Bachrach, Y., Rosenschein, J.S.: Minimal subsidies in expense sharing games. In: Kontogiannis, S., Koutsoupias, E., Spirakis, P.G. (eds.) SAGT 2010. LNCS, vol. 6386, pp. 347–358. Springer, Heidelberg (2010)
20. Myerson, R.B.: Graphs and cooperation in games. Math. Oper. Res. **2**(3), 225–229 (1977)
21. Roughgarden, T., Sundararajan, M.: Quantifying inefficiency in cost-sharing mechanisms. J. ACM **56**(4), 23 (2009)
22. Vondrák, J.: Optimal approximation for the submodular welfare problem in the value oracle model. In: Proceedings of STOC (2008)

The (Non)-Existence of Stable Mechanisms in Incomplete Information Environments

Nick Arnosti[1], Nicole Immorlica[2], and Brendan Lucier[2]([⊠])

[1] Department of Management Science and Engineering,
Stanford University, Stanford, USA
narnosti@stanford.edu
[2] Microsoft Research, Cambridge, USA
nicimm@gmail.com, brlucier@microsoft.com

Abstract. We consider two-sided matching markets, and study the incentives of agents to circumvent a centralized clearing house by signing binding contracts with one another. It is well-known that if the clearing house implements a stable match and preferences are known, then no group of agents can profitably deviate in this manner.

We ask whether this property holds even when agents have *incomplete information* about their own preferences or the preferences of others. We find that it does not. In particular, when agents are uncertain about the preferences of others, *every* mechanism is susceptible to deviations by groups of agents. When, in addition, agents are uncertain about their *own* preferences, every mechanism is susceptible to deviations in which a single pair of agents agrees in advance to match to each other.

1 Introduction

In entry-level labor markets, a large number of workers, having just completed their training, simultaneously seek jobs at firms. These markets are especially prone to certain failures, including unraveling, in which workers receive job offers well before they finish their training, and exploding offers, in which job offers have incredibly short expiration dates. In the medical intern market, for instance, prior to the introduction of the centralized clearing house (the *National Residency Matching Program*, or NRMP), medical students received offers for residency programs at US hospitals two years in advance of their employment date. In the market for law clerks, law students have reported receiving exploding offers in which they were asked to accept or reject the position on the spot (for further discussion, see Roth and Xing [17]).

In many cases, including the medical intern market in the United States and United Kingdom and the hiring of law students in Canada, governing agencies try to circumvent these market failures by introducing a centralized clearing house which solicits the preferences of all participants and uses these to recommend a matching. One main challenge of this approach is that of incentivizing

---

N. Arnosti — Work conducted at Microsoft Research.

E. Markakis and G. Schäfer (Eds.): WINE 2015, LNCS 9470, pp. 46–59, 2015.
DOI: 10.1007/978-3-662-48995-6_4

participation. Should a worker and firm suspect they each prefer the other to their assignment by the clearing house, then they would likely match with each other and not participate in the centralized mechanism. Roth [15] suggests that this may explain why clearing houses that fail to select a stable match have often had difficulty attracting participants.

Empirically, however, even clearing houses which produce stable matches may fail to prevent early contracting. Examples include the market for Canadian law students (discussed by Roth and Xing [17]) and the American gastroenterology match (studied by Niederle and Roth [12] and McKinney et al. [11]). This is perhaps puzzling, as selecting a stable match ensures that no group of participants can profitably circumvent the clearing house ex-post.

Our work offers one possible explanation for this phenomenon. While stable clearing houses ensure that for *fixed, known* preferences, no coalition can profitably deviate, in most natural settings, participants contemplating deviation do so without complete knowledge of others' preferences (and sometimes even their own preferences). Our main finding is that in the presence of such uncertainty, *no mechanism* can prevent agents from signing mutually beneficial side contracts.

We model uncertainty in preferences by assuming that agents have a common prior over the set of possible preference profiles, and may in addition know their own preferences. We consider two cases. In one, agents have no private information when contracting, and their decision of whether to sign a side contract depends only on the prior (and the mechanism used by the clearing house). In the second case, agents know their own preferences, but not those of others. When deciding whether to sign a side contract, agents consider their own preferences, along with the information revealed by the willingness (or unwillingness) of fellow agents to sign the proposed contract.

Note that with incomplete preference information, agents perceive the partner that they are assigned by a given mechanism to be a random variable. In order to study incentives for agents to deviate from the centralized clearing house, we must specify a way for agents to compare lotteries over match partners. One seemingly natural model is that each agent gets, from each potential partner, a utility from being matched to that partner. When deciding between two uncertain outcomes, agents simply compare their corresponding expected utilities. Much of the previous literature has taken this approach, and indeed, it is straightforward to discover circumstances under which agents would rationally contract early (see the full version of the paper for an example). Such cases are perhaps unsurprising; after all, the central clearing houses that we study solicit only ordinal preference lists, while the competing mechanisms may be designed with agents' cardinal utilities in mind.

For this reason, we consider a purely ordinal notion of what it means for an agent to prefer one allocation to another. In our model, an agent debating between two uncertain outcomes chooses to sign a side contract only if the rank that they assign their partner under the proposed contract strictly first-order stochastically dominates the rank that they anticipate if all agents participate in the clearing house. This is a strong requirement, by which we mean that it is easy for a mechanism to be stable under this definition, relative to a definition relying on expected utility. For instance, this definition rules out examples of beneficial deviations, where agents match to an acceptable, if sub-optimal, partner in order to avoid the possibility of a "bad" outcome.

Despite the strong requirements we impose on beneficial deviations, we show that every mechanism is vulnerable to side contracts when agents are initially uncertain about their preferences or the preferences of others. On the other hand, when agents are certain about their own preferences but not about the preferences of others, then there do exist mechanisms that resist the formation of side contracts, when those contracts are limited to involving only a pair of agents (i.e., one from each side of the market).

2   Related Work

Roth [14] and Roth and Rothblum [16] are among the first papers to model incomplete information in matching markets. These papers focus on the strategic implications of preference uncertainty, meaning that they study the question of whether agents should truthfully report to the clearinghouse. Our work, while it uses a similar preference model, assumes that the clearing house can observe agent preferences. While this assumption may be realistic in some settings, we adopt it primarily in order to separate the strategic manipulation of matching mechanisms (as studied in the above papers) from the topic of early contracting that is the focus of this work.

Since the seminal work of Roth and Xing [17], the relationship between stability and unraveling has been studied using observational studies, laboratory experiments, and theoretical models. Although work by Roth [15] and Kagel and Roth [5] concluded that stability plays an important role in encouraging participation, other papers note that uncertainty may cause unraveling even if a stable matching mechanism is used.

A common theme in these papers is that unraveling is driven by the motive of "insurance." For example, the closely related models of Li and Rosen [6], Suen [19], and Li and Suen [7,8] study two-sided assignment models with transfers in which binding contracts may be signed in one of two periods (before or after revelation of pertinent information). In each of these papers, unraveling occurs (despite the stability of the second-round matching) because of agents' risk-aversion: when agents are risk-neutral, no early matches form. Yenmez [20] also considers notions of interim and ex-ante stability in a matching market with transferable utility. He establishes conditions under which stable, incentive compatible, and budget-balanced mechanisms exist.

Even in models in which transfers are not possible (and so the notion of risk aversion has no obvious definition), the motive of insurance often drives early matching. The models presented by Roth and Xing [17], Halaburda [4], and Du and Livne [2] assume that agents have underlying cardinal utilities for each match, and compare lotteries over matchings by computing expected utilities. They demonstrate that unraveling may occur if, for example, workers are willing to accept an offer from their second-ranked firm (foregoing a chance to be matched to their top choice) in order to ensure that they do not match to a less-preferred option.[1]

While insurance may play a role in the early contracting observed by Roth and Xing [17], one contribution of our work is to show that it is not necessary to obtain such behavior. In this work, we show that even if agents are unwilling to forego top choices in order to avoid lower-ranked ones, they might rationally contract early with one another. Put another way, we demonstrate that some opportunities for early contracting may be identified on the basis of ordinal information alone (without making assumptions about agents' unobservable cardinal utilities).

Manjunath [10] and Gudmundsson [3] consider the stochastic dominance notion used in this paper; however they treat only the case (referred to in this paper as "ex-post") where the preferences of agents are fixed, and the only randomness comes from the assignment mechanism. One contribution of our work is to define a stochastic dominance notion of stability under asymmetric information. This can be somewhat challenging, as agents' actions signal information about their type, which in turn might influence the actions of others.[2]

Perhaps the paper that is closest in spirit to ours is that of Peivandi and Vohra [13], which considers the operation of a centralized exchange in a two-sided setting with transferable utility. One of their main findings is that every trading mechanism can be blocked by an alternative; our results have a similar flavor, although they are established in a setting with non-transferrable utility.

3   Model and Notation

In this section, we introduce our notation, and define what it means for a matching to be *ex-post*, *interim*, or *ex-ante* stable. There is a (finite, non-empty) set M of men and a (finite, non-empty) set W of women.

Definition 1. *Given M and W, a* **matching** *is a function $\mu : M \cup W \to M \cup W$ satisfying:*

1. *For each $m \in M$, $\mu(m) \in W \cup \{m\}$*

---

[1] In many-to-one settings, Sönmez [18] demonstrates that even in full-information environments, it may be possible for agents to profitably pre-arrange matches (a follow-up by Afacan [1] studies the welfare effects of such pre-arrangements). In order for all parties involved to strictly benefit, it must be the case that the firm hires (at least) one inferior worker in order to boost competition for their remaining spots (and thereby receive a worker who they would be otherwise unable to hire). Thus, the profitability of such an arrangement again relies on assumptions about the firm's underlying cardinal utility function.

[2] Liu et al. [9] have recently grappled with this inference procedure, and defined a notion of stable matching under uncertainty. Their model differs substantially from the one considered here: it takes a matching μ as given, and assumes that agents know the quality of their current match, but must make inferences about potential partners to whom they are not currently matched.

2. *For each* $w \in W$, $\mu(w) \in M \cup \{w\}$
3. *For each* $m \in M$ *and* $w \in W$, $\mu(m) = w$ *if and only if* $\mu(w) = m$.

We let $\mathcal{M}(M, W)$ be the set of matchings on M, W.

Given a set S, define $\mathcal{R}(S)$ to be the set of one-to-one functions mapping S onto $\{1, 2, \ldots, |S|\}$. Given $m \in M$, let $P_m \in \mathcal{R}(W \cup \{m\})$ be m's ordinal preference relation over women (and the option of remaining unmatched). Similarly, for $w \in W$, let $P_w \in \mathcal{R}(M \cup \{w\})$ be w's ordinal preference relation over the men. We think of $P_m(w)$ as giving the *rank* that m assigns to w; that is, $P_m(w) = 1$ implies that matching to w is m's most-preferred outcome.

Given sets M and W, we let $\mathcal{P}(M, W) = \prod_{m \in M} \mathcal{R}(W \cup \{m\}) \times \prod_{w \in W} \mathcal{R}(M \cup \{w\})$ be the set of possible preference profiles. We use P to denote an arbitrary element of $\mathcal{P}(M, W)$, and use ψ to denote a probability distribution over $\mathcal{P}(M, W)$. We use P_A to refer to the preferences of agents in the set A under profile P, and use P_a (rather than the more cumbersome $P_{\{a\}}$) to refer to the preferences of agent a.

Definition 2. *Given M and W, and $P \in \mathcal{P}(M, W)$, we say that matching μ is* **stable at preference profile** *P if and only if the following conditions hold.*

1. *For each* $a \in M \cup W$, $P_a(\mu(a)) \leq P_a(a)$.
2. *For each* $m \in M$ *and* $w \in W$ *such that* $P_m(\mu(m)) > P_m(w)$, *we have* $P_w(\mu(w)) < P_w(m)$.

This is the standard notion of stability; the first condition states that agents may only be matched to partners whom they prefer to going unmatched, and the second states that whenever m prefers w to his partner under μ, it must be that w prefers her partner under μ to m.

In what follows, we fix M and W, and omit the dependence of $\mathcal{M}$ and $\mathcal{P}$ on the sets M and W. We define a *mechanism* to be a (possibly random) mapping $\phi : \mathcal{P} \to \mathcal{M}$. We use A' to denote a subset of $M \cup W$.

We now define what it means for a coalition of agents to *block* the mechanism ϕ, and what it means for a *mechanism* (rather than a matching) to be stable. Because we wish to consider randomized mechanisms, we must have a way for agents to compare lotteries over outcomes. As mentioned in the introduction, our notion of blocking relates to stochastic dominance. Given random variables $X, Y \in \mathbb{N}$, say that X *first-order stochastically dominates* Y (denoted $X \succ Y$) if for all $n \in \mathbb{N}$, $\Pr(X \leq n) \geq \Pr(Y \leq n)$, with strict inequality for at least one value of n.

An astute reader will note that this definition reverses the usual inequalities; that is, $X \succ Y$ implies that X is "smaller" than Y. We adopt this convention because below, X and Y will represent the ranks assigned by each agent to their partner (where the most preferred option has a rank of one), and thus by our convention, $X \succ Y$ means that X is preferred to Y.

Definition 3 (Ex-Post Stability). *Given M, W and a profile $P \in \mathcal{P}(M, W)$, coalition A'* **blocks mechanism** *ϕ* **ex-post** *at P if there exists a mechanism ϕ' such that for each $a \in A'$,*

1. $\Pr(\phi'(P)(a) \in A') = 1$, *and*
2. $P_a(\phi'(P)(a)) \succ P_a(\phi(P)(a))$.

Mechanism ϕ is **ex-post stable at profile** P *if no coalition of agents blocks ϕ ex-post at P.*
Mechanism ϕ is **ex-post stable** *if it is ex-post stable at P for all $P \in \mathcal{P}(M, W)$.*
Mechanism ϕ is **ex-post pairwise stable** *if for all P, no coalition consisting of at most one man and at most one woman blocks ϕ ex post at P.*

Note that in the above setting, because P is fixed, the mechanism ϕ' is really just a random matching. The first condition in the definition requires that the deviating agents can implement this alternative (random) matching without the cooperation of the other agents; the second condition requires that for each agent, the random variable denoting the rank of his partner under the alternative ϕ' stochastically dominates the rank of his partner under the original mechanism.

Note that if the mechanism ϕ is deterministic, then it is ex-post pairwise stable if and only if the matching it produces is stable in the sense of Definition 2.

The above notions of blocking and stability are concerned only with cases where the preference profile P is fixed. In this paper, we assume that at the time of choosing between mechanisms ϕ and ϕ', agents have incomplete information about the profile P that will eventually be realized (and used to implement a matching). We model this incomplete information by assuming that it is common knowledge that P is drawn from a prior ψ over $\mathcal{P}$. Given a mechanism ϕ, each agent may use ψ to determine the ex-ante distribution of the rank of the partner that they will be assigned by ϕ. This allows us to define what it means for a coalition to block ϕ ex-ante, and for a mechanism ϕ to be ex-ante stable.

Definition 4 (Ex-Ante Stability). *Given M, W and a prior ψ over $\mathcal{P}(M, W)$, coalition A' blocks mechanism ϕ* **ex-ante** *at ψ if there exists a mechanism ϕ' such that if P is drawn from the prior ψ, then for each $a \in A'$,*

1. $\Pr(\phi'(P)(a) \in A') = 1$, *and*
2. $P_a(\phi'(P)(a)) \succ P_a(\phi(P)(a))$.

Mechanism ϕ is **ex-ante stable at prior** ψ *if no coalition of agents blocks ϕ ex-ante at ψ.*
Mechanism ϕ is **ex-ante stable** *if it is ex-ante stable at ψ for all priors ψ.*
Mechanism ϕ is **ex-ante pairwise stable** *if, for all priors ψ, no coalition consisting of at most one man and at most one woman blocks ϕ ex-ante at ψ.*

Note that the only difference between ex-ante and ex-post stability is that the randomness in Definition 4 is over both the realized profile P and the matching produced by ϕ, whereas in Definition 3, the profile P is deterministic. Put another way, the mechanism ϕ is ex-post stable if and only if it is ex-ante stable at all deterministic distributions ψ.

The notions of ex-ante and ex-post stability defined above are fairly straightforward because the information available to each agent is identical. In order to study the case where each agent knows his or her own preferences but not the preferences of others, we must define an appropriate notion of a blocking coalition. In particular, if man m decides to enter into a contract with woman w, m knows not only his own preferences, but also learns about those of w from the fact that she is willing to sign the contract. Our definition of what it means for a coalition to block ϕ in the interim takes this into account.

In words, given the common prior ψ, we say that a coalition A' *blocks ϕ in the interim* if there exists a preference profile P that occurs with positive probability under ψ such that when preferences are P, all members of A' agree that the outcome of ϕ' stochastically dominates that of ϕ, given their own preferences and the fact that other members of A' also prefer ϕ'. We formally define this concept below, where we use the notation $\psi(\cdot)$ to represent the probability measure assigned by the distribution ψ to the argument.

Definition 5 (Interim Stability). *Given M, W, and a prior ψ over $\mathcal{P}(M, W)$, coalition A' **blocks mechanism ϕ in the interim** if there exists a mechanism ϕ', and for each $a \in A'$, a subset of preferences $\mathcal{R}_a$ satisfying the following:*

1. *For each $P \in \mathcal{P}$, $\Pr(\phi'(P)(a) \in A') = 1$.*
2. *For each agent $a \in A'$ and each preference profile $\tilde{P}_a$, $\tilde{P}_a \in \mathcal{R}_a$ if and only if*
   (a) *$\psi(Y_a(\tilde{P}_a)) > 0$, where $Y_a(\tilde{P}_a) = \{P \colon P_a = \tilde{P}_a\} \cap \{P \colon P_{a'} \in \mathcal{R}_{a'} \ \forall a' \in A' \backslash \{a\}\}$*
   (b) *When P is drawn from the conditional distribution of ψ given $Y_a(\tilde{P}_a)$, we have $P_a(\phi'(P)(a)) \succ P_a(\phi(P)(a))$.*

*Mechanism ϕ is **interim stable at ψ** if no coalition of agents blocks ϕ in the interim at ψ.*
*Mechanism ϕ is **interim stable** if it is interim stable at ψ for all distributions ψ.*
*Mechanism ϕ is **interim pairwise stable** if, for all priors ψ, no coalition consisting of at most one man and at most one woman blocks ϕ in the interim at ψ.*

To motivate the above definition of an interim blocking coalition, consider a game in which a moderator approaches a subset A' of agents, and asks each whether they would prefer to be matched according to the mechanism ϕ (proposed by the central clearing house) or the alternative ϕ' (which matches agents in A' to each other). Only if all agents agree that they would prefer ϕ' is this mechanism used. Condition 1 simply states that the mechanism ϕ' generates matchings among the (potentially) deviating coalition A'.

We think of $\mathcal{R}_a$ as being a set of preferences for which agent a agrees to use mechanism ϕ'. The set $Y_a(\tilde{P}_a)$ is the set of profiles which agent a considers possible, conditioned on the events $P_a = \tilde{P}_a$ and the fact that all other agents in A' agree to use mechanism ϕ'. Condition 2 is a consistency condition on the preference subsets $\mathcal{R}_a$: (2a) states that agents in A' should agree to ϕ' only if they believe that there is a chance that the other agents in A' will also agree to ϕ' (that is, if ψ assigns positive mass to Y_a); moreover, (2b) states that in the cases when $P_a \in \mathcal{R}_a$ *and the other agents select ϕ'*, it should be the case that a "prefers" the mechanism ϕ' to ϕ (here and in the remainder of the paper, when we write that agent a prefers ϕ' to ϕ, we mean that *given the information available to* a, the rank of a's partner under ϕ' stochastically dominates the rank of a's partner under ϕ).

4   Results

We begin with the following observation, which states that the three notions of stability discussed above are comparable, in that ex-ante stability is a stronger requirement than interim stability, which is in turn a stronger requirement than ex-post stability.

Lemma 1. *If ϕ is ex-ante (pairwise) stable, then it is interim (pairwise) stable. If ϕ is interim (pairwise) stable, then it is ex-post (pairwise) stable.*

Proof. We argue the contrapositive in both cases. Suppose that ϕ is not ex-post stable. This implies that there exists a preference profile P, a coalition A', and a mechanism ϕ' that only matches agents in A' to each other, such that all agents in A' prefer ϕ' to ϕ, given P. If we take ψ to place all of its mass on profile P, then (trivially) A' also blocks ϕ in the interim, proving that ϕ is not interim stable.

Suppose now that ϕ is not interim stable. This implies that there exists a distribution ψ over $\mathcal{P}$, a coalition A', a mechanism ϕ' that only matches agents in A' to each other, and preference orderings $\mathcal{R}_a$ satisfying the following conditions: the set of profiles $Y = \{P : \forall a \in A', P_a \in \mathcal{R}_a\}$ has positive mass $\psi(Y) > 0$; and conditioned on the profile being in Y, agents in A' want to switch to ϕ', i.e., for all $a \in A'$ and *for all* $P_a \in \mathcal{R}_a$ agent a prefers ϕ' to ϕ conditioned on the profile being in Y. Thus, agent a must prefer ϕ' even ex ante (conditioned only on $P \in Y$).

If we take ψ' to be the conditional distribution of ψ given $P \in Y$, it follows that under ψ', all agents $a \in A'$ prefer mechanism ϕ' to mechanism ϕ ex-ante, so ϕ is not ex-ante stable.

4.1   Ex-Post Stability

We now consider each of our three notions of stability in turn, beginning with ex-post stability. By Lemma 1, ex-post stability is the easiest of the three conditions to satisfy. Indeed, we show there not only exist ex-post stable mechanisms, but that any mechanism that commits to always returning a stable matching is ex-post stable.

Theorem 1. *Any mechanism that produces a stable matching with certainty is ex-post stable.*

Note that if the mechanism ϕ is deterministic, then (trivially) it is ex-post stable if and only if it always produces a stable matching. Thus, for deterministic mechanisms, our notion of ex-post stability coincides with the "standard" definition of a stable mechanism. Theorem 1 states further that any mechanism that randomizes among stable matchings is also ex-post stable. This fact appears as Proposition 3 in [10].[3]

We next show in Example 1 that the converse of Theorem 1 does not hold. That is, there exist randomized mechanisms ϕ which sometimes select unstable matches but are nevertheless ex-post stable. In this and other examples, we use the notation $P_m : w_1, w_2, w_3$ as shorthand indicating that m ranks w_1 first, w_2 second, w_3 third, and considers going unmatched to be the least desirable outcome.

Example 1.

$$P_{m_1} : w_1, w_2, w_3 \qquad P_{w_1} : m_3, m_2, m_1$$
$$P_{m_2} : w_1, w_3, w_2 \qquad P_{w_2} : m_2, m_1, m_3$$
$$P_{m_3} : w_2, w_1, w_3 \qquad P_{w_3} : m_3, m_2, m_1$$

There is a unique stable match, given by $\{m_1w_2, m_2w_3, m_3w_1\}$.

Lemma 2. *For the market described in Example 1, no coalition blocks the mechanism that outputs a uniform random matching.*

Proof. Because the random matching gives each agent their first choice with positive probability, if agent a is in a blocking coalition, then it must be that the agent that a most prefers is also in this coalition. Furthermore, any blocking mechanism must always match all participants, and thus any blocking coalition must have an equal number of men and women. Thus, the only possible blocking coalitions are $\{m_2, m_3, w_1, w_2\}$ or all six agents. The first coalition cannot block; if the probability that m_2 and w_2 are matched exceeds $1/3$, m_2 will not participate. If the probability that m_3 and w_2 are matched exceeds $1/3$, then w_2 will not participate. But at least one of these quantities must be at least $1/2$.

Considering a mechanism that all agents participate in, for any set of weights on the six possible matchings, we can explicitly write inequalities saying that each agent must get their first choice with probability at least $1/3$, and their last with probability at most $1/3$. Solving these inequalities indicates that any random matching μ that (weakly) dominates a uniform random matching must satisfy

$$\Pr(\mu = \{m_1w_1, m_2w_2, m_3w_3\}) = \Pr(\mu = \{m_1w_2, m_2w_3, m_3w_1\})$$
$$= \Pr(\mu = \{m_1w_3, m_2w_1, m_3w_2\}),$$

$$\Pr(\mu = \{m_1w_1, m_2w_3, m_3w_2\}) = \Pr(\mu = \{m_1w_2, m_2w_1, m_3w_3\})$$
$$= \Pr(\mu = \{m_1w_3, m_2w_2, m_3w_1\}).$$

But any such mechanism gives each agent their first, second and third choices with equal probability, and thus does not strictly dominate the uniform random matching.

Finally, the following lemma establishes a simple necessary condition for ex-post incentive compatibility. This condition will be useful for establishing non-existence of stable outcomes under other notions of stability.

---

[3] We thank an anonymous reviewer for the reference.

Lemma 3. *If mechanism ϕ is ex-post pairwise stable, then if man m and woman w rank each other first under P, it follows that $\Pr(\phi(P)(m) = w) = 1$.*

Proof. This follows immediately: if $\phi(P)$ matches m and w with probability less than one, then m and w can deviate and match to each other, and both strictly benefit from doing so.

4.2 Interim Stability

The fact that a mechanism which (on fixed input) outputs a uniform random matching is ex-post stable suggests that our notion of a blocking coalition, which relies on ordinal stochastic dominance, is very strict, and that many mechanisms may in fact be stable under this definition even with incomplete information. We show in this section that this intuition is incorrect: despite the strictness of our definition of a blocking coalition, it turns out that *no* mechanism is interim stable.

Theorem 2. *No mechanism is interim stable.*

Proof. In the proof, we refer to *permutations* of a given preference profile P, which informally are preference profiles that are equivalent to P after a relabeling of agents. Formally, given a permutation σ on the set $M \cup W$ which satisfies $\sigma(M) = M$ and $\sigma(W) = W$, we say that P' is the **permutation of P obtained by** σ if for all $a \in M \cup W$ and a' in the domain of P_a, it holds that $P_a(a') = P'_{\sigma(a)}(\sigma(a'))$.

The proof of Theorem 2 uses the following example.

Example 2. Suppose that each agent's preferences are iid uniform over the other side, and consider the following preference profile, which we denote P:

$$
\begin{array}{ll}
P_{m_1} : w_1, w_2, w_3 & P_{w_1} : m_1, m_2, m_3 \\
P_{m_2} : w_1, w_3, w_2 & P_{w_2} : m_1, m_3, m_2 \\
P_{m_3} : w_3, w_1, w_2 & P_{w_3} : m_3, m_1, m_2
\end{array}
$$

Note that under profile P, m_1 and w_1 rank each other first, as do m_3 and w_3. By Lemma 1, if ϕ is interim stable, it must be ex-post stable. By Lemma 3, given this P, any ex-post stable mechanism must produce the match $\{m_1 w_1, m_2 w_2, m_3 w_3\}$ with certainty. Furthermore, if preference profile P' is a permutation of P, then the matching $\phi(P')$ must simply permute $\{m_1 w_1, m_2 w_2, m_3 w_3\}$ accordingly. Thus, on any permutation of P, ϕ gives four agents their first choices, and two agents their third choices.

Define the mechanism ϕ' as follows:

- If P' is the permutation of P obtained by σ, then

$$
\phi'(P') = \{\sigma(m_1)\sigma(w_2), \sigma(m_2)\sigma(w_1), \sigma(m_3)\sigma(w_3)\}.
$$

- On any profile that is not a permutation of P, ϕ' mimics ϕ.

Note that on profile P, ϕ' gives four agents their first choices, and two agents their second choices. If each agent's preferences are iid uniform over the other side, then each agent considers his or herself equally likely to play each role in the profile P (by symmetry, this is true even after agents observe their own preferences, as they know nothing about the preferences of others). Thus, conditioned on the preference profile being a permutation of P, all agents' interim expected allocation under ϕ offers a 2/3 chance of getting their first choice and a 1/3 chance of getting their third choice, while their interim allocation under ϕ' offers a 2/3 chance of getting their first choice and a 1/3 chance of getting their second choice. Because ϕ' and ϕ are identical on profiles which are not permutations of P, it follows that all agents strictly prefer ϕ' to ϕ ex-ante.

The intuition behind the above example is as follows. Stable matchings may be "inefficient", meaning that it might be possible to separate a stable partnership (m_1, w_1) at little cost to m_1 and w_1, while providing large gains to their new partners (say m_2 and w_2). When agents lack the information necessary to determine whether they are likely to play the role of m_1 or m_2, they will gladly go along with the more efficient (though ex-post unstable) mechanism.

In addition to proving that no mechanism is interim stable *for all priors*, Example 2 demonstrates that when the priori ψ is (canonically) taken to be uniform on $\mathcal{P}$, there exists no mechanism which is interim stable *at the prior ψ*. Indeed, if ϕ sometimes fails to match pairs who rank each other first, then such pairs have a strict incentive to deviate; if ϕ always matches mutual first choices, then all agents prefer to deviate to the mechanism ϕ' described above.

Theorem 2 establishes that it is impossible to design a mechanism ϕ that eliminates profitable deviations, but the deviating coalition in Example 2 involves six agents, and the contract ϕ' is fairly complex. In many settings, such coordinated action may seem implausible. One might ask whether there exist mechanisms that are at least immune to deviations by *pairs* of agents. The following theorem shows that the complexity of Example 2 is necessary: any mechanism that always produces a stable match is indeed interim pairwise stable.[4]

Lemma 4. *Any mechanism that produces a stable match with certainty is interim pairwise stable.*

Proof. Seeking a contradiction, suppose that ϕ always produces a stable match. Fix a man m, and a woman w with whom he might block ϕ in the interim. Note that m must prefer w to going unmatched; otherwise, no deviation with w can strictly benefit him. Thus, the best outcome (for m) from a contract with w is that they are matched with certainty. According to the definition of an interim blocking pair, m must believe that receiving w with certainty stochastically dominates the outcome of ϕ; that is to say, m must be certain that ϕ will give

---

[4] This result relies crucially on the fact that we're using the notion of stochastic dominance to determine blocking pairs. If agents instead evaluate lotteries over matches by computing expected utilities, it is easy to construct examples where two agents rank each other second, and both prefer matching with certainty to the risk of getting a lower-ranked alternative from ϕ (see the full version of the paper for an example).

him nobody better than w. Because ϕ produces a stable match, it follows that in cases where m chooses to contract with w, ϕ always assigns to w a partner that she (weakly) prefers to m, and thus she will not participate.

4.3   Ex-Ante Stability

In some settings, it is natural to model agents as being uncertain not only about the rankings of others, but also about their own preferences. One might hope that the result of Theorem 4 extends to this setting; that is, that if ϕ produces a stable match with certainty, it remains immune to pairwise deviations ex-ante. Theorem 3 states that this is not the case: ex-ante, no mechanism is even pairwise stable.

Theorem 3. *No mechanism is ex-ante pairwise stable.*

Proof. The proof of Theorem 3 uses the following example.

Example 3. Suppose that there are three men and three women, and fix $p \in (0, 1/4)$. The prior ψ is that preferences are drawn independently as follows:

$$
P_{m_1} = \begin{cases} w_1, w_3, w_2 \ w.p. \ 1 - 2p \\ w_2, w_1, w_3 \ w.p. \ p \\ w_3, w_2, w_1 \ w.p. \ p \end{cases} \qquad P_{w_1} = \begin{cases} m_1, m_3, m_2 \ w.p. \ 1 - 2p \\ m_2, m_1, m_3 \ w.p. \ p \\ m_3, m_2, m_1 \ w.p. \ p \end{cases}
$$

$$
P_{m_2} = \quad w_1, w_2 \qquad\qquad P_{w_2} = \quad m_1, m_2
$$
$$
P_{m_3} = \quad w_3 \qquad\qquad\quad P_{w_3} = \quad m_3
$$

Because m_3 and w_3 always rank each other first, we know by Lemmas 1 and 3 that if mechanism ϕ is ex-ante pairwise stable, it matches m_3 and w_3 with certainty. Applying Lemma 3 to the submarket $(\{m_1, m_2\}, \{w_1, w_2\})$, we conclude that

1. Whenever m_1 prefers w_2 to w_1, ϕ must match m_1 with w_2 (and m_2 with w_1) with certainty.
2. Whenever w_1 prefers m_2 to m_1, ϕ must match w_1 with m_2 (and m_1 with w_2) with certainty.
3. Whenever m_1 prefers w_1 to w_2 and w_1 prefers m_1 to m_2, ϕ must match m_1 with w_1.

After doing the relevant algebra, we see that w_1 and m_1 each get their first choice with probability $1 - 3p + 4p^2$, their second choice with probability p, and their third choice with probability $2p - 4p^2$. If w_1 and m_1 were to match to each other, they would get their first choice with probability $1 - 2p$, their second with probability p, and their third with probability p; an outcome that they both prefer. It follows that ϕ is not ex-ante pairwise stable, completing the proof.

The basic intuition for Example 3 is similar to that of Example 2. When m_1 ranks w_1 first and w_1 does not return the favor, it is unstable for them to match and m_1 will receive his third choice. In this case, it would (informally) be more "efficient" (considering only the welfare of m_1 and w_1) to match m_1 with w_1; doing so improves the ranking that m_1 assigns his partner by two positions, while only lowering the ranking that w_1 assigns her partner by one. Because men and women play symmetric roles in the above example, ex-ante, both m_1 and w_1 prefer the more efficient solution in which they always match to each other.

5   Discussion

In this paper, we extended the notion of stability to settings in which agents are uncertain about their own preferences and/or the preferences of others. We observed that when agents can sign contracts before preferences are fully known, every matching mechanism is susceptible to unraveling. While past work has reached conclusions that sound similar, we argue that our results are stronger in several ways.

First, previous results have assumed that agents are expected utility maximizers, and relied on assumptions about the utilities that agents get from each potential partner. Our work uses the stronger notion of stochastic dominance to determine blocking coalitions, and notes that there may exist opportunities for profitable circumvention of a central matching mechanism even when agents are unwilling to sacrifice the chance of a terrific match in order to avoid a poor one.

Second, not only can every mechanism be blocked under *some* prior, but also, for some priors, it is impossible to design a mechanism that is interim stable *at that prior*. This striking conclusion is similar to that of Peivandi and Vohra [13], who find (in a bilateral transferable utility setting) that for some priors over agent types, every potential mechanism of trade can be blocked.

In light of the above findings, one might naturally ask how it is that many centralized clearing houses have managed to persist. One possible explanation is that the problematic priors are "unnatural" and unlikely to arise in practice. We argue that this is not the case: Example 2 shows that blocking coalitions exist when agent preferences are independent and maximally uncertain, Example 3 shows that they may exist even when the preferences of most agents are known, and in the full version of the paper we show that they may exist even when one side has perfectly correlated (i.e. ex-post identical) preferences.

A more plausible explanation for the persistence of centralized clearing houses is that although mutually profitable early contracting opportunities may exist, agents lack the ability to identify and/or act on them. To take one example, even when profitable early contracting opportunities can be identified, agents may lack the ability to write binding contracts with one another (whereas our work assumes that they possess such commitment power). We leave a more complete discussion of the reasons that stable matching mechanisms might persist in some cases and fail in others to future work.

References

1. Afacan, M.O.: The welfare effects of pre-arrangements in matching markets. Econ. Theor. **53**(1), 139–151 (2013)
2. Du, S., Livne, Y.: Rigidity of transfers and unraveling in matching markets. Available at SSRN (2014)
3. Gudmundsson, J.: Sequences in Pairing Problems: A new approach to reconcile stability with strategy-proofness for elementary matching problems, November 2014. (Job Market Papers)
4. Halaburda, H.: Unravelling in two-sided matching markets and similarity of preferences. Games Econ. Behav. **69**(2), 365–393 (2010)
5. Kagel, J.H., Roth, A.E.: The dynamics of reorganization in matching markets: a laboratory experiment motivated by a natural experiment. Q. J. Econ. **115**(1), 201–235 (2000)
6. Li, H., Rosen, S.: Unraveling in matching markets. Am. Econ. Rev. **88**(3), 371–387 (1998)
7. Li, H., Suen, W.: Risk sharing, sorting, and early contracting. J. Polit. Econ. **108**(5), 1058–1091 (2000)
8. Li, H., Suen, W.: Self-fulfilling early-contracting rush. Int. Econ. Rev. **45**(1), 301–324 (2004)
9. Liu, Q., Mailath, G.J., Postlewaite, A., Samuelson, L.: Stable matching with incomplete information. Econometrica **82**(2), 541–587 (2014)
10. Manjunath, V.: Stability and the core of probabilistic marriage problems. Technical report, Working paper (2013)
11. McKinney, C.N., Niederle, M., Roth, A.E.: The collapse of a medical labor clearinghouse (and why such failures are rare). Am. Econ. Rev. **95**(3), 878–889 (2005)
12. Niederle, M., Roth, A.E.: The gastroenterology fellowship match: how it failed and why it could succeed once again. Gastroenterology **127**(2), 658–666 (2004)
13. Peivandi, A., Vohra, R.: On fragmented markets (2013)
14. Roth, A.E.: Two-sided matching with incomplete information about others' preferences. Games Econ. Behav. **1**(2), 191–209 (1989)
15. Roth, A.E.: A natural experiment in the organization of entry-level labor markets: regional markets for new physicians and surgeons in the united kingdom. Am. Econ. Rev. **81**(3), 415–440 (1991)
16. Roth, A.E., Rothblum, U.G.: Truncation strategies in matching markets in search of advice for participants. Econometrica **67**(1), 21–43 (1999)
17. Roth, A.E., Xing, X.: Jumping the gun: Imperfections and institutions related to the timing of market transactions. Am. Econ. Rev. **84**(4), 992–1044 (1994)
18. Sonmez, T.: Can pre-arranged matches be avoided in two-sided matching markets? J. Econ. Theor. **86**(1), 148–156 (1999)
19. Suen, W.: A competitive theory of equilibrium and disequilibrium unravelling in two- sided matching. RAND J. Econ. **31**(1), 101–120 (2000)
20. Yenmez, M.B.: Incentive-compatible matching mechanisms: consistency with various stability notions. Am. Econ. J. Microeconomics **5**(4), 120–141 (2013)

Fast Convergence in the Double Oral Auction

Sepehr Assadi, Sanjeev Khanna, Yang Li$^{(\boxtimes)}$, and Rakesh Vohra

University of Pennsylvania, Pennsylvania, USA
{sassadi,sanjeev,yangli2}@cis.upenn.edu, rvohra@seas.upenn.edu

Abstract. A classical trading experiment consists of a set of unit demand buyers and unit supply sellers with identical items. Each agent's value or opportunity cost for the item is their private information and preferences are quasi-linear. Trade between agents employs a double oral auction (DOA) in which both buyers and sellers call out bids or offers which an auctioneer recognizes. Transactions resulting from accepted bids and offers are recorded. This continues until there are no more acceptable bids or offers. Remarkably, the experiment consistently terminates in a Walrasian price. The main result of this paper is a mechanism in the spirit of the DOA that converges to a Walrasian equilibrium in a polynomial number of steps, thus providing a theoretical basis for the above-described empirical phenomenon. It is well-known that computation of a Walrasian equilibrium for this market corresponds to solving a maximum weight bipartite matching problem. The uncoordinated but rational responses of agents thus solve in a distributed fashion a maximum weight bipartite matching problem that is encoded by their private valuations. We show, furthermore, that every Walrasian equilibrium is reachable by some sequence of responses. This is in contrast to the well known auction algorithms for this problem which only allow one side to make offers and thus essentially choose an equilibrium that maximizes the surplus for the side making offers. Our results extend to the setting where not every agent pair is allowed to trade with each other.

1 Introduction

Chamberlin reported on the results of a market experiment in which prices failed to converge to a Walrasian equilibrium [5]. Chamberlin's market was an instance of the assignment model with homogeneous goods. There is a set of unit demand buyers and a set of unit supply sellers, and all items are identical. Each agent's value or opportunity cost for the good is their private information and preferences are quasi-linear. Chamberlin concluded that his results showed competitive theory to be inadequate. Vernon Smith, in an instance of insomnia, recounted in [18] demurred:

"The thought occurred to me that the idea of doing an experiment was right, but what was wrong was that if you were going to show that competitive

---

The full version of this paper can be found in [1]. Supported in part by National Science Foundation grants CCF-1116961, CCF-1552909, and IIS-1447470.

E. Markakis and G. Schäfer (Eds.): WINE 2015, LNCS 9470, pp. 60–73, 2015.
DOI: 10.1007/978-3-662-48995-6_5

equilibrium was not realizable ... you should choose an institution of exchange that might be more favorable to yielding competitive equilibrium. Then when such an equilibrium failed to be approached, you would have a powerful result. This led to two ideas: (1) why not use the double oral auction procedure, used on the stock and commodity exchanges? (2) why not conduct the experiment in a sequence of trading days in which supply and demand were renewed to yield functions that were daily flows?"

Instead of Chamberlin's unstructured design, Smith used a double oral auction (DOA) scheme in which both buyers and sellers call out bids or offers which an auctioneer recognizes [17]. Transactions resulting from accepted bids and offers are recorded. This continues until there are no more acceptable bids or offers. At the conclusion of trading, the trades are erased, and the market reopens with valuations and opportunity costs unchanged. The only thing that has changed is that market participants have observed the outcomes of the previous days trading and may adjust their expectations accordingly. This procedure was iterated four or five times. Smith was astounded: "I am still recovering from the shock of the experimental results. The outcome was unbelievably consistent with competitive price theory" [18](p. 156).

As noted by Daniel Friedman [8], the results in [17], replicated many times, are something of a mystery. How is it that the agents in the DOA overcome the impediments of both *private information* and *strategic uncertainty* to arrive at the Walrasian equilibrium? A brief survey of the various (early) theoretical attempts to do so can be found in Chap. 1 of [8]. Friedman concluded his survey of the theoretical literature with a two-part conjecture. "First, that competitive (Walrasian) equilibrium coincides with ordinary (complete information) Nash Equilibrium (NE) in interesting environments for the DOA institution. Second, that the DOA promotes some plausible sort of learning process which eventually guides the *both clever and not-so-clever* traders to a behavior which constitutes an 'as-if' complete-information NE."

Over the years, the first part of Friedman's conjecture has been well studied (see, e.g., [7]) but the second part of the conjecture is still left without a satisfying resolution. The focus of this paper is on the second part of Friedman's conjecture. More specifically, we design a mechanism which simulates the DOA, and prove that this mechanism always converges to a Walrasian equilibrium in polynomially many steps. Our mechanism captures the following four key properties of the DOA.

1. *Two-sided market:* Agents on either side of the market can make actions.
2. *Private information:* When making actions, agents have no other information besides their own valuations and the bids and offers submitted by others.
3. *Strategic uncertainty:* The agents have the freedom to choose their actions modulo mild rationality conditions.
4. *Arbitrary recognition:* The auctioneer (only) recognizes bids and offers in an *arbitrary* order.

Among these four properties, mechanisms that allow agents on either side to make actions (*two-sided market*) and/or limit the information each agent has (*private information*) have received more attention in the literature (see Sect. 1.1). However, very little is known for mechanisms that both work for strategically uncertain agents and recognize agents in an arbitrary order. Note that apart from resolving the second part of Friedman's conjecture, having a mechanism with these four properties itself is of great interest for multiple reasons. First, in reality, the agents are typically unwilling to share their private information to other agents or the auctioneer. Second, agents naturally prefer to act freely as oppose to being given a procedure and merely following it. Third, in large scale distributed settings, it is not always possible to find a real auctioneer who is trusted by every agent, and is capable of performing massive computation on the data collected from all agents. In the DOA (or in our mechanism) however, the auctioneer only recognizes actions in an arbitrary order, which can be replaced by any standard distributed token passing protocol, where an agent can take an action only when he is holding the token. In other words, our mechanism serves more like a platform (rather than a specific protocol) where rational agents always reach a Walrasian equilibrium no matter their actual strategy. To the best of our knowledge, no previous mechanism enables such a 'platform-like' feature. In the rest of this section, we summarize our results and discuss in more detail the four properties of the DOA in context of previous work.

1.1 Our Results and Related Work

We design a mechanism that simulates the DOA by simultaneously capturing two-sided market, private information, strategic uncertainty, and arbitrary recognition. More specifically, following the DOA, at each iteration of our mechanism, the auctioneer maintains a list of active price submission and a tentative assignment of buyers to sellers that 'clears' the market at the current prices (note that this can also be distributedly maintained by the agents themselves). Among the agents who wish to make or revise an earlier submission, an arbitrary one is recognized by the auctioneer and a new tentative assignment is formed. An agent can submit *any* price that strictly improves his payoff given the current submissions (rather than being forced to make a 'best' response, which is to submit the price that maximizes payoff). We show that as long as agents make myopically better responses, the market always converges to a Walrasian equilibrium in polynomial number of steps. Furthermore, *every* Walrasian equilibrium is the limit of some sequence of better responses. We should remark that the fact that an agent always improves his payoff does not imply that the total payoff of all agents always increases. For instance, a buyer can increase his payoff by submitting a higher price and 'stealing' the current match of some other buyer (whose payoff would drop).

To the best of our knowledge, no existing mechanism captures all four properties for the DOA that we proposed in this paper. For most of the early work on auction based algorithms (e.g., [2,6,7,12,16]), unlike the DOA, only one side of the market can make offers. By permitting only one side of the market to make offers, the auction methods essentially pick the Walrasian equilibrium (equilibria are not unique) that maximizes the total surplus of the side making the offers.

For two-sided auction based algorithms [3,4], along with the 'learning' based algorithms studied more recently [10,13], agents are required to follow a specific algorithm (or protocol) that determines their actions (and hence violates strategic uncertainty). For example, [4] requires that when an agent is activated, a buyer always matches to the 'best' seller and a seller always matches to the 'best' buyer (i.e., agents only make myopically *best* responses, which is not the case for the DOA). [10] has agents submit bids based on their current best alternative offer and prices are updated according to a common formula relying on knowledge of the agents opportunity costs and marginal values. [13], though not requiring agents to always make myopically best responses, has agents follow a specific (randomized) algorithm to submit conditional bids and choose matches. We should emphasize that agents acting based on some *random* process is different from agents being strategically uncertain. In particular, for the participants of the original DOA experiment (of [17]), there is no a priori reason to believe that they were following some specific random procedure during the experiment. On the contrary, as stated in Friedman's conjecture, there are *clever and not-so-clever* participants, and hence different agents could have completely different strategies and their strategies might even change when, for instance, seeing more agents matching with each other, or by observing the strategies of other agents. Therefore, analyzing a process where agents are strategically uncertain can be distinctly more complex than analyzing the case where agents behave in accordance with a well-defined stochastic process. In this paper, we consider an extremely general model of the agents: the agents are acting arbitrarily while only following some mild rationality conditions. Indeed, proving fast convergence (or even just convergence) for a mechanism with agents that are strategically uncertain is one of the main challenges of this work.

Arbitrary recognition is another critical challenge for designing our mechanism. For example, the work of [13,14] deploys randomization in the process of recognizing agents. This is again in contrast to the original DOA experiment, since the auctioneer did not use a randomized procedure when recognizing actions, and it is unlikely that the participants *decide* to make an action following some random process (in fact, some participants might be more 'active' than others, which could lead to the 'quieter' participants barely getting *any* chance to make actions, as long as the 'active' agents are still making actions).

The classical work on the *stable matching* problem [9] serves as a very good illustration for the importance of arbitrary recognition. Knuth [11] proposed the following algorithm for finding a stable matching. Start with an arbitrary matching; if it is stable, stop; otherwise, pick a blocking pair and match them; repeat this process until a stable matching is found. Knuth showed that the algorithm could cycle if the blocking pair is picked *arbitrarily*. Later, [15] showed that picking the blocking pairs at random suffices to ensure that the algorithm eventually converges to a stable matching, which suggests that it is the arbitrary selection of blocking pairs that causes Knuth's algorithm to not converge.

The setting of Knuth's algorithm is very similar to the process of the DOA in the sense that in any step of the DOA, a temporary matching is maintained and agents can make actions to (possibly) change the current matching. But perhaps surprisingly, we show that arbitrary recognition does not cause the DOA to suffer from the same cycling problem as Knuth's algorithm. The main reason, or the main difference between the two models is that our assignment model involves both matching and prices, while Knuth's algorithm only involves matchings. As a consequence, in our mechanism, the preferences of the agents change over time (since an agent always favors the better price submission, the preferences could change when new prices are submitted). In the instance that leads Knuth's algorithm to cycle (see [11]), the fundamental cause is that the preferences of *all* agents form a cycle. However, in our mechanism, preferences (though changing) are always consistent for all agents.

Based on this observation, we establish the limit of the DOA by introducing a small friction into the market: restricting the set of agents on the other side that each agent can trade with[1]. We show that in this case, there is an instance with a specific adversarial order of recognizing agents such that following this order, the preferences of the agents (over the entire order) form a cycle and the DOA may never converge. Finally, we complete the story by showing that if we change the mechanism to recognize agents randomly, with high probability, a Walrasian equilibrium will be reached in polynomial number of steps. This further emphasizes the distinction between random recognition and arbitrary recognition for DOA-like mechanisms.

Organization: The rest of the paper is organized as follows. In Sect. 2, we formally introduce the model of the market and develop some concepts and notation used throughout the paper. Our main results are presented in Sect. 3. We describe our DOA style mechanism and show that in markets with no trading restrictions, it converges in a number of steps polynomial in the number of agents. We then show that when each agent is restricted to trade only with an arbitrary subset of agents on the other side, the mechanism need not converge. A randomized variant of our mechanism is then presented to overcome this issue. Finally, we conclude with some directions for future work in Sect. 4.

2   Preliminaries

We will use the terms 'player' and 'agent' interchangeably throughout the paper. We use B to represent a buyer, S for a seller, and Z for either of them. Also, b is used as the bid submitted by a buyer and s as the offer from a seller.

Definition 1 (Market). *A market is denoted by* $G(\mathcal{B}, \mathcal{S}, E, val)$, *where* $\mathcal{B}$ *and* $\mathcal{S}$ *are the sets of buyers and sellers, respectively. Each buyer* $B \in \mathcal{B}$ *is endowed*

---

[1] In Chamberlin's experiment, buyers and sellers had to seek each other out to determine prices. This search cost meant that each agent was not necessarily aware of all prices on the other side of the market.

with a valuation of the item, and each seller $S \in \mathcal{S}$ has an opportunity cost for the item. We slightly abuse the terminology and refer to both of these values as the valuation of the agent for the item. The valuation of any agent Z is chosen from range $[0, 1]$, and denoted by $val(Z)$. Finally, E is the set of undirected edges between $\mathcal{B}$ and $\mathcal{S}$, which determines the buyer-seller pairs that may trade.

Let $m = |E|$ and $n = |\mathcal{B}| + |\mathcal{S}|$.

Definition 2 (Market State). *The state of a market at time t is denoted $\mathcal{S}^t(P^t, \Pi^t)$ ($\mathcal{S}(P, \Pi)$ for short, if time is clear or not relevant), where P is a price function revealing the price submission of each player and Π is a matching between $\mathcal{B}$ and $\mathcal{S}$, indicating which players are currently paired. In other words, the bid (offer) of a buyer B (seller S) is $P(B)$ ($P(S)$), and B, S are paired in Π iff $(B, S) \in \Pi$. In addition, we denote a player $Z \in \Pi$ iff Z is matched with some other player in Π, and denote his match by $\Pi(Z)$.*

Furthermore, the state where each buyer submits a bid of 0, each seller submits an offer of 1, and no player is matched is called the zero-information state.

We use the term zero-information because no player reveals non-trivial information about his valuation in this state.

Definition 3 (Valid State). *A state is called* valid *iff (a_1) the price submitted by each buyer (seller) is lower (higher) than his valuation, (a_2) two players are matched only when there is an edge between them, and (a_3) for any pair in the matching, the bid of the buyer is no smaller than the offer of the seller.*

In the following, we restrict attention to states that are valid.

Definition 4 (Utility). *For a market $G(\mathcal{B}, \mathcal{S}, E, val)$ at state $\mathcal{S}(P, \Pi)$, the utility of a buyer is defined as $val(B) - P(B)$, if B receives an item, and zero otherwise. Similarly, the utility of a seller is defined as $P(S) - val(S)$, if S trades his item, and zero otherwise.*

Note that what we have called utility is also called surplus.

Definition 5 (Stable State). *A stable state of a market $G(\mathcal{B}, \mathcal{S}, E, val)$ is a state $\mathcal{S}(P, \Pi)$ s.t. (a_1) for all $(B, S) \in E$, $P(B) \leq P(S)$ (a_2) if $Z \notin \Pi$, then $P(Z) = val(Z)$, and (a_3) if $(B, S) \in \Pi$, then $P(B) = P(S)$.*

Suppose $\mathcal{S}(P, \Pi)$ is not stable. Then, one of the following must be true.

1. There exists $(B, S) \in E$ such that $P(B) > P(S)$. Then, both B and S could strictly increase their utility by trading with each other using the average of their prices.
2. There exists $Z \notin \Pi$ such that $P(Z) \neq val(Z)$. This agent could raise his bid (if a buyer) or lower his offer (if a seller), without reducing his utility and having a better opportunity to trade.
3. There exists $(B, S) \in \Pi$ such that $P(B) > P(S)$ ($P(B) < P(S)$ results in an invalid state). One of the agents could do better by either raising his offer or lowering his bid.

Definition 6 (ε-Stable State). *For any $\varepsilon \geq 0$, a state $\mathcal{S}(P, \Pi)$ of a market $G(\mathcal{B}, \mathcal{S}, E, val)$ is ε-stable iff (a_1) for any $(B, S) \in E$, $P(B) - P(S) \leq \varepsilon$ (a_2) if player $Z \notin \Pi$, $P(Z) = val(Z)$, and (a_3) if $(B, S) \in \Pi$, $P(B) = P(S)$.*

Note that the only difference between a stable state and an ε-stable state lies in the first property. At any ε-stable state, no matched player will have a move to increase his utility by more than ε.

Definition 7 (Social Welfare). *For a market $G(\mathcal{B}, \mathcal{S}, E, val)$ with a matching Π, the social welfare (SW) of this matching is defined as the sum of the valuation of the matched buyers minus the total opportunity cost of the matched sellers. We denote by SW_Π the SW of matching Π.*

Definition 8 (ε-approximate SW). *For any market, a matching Π is said to give an ε-approximate SW if $SW_\Pi \geq SW_{\Pi^*} - n\varepsilon$ for any Π^* that maximizes SW. In other words, on average, the social welfare collected from each player using Π is at most ε less than that collected using Π^*.*

The following connection between (ε-)stable state and (ε-approximate) SW is well known in the literature (see the full version [1] Theorems 3.1 and 3.2 for a self contained proof.)

Theorem 2.1. *For any market $G(\mathcal{B}, \mathcal{S}, E, val)$, (a_1) a state is stable iff it maximizes SW, (a_2) for any $\varepsilon > 0$, any ε-stable state realizes an ε-approximate SW, and (a_3) if we define $\delta = \min\{|val(Z_1) - val(Z_2)| \mid Z_1, Z_2 \in \mathcal{B} \cup \mathcal{S}, val(Z_1) \neq val(Z_2)\}$, then for $0 \leq \varepsilon < \delta/n$, any ε-stable state maximizes SW.*

3   Convergence to a Stable State

We establish our main results in this section. We will start by describing a mechanism in the spirit of DOA, and show that for any *well-behaved* stable state, there is a sequence of agent moves that leads to this state. When the trading graph is a complete bipartite graph, i.e., the case of the DOA expriment, we show that convergence to a stable state occurs in number of steps that is polynomially bounded in the number agents. However, convergence to a stable state is not guaranteed when the trading graph is an incomplete bipartite graph. We propose a natural randomized extension of our mechanism, and show that with high probability, the market will converge to a stable state in number of steps that is polynomially bounded in the number of agents.

3.1   The Main Mechanism

To describe our mechanism, we need the notion of an ε-*interested* player.

Definition 9 (ε-Interested Player). *For a market at state $\mathcal{S}(P, \Pi)$ with any parameter $\varepsilon > 0$, a seller S is said to be ε-interested in his neighbor B iff either (a) $P(B) \geq P(S)$ and $S \notin \Pi$, or (b) $P(B) - P(S) \geq \varepsilon$ and $S \in \Pi$. The set of buyers interested in a seller S is defined analogously.*

When the parameter ε is clear from the context, we will simply refer to an $\varepsilon - interested$ player as an interested player.

Mechanism 1 *(with input parameter $\varepsilon > 0$)*

- **Activity Rule:** *Among the unmatched buyers, any buyer that neither submits a new higher bid nor has a seller that is interested in him, is labeled as inactive. All other unmatched buyers are labeled as active. An active (inactive) seller is defined analogously. An inactive player changes his status iff some player on the other side matches with him.*[2]
- **Minimum Increment:** *Each submitted price must be an integer multiple of ε.*[3]
- **Recognition:** *Among all active players, an arbitrary one is recognized.*
- **Matching:** *After a buyer B is recognized, B will choose an interested seller to match with if one exists. If the offer of the seller is lower than the bid b, it is immediately raised to b. The seller action is defined analogously.*
- **Tie Breaking:** *When choosing a player on the other side to match to, an unmatched player is given priority (the* unmatched first *rule).*

In each iteration, players are partitioned into two sets based on whether they are matched or not. The unmatched players are further partitioned into active players and inactive players. The only players with a myopic incentive to revise their submissions are those that are not matched.

Observe that since a buyer will never submit a bid higher than his valuation, and a seller will never make an offer below his own opportunity cost, by submitting only prices that are integer multiples of ε, an agent might not be able to submit his true valuation. However, since an agent can always submit a price at most ε away from the true valuation, if we pretend that the 'close to valuation' prices are true valuations, the maximum SW will decrease by at most $n\varepsilon$. By picking $\varepsilon' = \varepsilon/2$, if the market converges to an ε'-stable state, we still guarantee that the SW of the final state is at most $n\varepsilon$ away from the maximum SW.

When a buyer B chooses to increase his current bid: if s denotes the lowest offer in the neighborhood of B, and s' denotes the lowest offer of any unmatched seller in the neighborhood of B, then the new bid of B can be at most $\min\{s + \varepsilon, s'\}$. We refer to this as the *increment* rule. This may be viewed as a consequence of rationality – there is no incentive for a buyer to bid above the price needed to make a deal with some seller. A similar rule applies to sellers. With a slight abuse of the terminology, we call either rules increment rule. Notice, a player indifferent between submitting a new price and keeping his price unchanged will be assumed to break ties in favor of activity.

Note that the role of the auctioneer in Mechanism (1) is restricted to recognize agent actions, but never select actions for agents. In fact, the existence of an auctioneer is not even necessary for the mechanism to work. Minimum increment can be interpreted as setting the currency of the market to be ε. Arbitrary recognition can be achieved by a first come, first served principle. Activity rule

---

[2] This is common for eliminating no trade equilibria.

[3] This is part of many experimental implementations of the DOA.

and matching are both designed to ensure that players will keep making actions (submitting a new price or forming a valid match) if one exists.

If a state $\mathcal{S}(P, \Pi)$ satisfies $\forall (B, S) \in E, P(B) \leq P(S)$, then we call it a *valid starting state*. It is not difficult to verify the following property of Mechanism (1) (see the full version [1], Claim 4.2 for a detailed discussion).

Claim 3.1 *For any market, if we use Mechanism (1) and begin with a valid starting state, then any final state of the market is ε-stable.*

Note that by Theorem 2.1, if a market converges to an ε-stable state, it always realizes ε-approximate SW.

Definition 10 (Well-behaved). *A stable state $\mathcal{S}(P, \Pi)$, is* well-behaved *iff (a_1) for any $(B, S) \in E$, if $B \notin \Pi$ and $S \notin \Pi$, then $P(B) < P(S)$. An ε-stable state $\mathcal{S}(P, \Pi)$, is* well-behaved *iff not only property (a_1) is satisfied but also (a_2) for any $(B, S) \in E$, if either $B \notin \Pi$ or $S \notin \Pi$, then $P(B) \leq P(S)$.*

Note that the only states that are not well-behaved are the corner cases where a buyer-seller pair having the same valuation (thus having no contribution to SW) are not chosen in the matching, or players who can obtain utility at most ε stop attempting to match with others. We establish that any well-behaved (ε-stable) state is reachable (see the full version [1], Theorem 4.3 for a proof):

Theorem 3.2. *For any $\varepsilon > 0$, if we use Mechanism (1), and start from the zero-information state, any well-behaved ε-stable state can be reached via a sequence of at most n moves. Hence, any well-behaved stable state is also reachable.*

3.2 Complete Bipartite Graphs

We now prove that market with complete bipartite trading graph will always converge when using Mechanism (1).

Theorem 3.3. *For a market whose trading graph is a complete bipartite graph, if we use Mechanism (1) with any input parameter $\varepsilon > 0$, and begin with any valid starting state, then the market will converge to an ε-stable state after at most n^3/ε steps.*

We need the following lemma to prove Theorem 3.3 (see the full version [1], Lemma 4.1 for a proof).

Lemma 1. *For a market $G(\mathcal{B}, \mathcal{S}, E, val)$ whose trading graph is a complete bipartite graph, if we use Mechanism (1) with any input parameter $\varepsilon > 0$, then at any state $\mathcal{S}(P, \Pi)$ reached from a valid starting state, for any $(B, S) \in E$, if $P(B) > P(S)$, then both B and S are matched.*

Definition 11 (γ-feasible). *A market state $\mathcal{S}(P, \Pi)$ is said to be γ-feasible iff there are exactly γ matches in Π.*

Proof (of Theorem 3.3*)*. Assume at any time t, the state $\mathcal{S}^t$ of the market is γ^t-feasible. Define the following potential function

$$\Phi_P = \sum_{S_i \in \mathcal{S}} P(S_i) + \sum_{B_i \in \mathcal{B}} (1 - P(B_i))$$

Note that the value of Φ_P is always an integer multiple of ε. We will first show that γ^t forms a non-decreasing sequence over time, and then argue that, for any γ, the market can stay in a γ-feasible state for a bounded number of steps. Specifically, we will show that, if γ does not change, Φ_P is a non-increasing function and can stay unchanged for at most γ steps. Since the maximum value of Φ_P is bounded by n, it follows that after at most $(\gamma n)/\varepsilon$ steps, the market moves from a γ-feasible state to a $(\gamma + 1)$-feasible state (or converges).

We argue that γ^t forms a non-decreasing sequence over time. Since any recognized player is unmatched, if the action of an unmatched player Z results in a change in the matching, Z either matches with another unmatched player, or matches to a player that was already matched. In either case, the total number of matched pairs does not decrease.

Furthermore, we prove if γ does not change, then Φ_P is non-increasing. Moreover, the number of successive steps that Φ_P stay unchanged is at most γ.

To see that Φ_P is non-increasing, first note that Φ_P can increase only when either a buyer decreases his bid or a seller increases his offer. Assume an unmatched buyer B is recognized (seller case is analogous), and the price function before his move is P. To increase Φ_P, since B can only increase his bid, he must increase an offer by overbidding and matching with a seller S, resulting in the two of them submitting the same price b. The buyer bid increases by $b - P(B)$ and the seller offer increases by $b - P(S)$. Since B is unmatched, by Lemma 1, $P(B) \le P(S)$, and hence Φ_P will not increase.

We now bound the maximum number of steps for which Φ_P could remain unchanged. A move from a buyer B that does not change Φ_P occurs only when B overbids a matched seller S, where the bid and the offer are equal both before and after the move. We call this a *no-change* buyer move. By analogy, a no-change seller move can be defined.

In the remainder of the proof, we first argue that a no-change buyer move can never be followed by a no-change seller move, and vice versa. After that, we prove the upper bound on the number of consecutive no-change moves to show that Φ_P will eventually decrease (by at least ε).

Assume at time t_1, a buyer B_{t_1} made a no-change move and matched with a seller S_{t_1}, who was originally paired with the buyer B'_{t_1}.[4] We prove that no seller can make a no-change move at time $t_1 + 1$. The case that a seller makes a no-change move first can be proved analogously. Suppose at time $t_1 + 1$, a seller S_{t_1+1} is recognized and decreases his offer by ε. Since B_{t_1} made a no-change move, we have

$$P^{t_1}(B'_{t_1}) = P^{t_1}(B_{t_1}) \tag{1}$$

---

[4] An action at time time t will take effect at the time $t+1$, and P^t is the price function before any action is made at time t.

Denote the lowest seller offer (highest buyer bid) at any time t by s^t (b^t). By Lemma 1, $P^{t_1}(B_{t_1}) \leq P^{t_1}(S)$ for any seller S, hence $P^{t_1}(B_{t_1}) \leq s^{t_1}$. Moreover, since $P^{t_1}(B_{t_1}) = P^{t_1}(S_{t_1}) \geq s^{t_1}$, we have

$$P^{t_1}(B_{t_1}) = s^{t_1} \tag{2}$$

In other words, a buyer can make a no-change move, only if his bid is equal to the lowest offer. Similarly, if S_{t_1+1} can make a no-change move at time $t_1 + 1$, his offer is equal to the highest bid. Since the highest bid at time t_1 (b^{t_1}) is at most $s^{t_1} + \varepsilon$ (property (a_1) of ε-stable states), after B_{t_1} submits a bid of $s^{t_1} + \varepsilon$, he will be submitting the highest bid at time $t_1 + 1$. Hence

$$P^{t_1+1}(S_{t_1+1}) = b^{t+1} = P^{t_1+1}(B_{t_1}) = P^{t_1+1}(B'_{t_1}) + \varepsilon \tag{3}$$

Therefore, at time $t_1 + 1$, after S_{t_1+1} decreases his offer by ε, the unmatched buyer B'_{t_1} is interested in S_{t_1+1}. By the unmatched first rule, S_{t_1+1} will match with an unmatched player, hence this cannot be a no-change move.

This proves that a no-change seller move can never occur after a no-change buyer move and vice versa. We now prove the upper bound on the number of consecutive no-change buyer moves.

For any sequence of consecutive no-change buyer moves, if there exists a time t_2 such that $s^{t_2} > s^{t_2-1}$, for any unmatched buyer B at time t_2, $P^{t_2}(B) \leq s^{t_2-1} < s^{t_2}$. By Eq. (2), no buyer can make any more no-change move. Moreover, since any no-change buyer move will increase the submission of a matched seller who is submitting the lowest offer, after at most γ steps, the lowest offer must increase, implying that the length of the sequence is at most γ.

To conclude, the total number of steps that the market could stay in γ-feasible states is bounded by $(n/\varepsilon)\gamma$. As $\gamma \leq n$, the total number of steps before market converges is at most n^3/ε. $\qquad\square$

3.3   General Bipartite Graphs

In this section, we study the convergence of markets with an arbitrary bipartite trading graph. Although by Theorem 3.2, using Mechanism (1), the market can reach any well-behaved ε-stable state, when the trading graph of a market can be an arbitrary bipartite graph, there is no guarantee that Mechanism (1) will actually converge.

Claim 3.4. *In a market whose trading graph is an arbitrary bipartite graph, Mechanism (1) may never converge.*

Consider the market shown in Fig. 1. In this market, there are four buyers (B_1 to B_4) all with valuation 1 and four sellers (S_1 to S_4) all with opportunity cost 0. Moreover, the trading graph is a cycle of length 8, as illustrated by the first graph in Fig. 1. Assume at some time t, the market enters the state illustrated by the second graph, where B_1, B_2, S_1, B_3, S_2 are submitting 5ε, S_3, B_4, S_4 are submitting 6ε, and pairs (B_2, S_1), (B_3, S_2) and (B_4, S_4) are matched.

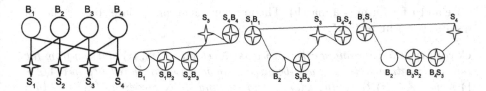

Fig. 1. Unstable market with general trading graph and Mechanism (1)

At time $t+1$, since B_1 is unmatched, he can be recognized and submit 6ε. S_1 is the only interested seller, hence B_1, S_1 will match and the offer of S_1 increases to 6ε, which leads to the state shown in the third graph. Similarly, at time $t+2$, since S_3 is unmatched, he can be recognized and submit 5ε. B_4 is the only interested buyer, hence B_4, S_3 will match and bid of B_4 increases to 6ε, which leads to the state shown in the fourth graph.

Notice that the states at time t and $t+2$ are isomorphic. By shifting the indices and repeating above two steps, the market will never converge.

Observe that the cycle described in Claim 3.4 is caused by an adversarial coordination between the actions of various agents. To break this pathological coordination, we introduce Mechanism (2) which is a natural extension of Mechanism (1) that uses randomization. We first define this mechanism, and then prove that on any trading graph, with high probability, the mechanism leads to convergence in a number of steps that is polynomially bounded in the number of agents.

Mechanism 2 *(with input parameter $\varepsilon > 0$)*

- **Activity Rule:** *Among the unmatched buyers, any buyer that neither submits a new higher bid nor has a seller that is interested in him, is labeled as inactive. All other unmatched buyers are labeled as active. An active (inactive) seller is defined analogously. An inactive player changes his status iff some player on the other side matches with him.*
- **Minimum Increment:** *Each submitted price must be an integer multiple of ε.*
- **Bounded Increment Rule:** *In each step, a player is only allowed to change his price by ε.*
- **Recognition:** *Among all players who are active, one is recognized uniformly at random.*
- **Matching:** *After a player, say a buyer B, is recognized, if B does not submit a new price, then B will match to an interested seller if one exists. If the offer of the seller is lower than the bid b, it is immediately raised to b. The seller action is defined analogously.*
- **Tie Breaking:** *When choosing a player on the other side to match to, an unmatched player is given priority (unmatched first rule).*

Notice that we ask players to move cautiously through the bounded increment rule. Players can either change the price by ε or match with an interested seller, and always favor being active. Note that, any move in Mechanism (1) can be simulated by at most $(1/\varepsilon + 1)$ moves in Mechanism (2) ($1/\varepsilon$ for submitting new price and 1 for forming a match). The following is an immediate consequence of results shown in Sect. 3.1.

Corollary 3.5. *For any market, if we use Mechanism (2), (i) starting from the zero-information state, any well-behaved ε-stable state can be reached in $n(1/\varepsilon + 1)$ steps, and (ii) beginning with a valid starting state, properties (a_1) and (a_3) of ε-stable states always hold, and the final state is ε-stable.*

We establish as our second main result that for any trading graph, with high probability, Mechanism (2) converges to a ε-stable state in a number of steps polynomial in the number of agents. We prove this result using the summation of the prices submitted by *all* agents as a potential function (denoted by Φ). Although Φ is clearly *not* monotone, the design of Mechanism (2) implies that Φ acts like a *random walk on a line*. Then, using a standard fact about random walks, we are able to analyze the behavior of Φ and argue fast convergence of the market (see the full version [1], Theorem 4.8 for a complete proof).

Theorem 3.6. *For any market $G(\mathcal{B}, \mathcal{S}, E, val)$, if we use Mechanism (2) with any input parameter $\varepsilon > 0$, and begin with a valid starting state, the market will converge to an ε-stable state after at most $O((n^3/\varepsilon^2) \log n)$ steps with high probability.*

4   Conclusions

We resolved the second part of Friedman's conjecture by designing a mechanism which simulates the DOA and proving that this mechanism always converges to a Walrasian equilibrium in polynomially many steps. Our mechanism captures four key properties of the DOA: agents on either side can make actions; agents only have limited information; agents can choose *any* better response (as opposed to the best response); and the submissions are recognized in an arbitrary order. An important aspect of our result is that, unlike previous models, *every* Walrasian equilibrium is reachable by some sequence of better responses.

For markets where only a restricted set of buyer-seller pairs are able to trade, we show that the DOA may never converge. However, if submissions are recognized randomly, and players only change their bids and offers by a small fixed amount, convergence is guaranteed. It is unclear that the latter condition is inherently necessary, and perhaps a convergence result can be established for a relaxed notion of bid and offer changes where players can make possibly large adjustments as long as they are consistent with the increment rule.

References

1. Assadi, S., Khanna, S., Li, Y., Vohra, R.: Fast convergence in the double oral auction. CoRR abs/1510.00086 (2015)
2. Bertsekas, D.: A distributed algorithm for the assignment problem. Laboratory for Information and Decision Systems, Working Paper, M.I.T., Cambridge (1979)

3. Bertsekas, D.P.: Linear Network Optimization: Algorithms and Codes. MIT Press, Cambridge (1991)
4. Bertsekas, D.P., Castaon, D.A.: A forward/reverse auction algorithm for asymmetric assignment problems. Comput. Optim. Appl. 1(3), 277–297 (1992). doi:10.1007/BF00249638
5. Chamberlin, E.H.: An experimental imperfect market. J. Polit. Econ. 56(2), 95–108 (1948)
6. Crawford, V.P., Knoer, E.M.: Job matching with heterogeneous firms and workers. Econometrica J. Econometric Soc. 49, 437–450 (1981)
7. Demange, G., Gale, D., Sotomayor, M.: Multi-item auctions. J. Polit. Econ. 94, 863–872 (1986)
8. Friedman, D.P., Rust, J.: The Double Auction Market: Institutions, Theories, and Evidence. Westview Press, Boulder (1993)
9. Gale, D., Shapley, L.S.: College admissions and the stability of marriage. Am. Math. Mon. 69, 9–15 (1962)
10. Kanoria, Y., Bayati, M., Borgs, C., Chayes, J., Montanari, A.: Fast convergence of natural bargaining dynamics in exchange networks. In: SODA (2011). http://dl.acm.org/citation.cfm?id=2133036.2133154
11. Knuth, D.: Mariages stables et leurs relations avec d&autres problèmes combinatoires: Collection de la Chaire Aisenstadt, Presses de l'Université de Montréal (1976). http://books.google.com/books?id=eAmFAAAAIAAJ
12. Kuhn, H.W.: The Hungarian method for the assignment problem. In: Jünger, M., Liebling, T.M., Naddef, D., Nemhauser, G.L. (eds.) 50 Years of Integer Programming 1958–2008 - From the Early Years to the State-of-the-Art, pp. 29–47. Springer, Heidelberg (2010)
13. Nax, H.H., Pradelski, B.S.R., Young, H.P.: Decentralized dynamics to optimal and stable states in the assignment game. In: CDC (2013)
14. Pradelski, B.S.: Decentralized dynamics and fast convergence in the assignment game: extended abstract. In: EC. ACM, New York (2015)
15. Roth, A.E., Vate, J.H.V.: Random paths to stability in two-sided matching. Econometrica J. Econometric Soc. 58, 1475–1480 (1990)
16. Shapley, L., Shubik, M.: The assignment game I: the core. Int. J. Game Theory 1(1), 111–130 (1971). doi:10.1007/BF01753437
17. Smith, V.L.: An experimental study of competitive market behavior. J. Polit. Econ. 70(2), 111–137 (1962)
18. Smith, V.L.: Papers in Experimental Economics. Cambridge University Press, New York (1991)

Minority Becomes Majority in Social Networks

Vincenzo Auletta[1], Ioannis Caragiannis[2], Diodato Ferraioli[1(✉)],
Clemente Galdi[3], and Giuseppe Persiano[1]

[1] Università degli Studi di Salerno, Fisciano, Italy
{auletta,dferraioli,pino.persiano}@unisa.it
[2] CTI "Diophantus" and University of Patras, Patras, Greece
caragian@ceid.upatras.gr
[3] Università di Napoli "Federico II", Napoli, Italy
clemente.galdi@unina.it

Abstract. It is often observed that agents tend to imitate the behavior of their neighbors in a social network. This imitating behavior might lead to the strategic decision of adopting a public behavior that differs from what the agent believes is the right one and this can subvert the behavior of the population as a whole.

In this paper, we consider the case in which agents express preferences over two alternatives and model social pressure with the *majority* dynamics: at each step an agent is selected and its preference is replaced by the majority of the preferences of her neighbors. In case of a tie, the agent does not change her current preference. A profile of the agents' preferences is *stable* if the each agent's preference coincides with the preference of at least half of the neighbors (thus, the system is in equilibrium).

We ask whether there are network topologies that are robust to social pressure. That is, we ask whether there are graphs in which the majority of preferences in an initial profile s always coincides with the majority of the preference in all stable profiles reachable from s. We completely characterize the graphs with this robustness property by showing that this is possible only if the graph has no edge or is a clique or very close to a clique. In other words, except for this handful of graphs, every graph admits at least one initial profile of preferences in which the majority dynamics can subvert the initial majority. We also show that deciding whether a graph admits a minority that becomes majority is NP-hard when the minority size is at most 1/4-th of the social network size.

This work was partially supported by the COST Action IC1205 "Computational Social Choice", by the Italian MIUR under the PRIN 2010–2011 project ARS TechnoMedia – Algorithmics for Social Technological Networks, by the European Social Fund and Greek national funds through the research funding program Thales on "Algorithmic Game Theory", by the EU FET project MULTIPLEX 317532, and by a Caratheodory basic research grant from the University of Patras.

E. Markakis and G. Schäfer (Eds.): WINE 2015, LNCS 9470, pp. 74–88, 2015.
DOI: 10.1007/978-3-662-48995-6_6

1   Introduction

Social scientists are greatly interested in understanding how social pressure can influence the behavior of agents in a social network. We consider the case in which agents connected through a social network must choose between two alternatives and, for concreteness, we consider two competing technologies: the current (or old) technology and a new technology. To make their decision, the agents take into account two factors: their personal relative valuation of the two technologies and the opinions expressed by their social neighbors. Thus, the public action taken by an agent (i.e., adopting the new technology or staying with the old) is the result of a mediation between her personal valuation and the social pressure derived from her neighbors.

The first studies concerning the adoption of new technologies date back to the middle of 20-th century, with the analysis of the adoption of hybrid seed corn among farmers in Iowa [16] and of tetracycline by physicians in US [6].

We assume that agents receive an initial signal about the quality of the new technology that constitutes the agent's initial preference. This signal is independent from the agent's social network; e.g., farmers acquired information about the hybrid corn from salesman and physicians acquired information about tetracycline from scientific publications. After the initial preference is formed, an agent tends to conform her preference to the one of her neighbors and thus to *imitate* their behavior, even if this disagrees with her own initial preference. This imitating behavior can be explained in several ways: an agent that sees a majority agreeing on an opinion might think that her neighbors have access to some information unknown to her and hence they have made the better choice; also agents can directly benefit from adopting the same behavior as their friends (e.g., prices going down).

Thus, the natural way of modeling the evolution of preferences in networks is through a majority dynamics: each agent has an initial preference and at each time step a subset of agents updates their opinion conforming to the majority of their neighbors in the network. As a tie-breaking rule it is usual to assume that when exactly half of the neighbors adopted the new technology, the agent decides to stay with her current choice to avoid the cost of a change. Thus, the network undergoes an opinion formation process where agents continue to update their opinions until a stable profile is reached, where each agent's behavior agrees with the majority of her neighbors. Notice that the dynamics does not take into account the relative merits of the two technologies and, without loss of generality, we adopt the convention that the technology that is preferred by the majority of the agents in the initial preference profile is the new technology.

In the setting described above, it is natural to ask whether and when the social pressure of conformism can change the opinion of some of the agents so that the initial majority is subverted. In the case of the adoption of a new technology, we are asking whether a minority of agents supporting the old technology can orchestrate a campaign and convince enough agents to reject the new technology, even if the majority of the agents had initially preferred the new technology.

This problem has been extensively studied in the literature. If we assume that updates occur *sequentially*, one agent at each time step, then it is easy to design graphs (e.g., a star) where the old technology, supported by an arbitrarily small minority of agents, can be adopted by most of the agents. Berger [2] proved that such a result holds even if at each time step all agents *concurrently* update their actions. However, Mossel et al. [13] and Tamuz and Tessler [17] proved that there are graphs for which, both with concurrent and sequential updates, at the end of the update process the new technology will be adopted by the majority of agents with high probability.

In [9,13] it is also proved that when the graph is an expander, agents will reach a *consensus* on the new technology with high probability for both sequential and concurrent updates (the probability is taken on the choice of initial configurations with a majority of new technology adopters). Thus, expander graphs are particularly efficient in aggregating opinions since, with high probability, social pressure does not prevent the diffusion of the new technology.

In this paper, we will extend this line of research by taking a worst-case approach instead of a probabilistic one. We ask whether there are graphs that are robust to social pressure, even when it is driven by a carefully and adversarially designed campaign. Specifically, we want to find out whether there are graphs in which no subset of the agents preferring the old technology (and thus consisting of less than half of the agents) can manipulate the rest of the agents and drive the network to a stable profile in which the majority of the agents prefers the old technology. This is easily seen to hold for two extreme graphs: the clique and the graph with no edge. In this paper, we prove that these are essentially[1] the only graphs where social pressure cannot subvert the majority.

In particular, our results highlight that even for expander graphs, where it is known that agents converge with high probability to consensus on the new technology, it is possible to fix a minority and orchestrate a campaign that brings the network into a stable profile where at least half of the agents decide to not adopt the new technology.

Overview of our contribution. We consider the following sequential dynamics. We have n agents and at any given point the system is described by the profile s in which $s(i) \in \{0, 1\}$ is the preference of the i-th agent. We say that agent i is *unhappy* in profile s if the majority of her neighbors have a preference different from $s(i)$. Profiles evolve according to the dynamics in which an *update* consists of non-deterministically selecting an unhappy agent and changing its preference. A profile in which no agent is unhappy is called *stable*.

In Sect. 2 (see Theorems 1 and 2), we characterize the set of social networks (graphs) where a majority can be subverted by social pressure. More specifically, we show that for each of these graphs it is possible to select a minority of agents not supporting the new technology and a sequence of updates (a campaign) that leads the network to a stable profile where the majority of the agents prefers the old technology. As described above, we will prove that this class is very large

---

[1] It turns out that for an even number of nodes, there are a few more very dense graphs enjoying such a property.

and contains all graphs except a small set of forbidden graphs, consisting of the graph with no edges and of other graphs that are almost cliques. Proving this fact turned out to be a technically challenging task and it is heavily based on properties of local optima of graph bisections.

Then we turn our attention to related computational questions. First we show that we can compute in polynomial time an initial preference profile, where the majority of the agents supports the new technology, and a sequence of update that ends in a stable profile where at least half of the agents do not adopt the new technology. This is done through a polynomial-time local-search computation of a bisection of locally minimal width.

We actually prove a stronger result. In principle, it could be that from the starting profile the system needs to undergo a long sequence of updates, in which the minority gains and loses member to eventually reach a stable profile in which the minority has become a majority. Our algorithm shows that this can always be achieved by means of a short sequence of at most two updates after which any sequence of updates will bring the system to a stable profile in which the initial minority has become majority. This makes the design of an adversarial campaign even more realistic, since such a campaign only has to identify the few "swing" agents and thus it turns out to be very simple to implement.

However, the simplicity of the subverting campaign comes at a cost. Indeed, our algorithm always computes an initial preferences profile that has very large minorities, consisting of $\left\lfloor \frac{n-1}{2} \right\rfloor$ agents. We remark that, even in case of large minorities, it is not trivial to give a sequence of update steps that ends in a stable profile where the majority is subverted. Indeed, even if the large minority of the original profile makes it easy to find a few agents of the original majority that prefer to change their opinions, this is not sufficient in order to prove that the majority has been subverted, since we have also to prove that there are no other nodes in the original minority that prefer to change their preference.

Moreover, we observe that, even if there are cases in which such a large minority is necessary, the idea behind our algorithm can be easily turned into an heuristic that checks whether the majority can be subverted by a smaller minority (e.g., by considering unbalanced partitions in place of bisections).

On the other side, we show that a large size of the minority in the initial preference profile seems to be necessary in order to quickly compute a subverting minority and its corresponding sequence of updates. Indeed, given a n-node social network, deciding whether there exists a minority of less than $n/4$ nodes and a sequence of update steps that bring the system to a stable profile in which the majority has been subverted is an NP-hard problem (see Theorem 4).

The main source of computational hardness seems to arise from the computation of the initial preference profile. Indeed, if this profile is given, computing the maximum number of adopters of the new technology (and, hence, deciding whether majority can be subverted) and the corresponding sequence of updates turns out to be possible in polynomial time (see Theorem 5).

Related work. There is a vast literature on the effect that social pressure has on the behavior of a system as a whole. In many works, influence is modeled by agents simply following the majority [2,9,13,17]. A generalization of this imitating behavior is discussed in [13].

A different approach is taken in [14], where each agent updates her behavior according to a Bayes rule that takes in account its own initial preference and what is declared by neighbors on the network.

Yet another approach assumes that agents are strategic and rational. That is, they try to maximize some utility function that depends on the level of coordination with the neighbors on the network. Here, the updates occur according to a best response dynamics or some other more complex game dynamics. Along this direction, particularly relevant to our works are the ones considering best-response dynamics from truthful profiles in the context of iterative voting, e.g., see [4,12]. In particular, closer to our current work is the paper of Brânzei et al. [4] who present bounds on the quality of equilibria that can be reached from a truthful profile using best-response play and different voting rules. The important difference is that there is no underlying network in their work.

Our work is also strictly related with a line of work in social sciences that aims to understand how opinions are formed and expressed in a social context. A classical simple model in this context has been proposed by Friedkin and Johnsen [11] (see also [7]). Its main assumption is that each individual has a private initial belief and that the opinion she eventually expresses is the result of a repeated averaging between her initial belief and the opinions expressed by other individuals with whom she has social relations. The recent work of Bindel et al. [3] assumes that initial beliefs and opinions belong to [0, 1] and interprets the repeated averaging process as a best-response play in a naturally defined game that leads to a unique equilibrium.

An obvious refinement of this model is to consider discrete initial beliefs and opinions by restricting them, for example, to two discrete values (see [5,10]). Clearly, the discrete nature of the opinions does not allow for averaging anymore and several nice properties of the opinion formation models mentioned above — such as the uniqueness of the outcome — are lost. In contrast, in [10] and in [5], it is assumed that each agent is strategic and aims to pick the most beneficial strategy for her, given her internal initial belief and the strategies of her neighbors. Interestingly, it turns out that the majority rule used in this work for describing how agents update their behavior can be seen as a special case of the discrete model of [5,10], in which agents assign a weight to the initial preference smaller than the one given to the opinion of the neighbors.

Studies on social networks consider several phenomena related to the spread of social influence such as information cascading, network effects, epidemics, and more. The book of Easley and Kleinberg [8] provides an excellent introduction to the theoretical treatment of such phenomena. From a different perspective, problems of this type have also been considered in the distributed computing literature, motivated by the need to control and restrict the influence of failures in distributed systems; e.g., see the survey by Peleg [15] and the references therein.

Preliminaries. We formally describe our model as follows. There are n agents; we use $[n] = \{1, 2, ..., n\}$ to denote their set. Each agent corresponds to a distinct node of a graph $G = (V, E)$ that represents the *social network*; i.e., the network of social relations between the agents. Agent i has an initial preference $s_0(i) \in \{0, 1\}$. At each time step, agent i can update her preference to $s(i) \in \{0, 1\}$. A *profile* is a vector of preferences, with one preference per agent. We use bold symbols for profiles; i.e., $s = (s(1), \ldots, s(n))$. In particular, we sometimes call the profile of initial preferences $(s_0(1), \ldots, s_0(n))$ as the *truthful profile*. Moreover, for any $y \in \{0, 1\}$, we denote as $\overline{y}$ the negation of y; i.e., $\overline{y} = 1 - y$.

A graph G is mbM (*minority becomes majority*) if there exists a profile s_0 of initial preferences such that: the number of nodes that prefer 0 is a strict majority, i.e., $|\{x \in V : s_0(x) = 0\}| > n/2$; and there is a *subverting* sequence of updates that starts from s_0 and reaches a stable profile s in which the number of nodes that prefer 0 is not a majority, i.e., $|\{x \in V : s(x) = 0\}| \leq n/2$. A profile of initial preferences that witnesses a graph being mbM will be also termed mbM.

2 Characterizing the mbM Graphs

The main result of this section is a characterization of the mbM graphs. More formally, we have the following definition. A graph G with n nodes is *forbidden* if one of the following conditions is satisfied:

F1: G has no edge;

oF2: G has an odd number of nodes, all of degree $n - 1$ (that is, G is a clique);

eF2: G has an even number of nodes and all its nodes have degree at least $n - 2$;

eF3: G has an even number of nodes, $n - 1$ nodes of G form a clique, and the remaining node has degree at most 2;

eF4: G has an even number of nodes, $n - 1$ nodes of G have degree $n - 2$ but they do not form a clique, and the remaining node has degree at most 4.

We begin by proving the following statement.

Theorem 1. *No forbidden graph is* mbM.

Proof. We will distinguish between cases for a forbidden graph G. Clearly, if G is F1, then it is not mbM since no node can change its preference. Now assume that G is eF2 (respectively, oF2) and consider a profile in which there are at least $\frac{n}{2} + 1$ (respectively, $\frac{n+1}{2}$) agents with preference 0. Then, every node x with initial preference 0 has at most $\frac{n}{2} - 1$ neighbors with initial preference 1 and at least $\frac{n}{2} - 1$ neighbors with initial preference 0 (respectively, at most $\frac{n-1}{2}$ neighbors with initial preference 1 and at least $\frac{n-1}{2}$ neighbors with initial preference 0). Hence, x is not unhappy and stays with preference 0.

Now, consider the case where G is eF3 and let u be the node of degree at most 2. Consider profile s_0 of initial preferences in which there are at least $\frac{n}{2} + 1$ agents with preference 0. First observe that in the truthful profile s_0 any node x other than u that has preference 0 is adjacent to at most $\frac{n}{2} - 1$ nodes with initial preference 1 and to at least $\frac{n}{2} - 1$ nodes with initial preference 0. Then, x is not unhappy and stays with preference 0. Hence, u is the only node that may want to switch from 0 to 1. But this is possible only if all nodes in the neighborhood of u have preference 1, which implies that the neighborhood of any node with initial

preference 0 does not change after the switch of u, i.e., nodes with preference 0
still are not unhappy and thus they have no incentive to switch to 1. Then, any
node with preference 1 that is not adjacent to u has at most $\frac{n}{2} - 2$ neighbors
with preference 1 and at least $\frac{n}{2}$ neighbors with preference 0. Also, any node with
preference 1 that is adjacent to u has $\frac{n}{2} - 1$ neighbors with preference 1 and $\frac{n}{2}$
neighbors with preference 0. So, every node with preference 1 will eventually
switches to 0.

It remains to consider the case where G is eF4; let u be the node of degree
at most 4. Actually, it can be verified that u can have degree either 2 or 4 and
its neighbors form pair(s) of non-adjacent nodes. Consider a truthful profile in
which there are at least $\frac{n}{2} + 1$ agents with preference 0. Observe that a node
different from u that has initial preference 0 has at most $\frac{n}{2} - 1$ neighbors with
preference 1 and at least $\frac{n}{2} - 1$ neighbors with preference 0. So, it is not unhappy
and has no incentive to switch to preference 1. The only node that might do so
is u, provided that the strict majority of its neighbors (i.e., both of them if u has
degree 2 and at least three of them if u has degree 4) have preferences 1. This
switch cannot trigger another switch of the preference of an agent from 0 node
to 1. Indeed, there is at most one agent with preference 0 that can be adjacent
to u. Since this node is not adjacent to one of the neighbors of u with preference
1, it has at most $\frac{n}{2} - 1$ neighbors with preference 1 (and at least $\frac{n}{2} - 1$ neighbors
with preference 0). Hence, it has no incentive to switch to preference 1 either.
Now, consider two neighbors of u with preference 1 that are not adjacent (these
nodes certainly exist). Each of them is adjacent to $\frac{n}{2} - 2$ nodes with preference
1 and $\frac{n}{2}$ nodes with preference 0. Hence, they have an incentive to switch to 0.
Then, the number of nodes with preference 1 is at most $\frac{n}{2} - 2$ and eventually all
nodes will switch to preference 0.   □

The following is the main result of this section.

Theorem 2. *Every non-forbidden graph is* mbM.

We next give the proof for the simpler case of graphs with an odd number of
nodes and postpone the full proof to the full version [1]. Let us start with the
following definitions. A *bisection* $\mathcal{S} = (S, \overline{S})$ of a graph $G = (V, E)$ with n nodes
is simply a partition of the nodes of V into two sets S and $\overline{S}$ of sizes $\lceil n/2 \rceil$ and
$\lfloor n/2 \rfloor$, respectively. We will refer to S and $\overline{S}$ as the *sides* of bisection $\mathcal{S}$. The *width*
$W(S, \overline{S})$ of a bisection $\mathcal{S}$ is the number of edges of G whose endpoints belong
to different sides of the partition. The *minimum* bisection $\mathcal{S}$ of G has minimum
width among all partitions of G. We extend notation $W(A, B)$ to any pair (A, B)
of subsets of nodes of G in the obvious way. When $A = \{x\}$ is a singleton we will
write $W(x, B)$ and similarly for B. Thus, if nodes x and y are adjacent, then
$W(x, y) = 1$; otherwise $W(x, y) = 0$. For a bisection $\mathcal{S} = (S, \overline{S})$, we define the
deficiency $\mathsf{def}_{\mathcal{S}}(x)$ *of node* x *w.r.t. bisection* $\mathcal{S}$ as $\mathsf{def}_{\mathcal{S}}(x) = W(x, S) - W(x, \overline{S})$ if
$x \in S$, and $\mathsf{def}_{\mathcal{S}}(x) = W(x, \overline{S}) - W(x, S)$ if $x \in \overline{S}$. Let $\mathcal{S} = (S, \overline{S})$ be a minimum
bisection of a graph G with n nodes. Then it is not hard to see that minimality
of $\mathcal{S}$ implies that for every $x \in S$ and $y \in \overline{S}$, $\mathsf{def}_{\mathcal{S}}(x) + \mathsf{def}_{\mathcal{S}}(y) + 2W(x, y) \geq 0$.
Moreover if n is odd, $\mathsf{def}_{\mathcal{S}}(x) \geq 0$.

We have the following technical lemma.

Lemma 1. *Suppose that a graph G admits a bisection $\mathcal{S} = (S, \overline{S})$ in which S consists of nodes with non-negative deficiency and includes at least one node with positive deficiency. Then G is* mbM.

Proof. Let v be the node with positive deficiency in S and consider profile $\mathbf{s}_0$ of initial preferences in which any node in S except v has preference 1 and remaining nodes have preference 0. Hence, in $\mathbf{s}_0$ there is a majority of $\lceil n/2 \rceil$ agents with preference 0. Observe also that in $\mathbf{s}_0$, v is adjacent to $W(v, S)$ nodes with preference 1 and to $W(v, \overline{S})$ nodes with preference 0. Since $\mathsf{def}_S(v) > 0$ then v is unhappy with preference 0 and updates her preference to 1. We thus reach a profile $\mathbf{s}_1$ in which $\lceil n/2 \rceil$ nodes have preference 1 (that is, all nodes in S). We conclude the proof of the lemma by showing that every node of S is not unhappy and thus it stays with preference 1^2. This is obvious for v. Let us consider $u \in S$ and $u \neq v$. Then u has $W(u, S)$ neighbors with preference 1 and $W(u, \overline{S})$ neighbors with preference 0. Since $\mathsf{def}_S(u) \geq 0$, we have that $W(u, S) \geq W(u, \overline{S})$. Hence, the number of neighbors of u with preference 0 is not a majority. Then, u is not unhappy, and thus stays with preference 1.   □

We are now ready to prove Theorem 2 for odd-sized graphs. We remind the reader that the (more complex) proof for even-size graphs is in the full version [1].

Proposition 3. *Non-forbidden graphs with an odd number of nodes are* mbM.

Proof. Let G be a non-forbidden graph with an odd number of nodes and let $\mathcal{S} = (S, \overline{S})$ be a minimum bisection for G. By minimality of $\mathcal{S}$, we have that $\mathsf{def}_S(x) \geq 0$, for all $x \in S$. If S contains at least a node v with $\mathsf{def}_S(v) > 0$ then, by minimality of $\mathcal{S}$, G is mbM. So assume that $\mathsf{def}_S(x) = 0$ for all $x \in S$.

Minimality of $\mathcal{S}$ implies that if $\mathsf{def}_S(v) < 0$ for $v \in \overline{S}$ then $\mathsf{def}_S(v) \geq -2$ and v is connected to all nodes in S. Therefore $W(v, S) = \lceil n/2 \rceil$ and, since $W(v, \overline{S}) \leq \lfloor n/2 \rfloor - 1$, $\mathsf{def}_S(v) = -2$. We denote by A the set of all the nodes $y \in \overline{S}$ with $\mathsf{def}_S(x) = -2$; therefore, all nodes $y \in \overline{S} \setminus A$ have $\mathsf{def}_S(y) \geq 0$.

Let us first consider the case in which $A \neq \emptyset$ and there are two non-adjacent nodes $u, w \in S$. Then pick any node $v \in A$ and consider partition $\mathcal{T} = (T, \overline{T})$ with $T = S \cup \{v\} \setminus \{u\}$. We have that $W(v, T) = W(v, S) - 1 = \lceil n/2 \rceil - 1$ and $W(v, \overline{T}) = W(v, \overline{S}) + 1 = \lfloor n/2 \rfloor + 1$ and hence $\mathsf{def}_T(v) = 0$. For any $x \in T \setminus \{v, w\}$, we have $\mathsf{def}_T(x) \geq \mathsf{def}_S(x) = 0$. Node w is connected to v but not to u and, thus, $\mathsf{def}_T(w) \geq \mathsf{def}_S(w) + 2 = 2$. Then, by Lemma 1, G is mbM.

Assume now that $A \neq \emptyset$ and S is a clique. That is, $W(x, S) = \lceil n/2 \rceil - 1$ for every $x \in S$, and, since $\mathsf{def}_S(x) = 0$, it must be that $W(x, \overline{S}) = W(x, S)$ and thus x is connected to all nodes in $\overline{S}$. Therefore, for all $y \in \overline{S}$, $W(y, S) = \lceil n/2 \rceil$

---

^{2} This is sufficient since the switch of nodes in $\overline{S}$ that are unhappy with preference 0 only increases the number of nodes with preference 1. Moreover, if some nodes in $\overline{S}$ switch their preferences, then the number of nodes with preference 1 in the neighborhood of any node in S can only increase.

and, since $\mathsf{def}_S(y) \geq -2$ it must be that $W(y, \overline{S}) \geq \lceil n/2 \rceil - 2 = |\overline{S}| - 1$. In other words, every node of $\overline{S}$ is connected to every node of $\overline{S}$ and thus G is a clique.

Finally, assume that $A = \emptyset$; that is, $\mathsf{def}_S(y) \geq 0$ for any $y \in \overline{S}$. If for some $v \in \overline{S}$, we have $\mathsf{def}_S(v) > 0$, then consider partition $\mathcal{T} = (T, \overline{T})$ with $T = \overline{S} \cup \{u\}$, where u is any node from S. For any $x \in T \cap \overline{S}$, $\mathsf{def}_T(x) \geq \mathsf{def}_S(x) \geq 0$, $\mathsf{def}_T(u) = -\mathsf{def}_S(u) = 0$ and $\mathsf{def}_T(v) \geq \mathsf{def}_S(v) \geq 1$. By Lemma 1, G is mbM.

Finally, we consider the case in which $\mathsf{def}_S(y) = 0$ for every node x of G. Since G is not empty, there exists at least one edge in G and, since the endpoints of this edge have $\mathsf{def}_S = 0$ there must be at least node $v \in S$ with a neighbor $w \in \overline{S}$. Now, consider partition $\mathcal{T} = (T, \overline{T})$ with $T = S \cup \{w\}$. We have that every node $x \in T \cap S$, has $\mathsf{def}_T(x) \geq \mathsf{def}_S(x) = 0$, $\mathsf{def}_T(w) = -\mathsf{def}_S(w) = 0$, and $\mathsf{def}_T(w) > \mathsf{def}_S(w) = 0$. The claim again follows by Lemma 1. □

We note that the only property required is local minimality. Since a local-search algorithm can compute a locally minimal bisection in polynomial time, we can make constructive the proof of Proposition 3, and quickly compute the subverting minority and the corresponding updates.

3   Hardness for Weaker Minorities

We next show that deciding if it is possible to subvert the majority starting from a weaker minority is a computationally hard problem.

Theorem 4. *For every constant* $0 < \varepsilon < \frac{1}{8}$, *given a graph G with n nodes, it is NP-hard to decide whether there exists an* mbM *profile of initial preferences with at most $n\left(\frac{1}{4} - \varepsilon\right)$ nodes with initial preference 1.*

Proof. We will use a reduction from the NP-hard problem 2P2N-3SAT, the problem of deciding whether a 3SAT formula in which every variable appears as positive in two clauses and as negative in two clauses has a truthful assignment or not (the NP-hardness follows by the results of [18]).

Given a Boolean formula ϕ with C clauses and V variables that is an instance of 2P2N-3SAT (thus $3C = 4V$ and C is a multiple of 4), we will construct a graph $G(\phi)$ with n nodes such that there exists a profile of initial preferences with at most $n\left(\frac{1}{4} - \varepsilon\right)$ nodes of $G(\phi)$ with preference 1 such that a sequence of updates can lead to a stable profile in which at least $n/2$ nodes have preference 1 if and only if ϕ has a satisfying assignment.

The graph $G(\phi)$ has the following nodes and edges. For each variable x of ϕ, $G(\phi)$ includes a *variable* gadget for x consisting of 25 nodes and 50 edges. The nodes of the variable gadget for x are the *literal nodes*, x and $\overline{x}$, nodes $v_1(x), \ldots, v_7(x)$, nodes $v_1(\overline{x}), \ldots, v_7(\overline{x})$, nodes $v_0(x)$ and $w_0(x)$, and nodes $w_1(x)$, $\ldots, w_7(x)$. The edges are $(x, v_i(x))$ and $(\overline{x}, v_i(\overline{x}))$ for $i = 1, \ldots, 7$, $(v_i(x), v_{i+1}(x))$ and $(v_i(\overline{x}), v_{i+1}(\overline{x}))$ for $i = 1, \ldots, 6$, $(v_0(x), v_7(x))$, $(v_0(x), v_7(\overline{x}))$, $(v_0(x), w_0(x))$, $(w_0(x), v_i(x))$, $(w_0(x), v_i(\overline{x}))$ and $(w_0(x), w_i(x))$ for $i = 1, \ldots, 7$. For each clause c of ϕ, graph $G(\phi)$ includes a *clause gadget* for c consisting of 18 nodes and 32 edges. The nodes of the gadget are the *clause* node c, nodes $u_1(c)$, $u_2(c)$, and nodes $v_1(c), \ldots, v_{15}(c)$. The 32 edges are $(c, u_1(x)), (c, u_2(x))$, and $(u_i(c), v_j(c))$ with $i = 1, 2$ and $j = 1, \ldots, 15$. In $G(\phi)$, for every clause c, the clause node c is connected to the three literal nodes corresponding to the literals that appear in clause c in ϕ. Therefore, each literal node is connected to the two clauses in which it appears. Graph $G(\phi)$ includes a *clique* of even size N, with $12C \le N \le \frac{95C}{16\varepsilon} - \frac{123C}{4}$; the clique is disconnected from the rest of the graph. Graph $G(\phi)$ includes $N + \frac{99C}{4}$ additional *isolated* nodes. Overall, the total number of nodes in $G(\phi)$ is $n = 2N + \frac{99C}{4} + 25V + 18C = 2N + \frac{123C}{2}$.

A profile of initial preferences to the nodes of $G(\phi)$ is called *proper* if: for every variable x, it assigns preference 1 to node $w_0(x)$ and to exactly one literal node of the gadget of x; for every clause c, it assigns preference 1 to nodes $u_1(c)$ and $u_2(c)$; it assigns preference 1 to exactly $\frac{N}{2}$ nodes of the clique; it assigns preference 0 to all the remaining nodes. Hence, in a proper profile the number of nodes with preference 1 is $2V + 2C + \frac{N}{2} = \frac{7C}{2} + \frac{N}{2} \le n(\frac{1}{4} - \varepsilon)$; the inequality follows from the upper bound in the definition of N.

We now prove that $G(\phi)$ has a proper profile of initial preferences that leads to a majority of nodes with preference 1 if and only if ϕ is satisfiable. First observe that every clique node switches her preference to 1 (as the strict majority of its neighbors has initially preference 1 and this number gradually increases until all clique nodes switch to 1).

We next prove that starting from a proper profile of initial preferences, there is a sequence of updates that leads to a stable profile in which 17 nodes of every variable gadget have preference 1. To see this, consider a proper profile that assigns preference 1 to x (and to $w_0(x)$) and the following sequence of updates: node $v_1(x)$ switches from 0 to 1; then, for $i = 1, \ldots, 6$, node $v_{i+1}(x)$ switches to 1 immediately after node $v_i(x)$; node $v_0(x)$ switches to 1 after node $v_7(x)$; finally, $w_1(x), \ldots, w_7(x)$ can switch in any order. Observe that in this sequence any switching node is unhappy since it has a strict majority of nodes with preference 1 in its neighborhood. Also, the resulting profile where the 17 nodes $w_0(x), w_1(x), \ldots, w_7(x), v_0(x), v_1(x), \ldots, v_7(x)$, and x have preference 1 is stable, i.e., no node in the gadget is unhappy. Indeed, for each node with preference 1, the strict majority of the preferences of its neighbors is 1. Hence, the node has no incentive to switch to preference 0. For each of the remaining nodes (with preference 0), at least half of its neighbors is 0. Hence, this node has no incentive to switch to preference 1 either. A similar sequence can be constructed for a proper profile that assigns preference 1 to node $\bar{x}$ (and $w_0(x)$) of the gadget for variable x. Intuitively, the two proper profiles of initial preferences simulate the assignment of values TRUE and FALSE to variable x, respectively.

In addition, it is easy to see that starting from a proper profile, there is no sequence of updates that reaches a stable profile where more than 17 nodes in a variable gadget have preference 1 (the observation needed here is the same that guarantees that we reach a stable profile above).

Let us now consider the clause gadgets associated with clause c of ϕ. We observe that, starting from a proper profile of initial preferences, there exists a sequence of updates that leads to a stable profile in which 17 nodes of the clause gadget have preference 1. Indeed, starting from the proper assignment of preference 1 to nodes $u_1(c)$ and $u_2(c)$, nodes $v_1(c), \ldots, v_{15}(c)$ will switch from 0 to 1 in arbitrary order (for each of them both neighbors have preference 1). After these updates, at least 15 out of 17 neighbors of $u_1(c)$ and $u_2(c)$ have preference 1 and both neighbors of the nodes $v_1(c), \ldots, v_{15}(c)$ have preference 0. Hence, none among these nodes have any incentive to switch to preference 0.

Let us now focus on the clause nodes and observe that node c in the corresponding clause gadget will switch to 1 if and only if at least one of the literal nodes corresponding to literals that appear in c have preference 1 (since the degree of a clause node is five and nodes $u_1(c)$ and $u_2(c)$ have preference 1). This switch cannot trigger any other switch in literal nodes or in nodes of clause gadgets since the preference of these nodes coincides with a strict majority of preferences in its neighborhood. Hence, the fact that a clause node has preference 1 (respectively, 0) corresponds to the clause being satisfied (respectively, not satisfied) by the Boolean assignment induced by the proper profile of initial preferences. Eventually, the updates lead to an additional number of C clause nodes adopting preference 1 in the stable profile if and only if ϕ is satisfiable.

In conclusion, we have that if ϕ is satisfiable there is a sequence of updates converging to a profile with $17V + 17C + N + C = N + 123C/4 = n/2$ nodes with preference 1. Otherwise, if ϕ is not satisfiable, any sequence of updates converges to a stable profile with strictly less than $n/2$ nodes having preference 1.

We conclude the proof by showing that it is sufficient to restrict to proper assignments as non-proper assignments will never lead to a stable profile with a majority of nodes with preference 1. First observe that if the total number of clique and isolated nodes with preference 1 is strictly less than $\frac{N}{2}$, then no clique and isolated nodes with preference 0 will adopt preference 1. Thus, in this case, even counting all nodes in variable and clause gadgets, any sequence of updates converges to a stable profile with at most $25V + 18C + \frac{N}{2} - 1 < \frac{n}{2}$ nodes with preference 1 (where we used that $N \geq 12C$).

Let us now focus on a profile of initial preferences that assigns preference 1 to at most $7C/2 = 2C + 2V$ nodes from variable and clause gadgets. Suppose that this profile of initial preferences is such that a sequence of updates leads to a stable profile with at least $n/2$ nodes with preference 1. We will show that this profile of initial preferences must be proper.

First, observe that if at most one node in a variable gadget is assigned preference 1, then all nodes in the gadget will eventually adopt preference 0 after a sequence of updates. Indeed, a literal node will have at least six neighbors with preference 0 and at most three with preference 1, and any non-literal node will have at most one out of at least three of its neighbors with preference 0.

Consider now profiles of initial preferences that assign preference 1 to two nodes of the variable gadget of x in a non-proper way. We show that any sequence of updates leads to a profile in which all nodes of the gadget adopt preference 0.

Indeed, assume that $w_0(x)$ has preference 1 and both x and $\bar{x}$ have preference 0. Clearly, the nodes $w_1(x), \ldots, w_7(x)$ can switch from 0 to 1 in any order. Among the non-literal nodes $v_i(x)$ and $v_i(\bar{x})$, only one among the degree-3 nodes $v_0(x)$, $v_1(x)$, and $v_1(\overline{x})$ can switch from 0 to 1; this can only happen if the second node with preference 1 is in the neighborhood of one of these nodes (i.e., some of the nodes $v_7(x)$, $v_7(\overline{x})$, $v_2(x)$, or $v_2(\overline{x})$). But then, the literal nodes will have at most four neighbors with preference 1 and they cannot switch to 1. So, no other node has any incentive to switch from 0 to 1. Then, $w_0(x)$ has at least 13 among its 22 neighbors with preference 0 and will switch from 1 to 0, followed by the nodes $w_1(x), ..., w_7(x)$ that will switch back to 0 as well. Then, there are at most two nodes with preference 1 among the nodes $v_i(x)$ and $v_i(\overline{x})$ that will eventually switch to 0 as well (since they have degree at least three).

Assume now that literal node x has preference 1 (the case for $\overline{x}$ is symmetric) and that $w_0(x)$ has preference 0. Then, only the degree-3 node $v_1(x)$ that is adjacent to x can switch to 1 provided that the second node with preference 1 is node $v_2(x)$. Now notice that no other node can switch from 0 to 1. Even worse, the literal node x has at least five (out of nine) neighbors of preference 0 and will switch from 1 to 0. And then, we are left with at most two nodes with preference 1 among the nodes $v_i(x)$ and $v_i(\overline{x})$ that will eventually switch to 0 as well.

Finally, we consider the case in which $w_0(x)$ and the two literal nodes have preference 0. Now the only node that can initially switch from 0 to 1 is $v_0(x)$ provided that the two nodes with preference 1 are $v_7(x)$ and $v_7(\overline{x})$. But then, there is no other node that can switch from 0 to 1 and, eventually, nodes $v_7(x)$ and $v_7(\overline{x})$ will switch to 0 and finally node $v_0(x)$ will switch back to 0.

We have covered all possible cases in which a variable gadget has a nonproper assignment of preference 1 to two nodes and shown that in all of these cases, all nodes of the gadget will switch to preference 0. On the other hand, as discussed above, a proper profile of initial preferences can end up with preference 1 in 17 nodes of the variable gadget.

Now, observe that if at most one node in a clause gadget has preference 1 (or two nodes are assigned preference 1 in a non-proper way), then all the 17 non-clause nodes in the gadget will end up with preference 0. This is due to the fact that none among the nodes $v_1(c), ..., v_{15}(c)$ can switch from 0 to 1 since at least one of their neighbors will have preference 0. But this means that nodes $u_1(c)$ and $u_2(c)$ are adjacent to many (i.e., at least 13) nodes with preference 0; so, they will also switch to 0. And then, if there is still some node $v_i(c)$ with preference 1, it will switch to 0 since both its neighbors have preference 1.

Now, by denoting with V_0, V_1, V_3 the number of variable gadgets that have 0, 1 or at least 3 nodes with preference 1 and by V_{2p} and V_{2n} the number of variable gadgets with proper and non-proper assignment of preference 1 to exactly two nodes, we have $V = V_0 + V_1 + V_{2n} + V_{2p} + V_3$ and, by denoting with C_0, C_1, and C_3 the number of clause gadgets with 0, 1, and at least 3 nodes with preference 1 in nodes other than the clause node and by C_{2p} and C_{2n} the number of clause gadgets with two nodes with preference 1 assigned in a proper and non-proper way, we have $C = C_0 + C_1 + C_{2n} + C_{2p} + C_3$. Since the total number of nodes with preference 1 does not exceed $2V + 2C$, we have $V_1 + 2V_{2n} + 2V_{2p} + 3V_3 + C_1 + 2C_{2n} + 2C_{2p} + 3C_3 \leq 2V + 2C$ from which we get $V_3 + C_3 \leq 2C_0 + C_1 + 2V_0 + V_1$. Now consider the difference between the number of nodes with preference 1 in any stable profile reached after a sequence of updates and the quantity $17V + 18C$. It is at most $17V_{2p} + 25V_3 + C_0 + C_1 + C_{2n} + 18C_{2p} + 18C_3 - 17V - 18C = -17V_0 - 17V_1 - 17V_{2n} + 8V_3 - 17C_0 - 17C_1 - 17C_{2n} \leq -V_0 - 9V_1 - 17V_{2n} - C_0 - 9C_1 - 17C_{2n} - 8C_3$. Hence, if at least one of V_0, V_1, V_{2n}, C_0, C_1, C_{2n}, and C_3 is positive, the proof follows since the number of nodes with preference 1 will be strictly less than $N + 17V + 18C = n/2$. Otherwise, i.e., if all these quantities are 0, this implies that $C = C_{2p}$ and $V = V_{2p} + V_3$ which in turn implies that $V_3 = 0$ since the number of nodes with preference 1 cannot exceed $2C + 2V$. Hence, the only case where a sequence of updates may lead to a stable profile with at least $N + 17V + 18C$ nodes having preference 1 is when the profile of initial preferences is proper. The claim follows.   □

Checking whether minority can become majority. We next show that, given a graph G and a profile of initial preferences s_0, it is possible to decide whether s_0 is mbM for G in polynomial time. Moreover, if this is the case, then there is an efficient algorithm that computes the subverting sequence of updates. This algorithm was used in [5] for bounding the price of stability. Due to page limit, the proof of Theorem 5 is only sketched. We refer the interested reader to the full version of the paper [1].

Theorem 5. *There is a polynomial time algorithm that, given a graph $G = (V, E)$ and a profile of initial preferences s_0, decides whether s_0 is mbM for graph G and, if it is, it outputs a subverting sequence of updates.*

Proof (Sketch). Consider the algorithm used in [5] for bounding the price of stability. The running time of the algorithm is polynomial in the size of the input graph, since each node updates its preference at most twice. The fact that the profile s_0' returned by the algorithm is a stable profile is proved in [5, Lemma 3.3]. We next show that s_0' is actually the stable profile that maximizes the number of nodes with preference 1. Specifically, consider a sequence σ of updates leading to a stable profile s that maximizes the number of nodes with preference 1. We will show that there is another sequence of updates that has the form computed by the algorithm described above and converges to a stable profile in which the agents with preference 1 are at least as many as in s.   □

4   Conclusions and Open Problems

In this work we showed that, for any social network topology except very few and extreme cases, social pressure can subvert a majority. We proved this with respect to a very natural *majority* dynamics in the case in which agents must express preferences. We also showed that, for each of these graphs, it is possible to compute in polynomial time an initial majority and a sequence of updates that subverts it. The initial majority constructed in this way consists of only $\lceil (n+1)/2 \rceil$ agents. On the other hand, our hardness result proves that it may be hard to compute an initial majority of size at least $3n/4$ that can be subverted by the social pressure. The main problem that this work left open is to close this gap.

Even if computational considerations rule out a simple characterization of the graphs for which a large majority can be subverted, it would be still interesting to gain knowledge on these graphs. Specifically, can we prove that the set of graphs for which large majority can be subverted can be easily described by some simple (but hard to compute) graph-theoretic measure? We believe that our ideas can be adapted (e.g., by considering unbalanced partitions in place of bisections), for gaining useful hints in this direction.

References

1. Auletta, V., Caragiannis, I., Ferraioli, D., Galdi, C., Persiano, G.: Minority becomes majority in social networks. CoRR, abs/1402.4050 (2014)
2. Berger, E.: Dynamic monopolies of constant size. J. Comb. Theory Ser. B **83**(2), 191–200 (2001)
3. Bindel, D., Kleinberg, J.M., Oren, S.: How bad is forming your own opinion? In: Proceedings of the 52nd Annual IEEE Symposium on Foundations of Computer Science (FOCS), pp. 55–66 (2011)
4. Brânzei, S., Caragiannis, I., Morgenstern, J., Procaccia, A.D.: How bad is selfish voting? In: Proceedings of the 27th AAAI Conference on Artificial Intelligence (AAAI), pp. 138–144 (2013)
5. Chierichetti, F., Kleinberg, J.M., Oren, S.: On discrete preferences and coordination. In: Proceedings of the 14th ACM Conference on Electronic Commerce (EC), pp. 233–250 (2013)
6. Coleman, J.S., Katz, E., Menzel, H.: Medical Innovation: A Diffusion Study. Advanced Study in Sociology. Bobbs-Merrill Co., Indianapolis (1966)
7. DeGroot, M.H.: Reaching a consensus. J. Am. Stat. Assoc. **69**(345), 118–121 (1974)
8. Easley, D., Kleinberg, J.: Networks, Crowds, and Markets: Reasoning about A Highly Connected World. Cambridge University Press, New York (2010)
9. Feldman, M., Immorlica, N., Lucier, B., Weinberg, S.M.: Reaching consensus via non-Bayesian asynchronous learning in social networks. In: Approximation, Randomization, and Combinatorial Optimization. Algorithms and Techniques (APPROX/RANDOM 2014), vol. 28, pp. 192–208 (2014)
10. Ferraioli, D., Goldberg, P.W., Ventre, C.: Decentralized dynamics for finite opinion games. In: Serna, M. (ed.) SAGT 2012. LNCS, vol. 7615, pp. 144–155. Springer, Heidelberg (2012)
11. Friedkin, N.E., Johnsen, E.C.: Social influence and opinions. J. Math. Sociol. **15**(3–4), 193–205 (1990)
12. Meir, R., Polukarov, M., Rosenschein, J.S., Jennings, N.R.: Convergence to equilibria in plurality voting. In: Proceedings of the 24th AAAI Conference on Artificial Intelligence (AAAI), pp. 823–828 (2010)
13. Mossel, E., Neeman, J., Tamuz, O.: Majority dynamics and aggregation of information in social networks. Auton. Agent. Multi-Agent Syst. **28**(3), 408–429 (2014)
14. Mossel, E., Sly, A., Tamuz, O.: Asymptotic learning on Bayesian social networks. Probab. Theory Relat. Fields **158**(1–2), 127–157 (2014)
15. Peleg, D.: Local majorities, coalitions and monopolies in graphs: a review. Theoret. Comput. Sci. **282**, 231–257 (2002)

16. Ryan, B., Gross, N.G.: Acceptance and diffusion of hybrid corn seed in two Iowa communities, vol. 372. Agricultural Experiment Station, Iowa State College of Agriculture and Mechanic Arts (1950)

17. Tamuz, O., Tessler, R.J.: Majority dynamics and the retention of information. Isr. J. Math. **206**(1), 483–507 (2013)

18. Yoshinaka, R.: Higher-order matching in the linear lambda calculus in the absence of constants is NP-complete. In: Giesl, J. (ed.) RTA 2005. LNCS, vol. 3467, pp. 235–249. Springer, Heidelberg (2005)

New Complexity Results and Algorithms
for the Minimum Tollbooth Problem

Soumya Basu[✉], Thanasis Lianeas, and Evdokia Nikolova

University of Texas at Austin, Austin, USA
basusoumya@utexas.edu

Abstract. The inefficiency of the Wardrop equilibrium of nonatomic routing games can be eliminated by placing tolls on the edges of a network so that the socially optimal flow is induced as an equilibrium flow. A solution where the minimum number of edges are tolled may be preferable over others due to its ease of implementation in real networks. In this paper we consider the minimum tollbooth ($MINTB$) problem, which seeks social optimum inducing tolls with minimum support. We prove for single commodity networks with linear latencies that the problem is NP-hard to approximate within a factor of 1.1377 through a reduction from the minimum vertex cover problem. Insights from network design motivate us to formulate a new variation of the problem where, in addition to placing tolls, it is allowed to remove unused edges by the social optimum. We prove that this new problem remains NP-hard even for single commodity networks with linear latencies, using a reduction from the partition problem. On the positive side, we give the first exact polynomial solution to the $MINTB$ problem in an important class of graphs—series-parallel graphs. Our algorithm solves $MINTB$ by first tabulating the candidate solutions for subgraphs of the series-parallel network and then combining them optimally.

1 Introduction

Traffic congestion levies a heavy burden on millions of commuters across the globe. The congestion cost to the U.S. economy was measured to be \$126 billion in the year 2013 with an estimated increase to \$186 billion by year 2030 [16]. Currently the most widely used method of mitigating congestion is through congestion pricing, and one of the most common pricing schemes is through placing tolls on congested roads that users have to pay, which makes these roads less appealing and diverts demand, thereby reducing congestion.

Mathematically, an elegant theory of traffic congestion was developed starting with the work of Wardrop [18] and Beckman et al. [6]. This theory considered a network with travel time functions that are increasing in the network flow, or the number of users, on the corresponding edges. Wardrop differentiated between two main goals: (1) user travel time is minimized, and (2) the total travel time of all users is minimized. This led to the investigation of two different resulting traffic assignments, or flows, called a Wardrop equilibrium and a social or system

© Springer-Verlag Berlin Heidelberg 2015
E. Markakis and G. Schäfer (Eds.): WINE 2015, LNCS 9470, pp. 89–103, 2015.
DOI: 10.1007/978-3-662-48995-6_7

optimum, respectively. It was understood that these two flows are unfortunately often not the same, leading to tension between the two different objectives. Remarkably, the social optimum could be interpreted as an equilibrium with respect to modified travel time functions, that could in turn be interpreted as the original travel time functions plus tolls.

Consequently, the theory of congestion games developed a mechanism design approach to help users routing along minimum cost paths reach a social optimum through a set of optimal tolls that would be added to (all) network edges. Later, through the works of Bergendorff et al. [7] and Hearn & Ramana [12], it was understood that the set of optimal tolls is not unique and there has been work in diverse branches of literature such as algorithmic game theory, operations research and transportation on trying to limit the toll cost paid by users by limiting the number of tolls placed on edges.

Related Work. The natural question of what is the minimum number of edges that one needs to place tolls on so as to lead selfish users to a social optimum, was first raised by Hearn and Ramana [12]. The problem was introduced as the minimum tollbooth ($MINTB$) problem and was formulated as a mixed integer linear program. This initiated a series of works which led to new heuristics for the problem. One heuristic approach is based on genetic algorithms [3,8]. In 2009, a combinatorial benders cut based heuristic was proposed by Bai and Rubin [2]. The following year, Bai et al. proposed another heuristic algorithm based on LP relaxation using a dynamic slope scaling method [1]. More recently, Stefanello et al. [15] have approached the problem with a modified genetic algorithm technique.

The first step in understanding the computational complexity of the problem was by Bai et al. [1] who proved that $MINTB$ in multi commodity networks is NP-hard via a reduction from the minimum cardinality multiway cut problem [10]. In a related direction, Harks et al. [11] addressed the problem of inducing a predetermined flow, not necessarily the social optimum, as the Wardrop equilibrium, and showed that this problem is APX-hard, via a reduction from length bounded edge cuts [4]. Clearly, $MINTB$ is a special case of that problem and it can be deduced that the hardness results of Harks et al. [11] do not carry forward to the $MINTB$ problem. A related problem is imposing tolls on a constrained set of edges to minimize the social cost under equilibrium [13].

The latest work stalls at this point leaving open both the question of whether approximations for multi commodity networks are possible, and what the hardness of the problem is for single commodity networks or for any meaningful subclass of such networks.

Our Contribution. In this work, we make progress on this difficult problem by deepening our understanding on what can and cannot be computed in polynomial time. In particular, we make progress in both the negative and positive directions by providing NP-hardness and hardness of approximation results for the single commodity network, and a polynomial-time exact algorithm for computing the minimum cardinality tolls on series-parallel graphs.

Specifically, we show in Theorem 1 that the minimum tollbooth problem for single commodity networks and linear latencies is hard to approximate to within a factor of 1.1377, presenting the first hardness of approximation result for the $MINTB$ problem.

Further, motivated by the observation that removing or blocking an edge in the network bears much less cost compared to the overhead of toll placement, we ask: if all unused edges under the social optimum are removed, can we solve $MINTB$ efficiently? The NP-hardness result presented in Theorem 2 for $MINTB$ in single commodity networks with only used edges, settles it negatively, yet the absence of a hardness of approximation result creates the possibility of a polynomial time approximation scheme upon future investigation.

Observing that the Braess structure is an integral part of both NP-hardness proofs, we seek whether positive progress is possible for the problem in series-parallel graphs. We propose an exact algorithm for series-parallel graphs with $\mathcal{O}(m^3)$ runtime, m being the number of edges. Our algorithm provably (see Theorem 4) solves the $MINTB$ problem in series-parallel graphs, giving the first exact algorithm for $MINTB$ on an important class of graphs.

2   Preliminaries and Problem Definition

We are given a directed graph $G(V, E)$ with edge delay or latency functions $(\ell_e)_{e \in E}$ and demand r that needs to be routed between a source s and a sink t. We will abbreviate an instance of the problem by the tuple $\mathcal{G} = (G(V, E), (\ell_e)_{e \in E}, r)$. For simplicity, we usually omit the latency functions, and refer to the instance as (G, r). The function $\ell_e : \mathbb{R}_{\geq 0} \to \mathbb{R}_{\geq 0}$ is a non-decreasing cost function associated with each edge e. Denote the (non-empty) set of simple $s - t$ paths in G by $\mathcal{P}$.

Flows. Given an instance (G, r), a (feasible) *flow* f is a non-negative vector indexed by the set of feasible $s - t$ paths $\mathcal{P}$ such that $\sum_{p \in \mathcal{P}} f_p = r$. For a flow f, let $f_e = \sum_{p : e \in p} f_p$ be the amount of flow that f routes on each edge e. An edge e is used by flow f if $f_e > 0$, and a path p is used by flow f if it has strictly positive flow on all of its edges, namely $\min_{e \in p}\{f_e\} > 0$. Given a flow f, the cost of each edge e is $\ell_e(f_e)$ and the cost of path p is $\ell_p(f) = \sum_{e \in p} \ell_e(f_e)$.

Nash Flow. A flow f is a *Nash (equilibrium) flow*, if it routes all traffic on minimum latency paths. Formally, f is a Nash flow if for every path $p \in \mathcal{P}$ with $f_p > 0$, and every path $p' \in \mathcal{P}$, $\ell_p(f) \leq \ell_{p'}(f)$. Every instance (G, r) admits at least one Nash flow, and the players' latency is the same for all Nash flows (see e.g., [14]).

Social Cost and Optimal Flow. The *Social Cost* of a flow f, denoted $C(f)$, is the total latency $C(f) = \sum_{p \in \mathcal{P}} f_p \ell_p(f) = \sum_{e \in E} f_e \ell_e(f_e)$. The *optimal* flow of an instance (G, r), denoted o, minimizes the total latency among all feasible flows.

In general, the Nash flow may not minimize the social cost. As discussed in the introduction, one can improve the social cost at equilibrium by assigning tolls to the edges.

Tolls and Tolled Instances. A set of *tolls* is a vector $\Theta = \{\theta_e\}_{e \in E}$ such that the toll for each edge is nonnegative: $\theta_e \geq 0$. We call *size* of Θ the size of the support of Θ, i.e., the number of edges with strictly positive tolls, $|\{e : \theta_e > 0\}|$. Given an instance $\mathcal{G} = (G(V, E), (\ell_e)_{e \in E}, r)$ and a set of tolls Θ, we denote the tolled instance by $\mathcal{G}^\theta = (G(V, E), (\ell_e + \theta_e)_{e \in E}, r)$. For succinctness, we may also denote the tolled instance by (G^θ, r). We call a set of tolls, Θ, *opt-inducing* for an instance $\mathcal{G}$ if the optimal flow in $\mathcal{G}$ and the Nash flow in $\mathcal{G}^\theta$ coincide.

Opt-inducing tolls need not be unique. Consequently, a natural problem is to find a set of optimal tolls of minimum size, which is the problem we consider here.

Definition 1 (Minimum Tollbooth problem ($MINTB$)). *Given instance $\mathcal{G}$ and an optimal flow o, find an opt-inducing toll vector Θ such that the support of Θ is less than or equal to the support of any other opt-inducing toll vector.*

The following definitions are needed for Sect. 4.

Series-Parallel Graphs. A directed $s-t$ multi-graph is *series-parallel* if it consists of a single edge (s, t) or of two series-parallel graphs with terminals (s_1, t_1) and (s_2, t_2) composed either in series or in parallel. In a *series composition*, t_1 is identified with s_2, s_1 becomes s, and t_2 becomes t. In a *parallel composition*, s_1 is identified with s_2 and becomes s, and t_1 is identified with t_2 and becomes t.

A series-parallel (SP) graph G with n nodes and m edges can be efficiently represented using a parse tree decomposition of size $\mathcal{O}(m)$, which can be constructed in time $\mathcal{O}(m)$ due to Valdes et al. [17].

Series-Parallel Parse Tree. A series-parallel parse tree T is a rooted binary tree representation of a given SP graph G that is defined using the following properties:

1. Each node in the tree T represents a SP subgraph H of G, with the root node representing the graph G.
2. There are three type of nodes: 'series' nodes, 'parallel' nodes, which have two children each, and the 'leaf' nodes which are childless.
3. A 'series' ('parallel') node represents the SP graph H formed by the 'series combination' ('parallel combination') of its two children H_1 and H_2.
4. The 'leaf' node represents a parallel arc network, namely one with two terminals s and t and multiple edges from s to t.

For convenience, when presenting the algorithm, we allow 'leaf' nodes to be multi-edge/parallel-arc networks. This will not change the upper bounds on the time complexity or the size of the parse tree.

3   Hardness Results for $MINTB$

In this section we provide hardness results for $MINTB$. We study two versions of the problem. The first one considers arbitrary instances while the second considers arbitrary instances where the optimal solution uses all edges, i.e. $\forall e \in$

$E : o_e > 0$. Recall that the motivation for separately investigating the second version comes as a result of the ability of the network manager to make some links unavailable.

3.1 Single-Commodity Network with Linear Latencies

We give hardness results on finding and approximating the solution of $MINTB$ in general instances with linear latencies. In Theorem 1 we give an inapproximability result by a reduction from a Vertex Cover related NP-hard problem and as a corollary (Corollary 1) we get the NP-hardness of $MINTB$ on single commodity networks with linear latencies. The construction of the network for the reduction is inspired by the NP-hardness proof of the length bounded cuts problem in [4].

Theorem 1. *For instances with linear latencies, it is NP-hard to approximate the solution of $MINTB$ by a factor of less than 1.1377.*

Proof. The proof is by a reduction from an NP-hard variant of Vertex Cover (VC) due to Dinur and Safra [9]. Reminder: a Vertex Cover of an undirected graph $G(V, E)$ is a set $S \subseteq V$ such that $\forall \{u, v\} \in E : S \cap \{u, v\} \neq \emptyset$.

Given an instance $\mathcal{V}$ of VC we are going to construct an instance $\mathcal{G}$ of $MINTB$ which will give a one-to-one correspondence (Lemma 1) between Vertex Covers in $\mathcal{V}$ and opt-inducing tolls in $\mathcal{G}$. The inapproximability result will follow from that correspondence and an inapproximability result concerning Vertex Cover by [9]. We note that we will not directly construct the instance of $MINTB$. First, we will construct a graph with edge costs that are assumed to be the costs of the edges (used or unused) under the optimal solution and then we are going to assign linear cost functions and demand that makes the edges under the optimal solution to have costs equal to the predefined costs.

We proceed with the construction. Given an instance $G_{vc}(V_{vc}, E_{vc})$ of VC, with n_{vc} vertices and m_{vc} edges, we construct a directed single commodity network $G(V, E)$ with source s and sink t as follows:

1. For every vertex $v_i \in V_{vc}$ create gadget graph $G_i(V_i, E_i)$, with $V_i = \{a_i, b_i, c_i, d_i\}$ and $E_i = \{(a_i, b_i), (b_i, c_i), (c_i, d_i), (a_i, d_i)\}$, and assign costs equal to 1 for edges $e_{1,i} = (a_i, b_i)$ and $e_{3,i} = (c_i, d_i)$, 0 for edge $e_{2,i} = (b_i, c_i)$, and 3 for edge $e_{4,i} = (a_i, d_i)$. All edges $e_{1,i}, e_{2,i}, e_{3,i}$ and $e_{4,i}$ are assumed to be used.
2. For each edge $e_k = \{v_i, v_j\} \in E_{vc}$ add edges $g_{1,k} = (b_i, c_j)$ and $g_{2,k} = (b_j, c_i)$ with cost 0.5 each. Edges $g_{1,k}$ and $g_{2,k}$ are assumed to be unused.
3. Add source vertex s and sink vertex t and for all $v_i \in V_{vc}$ add edges $s_{1,i} = (s, a_i)$ and $t_{1,i} = (d_i, t)$ with 0 cost, and edges $s_{2,i} = (s, b_i)$ and $t_{2,i} = (c_i, t)$ with cost equal to 1.5. Edges $s_{1,i}$ and $t_{1,i}$ are assumed to be used and edges $s_{2,i}$ and $t_{2,i}$ are assumed to be unused.

The construction is shown in Fig. 1 where the solid lines represent used edges and dotted lines represent unused edges. The whole network consists of $(2+4n_{vc})$

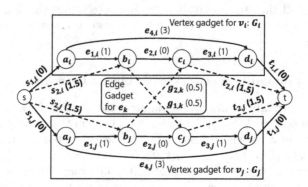

Fig. 1. Gadgets for the reduction from VC to $MINTB$. The pair of symbols on each edge corresponds to the name and the cost of the edge respectively. Solid lines represent used edges and dotted lines represent unused edges.

nodes and $(8n_{vc} + 2m_{vc})$ edges, therefore, it can be constructed in polynomial time, given G_{vc}.

We go on to prove the one-to-one correspondence lemma.

Lemma 1. *(I) If there is a Vertex Cover in G_{vc} with cardinality x, then there are opt-inducing tolls for G of size $n_{vc} + x$.*
(II) If there are opt-inducing tolls for G of size $n_{vc} + x$, then there is a Vertex Cover in G_{vc} with cardinality x.

Proof. Readers are referred to the proof of Lemma 1 in [5]. ∎

Statement (I) in the above lemma directly implies that if the minimum Vertex Cover of G_{vc} has cardinality x then the optimal solution of the $MINTB$ instance has size at most $n_{vc} + x$.

From the proof of Theorem 1.1 in [9] we know that there exist instances G_{vc} where it is NP-hard to distinguish between the case where we can find a Vertex Cover of size $n_{vc} \cdot (1 - p + \epsilon)$, and the case where any vertex cover has size at least $n_{vc} \cdot (1 - 4p^3 + 3p^4 - \epsilon)$, for any positive ϵ and $p = (3 - \sqrt{5})/2$. We additionally know that the existence of a Vertex Cover with cardinality in between the gap implies the existence of a Vertex Cover of cardinality $n_{vc} \cdot (1 - p + \epsilon)$.[1]

Assuming that we reduce from such an instance of VC, the above result implies that it is NP-hard to approximate $MINTB$ within a factor of $1.1377 < \frac{2 - 4p^3 + 3p^4 - \epsilon}{2 - p + \epsilon}$ (we chose an ϵ for inequality to hold). To reach a contradiction assume the contrary, i.e. there exists a β-approximation algorithm *Algo* for $MINTB$, where $\beta \leq 1.1377 < \frac{2 - 4p^3 + 3p^4 - \epsilon}{2 - p + \epsilon}$. By Lemma 1 statement (I), if there

---

[1] The instance they create will have either a Vertex Cover of cardinality $n_{vc} \cdot (1 - p + \epsilon)$ or all Vertex Covers with cardinality $\geq n_{vc} \cdot (1 - 4p^3 + 3p^4 - \epsilon)$.

exists a Vertex Cover of cardinality $\hat{x} = n_{vc} \cdot (1 - p + \epsilon)$ in G_{vc}, then the cardinality in an optimal solution to $MINTB$ on the corresponding instance is $OPT \leq n_{vc} + \hat{x}$. Further, $Algo$ produces opt-inducing tolls with size $n_{vc} + y$, from which we can get a Vertex Cover of cardinality y in the same way as we did inside the proof of statement (II) of Lemma 1. Then by the approximation bounds and using $\hat{x} = n_{vc} \cdot (1 - p + \epsilon)$ we get

$$\frac{n_{vc} + y}{n_{vc} + \hat{x}} \leq \frac{n_{vc} + y}{OPT} \leq \beta < \frac{2 - 4p^3 + 3p^4 - \epsilon}{2 - p + \epsilon} \Rightarrow y < (1 - 4p^3 + 3p^4 - \epsilon)n_{vc}$$

The last inequality would answer the question whether there exists a Vertex Cover with size $n_{vc} \cdot (1 - p + \epsilon)$, as we started from an instance for which we additionally know that the existence of a Vertex Cover with cardinality $y < (1 - 4p^3 + 3p^4 - \epsilon)n_{vc}$ implies the existence of a Vertex Cover of cardinality $n_{vc} \cdot (1 - p + \epsilon)$.

What is left for concluding the proof is to define the linear cost functions and the demand so that at an optimal solution all edges have costs equal to the ones defined above.

Define the demand to be $r = 2n_{vc}$ and assign: for every i the cost functions $\ell_0(x) = 0$ to edges $s_{1,i}$, $t_{1,i}$ and $e_{2,i}$, the cost function $\ell_1(x) = \frac{1}{2}x + \frac{1}{2}$ to edges $e_{1,i}$ and $e_{3,i}$, the cost function $\ell_2(x) = 1.5$ to edges $s_{2,i}$ and $t_{2,i}$, and the cost function $\ell_3(x) = 3$ to edge $e_{4,i}$, and for each k, the cost function $\ell_4(x) = 0.5$ to edges $g_{1,k}$ and $g_{2,k}$. The optimal solution will assign for each G_i one unit of flow to path $s - a_i - b_i - c_i - d_i - t$ and one unit of flow to $s - a_i - d_i - t$. This makes the costs of the edges to be as needed, as the only non constant cost is ℓ_1 and $\ell_1(1) = 1$.

To verify that this is indeed an optimal flow, one can assign to each edge e instead of its cost function, say $\ell_e(x)$, the cost function $\ell_e(x) + x\ell'_e(x)$. The optimal solution in the initial instance should be an equilibrium for the instance with the pre-described change in the cost functions (see e.g. [14]). This will hold here as under the optimal flow and with respect to the new cost functions the only edges changing cost will be $e_{1,i}$ and $e_{3,i}$, for each i, and that new cost will be 1.5 $(\ell_1(1) + 1\ell'_1(1) = 1.5)$. $\qquad\square$

Consequently, we obtain the following corollary.

Corollary 1. *For single commodity networks with linear latencies, $MINTB$ is NP-hard.*

Proof. By following the same reduction, by Lemma 1 we get that solving MINTB in G gives the solution to VC in G_{vc} and vice versa. Thus, $MINTB$ is NP-hard. $\qquad\square$

3.2 Single-Commodity Network with Linear Latencies and All Edges Under Use

In this section we turn to study $MINTB$ for instances where all edges are used by the optimal solution. Note that this case is not captured by Theorem 1, as in the reduction given for proving the theorem, the existence of unused paths in network G was crucially exploited. Nevertheless, $MINTB$ remains NP hard for this case.

Theorem 2. *For instances with linear latencies, it is NP-hard to solve $MINTB$ even if all edges are used by the optimal solution.*

Proof. The proof comes by a reduction from the partition problem (PARTITION) which is well known to be NP-complete (see e.g. [10]). $PARTITION$ is: Given a multiset $S = \{\alpha_1, \alpha_2, \ldots, \alpha_n\}$ of positive integers, decide (YES or NO) whether there exists a partition of S into sets S_1 and S_2 such that $S_1 \cap S_2 = \emptyset$ and $\sum_{\alpha_i \in S_1} \alpha_i = \sum_{\alpha_j \in S_2} \alpha_j = \frac{\sum_{i=1}^n \alpha_i}{2}$.

Given an instance of $PARTITION$ we will construct an instance of $MINTB$ with used edges only and show that getting the optimal solution for $MINTB$ solves $PARTITION$. Though, we will not directly construct the instance. First we will construct a graph with edge costs that are assumed to be the costs of the edges under the optimal solution and then we are going to assign linear cost functions and demand that makes the edges under the optimal solution to have costs equal to the predefined costs. For these costs we will prove that if the answer to $PARTITION$ is YES, then the solution to $MINTB$ puts tolls to $2n$ edges and if the answer to $PARTITION$ is NO then the solution to $MINTB$ puts tolls to more than $2n$ edges. Note that the tolls that will be put on the edges should make all s-t paths of the $MINTB$ instance having equal costs, as all of them are assumed to be used.

Next, we construct the graph of the reduction together with the costs of the edges. Given the multi-set $S = \{\alpha_1, \alpha_2, \ldots, \alpha_n\}$ of $PARTITION$, with $\sum_{i=1}^n \alpha_i = 2B$, construct the $MINTB$ instance graph $G(V, E)$, with source s and sink t, in the following way:

1. For each i, construct graph $G_i = (V_i, E_i)$, with $V_i = \{u_i, w_i, x_i, v_i\}$ and $E_i = \{(u_i, w_i), (w_i, v_i), (u_i, x_i), (x_i, v_i), (w_i, x_i), (w_i, x_i)\}$. Edges $a_i = (u_i, w_i)$ and $b_i = (x_i, v_i)$ have cost equal to α_i, edges $c_{1,i} = (w_i, x_i)$ and $c_{2,i} = (w_i, x_i)$ have cost equal to $2\alpha_i$ and edges $q_i = (w_i, v_i)$ and $g_i = (u_i, x_i)$ have cost equal to $4\alpha_i$.
2. For $i = 1$ to $n - 1$ identify v_i with u_{i+1}. Let the source vertex be $s = u_1$ and the sink vertex be $t = v_n$.
3. Add edge $h = (s, t)$ to connect s and t directly with cost equal to $11B$.

The constructed graph is presented in Fig. 2. It consists of $(3n + 1)$ vertices and $6n + 1$ edges and thus can be created in polynomial time, given S.

We establish the one-to-one correspondence between the two problems in the following lemma.

Lemma 2. *(I) If the answer to $PARTITION$ on S is YES then the size of opt-inducing tolls for G is equal to $2n$.*
(II) If the answer to $PARTITION$ on S is NO then the size of opt-inducing tolls for G is strictly greater than $2n$.

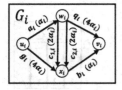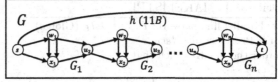

Fig. 2. The graph for $MINTB$, as it arises from $PARTITION$. The pair of symbols on each of G_i's edges correspond to the name and the cost of the edge respectively.

Proof. The proof is identical to the proof of Lemma 2 in [5].

What is left for concluding the proof is to define the linear cost functions and the demand so that at the optimal solution all edges have costs equal to the ones defined above.

Define the demand to be $r = 4$ and assign the cost function $\ell_h = 11B$ to edge h and for each i, the cost function $\ell_i^1(x) = \frac{1}{4}\alpha_i x + \frac{1}{2}\alpha_i$ to edges a_i, b_i, the cost function $\ell_i^2(x) = \alpha_i x + \frac{3}{2}\alpha_i$ to edges $c_{1,i}$ and $c_{2,i}$, and the constant cost function $\ell_i^3(x) = 4\alpha_i$ to edges q_i and g_i. The optimal flow then assigns 1 unit of flow to edge h which has cost $11B$, and the remaining 3 units to the paths through G_i. In each G_i, 1 unit will pass through $a_i - q_i$, 1 unit will pass through $g_i - b_i$, $1/2$ units will pass through $a_i - c_{1,i} - b_i$, and $1/2$ unit will pass through $a_i - c_{2,i} - b_i$. This result to a_i and b_i costing α_i, to $c_{1,i}$ and $c_{2,i}$ costing $2\alpha_i$, and to q_i and g_i costing $4\alpha_i$, as needed. We can easily verify that it is indeed an optimal flow using a technique similar to the one used in Theorem 1.

4   Algorithm for MINTB on Series-Parallel Graphs

In this section we propose an exact algorithm for $MINTB$ in series-parallel graphs. We do so by reducing it to a solution of an equivalent problem defined below.

Consider an instance $\mathcal{G} = \{G(V,E),(\ell_e)_{e\in E},r\}$ of $MINTB$, where $G(V,E)$ is a SP graph with terminals s and t. Since the flow we want to induce is fixed, i.e. the optimal flow o, by abusing notation, let *length* ℓ_e denote $\ell_e(o_e)$, for each $e \in E$, and let *used edge-set*, $E_u = \{e \in E : o_e > 0\}$, denote the set of used edges under o. For $\mathcal{G}$, we define the corresponding ℓ-*instance* (length-instance) to be $\mathcal{S}(\mathcal{G}) = \{G(V,E),\{\ell_e\}_{e\in E},E_u\}$. We may write simply $\mathcal{S}$, if $\mathcal{G}$ is clear form the context. By the definition below and the equilibrium definition, Lemma 3 easily follows.

Definition 2. *Given an ℓ-instance $\mathcal{S} = \{G(V,E),\{\ell_e\}_{e\in E},E_u\}$, inducing a length L in G is defined as the process of finding $\ell'_e \geq \ell_e$, for all $e \in E$, such that when replacing ℓ_e with ℓ'_e: (i) all used $s - t$ paths have length L and (ii) all unused $s - t$ paths have length greater or equal to L, where a path is used when all of its edges are used, i.e. they belong to E_u.*

Algorithm 1. MAKELISTPL

Input: Parallel link network: P, List: lst_P (Global)
Output: Processed list: lst_P
1 Reorder the m edges such that
   $\ell_1 \leq \ell_2 \leq \cdots \leq \ell_m$;
2 Append $\ell_{m+1} = \infty$ to the lengths;
3 Let ℓ_{max} be the max length of used edges;
4 The minimum number of edges to be tolled,
   $i_0 \leftarrow \min\{i : \ell_{i+1} \geq \ell_{max}, 0 \leq i \leq m+1\}$;
5 **for** $i \leftarrow i_0$ **to** m **do**
6 | Create the new element α
7 | $(\alpha \cdot \eta, \alpha \cdot \ell) \leftarrow (i, \ell_{i+1})$;
8 | Insert α in lst_P

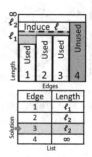

Fig. 3. Example of list

Lemma 3. *Consider an instance $\mathcal{G}$ on a SP graph $G(V, E)$ with corresponding ℓ-instance $\mathcal{S}$. L is induced in G with modified lengths ℓ'_e if and only if $\{\ell'_e - \ell_e\}_{e \in E}$ is an opt-inducing toll vector for $\mathcal{G}$.*

We call edges with $\ell'_e > \ell_e$ tolled edges as well. Under these characterizations, observe that finding a toll vector Θ that solves $MINTB$ for instance $\mathcal{G}$ with graph G, is equivalent to inducing length L in G with minimum number of tolled edges, where L is the common equilibrium cost of the used paths in G^θ. In general, this L is not known in advance and it might be greater than ℓ_{max}, i.e. the cost of the most costly used path in G, see e.g. Fig. 4. Though, for SP graphs we prove (Lemma 4) a monotonicity property that ensures that inducing length ℓ_{max} results in less or equal number of tolled edges than inducing any $\ell' > \ell_{max}$. Our algorithm relies on the above equivalence and induces ℓ_{max} with minimum number of tolled edges.

Algorithm for Parallel Link Networks: Before introducing the algorithm for $MINTB$ on SP graphs, we consider the problem of inducing a length L in a parallel link network P using minimum number of edges. It is easy to see that all edges with length less than the maximum among used edges, say ℓ_{max}, should get a toll. Similarly, to induce any length $\ell > \ell_{max}$, all edges with cost less than ℓ are required to be tolled.

Define an 'edge-length' pair as the pair (η, ℓ) such that by using at most η edges a length ℓ can be induced in a given graph. Based on the above observations we create the 'edge-length' pair list, lst_P, in Algorithm 1. By reordering the edges in increasing length order, let edge k have length ℓ_k for $k = 1$ to m. Also let there be i_0 number of edges with length less than ℓ_{max}. The list gets the first entry (i_0, ℓ_{max}) and subsequently for each $i = i_0 + 1$ to m, gets the entry (i, ℓ_{i+1}), where $\ell_{m+1} = \infty$.

To induce any length ℓ, starting from the first 'edge-length' pair in list lst_P we linearly scan the list until for the first time we encounter the 'edge-length' pair with η edges and length strictly greater than ℓ. Clearly $(\eta - 1)$ is the minimum number of edges required to induce ℓ as illustrated in Fig. 3.

Algorithm Structure: The proposed algorithm for $MINTB$ proceeds in a recursive manner on a given parse tree T of the SP graph G of an ℓ-*instance* S, where we create S given instance G and optimal flow o. Recall that for each node v of the parse tree we have an associated SP subgraph G_v with the terminals s_v and t_v. The two children of node v, whenever present, represent two subgraphs of G_v, namely G_1 and G_2. Similar to the parallel link graph our algorithm creates an 'edge-length' pair list for each node v. Due to lack of space the algorithms are presented in the full version of this paper [5]. From hereon Algorithm i in this paper will refer to Algorithm i in [5], for all $i \geq 2$.

Central Idea. Beginning with the creation of a list for each leaf node of the parse tree using Algorithm 1 we keep on moving up from the leaf level to the root level. At every node the list of its two children, lst_1 and lst_2, are optimally combined to get the current list lst_v. For each 'edge-length' pair (η, ℓ) in a current list we maintain two pointers $(p1, p2)$ to point to the two specific pairs, one each from its descendants, whose combination generates the pair (η, ℓ). Hence each element in the list of a 'series' or 'parallel' node v is given by a tuple, $(\eta, \ell, p1, p2)$.

The key idea in our approach is that the size of the list lst_v for every node v, is upper bounded by the number of edges in the subgraph G_v. Furthermore, for each series or parallel node, we devise polynomial time algorithms, Algorithm 5 and Algorithm 6 respectively, which carry out the above combinations optimally.

Optimal List Creation. Specifically, we first compute the number of edges necessary to induce the length of maximum used path between s_v and t_v, which corresponds to the first 'edge-length' pair in lst_v. Moreover, the size of the list is limited by the number of edges necessary for inducing the length ∞, as computed next. Denoting the first value by s and the latter by f, for any 'edge-length' pair (η, ℓ) in lst_v, $\eta \in \{s, s+1, \ldots, f\}$.

Considering an η in that range we may use η' edges in subgraph G_1 and $\eta - \eta'$ edges in subgraph G_2 to induce some length, which gives a feasible division of η. Let η' induce ℓ_1 in G_1 and $\eta - \eta'$ induce ℓ_2 in G_2. In a 'series' node the partition induces $\ell = \ell_1 + \ell_2$ whereas in a 'parallel' node it induces $\ell = \min\{\ell_1, \ell_2\}$.

Next we fix the number of edges to be η and find the feasible division that maximizes the induced length in G and subsequently a new 'edge-length' pair is inserted in lst_v. We repeat for all η, starting from s and ending at f. This gives a common outline for both Algorithms 5 and 6. A detailed description is provided in Theorem 3.

Placing Tolls on the Network. Once all the lists have been created, Algorithm 4 traverses the parse tree starting from its root node and optimally induces the necessary lengths at every node. At the root node the length of the maximum used path in G is induced. At any stage, due to the optimality of the current list, given a length ℓ that can be induced there exists a unique 'edge-length' pair that gives the optimal solution. In the recursive routine after finding this specific pair, we forward the length required to be induced on its two children. For a 'parallel' node the length ℓ is forwarded to both of its children, whereas in a 'series' node the length is appropriately split between the two. Following the tree traversal the algorithm eventually reaches the leaf nodes, i.e. the parallel link graphs, where given a length ℓ the optimal solution is to make each edge e with length $\ell_e < \ell$ equal to length ℓ by placing toll $\ell - \ell_e$. A comprehensive explanation is presented under Lemma 5 in [5].

4.1   Optimality and Time Complexity of Algorithm SolMINTB

Proof Outline: The proof of Theorem 4 which states that the proposed algorithm solves the $MINTB$ problem in SP graphs in polynomial time, is split into Lemmas 4 and 5 and Theorem 3. The common theme in the proofs is the use of an inductive reasoning starting from the base case of parallel link networks, which is natural given the parse tree decomposition. Lemma 4 gives a monotonicity property of the number of edges required to induce length ℓ in a SP graph guiding us to induce the length of maximum used path to obtain an optimal solution.

The key Theorem 3 is essentially the generalization of the ideas used in the parallel link network to SP graphs. It proves that the lists created by Algorithm 2 follow three desired properties. (1) The maximality of the 'edge-length' pairs in a list, i.e. for any 'edge-length' pair (η, ℓ) in lst_v it is not possible to induce a length greater than ℓ in G_v using at most η edges. (2) The 'edge-length' pairs in a list follows an increasing length order which makes it possible to locate the optimal solution efficiently. (3) Finally the local optimality of a list at any level of the parse tree ensures that the 'series' or 'parallel' combination preserves the same property in the new list.

In Lemma 5 we prove that the appropriate tolls on the edges can be placed provided the correctness of Theorem 3. The basic idea is while traversing down the parse tree at each node we induce the required length in a locally optimal manner. Finally, in the leaf nodes the tolls are placed on the edges and the process inducing a given length is complete. Exploiting the linkage between the list in a specific node and the lists in its children we can argue that these local optimal solutions lead to a global optimal solution.

Finally, in our main theorem, Theorem 4, combining all the elements we prove that the proposed algorithm solves $MINTB$ optimally. In the second part of the proof of Theorem 4, the analysis of running time of the algorithm is carried out. The creation of the list in each node of the parse tree takes $\mathcal{O}(m^2)$ time, whereas the number of nodes is bounded by $\mathcal{O}(m)$, implying that Algorithm 2 terminates in $\mathcal{O}(m^3)$ time. Here m is the number of edges in the SP graph G.

Proof of Correctness: In what follows we state the key theorems and lemmas, while interested readers are referred to [5] for the complete proofs.

Lemma 4. *In an ℓ-instance S, with SP graph G and maximum used (s,t) path length ℓ_{max}, any length L can be induced in G if and only if $L \geq \ell_{max}$. Moreover if length L is induced optimally with T edges then length $\ell_{max} \leq \ell \leq L$ can be induced optimally with $t \leq T$ edges.*

Note: The above lemma breaks in general graphs. As an example, in the graph in Fig. 4 to induce a length of 3 we require 3 edges, whereas to induce a length of 4 only 2 edges are sufficient.

Fig. 4. Counter Example

Theorem 3. *Let S be an ℓ-instance and G be the associated SP graph with parse tree representation T. For every node v in T, let the corresponding SP network be G_v and $\ell_{max,v}$ be the length of the maximum used path from s_v to t_v. Algorithm 3 creates the list, lst_v, with the following properties.*

1. *For each 'edge-length' pair (η_i, ℓ_i), $i = 1$ to m_v, in lst_v, ℓ_i is the maximum length that can be induced in the network G_v using at most η_i edges.*
2. *For each 'edge-length' pair (η_i, ℓ_i) in the list lst_v, we have the total ordering, i.e. $\eta_{i+1} = \eta_i + 1$ for all $i = 1$ to $m_v - 1$, and $\ell_{max,v} = \ell_1 \leq \ell_2 \leq \cdots \leq \ell_{m_v} = \infty$.*
3. *In G_v, length ℓ is induced by minimum $\eta_{\hat{i}}$ edges if and only if $\ell \geq \ell_1$ and $\hat{i} = \arg\min\{\eta_j : (\eta_j, \ell_j) \in lst_v \wedge \ell_j \geq \ell\}$.*

Lemma 5. *In an ℓ-instance S with SP graph G, suppose we are given lists lst_v, for all nodes v in the parse tree T of G, all of which satisfy properties 1, 2 and 3 in Theorem 3. Algorithm 4 induces any length $\ell_{in} \geq \ell_1$ optimally in G, where ℓ_1 is the length of the first 'edge-length' pair in lst_r, r being the root node of T. Moreover, it specifies the appropriate tolls necessary for every edge.*

Theorem 4. *Algorithm 2 solves the $MINTB$ problem optimally in time $\mathcal{O}\left(m^3\right)$ for the instance $\mathcal{G} = \{G(V,E), (\ell_e)_{e \in E}, r\}$, where $G(V,E)$ is a SP graph with $|V| = n$ and $|E| = m$.*

5   Conclusion

In this paper we consider the problem of inducing the optimal flow as network equilibrium and show that the problem of finding the minimum cardinality toll, i.e. the $MINTB$ problem, is NP-hard to approximate within a factor of 1.1377. Furthermore we define the minimum cardinality toll with only used edges left in the network and show in this restricted setting the problem remains NP-hard even for single commodity instances with linear latencies. We leave the hardness of approximation results of the problem open. Finally, we propose a polynomial time algorithm that solves $MINTB$ in series-parallel graphs, which exploits the parse tree decomposition of the graphs. The approach in the algorithm fails to generalize to a broader class of graphs. Specifically, the monotonicity property proved in Lemma 4 holds in series-parallel graphs but breaks down in general graphs revealing an important structural difficulty inherent to $MINTB$ in general graphs. Future work involves finding approximation algorithms for $MINTB$. The improvement of the inapproximability results presented in this paper provides another arena to this problem, e.g. finding stronger hardness of approximation results for $MINTB$ in multi-commodity networks.

Acknowledgement. We would like to thank Steve Boyles and Sanjay Shakkottai for helpful discussions. This work was supported in part by NSF grant numbers CCF-1216103, CCF-1350823 and CCF-1331863.

References

1. Bai, L., Hearn, D.W., Lawphongpanich, S.: A heuristic method for the minimum toll booth problem. J. Global Optim. **48**(4), 533–548 (2010)
2. Bai, L., Rubin, P.A.: Combinatorial benders cuts for the minimum tollbooth problem. Oper. Res. **57**(6), 1510–1522 (2009)
3. Bai, L., Stamps, M.T., Harwood, R.C., Kollmann, C.J., Seminary, C.: An evolutionary method for the minimum toll booth problem: The methodology. Acad. Info. Manage. Sci. J. **11**(2), 33 (2008)
4. Baier, G., Erlebach, T., Hall, A., Köhler, E., Kolman, P., Pangrác, O., Schilling, H., Skutella, M.: Length-bounded cuts and flows. ACM Trans. Algorithms (TALG) **7**(1), 4 (2010)
5. Basu, S., Lianeas, T., Nikolova, E.: New complexity results and algorithms for the minimum tollbooth problem (2015). http://arxiv.org/abs/1509.07260
6. Beckmann, M., McGuire, C., Weinstein, C.: Studies in the Economics of Transportation. Yale University Press, New Haven (1956)
7. Bergendorff, P., Hearn, D.W., Ramana, M.V.: Congestion toll pricing of traffic networks. In: Pardalos, P.M., Heam, D.W., Hager, W.W. (eds.) Network Optimization. LNEMS, vol. 450, pp. 51–71. Springer, Heidelberg (1997)
8. Buriol, L.S., Hirsch, M.J., Pardalos, P.M., Querido, T., Resende, M.G., Ritt, M.: A biased random-key genetic algorithm for road congestion minimization. Optim. Lett. **4**(4), 619–633 (2010)
9. Dinur, I., Safra, S.: On the hardness of approximating minimum vertex cover. Ann. Math. **162**, 439–485 (2005)
10. Garey, M.R., Johnson, D.S.: Computers and intractability, vol. 29. W.H. Freeman (2002)
11. Harks, T., Schäfer, G., Sieg, M.: Computing flow-inducing network tolls. Technical report, 36–2008, Institut für Mathematik, Technische Universität Berlin, Germany (2008)
12. Hearn, D.W., Ramana, M.V.: Solving congestion toll pricing models. In: Marcotte, P., Nguyen, S., (eds.) Equilibrium and Advanced Transportation Modelling. Centre for Research on Transportation, pp. 109–124. Springer, New York (1998)
13. Hoefer, Martin, Olbrich, Lars, Skopalik, Alexander: Taxing subnetworks. In: Papadimitriou, Christos, Zhang, Shuzhong (eds.) WINE 2008. LNCS, vol. 5385, pp. 286–294. Springer, Heidelberg (2008)
14. Roughgarden, T.: Selfish Routing and the Price of Anarchy. MIT Press, Cambridge (2005)

15. Stefanello, F., Buriol, L., Hirsch, M., Pardalos, P., Querido, T., Resende, M., Ritt, M.: On the minimization of traffic congestion in road networks with tolls. Ann. Oper. Res., 1–21 (2013)
16. The Centre for Economics and Business Research: The future economic and environmental costs of gridlock in 2030. Technical report. INRIX, Inc. (2014)
17. Valdes, J., Tarjan, R.E., Lawler, E.L.: The recognition of series parallel digraphs. In: Proceedings of the Eleventh Annual ACM Symposium on Theory of Computing, pp. 1–12. ACM (1979)
18. Wardrop, J.G.: Road Paper. Some Theoretical Aspects of Road Traffic Research. In: ICE Proceedings: Engineering Divisions, vol. 1, pp. 325–362. Thomas Telford (1952)

Ad Exchange: Envy-Free Auctions
with Mediators

Oren Ben-Zwi[1], Monika Henzinger[2], and Veronika Loitzenbauer[2]([✉])

[1] Emarsys Labs, Emarsys eMarketing Systems AG, Vienna, Austria
oren.ben-zwi@emarsys.com
[2] Faculty of Computer Science, University of Vienna, Vienna, Austria
{monika.henzinger,veronika.loitzenbauer}@univie.ac.at

Abstract. Ad exchanges are an emerging platform for trading advertisement slots on the web with billions of dollars revenue per year. Every time a user visits a web page, the publisher of that web page can ask an ad exchange to auction off the ad slots on this page to determine which advertisements are shown at which price. Due to the high volume of traffic, ad networks typically act as mediators for individual advertisers at ad exchanges. If multiple advertisers in an ad network are interested in the ad slots of the same auction, the ad network might use a "local" auction to resell the obtained ad slots among its advertisers.

In this work we want to deepen the theoretical understanding of these new markets by analyzing them from the viewpoint of combinatorial auctions. Prior work studied mostly single-item auctions, while we allow the advertisers to express richer preferences over multiple items. We develop a game-theoretic model for the entanglement of the *central* auction at the ad exchange with the *local* auctions at the ad networks. We consider the incentives of all three involved parties and suggest a *three-party competitive equilibrium*, an extension of the Walrasian equilibrium that ensures envy-freeness for all participants. We show the existence of a three-party competitive equilibrium and a polynomial-time algorithm to find one for gross-substitute bidder valuations.

Keywords: Ad-exchange · Combinatorial auctions · Gross substitute · Walrasian equilibrium · Three-party equilibrium · Auctions with mediators

1 Introduction

As advertising on the web becomes more mature, *ad exchanges* (AdX) play a growing role as a platform for selling advertisement slots from publishers to

This work was funded by the Vienna Science and Technology Fund (WWTF) through project ICT10-002. Additionally the research leading to these results has received funding from the European Research Council under the European Union's Seventh Framework Programme (FP/2007-2013)/ERC Grant Agreement no. 340506. A full version of the paper is available at https://eprints.cs.univie.ac.at/4456/.

O. Ben-Zwi—Work done while at the University of Vienna.

E. Markakis and G. Schäfer (Eds.): WINE 2015, LNCS 9470, pp. 104–117, 2015.
DOI: 10.1007/978-3-662-48995-6_8

advertisers. Following the Yahoo! acquisition of Right Media in 2007, all major web companies, such as Google, Facebook, and Amazon, have created or acquired their own ad exchanges. Other major ad exchanges are provided by the Rubicon Project, OpenX, and AppNexus. In 2012 the total revenue at ad exchanges was estimated to be around two billion dollars [4]. Every time a user visits a web page, the publisher of that web page can ask an ad exchange to auction off the ad slots on this page. Thus, the goods traded at an ad exchange are *ad impressions*. This process is also known as *real-time bidding* (RTB). A web page might contain multiple ad slots, which are currently modeled to be sold separately in individual auctions. Individual advertisers typically do not directly participate in these auctions but entrust some ad network to bid on their behalf. When a publisher sends an ad impression to an exchange, the exchange usually contacts several ad networks and runs a (variant of a) second-price auction [13] between them, potentially with a reserve price under which the impression is not sold. An ad network (e.g. Google's Display Network [6]) might then run a second, "local" auction to determine the allocation of the ad slot among its advertisers. We study this interaction of a *central auction* at the exchange and *local auctions* at the ad networks.[1]

We develop a game-theoretic model that considers the incentives of the following three parties: (1) the ad exchange, (2) the ad networks, and (3) the advertisers. As the ad exchange usually charges a fixed percentage of the revenue and hands the rest to the publishers, the ad exchange and the publishers have the same objective and can be modeled as one entity. We then study equilibrium concepts of this new model of a three-party exchange. Our model is described as an ad exchange, but it may also model other scenarios with mediators that act between bidders and sellers, as noted already by Feldman et al. [5]. The main differences between our model and earlier models (discussed in detail at the end of this section) are the following: (a) We consider the incentives of all three parties *simultaneously*. (b) While most approaches in prior work use Bayesian assumptions, we apply *worst-case analysis*. (c) We allow auctions with *multiple heterogeneous items*, namely combinatorial auctions, in contrast to the single-item auctions studied so far. Multiple items arise naturally when selling ad slots on a per-impression basis, since there are usually multiple advertisement slots on a web page.

To motivate the incentives of ad networks and exchanges, we compare next their short and long-term revenue considerations, following Mansour et al. [13] and Muthukrishnan [14]. Ad exchanges and ad networks generate revenue as follows: (1) An ad exchange usually receives some percentage of the price paid by the winner(s) of the central auction. (2) An ad network can charge a higher price to its advertisers than it paid to the exchange or it can be paid via direct contracts with its advertisers. Thus both the ad exchange and the ad networks (might) profit from higher prices in their auctions. However, they also have a motivation not to charge too high prices as (a) the advertisers could stick to

---

[1] In this work an auction is an algorithm to determine prices of items and their allocation to bidders.

alternative advertising channels such as long-term contracts with publishers, and (b) there is a significant competition between the various ad exchanges and ad networks, as advertisers can easily switch to a competitor. Thus, lower prices (might) increase advertiser participation and, hence, the long-term revenue of ad exchanges and ad networks. We only consider a single auction (of multiple items) and leave it as an open question to study changes over time. We still take the long-term considerations outlined above into account by assuming that the ad exchange aligns its strategic behavior with its long-term revenue considerations and only desires for each central auction to sell *all* items.[2] In our model the incentive of an ad network to participate in the exchange comes from the opportunity to purchase some items at a low price and then resell them at a higher price. However, due to long-term considerations, our model additionally requires the ad networks to "satisfy their advertisers" by faithfully representing the advertisers' preferences towards the exchange, while still allowing the ad networks to extract revenue from the competition between the advertisers in their network. An example for this kind of restriction for an ad network is Google's Display Network [6] that guarantees its advertisers that each ad impression is sold via a second-price auction, independent of whether an ad exchange is involved in the transaction or not [13].

To model a *stable* outcome in a three-party exchange, we use the equilibrium concept of *envy-freeness* for all three types of participants. A participant is envy-free if he receives his most preferred set of items under the current prices. Envy-freeness for all participants is a natural notion to express stability in a market, as it implies that no coalition of participants would strictly profit from deviating from the current allocation and prices (assuming truthfully reported preferences). Thus an envy-free equilibrium supports stability in the market prices, which in turn facilitates, for example, revenue prediction for prospective participants and hence might increase participation and long-term revenue. For only two parties, i.e., sellers and buyers, where the sellers have no intrinsic value for the items they sell, envy-freeness for all participants is equal to a *competitive* or *Walrasian* equilibrium [20], a well established notion in economics to characterize an equilibrium in a market where demand equals supply. We provide a generalization of this equilibrium concept to three parties.

Our Contribution. We introduce the following model for ad exchanges. A *central seller* wants to sell k *items*. There are m *mediators* $\mathcal{M}_i$, each with her own n_i *bidders*. Each bidder has a valuation function over the items. In the ad exchange setting, the central seller is the ad exchange, the items are the ad slots shown to a visitor of a web page, the mediators are the ad networks, and the bidders are the advertisers. A bidder does not have any direct "connection" to the central seller. Instead, all communication is done through the mediators. A mechanism for allocating the items to the bidders is composed of a *central auction* with mediators acting as bidders, and then *local auctions*, one per mediator, in which

---

[2] Our model and results can be adapted to include reserve prices under which the ad exchange is not willing to sell an item.

every mediator allocates the set of items she bought in the central auction; that is, an auction where the bidders of that mediator are the only participating bidders and the items that the mediator received in the central auction are the sole items. The prices of the items obtained in the central auction provide a lower bound for the prices in the local auctions, i.e., they act as reserve prices in the local auctions. We assume that the central seller and the bidders have quasi-linear utilities, i.e., utility functions that are linear in the price, and that their incentive is to maximize their utility. For the central seller this means that his utility from selling a set of slots is just the sum of prices of the items in the set. The utility of a bidder on receiving a set of items S is his value for S minus the sum of the prices of the items in S.

The incentive of a mediator, however, is not so straightforward and needs to be defined carefully. In our model, to "satisfy" her bidders, each mediator guarantees her bidders that the outcome of the local auction will be *minimal envy free*, that is, for the final local price vector, the item set that is allocated to any bidder is one of his most desirable sets over *all* possible item sets (even sets that contain items that were not allocated to his mediator, i.e., each bidder is not only *locally*, but *globally envy-free*) and there is no (item-wise) smaller price vector that fulfills this requirement. We assume that each mediator wants to maximize her revenue[3] and define the revenue of a mediator for a set of items S as the difference between her earnings when selling S with this restriction and the price she has to pay for S at the central auction.

For this model we define a new equilibrium concept, namely the *three-party competitive equilibrium*. At this equilibrium all three types of participants are envy-free. Envy-free solutions for the bidders always exist, as one can set the prices of all items high enough so that no bidder will demand any item. Additionally, we require that there is no envy for the central seller, meaning that all items are sold. If there were no mediators, then a two-party envy-free solution would be exactly a *Walrasian equilibrium*, which for certain scenarios can be guaranteed [11]. However, with mediators it is not a-priori clear that a three-party competitive equilibrium exists as, additionally, the mediators have to be envy-free. We show that for our definition of a mediator's revenue (a) the above requirements are fulfilled and (b) a three-party competitive equilibrium exists whenever a Walrasian equilibrium for the central auction exists or whenever a two-party equilibrium exists for the bidders and the central seller without mediators. Interestingly, we show that for gross-substitute bidder valuations the incentives of this kind of mediator can be represented with an OR-valuation over the valuations of her bidders. This then leads to the following result: *For gross-substitute bidder valuations a three-party competitive equilibrium can be computed in polynomial time.* In particular, we will show how to compute the three-party competitive equilibrium with minimum prices.

Related Work. The theoretical research on ad exchanges was initialized by a survey of Muthukrishnan [14] that lists several interesting research directions.

---

[3] For the purpose of this paper, the terms revenue and utility are interchangeable.

Our approach specifically addresses his 9th problem, namely to enable the advertisers to express more complex preferences that arise when multiple advertisement slots are auctioned off at once as well as to design suitable auctions for the exchange and the ad networks to determine allocation and prices given these preferences.

The most closely related work with respect to the model of the ad exchange is Feldman et al. [5]. It is similar to our work in two aspects: (1) The mediator bids on behalf of her bidders in a central auction and the demand of the mediator as well as the tentative allocation and prices for reselling to her bidders are determined via a local auction. (2) The revenue of the mediator is the price she can obtain from reselling minus the price she paid in the central auction. The main differences are: (a) Only one item is auctioned at a time and thus the mediator can determine her valuation with a single local auction. (b) Their work does not consider the incentives of the bidders, only of the mediators and the central seller. (c) A Bayesian setting is used where the mediators and the exchange know the probability distributions of the bidders' valuations. Based on this information, the mediators and the exchange choose reserve prices for their second-price auctions to maximize their revenue. The work characterizes the equilibrium strategies for the selection of the reserve prices.

Mansour et al. [13] (mainly) describe the auction at the DoubleClick exchange. Similar to our work advertisers use ad networks as mediators for the central auction. They observe that if mediators that participate in a single-item, second-price central auction are only allowed to submit a single bid, then it is not possible for the central auction to correctly implement a second-price auction over *all* bidders as the bidders with the highest and the second highest value might use the same mediator. Thus they introduce the *Optional Second Price* auction, where every mediator is allowed to optionally submit the second highest bid with the highest bid. In such an auction each mediator can guarantee to her bidders that if one of them is allocated the item, then he pays the (global) second-price for it. For the single-item setting, the bidders in their auction and in our auction pay the same price. If the mediator of the winning bidder did *not* specify an optional second price, then her revenue will equal the revenue of our mediator. If she did, her revenue will be zero and the central seller will receive the gain between the prices in the local and the central auction.

Stavrogiannis et al. [18] consider a game between bidders and mediators, where the bidders can select mediators (based on Bayesian assumptions of each other's valuations) and the mediators can set the reserve prices in the second-price local auction. The work presents mixed Nash equilibrium strategies for the bidders to select their mediator. In [19] the same authors compare different single-item local auctions with respect to the achieved social welfare and the revenue of the mediators and the exchange.

Balseiro et al. (2013) introduced a setting that does *not* include mediators [1]. Instead, they see the ad exchange as a game between publishers, who select parameters such as reserve prices for second-price auctions, and advertisers, whose budget constraints link different auctions over time. They introduced a new equilibrium concept for this game and used this to analyze the impact of auction design questions such as the selection of a reserve price. Balseiro et al. (2014) studied a publisher's trade-off between using an ad exchange versus fulfilling long-term contracts with advertisers [2].

Equilibria in trading networks (such as ad exchanges) are also addressed in the "matching with contracts" literature. Hatfield and Milgrom [10] presented a new model where instead of bidders and items there are *agents* and *trades* between pairs of agents. The potential trades are modeled as edges in a graph where the agents are represented by the nodes. Agent valuations are then defined over the potential trades and assumed to be monotone substitute. They proved the existence of an (envy-free) equilibrium when the agent-trades graph is bipartite. Later this was improved to directed acyclic graphs by Ostrovsky [16] and to arbitrary graphs by Hatfield et al. [9]. They did not show (polynomial-time) algorithms to reach equilibria. Our model can be reduced to this model, hence a three-party equilibrium exists when all bidders are monotone gross substitute. The result of this reduction (not stated here) is not polynomial in the number of bidders and items.

2 Preliminaries

Let Ω denote a set of k items. A *price vector* is an assignment of a non-negative price to every element of Ω. For a price vector $p = (p_1, ..., p_k)$ and a set $S \subseteq \Omega$ we use $p(S) = \sum_{j \in S} p_j$. For any two price vectors p, r an inequality such as $p \geq r$ as well as the operations $\min(p, r)$ and $\max(p, r)$ are meant item-wise.

We denote with $\langle \Omega_b \rangle = \langle \Omega_b \rangle_{b \in \mathcal{B}}$ an *allocation* of the items in Ω such that for all bidders $b \in \mathcal{B}$ the set of items allocated to b is given by Ω_b and we have $\Omega_b \subseteq \Omega$ and $\Omega_b \cap \Omega_{b'} = \emptyset$ for $b' \neq b$, $b' \in \mathcal{B}$. Note that some items might not be allocated to any bidder.

A *valuation* function v_b of a bidder b is a function from 2^Ω to $\mathbb{R}$, where 2^Ω denotes the set of all subsets of Ω. We assume throughout the paper $v_b(\emptyset) = 0$. Unless specified otherwise, for this work we assume *monotone* valuations, that is, for $S \subseteq T$ we have $v_b(S) \leq v_b(T)$. This assumption is made for ease of presentation. We use $\{v_b\}$ to denote a collection of valuation functions. The (quasi-linear) *utility* of a bidder b from a set $S \subseteq \Omega$ at prices $p \geq 0$ is defined as $u_{b,p}(S) = v_b(S) - p(S)$. The *demand* $D_b(p)$ of a bidder b for prices $p \geq 0$ is the set of subsets of items $S \subseteq \Omega$ that maximize the bidder's utility at prices p. We call a set in the demand a *demand representative*. Throughout the paper we omit subscripts if they are clear from the context.

Definition 1 (Envy Free). *An allocation $\langle \Omega_b \rangle$ of items Ω to bidders $\mathcal{B}$ is envy free (on Ω) for some prices p if for all bidders $b \in \mathcal{B}$, $\Omega_b \in D_b(p)$. We say that prices p are envy free (on Ω) if there exists an envy-free allocation (on Ω) for these prices.*

There exist envy-free prices for any valuation functions of the bidders, e.g., set all prices to $\max_{b,S} v_b(S)$. For these prices the allocation which does not allocate any item is envy free. Thus also minimal envy-free prices always exist, but are in general not unique.

Definition 2 (Walrasian Equilibrium (WE)). *A Walrasian equilibrium (on* Ω*) is an envy-free allocation* $\langle \Omega_b \rangle$ *(on* Ω*) with prices p such that all prices are non-negative and the price of unallocated items is zero. We call the allocation* $\langle \Omega_b \rangle$ *a Walrasian allocation (on* Ω*) and the prices p Walrasian prices (on* Ω*).*

We assume that the central seller has a value of zero for every subset of the items; thus (with quasi-linear utility functions) selling all items makes the seller envy free. In this case a Walrasian equilibrium can be seen as an *envy-free two-party equilibrium*, i.e., envy free for the buyers and the seller. Note that for a Walrasian price vector there might exist multiple envy-free allocations.

2.1   Valuation Classes

A *unit demand* valuation assigns a value to every item and defines the value of a set as the *maximum* value of an item in it. An *additive* valuation also assigns a value to every item but defines the value of a set as the *sum* of the values of the items in the set. Non-negative unit demand and non-negative additive valuations both have the gross-substitute property (defined below) and are by definition monotone.

Definition 3 (Gross Substitute (GS)). *A valuation function is* gross substitute *if for every two price vectors* $p^{(2)} \geq p^{(1)} \geq 0$ *and every set* $D^{(1)} \in D(p^{(1)})$, *there exists a set* $D^{(2)} \in D(p^{(2)})$ *with* $j \in D^{(2)}$ *for every* $j \in D^{(1)}$ *with* $p_j^{(1)} = p_j^{(2)}$.

For *gross-substitute* valuations of the bidders a Walrasian equilibrium is guaranteed to exist in a two-sided market [11] and can be computed in polynomial time [15,17]. Further, gross substitute is the maximal valuation class containing the unit demand class for which the former holds [7]. Several equivalent definitions are known for this class [7,17]. We will further use that for gross-substitute valuations the Walrasian prices form a complete lattice [7].

We define next an OR-valuation. Lehmann et al. [12] showed that the OR of gross-substitute valuations is gross substitute.

Definition 4 (OR-player). *The* OR *of two valuations v and w is defined as* $(v \text{ OR } w)(S) = \max_{R,T \subseteq S, R \cap T = \emptyset}(v(R) + w(T))$. *Given a set of valuations* $\{v_b\}$ *for bidders* $b \in \mathcal{B}$ *we say that the* OR*-player is a player with valuation* $v_{\text{OR}}(S) = \max_{\langle S_b \rangle} \sum_{b \in \mathcal{B}} v_b(S_b)$.

3   Model and Equilibrium

There are k items to be allocated to m mediators. Each mediator $\mathcal{M}_i$ represents a set $\mathcal{B}_i$ of bidders, where $|\mathcal{B}_i| = n_i$. Each bidder is connected to a unique mediator. Each bidder has a valuation function over the set of items and a quasi-linear utility function. A *central auction* is an auction run on all items with mediators as bidders. After an allocation $\langle \Omega_i \rangle$ and prices r at the central auction are set, another m *local auctions* are conducted, one by each mediator. In the local auction for mediator $\mathcal{M}_i$ the items Ω_i that were allocated to her in the central auction are the sole items and the bidders $\mathcal{B}_i$ are the sole bidders. A solution is an assignment of central-auction and local-auction prices to items and an allocation of items to bidders and hence, by uniqueness, also to mediators. We define next a three-party equilibrium based on envy-freeness.

Definition 5 (Equilibrium). *A three-party competitive equilibrium is an allocation of items to bidders and a set of $m + 1$ price vectors $r, p^1, p^2, \ldots, p^m$ such that the following requirements hold. For $1 \leq i \leq m$*

1. *every mediator[4] $\mathcal{M}_i$ is allocated a set Ω_i in her demand at price r,*
2. *every item j with non-zero price r is allocated to a mediator,*
3. *the price p^i coincides with r for all items not in Ω_i,*
4. *and every bidder $b \in \mathcal{B}_i$ is allocated a subset of Ω_i that is in his demand at price p^i.*

In other words, the allocation to the bidders in $\mathcal{B}_i$ with prices p^i must be envy-free for the bidders, the allocation to the mediators with prices r must be envy free for the mediators and for the central seller, i.e., must be a Walrasian equilibrium; and the prices p^i must be equal to the prices r for all items not assigned to mediator $\mathcal{M}_i$.

Note that the allocation of the items to the mediators and prices r are the outcome of a *central* auction run by the central seller, while the allocation to the bidders in $\mathcal{B}_i$ and prices p^i correspond to the outcome of a *local* auction run by mediator $\mathcal{M}_i$. These auctions are connected by the demands of the mediators and Requirement 3.

We next present our mediator model. The definition of an Envy-Free Mediator, or EF-mediator for short, reflects the following idea: To determine her revenue for a set of items S at central auction prices r, the mediator simulates the local auction she would run if she would obtain the set S at prices r. Given the outcome of this "virtual auction", she can compute her potential revenue for S and r as the difference between the virtual auction prices of the items sold in the virtual auction and the central auction prices for the items in S. However, as motivated in the introduction, the mediator is required to represent the preferences of her bidders and therefore not every set S is "allowed" for the mediator, that is, for some sets the revenue of the mediator is set to -1. The sets that maximize the revenue are then in the demand of the mediator at central auction prices r. To make the revenue of a mediator well-defined and to follow our motivation that a mediator should satisfy her bidders, the virtual auctions specifically compute minimal envy-free price vectors.

Definition 6 (Envy-Free Mediator). *An EF-mediator $\mathcal{M}_i$ determines her demand for a price vector $r \geq 0$ as follows. For each subset of items $S \subseteq \Omega$ she*

---

[4] Regardless of any demand definition.

runs a virtual auction with items S, her bidders $\mathcal{B}_i$, and reserve prices r. We assume that the virtual auction computes minimal envy-free prices $p^S \geq r$ and a corresponding envy-free allocation $\langle S_b \rangle$.[5] We extend the prices p^S to all items in Ω by setting $p_j^S = r_j$ for $j \in \Omega \setminus S$, and define the revenue $R_{i,r}(S)$ of the mediator for a set S as follows. If the allocation $\langle S_b \rangle$ is envy free for the bidders $\mathcal{B}_i$ and prices p^S on Ω, then $R_{i,r}(S) = \sum_{b \in \mathcal{B}_i} p^S(S_b) - r(S)$; otherwise, we set $R_{i,r}(S) = -1$. The demand $D_i(r)$ of $\mathcal{M}_i$ is the set of all sets S that maximize the revenue of the mediator for the reserve prices r. The local auction of $\mathcal{M}_i$ for a set Ω_i allocated to her in the central auction at prices r is equal to her virtual auction for Ω_i and r.

Following the above definition, we say that a price vector is *locally envy free* if it is envy free for the bidders $\mathcal{B}_i$ on the subset $\Omega_i \subseteq \Omega$ assigned to mediator $\mathcal{M}_i$ and *globally envy free* if it is envy free for the bidders $\mathcal{B}_i$ on Ω. Note that if p^S is envy free on Ω, then it is minimal envy free $\geq r$ on Ω for the bidders $\mathcal{B}_i$.

An interesting property of EF-mediators is that every Walrasian equilibrium in the central auction can be combined with the outcome of the local auctions of EF-mediators to form a three-party competitive equilibrium.

Theorem 1. *Assume all mediators are EF-mediators. Then a Walrasian equilibrium in the central auction with allocation $\langle \Omega_i \rangle$ together with the allocation and prices computed in the local auctions of the mediators $\mathcal{M}_i$ on their sets Ω_i (not necessarily Walrasian) form a three-party competitive equilibrium.*

Further, with EF-mediators a three-party competitive equilibrium exists whenever a Walrasian eq. exists for the bidders and items without the mediators.

Theorem 2. *Assume all mediators are EF-mediators and a Walrasian equilibrium exists for the set of bidders and items (without mediators). Then there exists a three-party competitive equilibrium.*

The proof of Theorem 2 only shows the existence of trivial three-party equilibria that basically ignores the presence of mediators. However, three-party equilibria and EF-mediators allow for richer outcomes that permit the mediators to gain revenue from the competition between their bidders while still representing the preferences of their bidders towards the central seller. In the next section we show how to find such an equilibrium provided that the valuations of all bidders are gross substitute. Recall that gross-substitute valuations are the most general valuations that include unit demand valuations for which a Walrasian equilibrium exists [7]; and that efficient algorithms for finding a Walrasian equilibrium are only known for this valuation class.

4   An Efficient Algorithm for Gross-substitute Bidders

In this section we will show how to find, in polynomial time, a three-party competitive equilibrium if the valuations of all bidders are gross substitute.

---

[5] If there are multiple envy-free allocations on S for the prices p^S, the mediator chooses one that maximizes $\sum_{b \in \mathcal{B}_i} p^S(S_b)$.

The prices the bidders have to pay at equilibrium, and thus the utilities they achieve, will be the same as in a Walrasian equilibrium (between bidders and items) with minimum prices (see full version). The price the bidders pay is split between the mediators and the exchange. We show how to compute an equilibrium where this split is best for the mediators and worst for the exchange. In turn the computational load can be split between the mediators and the exchange as well. The algorithm will be based on existing algorithms to compute Walrasian equilibria for gross-substitute bidders.

The classical (two-party) *allocation problem* is the following: We are given k items and n valuation functions and we should find an equilibrium allocation (with or without equilibrium prices) if one exists. Recall that in general a valuation function has a description of size exponential in k. Therefore, the input valuation functions can only be accessed via an *oracle*, defined below. An *efficient* algorithm runs in time polynomial in n and k (where the oracle access is assumed to take constant time).

Given an algorithm that computes a Walrasian allocation for gross-substitute bidders, by a result of Gul and Stacchetti [7] minimum Walrasian prices can be computed by solving the allocation problem $k + 1$ times. A Walrasian allocation can be combined with any Walrasian prices to form a Walrasian equilibrium [7]. Thus we can assume for gross-substitute valuations that a polynomial-time algorithm for the allocation problem also returns a vector of minimum prices that support the allocation.

Two main oracle definitions that were considered in the literature are the *valuation oracle*, where a query is a set of items S and the oracle replies with the exact value of S; and the *demand oracle*, where a query is a price vector p and the oracle replies with a demand representative D [3].

It is known that a demand oracle is strictly stronger than a valuation oracle, i.e., a valuation query can be simulated by a polynomial number of demand queries but not vice versa. For gross-substitute valuations, however, these two query models are polynomial-time equivalent, see Paes Leme [17]. The two-party allocation problem is efficiently solvable for gross-substitute valuations [15,17].

We define the *three-party allocation problem* in the same manner. We are given k items, n valuation functions over the items and m mediators, each associated with a set of unique bidders. We are looking for a three-party equilibrium allocation (and equilibrium prices) if one exists. We will assume that the input valuations are given through a valuation oracle.

The algorithm will be based on the following central result: For gross substitute valuations of the bidders an EF-mediator and an OR-player over the valuations of the same bidders are equivalent with respect to their demand and their allocation of items to bidders. Thus in this case EF-mediators can be considered as if they have a gross-substitute valuation. Note that for general valuations this equivalence does not hold.

Theorem 3. *If the valuation functions of a set of bidders $\mathcal{B}_i$ are gross substitute, then the demand of an EF-mediator for $\mathcal{B}_i$ is equal to the demand of an OR-player for $\mathcal{B}_i$. Moreover, the allocation in a virtual auction of the EF-mediator* *for reserve prices r and a set of items S in the demand is an optimal allocation for the* OR-*player for S and r and vice versa.*

To this end, we will first show for the virtual (and local) auctions that a modified Walrasian equilibrium, the RESERVE-WE(r), exists for gross-substitute valuations with reserve prices. For this we will use yet another reduction to a (standard) Walrasian equilibrium without reserve prices but with an additional additive player[6].

Definition 7 (Walrasian Equilibrium with Reserve Prices r (RESERVE-WE(r))) [8]. *A Walrasian equilibrium with reserve prices $r \geq 0$ (on Ω) is an envy-free allocation $\langle \Omega_b \rangle$ (on Ω) with prices p such that $p \geq r$, and the price of every unallocated item is equal to its reserve price, i.e., $p_j = r_j$ for $j \notin \cup_b \Omega_b$. We say that $\langle \Omega_b \rangle$ is a RESERVE-WE(r) allocation (on Ω) and p are RESERVE-WE(r) prices (on Ω).*

4.1 Properties of Walrasian Equilibria with Reserve Prices

In this section we generalize several results about Walrasian equilibria to Walrasian equilibria with reserve prices. Similar extensions were shown for unit demand valuations in [8].

We first define a suitable linear program. The RESERVE-LP(r) is a linear program obtained from a reformulation of the dual of the LP-relaxation of the welfare maximization integer program after adding reserve prices $r \geq 0$. More details on this reformulation are given in the full version of the paper.

For an integral solution to the RESERVE-LP(r) we can interpret this reformulation as a solution to a WELFARE-LP with an additional additive player whose value for an item is equal to that item's reserve price. We will use this interpretation to extend known results for Walrasian equilibria to Walrasian equilibria with reserve prices. The results are summarized in Theorem 4 below. We use the following definition.

Definition 8 (Additional Additive Player). *Let $\{v_b\}$ be a set of valuation functions over Ω for bidders $b \in \mathcal{B}$, and let $r \geq 0$ be reserve prices for the items in Ω. Let $\{v'_{b'}\}$ be the set of valuation functions when an additive bidder a is added, i.e., for the bidders $b' \in \mathcal{B}' = \mathcal{B} \cup \{a\}$ with $v'_{b'}(S) = v_{b'}(S)$ for $b' \neq a$ and $v'_a(S) = \sum_{j \in S} r_j$ for all sets $S \subseteq \Omega$. For an allocation $\langle \Omega_b \rangle_{b \in \mathcal{B}}$ we define $\langle \Omega'_{b'} \rangle_{b' \in \mathcal{B}'}$ with $\Omega'_{b'} = \Omega_{b'}$ for $b' \neq a$ and $\Omega'_a = \Omega \setminus \cup_b \Omega_b$.*

Theorem 4. *(a) The allocation $\langle \Omega_b \rangle$ and the prices p are a RESERVE-WE(r) for $r \geq 0$ and bidders $\mathcal{B}$ if and only if the allocation $\langle \Omega'_{b'} \rangle$ and prices p' are a WE for the bidders $\mathcal{B}'$, where we have $p_j = p'_j$ for $j \in \cup_{b \in \mathcal{B}} \Omega_b$ and $p_{j'} = r_{j'}$ for $j' \in \Omega \setminus \cup_{b \in \mathcal{B}} \Omega_b$ (a1). The allocation $\langle \Omega_b \rangle$ is a RESERVE-WE(r) allocation if and only if $\langle \Omega_b \rangle$ is an integral solution to the RESERVE-LP(r) (a2).*

---

[6] Such a player was introduced by Paes Leme [17] to find the demand of an OR-player (with a slightly different definition of OR).

(b) If the valuations $\{v\}$ are gross substitute, then (b1) there exists a RESERVE-WE(r) for $\{v\}$ and (b2) the RESERVE-WE(r) price vectors form a complete lattice.

Theorem 4 will be used in the next section to characterize the outcome of the virtual auctions of an EF-mediator. It also provides a polynomial-time algorithm to compute a RESERVE-WE(r) when the bidders in $\mathcal{B}$ have gross-substitute valuations, given a polynomial-time algorithm for a WE for gross-substitute bidders.

4.2 The Equivalence of the EF-mediator and the OR-player for Gross-substitute Valuations—Proof Outline

In this section we outline the proof of Theorem 3, the complete proof can be found in the full version of the paper. The proof proceeds as follows. We first characterize the demand of an EF-mediator for bidders with gross-substitute valuations. As a first step we show that for such bidders an EF-mediator actually computes a RESERVE-WE(r) with minimum prices in each of her virtual auctions. The minimality of the prices implies that whenever the virtual auction prices for an item set S are globally envy-free, they are also minimum RESERVE-WE(r) prices for the set of all items Ω and the bidders in $\mathcal{B}_i$. Thus, given reserve prices r, all virtual auctions of an EF-mediator result in the same price vector p as long as they are run on a set S with non-negative revenue. With the help of some technical lemmata we then completely characterize the demand of an EF-mediator and show that the mediator does not have to run *multiple* virtual auctions to determine her demand; it suffices to run *one* virtual auction on Ω where the set of allocated items is a set in the demand of the EF-mediator. Thus for gross-substitute bidders the mediator can efficiently answer demand queries and compute the outcome of her local auction.

Finally we compare the utility function of the OR-player to the optimal value of the RESERVE-LP(r) to observe that they have to be equal (up to an additive constant) for item sets that are in the demand of the OR-player. Combined with the above characterization of the demand of the mediator, we can then relate both demands at central auction prices r to optimal solutions of the RESERVE-LP(r) for r and Ω and hence show the equality of the demands for these two mediator definitions for gross-substitute valuations of the bidders. Recall that an OR-player over gross-substitute valuations has a gross-substitute valuation [12]. Thus in this case we can regard the EF-mediator as having a gross-substitute valuation. This implies that a Walrasian equilibrium for the central auction exists and, with the efficient demand oracle defined above, can be computed efficiently when all bidders have gross-substitute valuations and all mediators are EF-mediators.

4.3 Computing an Equilibrium

The basic three-party auction is simple: First run the central auction at the exchange, then the local auctions at the mediators. In this section we summarize the details and analyze the time needed to compute a three-party competitive equilibrium. We assume that all bidders have gross-substitute valuations and that their valuations can be accessed via a demand oracle. We assume, for simplicity, that there are m EF-mediators, each with n/m distinct bidders. We will use known polynomial-time auctions for the two-party allocation problem, see [17] for a recent survey. Theorem 4 shows how such an auction can be modified to yield a RESERVE-WE(r) instead of a Walrasian equilibrium.

Let A be a polynomial-time algorithm that can access n gross-substitute valuations over k items Ω via a demand oracle and outputs a Walrasian price vector $p \in \mathbb{R}^k$ and a Walrasian allocation $\langle \Omega_i \rangle_{i \in [n]}$. Let the runtime of A be $T(n, k) = O(n^\alpha k^\beta)$ for constants α, β.

Although we can assume oracle access to the bidders' valuations, we cannot assume it for the mediators' (gross-substitute) valuations, as they are not part of the input. However, as outlined in the previous section, a mediator can determine a set in her demand by running a single virtual auction to compute a RESERVE-WE(r), i.e., there is an efficient demand oracle for the mediators. Hence, solving the allocation problem for the central auction can be done in time $T(m, k) \cdot T(n/m, k) = O(n^\alpha k^{2\beta})$. Further, the local auctions for all mediators take time $O(m \cdot T(n/m, k))$ and thus the total time to compute a three-party competitive equilibrium is $O(n^\alpha k^{2\beta})$.[7]

5   Short Discussion

We proposed a new model for auctions at ad exchanges. Our model is more general than previous models in the sense that it takes the incentives of all three types of participants into account and that it allows to express preferences over multiple items. Interestingly, at least when gross-substitute valuations are considered, this generality does not come at the cost of tractability, as shown by our polynomial-time algorithm. Note that this is the most general result we could expect in light of the classical (two-sided) literature on combinatorial auctions.[8]

Acknowledgment. We wish to thank Noam Nisan for helpful discussions.

References

1. Balseiro, S., Besbes, O., Weintraub, G.Y.: Auctions for online display advertising exchanges: approximations and design. In: Proceedings of the Fourteenth ACM Conference on Electronic Commerce (EC 2013), pp. 53–54. ACM, New York (2013)
2. Balseiro, S.R., Feldman, J., Mirrokni, V.S., Muthukrishnan, S.: Yield optimization of display advertising with ad exchange. Manage. Sci. **60**(12), 2886–2907 (2014). Preliminary version in EC 2011

---

[7] We additionally discuss the case $k = o(\log n)$ in the full version of the paper.

[8] We elaborate and sketch new directions in the full version of the paper.

3. Blumrosen, L., Nisan, N.: Combinatorial auctions. In: Nisan, N., Roughgarden, T., Tardos, É., Vazirani, V.V. (eds.) Algorithmic Game Theory, Chap. 11, pp. 267–299. Cambridge University Press, Cambridge (2007)
4. Business Insider: How Massive Real-Time Bidding Has Gotten in Online Advertising. http://www.businessinsider.com/dollar-numbers-for-real-time-bidding-in-digital-advertising-2012-6. Accessed 28 July 2014
5. Feldman, J., Mirrokni, V., Muthukrishnan, S., Pai, M.M.: Auctions with intermediaries: extended abstract. In: Proceedings of the 11th ACM Conference on Electronic Commerce (EC 2010), pp. 23–32. ACM, New York (2010)
6. GDN: About the Display Network ad auction. https://support.google.com/adwords/answer/2996564. Accessed 30 January 2014
7. Gul, F., Stacchetti, E.: Walrasian equilibrium with gross substitutes. J. Econ. Theor. **87**(1), 95–124 (1999)
8. Guruswami, V., Hartline, J.D., Karlin, A.R., Kempe, D., Kenyon, C., McSherry, F.: On profit-maximizing envy-free pricing. In: Sixteenth Annual ACM-SIAM Symposium on Discrete Algorithms (SODA), pp. 1164–1173. SIAM, Philadelphia (2005)
9. Hatfield, J.W., Kominers, S.D., Nichifor, A., Ostrovsky, M., Westkamp, A.: Stability and competitive equilibrium in trading networks. J. Polit. Econ. **121**(5), 966–1005 (2013)
10. Hatfield, J.W., Milgrom, P.R.: Matching with contracts. Am. Econ. Rev. **95**(4), 913–935 (2005)
11. Kelso, A.S.J., Crawford, V.P.: Job matching, coalition formation, and gross substitutes. Econometrica **50**(6), 1483–1504 (1982)
12. Lehmann, B., Lehmann, D.J., Nisan, N.: Combinatorial auctions with decreasing marginal utilities. Games Econ. Behav. **55**(2), 270–296 (2006)
13. Mansour, Y., Muthukrishnan, S., Nisan, N.: Doubleclick ad exchange auction (2012). http://arxiv.org/abs/1204.0535
14. Muthukrishnan, S.: Ad exchanges: research issues. In: Leonardi, S. (ed.) WINE 2009. LNCS, vol. 5929, pp. 1–12. Springer, Heidelberg (2009)
15. Nisan, N., Segal, I.: The communication requirements of efficient allocations and supporting prices. J. Econ. Theor. **129**, 192–224 (2006)
16. Ostrovsky, M.: Stability in supply chain networks. Am. Econ. Rev. **98**(3), 897–923 (2008)
17. Paes Leme, R.: Gross substitutability: an algorithmic survey (2014). http://www.renatoppl.com/papers/gs-survey-jul-14.pdf
18. Stavrogiannis, L.C., Gerding, E.H., Polukarov, M.: Competing intermediary auctions. In: International Conference on Autonomous Agents and Multi-Agent Systems (AAMAS 2013), pp. 667–674. International Foundation for Autonomous Agents and Multiagent Systems, Richland (2013)
19. Stavrogiannis, L.C., Gerding, E.H., Polukarov, M.: Auction mechanisms for demand-side intermediaries in online advertising exchanges. In: International Conference on Autonomous Agents and Multi-Agent Systems (AAMAS 2014), pp. 1037–1044. International Foundation for Autonomous Agents and Multiagent Systems, Richland (2014)
20. Walras, L.: Éléments d'Économie Politique Pure, ou Théorie de la Richesse Sociale. Corbaz, Lausanne (1874)

Computing Approximate Nash Equilibria in Network Congestion Games with Polynomially Decreasing Cost Functions

Vittorio Bilò[1], Michele Flammini[2,3], Gianpiero Monaco[2],
and Luca Moscardelli[4(✉)]

[1] Department of Mathematics and Physics, University of Salento, Lecce, Italy
vittorio.bilo@unisalento.it
[2] DISIM - University of L'Aquila, L'Aquila, Italy
{michele.flammini,gianpiero.monaco}@univaq.it
[3] Gran Sasso Science Institute, L'Aquila, Italy
[4] Department of Economic Studies, University of Chieti-Pescara, Pescara, Italy
luca.moscardelli@unich.it

Abstract. We consider the problem of computing approximate Nash equilibria in monotone congestion games with polynomially decreasing cost functions. This class of games generalizes the one of network congestion games, while polynomially decreasing cost functions also include the fundamental Shapley cost sharing value. We design an algorithm that, given a parameter $\gamma > 1$ and a subroutine able to compute ρ-approximate best-responses, outputs a $\gamma(1/p + \rho)$-approximate Nash equilibrium, where p is the number of players. The computational complexity of the algorithm heavily depends on the choice of γ. In particular, when $\gamma \in O(1)$, the complexity is quasi-polynomial, while when $\gamma \in \Omega(p^\epsilon)$, for a fixed constant $\epsilon > 0$, it becomes polynomial. Our algorithm provides the first non-trivial approximability results for this class of games and achieves an almost tight performance for network games in directed graphs. On the negative side, we also show that the problem of computing a Nash equilibrium in Shapley network cost sharing games is PLS-complete even in undirected graphs, where previous hardness results where known only in the directed case.

1 Introduction

In the last years, with the advent of the Internet, considerable research interest has been devoted to modeling and analysing behavioral dynamics in non-cooperative networks, where selfish users compete for limited resources. One of the most natural and investigated frameworks in this setting is that of the congestion games introduced by Rosenthal [16], in which the set of the strategies of each player corresponds to some collection of subsets of a given set of common resources (when the set of resources is given by the set of edges of a graph, we speak of network congestion games). The cost of a strategy is the sum of the costs of the selected resources, where the cost of each single resource depends

© Springer-Verlag Berlin Heidelberg 2015
E. Markakis and G. Schäfer (Eds.): WINE 2015, LNCS 9470, pp. 118–131, 2015.
DOI: 10.1007/978-3-662-48995-6_9

on its congestion, i.e., the number of players using it. These games are particularly suited to model selfish routing in unregulated networks as well as the phenomenon of spontaneous network formation. In the former case, the cost of each resource is assumed to increase with its congestion, whereas in the latter the cost decreases. Rosenthal proved that each congestion game admits an exact potential function, that is, a real-valued function on the set of all possible strategy profiles such that the difference in the potential between two strategy profiles that differ in the strategic choice of a single player is equal to the difference of the costs experienced by this player in the two profiles. This implies that each congestion game has a pure Nash equilibrium [14] (from now on, simply, Nash equilibrium), that is a strategy profile in which no player can decrease her cost by unilaterally deviating to a different strategy. Moreover, it also follows that a Nash equilibrium can always be reached by the so called Nash dynamics, that is, the iterative process in which, at each step, an unsatisfied player is allowed to switch to a better strategy, i.e. lowering her cost. Monderer and Shapley [13] proved that the class of congestion games and that of exact potential games are equivalent. Questions related to the performance of Nash equilibria, such as price of anarchy [12] and price of stability [3], have been extensively addressed in the context of congestion games and some of their variations. In this paper, we focus on the problem of computing (approximate) Nash equilibria in a variant of congestion games, that we name *monotone congestion games*, which includes the class of the network congestion games.

Related Work. The complexity of computing (approximate) Nash equilibria in (network) congestion games with increasing cost functions is fairly well-understood. Fabrikant et al. [8] prove that computing a Nash equilibrium is PLS-complete in both network games and general games in which the players share the same set of strategies (symmetric games). Ackermann et al. [1] shows that PLS-completeness holds even for network games with linear cost functions. For the special case of symmetric network congestion games, instead, Fabrikant et al. [8] show how to efficiently compute a Nash equilibrium by solving a min-cost flow problem. These impossibility results together with the fact that in many real-life applications it may be the case that a player incurs a cost for changing her strategy, have naturally led researchers to consider the notion of approximate Nash equilibria. An ϵ-approximate Nash equilibrium, for some $\epsilon \geq 1$, is a strategy profile in which no player can improve her payoff by a multiplicative factor of at least ϵ by unilaterally deviating to a different strategy. Skopalik and Vöcking [17] prove that computing an ϵ-approximate Nash equilibrium is PLS-complete for any polynomial time computable $\epsilon \geq 1$. Moreover they also show that ϵ-approximate better-response dynamics can require an exponential number of steps to converge. This PLS-hardness result implies that any dynamics for reaching an approximate equilibrium that requires only polynomial-time computation per iteration does not always converge in a polynomial number of iterations, unless PLS $\subseteq$ P. On the positive side, Chien and Sinclair [7] show that, in symmetric games with positive delays, an ϵ-approximate Nash equilibrium can be reached after a polynomial number of steps of the Nash dynamics.

Chen et al. [6] extended the result of [7] to symmetric games with only negative delays, while showed that the problem becomes PLS-complete if both positive and negative delays are present. Finally, Caragiannis et al. [4] provide a polynomial time algorithm that computes a $(2 + \gamma)$-approximate Nash equilibrium (for arbitrary small $\gamma > 0$) for games with linear cost functions and a $d^{O(d)}$-approximate Nash equilibrium for games with polynomial cost functions with constant maximum degree d.

Much less is known about the complexity of computing (approximate) Nash equilibria in congestion games with decreasing cost functions. Indeed, there are results only for special cases of network games under the **Shapley cost sharing protocol** [3], i.e., games in which each resource has a cost which is equally split among all the players using it. Specifically, given a graph each player i wants to connect a pair of nodes (s_i, t_i). Syrgkanis [18] shows that computing a Nash equilibrium in these games is PLS-complete if the underlying graph is directed. Albers and Lenzner [2] consider the case in which the graph is undirected and $s_i = s$ for each player i and prove that any optimal network (i.e., a minimum cost Steiner tree whose set of terminals is given by the union of the source-destination nodes of all the players) is an H_p-approximate Nash equilibrium, where p is the number of players. We stress that the algorithm proposed by Caragiannis et al. [4] cannot be extended to deal with decreasing cost functions since it strongly exploits the following facts (here, linear functions are considered, but a similar discussion can be made for polynomial functions): *(i)* if, given a game, some players are frozen on some strategies, the resulting game is a congestion game with less players and different cost functions, but always belonging the the class of linear cost functions; *(ii)* the potential value of an ϵ-approximate Nash equilibrium approximates the minimum of the potential function by a factor or order ϵ. It can be easily verified as neither property *(i)* nor property *(ii)* are verified by congestion games with decreasing cost functions. Thus, there exists a tremendous theoretical gap between positive and negative results regarding the computability of (approximate) Nash equilibria in network games under the Shapley cost sharing protocol.

Our Contribution. Our aim is to address the class of network congestion games with polynomially decreasing cost functions. Since network games have the property that, whenever a given strategy (subset of edges) s fulfills the connectivity requirement of a player, then any superset $s' \supset s$ with redundant edges also does, we introduce the general class of the **monotone congestion games**, that properly includes that of network congestion games. In these games, the strategy set of each player is closed under resource addition. We remark that this feature, while indifferent for plain Nash equilibria, as no strategy at equilibrium will contain redundant resources, can be profitably exploited in order to get low factor approximate Nash equilibria.

Building upon this observation, we consider the complexity of computing (approximate) Nash Equilibria in monotone congestion games with polynomially decreasing cost functions, providing both positive and negative results. In particular, we consider the case in which each resource j has a cost c_j and the cost that each player incurs when using j is given by c_j/x^α for some constant $\alpha > 0$, where x is the number of players using j. Observe that, for $\alpha = 1$, we recover the Shapley cost sharing protocol, so that our model widely generalizes the models previously studied in the literature.

On the positive side, we design an algorithm that computes an approximate Nash equilibrium with provable performance guarantee. In particular, assuming that a ρ-approximate best-response for the game can be computed in time t, for a fixed parameter $\gamma \in (1, rp)$, our algorithm returns a $\gamma(p^{-1} + \rho)$-approximate Nash equilibrium in time $t \cdot (rp)^{O\left(\frac{\alpha \log p}{\log \gamma}\right)}$, where r is the number of resources and p the number of players. Such a result has several implications for network congestion games, that we describe in detail in Sect. 4. On the negative side, we prove that the problem of computing a Nash equilibrium in network games under the Shapley cost sharing protocol is PLS-complete even for undirected graphs, thus closing a long standing open question which had been answered before by Syrgkanis [18] only for the case of directed graphs.

2 Model

A *congestion game with polynomially decreasing latency functions* is a tuple $(P, R, (S_i)_{i \in P}, (c_j)_{j \in R}, \alpha)$, where P is a set of p players, R a set of r resources, $S_i \subseteq 2^R \setminus \{\emptyset\}$ is the set of strategies of player $i \in P$, $c_j > 0$ is the cost of resource $j \in R$ and $\alpha > 0$ is a real value. A strategy profile $\boldsymbol{\sigma} = (\sigma_1, \ldots, \sigma_p)$ is the outcome in which each player $i \in P$ chooses strategy $\sigma_i \in S_i$. The cost that player i experiences in strategy profile $\boldsymbol{\sigma}$ is defined as $cost_i(\boldsymbol{\sigma}) = \sum_{j \in \sigma_i} \frac{c_j}{p_j(\boldsymbol{\sigma})^\alpha}$, where $p_j(\boldsymbol{\sigma}) = |\{i \in P : j \in \sigma_i\}|$ denotes the number of players using resource j in $\boldsymbol{\sigma}$.

Given a strategy profile $\boldsymbol{\sigma}$, a player $i \in P$ and a strategy $S \in S_i$, we denote with $\boldsymbol{\sigma}_{-i} \diamond S$ the strategy profile obtained from $\boldsymbol{\sigma}$ when i switches from strategy σ_i to strategy S. A strategy profile $\boldsymbol{\sigma}$ is a Nash equilibrium if, for each $i \in P$ and $S \in S_i$, $cost_i(\boldsymbol{\sigma}) \leq cost_i(\boldsymbol{\sigma}_{-i} \diamond S)$, that is, no player can lower her cost by unilaterally switching to a different strategy. More generally, given a value $\epsilon \geq 1$, $\boldsymbol{\sigma}$ is an ϵ-approximate Nash equilibrium if, for each $i \in P$ and $S \in S_i$, $cost_i(\boldsymbol{\sigma}) \leq \epsilon \cdot cost_i(\boldsymbol{\sigma}_{-i} \diamond S)$, that is, no player can lower her cost of a factor of more than ϵ by unilaterally switching to a different strategy. Clearly, a Nash equilibrium is an ϵ-approximate Nash equilibrium with $\epsilon = 1$.

For a fixed strategy profile $\boldsymbol{\sigma}$ and a player $i \in P$, a *best-response* for i in $\boldsymbol{\sigma}$ is a strategy $br_i(\boldsymbol{\sigma}) \in S_i$ such that $cost_i(\boldsymbol{\sigma}_{-i} \diamond br_i(\boldsymbol{\sigma})) = \min_{S \in S_i} \{cost_i(\boldsymbol{\sigma}_{-i} \diamond S)\}$. A strategy $S \in S_i$ is an ϵ-approximate best-response for i in $\boldsymbol{\sigma}$ if $cost_i(\boldsymbol{\sigma}_{-i} \diamond S) \leq \epsilon \cdot cost_i(\boldsymbol{\sigma}_{-i} \diamond br_i(\boldsymbol{\sigma}))$. It follows that $\boldsymbol{\sigma}$ is an ϵ-approximate Nash equilibrium if and only if each player $i \in P$ is playing an ϵ-approximate best-response. Given a game $\mathcal{G}$, we denote with $\Pi_{br}(\mathcal{G})$ the problem of computing a best-response for a generic player $i \in P$ in a generic strategy profile $\boldsymbol{\sigma}$ of $\mathcal{G}$. As we will see, there are games for which such a problem is in P and others for which it is NP-hard.

Let $\Phi : \times_{i \in P} S_i \to \mathbb{R}_+$ be the function such that $\Phi(\boldsymbol{\sigma}) = \sum_{j \in R} \sum_{i=1}^{p_j(\boldsymbol{\sigma})} \frac{c_j}{i^\alpha}$. Function Φ, originally defined by Rosenthal [16] for the general class of the congestion games, is an *exact potential function*, that is $\Phi(\sigma_{-i} \diamond S) - \Phi(\sigma) = cost_i(\sigma_{-i} \diamond S) - cost_i(\sigma)$ for each strategy profile σ and strategy $S \in \mathcal{S}_i$.

We say that a congestion game is *monotone* if, for each player $i \in P$, the set of strategies $\mathcal{S}_i$ is closed under resource addition, that is, for each $S \in \mathcal{S}_i$ and $R' \subseteq R$, it holds that $S \cup R' \in \mathcal{S}_i$. In such a case, given two strategies $S, S' \in \mathcal{S}_i$, we say that S' is *dominated* by S if $S \subset S'$. Note that, by definition, no player ever plays a dominated strategy in a Nash equilibrium, whereas, for suitable values of ϵ, this cannot be a priori excluded in an ϵ-approximate Nash equilibrium.

An interesting and well-studied class of monotone congestion games is represented by *network congestion games*. In these games, the set of resources is represented by the set of edges of a weighted graph $G = (V, E, c)$ (thus $|E| = r$), where $|V| = n$, which may be either directed or undirected, and the set of strategies for each player is defined in intensive form as follows: each player $i \in P$ is associated with a set of k_i subsets of nodes $V_{ij} \subseteq V$ for $j = 1, \ldots, k_i$ that she wishes to serve in some way (for instance, by providing them a connection, a fault-tolerant connection, a point-to-point connection, etc.). Note that differently from classical congestion game, here the representation of the game does not need to keep the whole set of strategies explicitly. Several interesting subcases, all of them assuming $k_i = 1$ for any $i \in P$, have been considered so far in the literature, such as *multi-source games* in which $|V_{i1}| = |V_i| = 2$ for each $i \in P$, *multicast games* in which, in addition, $V_i \cap V_{i'} = \{s\}$ for each $i, i' \in P$ and *broadcast games* in which, as a further restriction, one has $\bigcup_{i \in P} V_i = V$. Finally, *Shapley games* are congestion games with polynomially decreasing latency functions in which $\alpha = 1$. Note that $\Pi_{br}(\mathcal{G})$ is NP-hard for several network congestion games $\mathcal{G}$.

3   The Approximation Algorithm

Fix a monotone congestion game with polynomially decreasing latency functions $\mathcal{G}$. Throughout this section we assume, without loss of generality, that $\mathcal{G}$ is normalized in such a way that $c_j \geq 1$ for each $j \in R$. For any set of resources $R' \subseteq R$, define $c(R') = \sum_{j \in R'} c_j$ and, for each player $i \in P$, let $\mathsf{S}_i^* = \operatorname{argmin}_{S \in \mathcal{S}_i} \{c(S)\}$ be the minimum cost strategy for player i. Finally, let $c_{max} = \max_{j \in R}\{c_j\}$. For a fixed parameter $\gamma \in (1, rp)$, set $\Delta = \frac{rp^{\alpha+1}}{\gamma}$. Since both $\gamma > 1$ and $\Delta > 1$, it follows that $\log_\gamma x = O(\log x)$ and $\log_\Delta x = O(\log x)$ for each $x \in \mathbb{R}$.

Definition 1. *A player i belongs to class cl_ℓ^P (or is of class cl_ℓ^P) if $c(\mathsf{S}_i^*) \in (\Delta^{\ell-1}, \Delta^\ell]$, while a resource j belongs to class cl_ℓ^R if $c_j \in (\Delta^{\ell-1}, \Delta^\ell]$.*

Note that the maximum number of player classes is $L = \lceil \log_\Delta(r \cdot c_{max}) \rceil = O(\log r + \log c_{max})$ which is polynomial in the dimension of $\mathcal{G}$. We denote by $p_\ell = |cl_\ell^P|$ the number of players of class cl_ℓ^P, by $\overline{p}_\ell = \sum_{\ell'=\ell+3}^{L} p_{\ell'}$ the number of players of class at least $\ell + 3$ and set $P_\ell = \sum_{\ell'=\ell-2}^{\ell+2} p_{\ell'}$, where we use the convention $p_{-1} = p_0 = p_{L+1} = p_{L+2} = 0$.

Our approximation algorithm is based on the following idea: the strategy σ_i chosen by each player $i \in P$ is composed of two parts: a fixed strategy, denoted as FS_i, which is independent of what the other players do and is kept fixed throughout the whole execution of the algorithm, and a variable strategy, denoted as VS_i, which may change over time and expresses the player's reaction to the choices of the others.

For a player $i \in cl_\ell^P$, let $cl(i)$ denote the index of the player class containing i, i.e., $cl(i) = \ell$. Similarly, for a resource $j \in cl_\ell^R$, let $cl(j)$ denote the index of the resource class containing j, i.e., $cl(j) = \ell$. The fixed strategy of each player $i \in P$ is defined as follows:

$$\mathsf{FS}_i = \begin{cases} \emptyset & \text{if } cl(i) \in \{1,2\}, \\ \bigcup_{\ell=1}^{cl(i)-2} cl_\ell^R & \text{if } cl(i) \geq 3. \end{cases}$$

The set of variable strategies available to player $i \in P$ is

$$\mathcal{VS}_i = \{S \subseteq R \mid \exists S' \in \mathcal{S}_i : S = S' \setminus \mathsf{FS}_i\}.$$

We denote with $\mathsf{VS}_i^* = \mathrm{argmin}_{S \in \mathcal{VS}_i}\{c(S)\}$ the minimum cost variable strategy for player i, while, for a given strategy profile σ, we denote with

$$br_i^V(\sigma) = \mathrm{argmin}_{S \in \mathcal{VS}_i}\{cost_i(\sigma_{-i} \diamond (S \cup \mathsf{FS}_i))\}$$

the variable strategy for player i of minimum shared cost. Since $\Pi_{br}(\mathcal{G})$ may be NP-hard, we assume that each player $i \in P$ can compute a ρ-approximate best-response, denoted with $\widetilde{br}_i^V(\sigma)$, by using an algorithm of time complexity $t := t(\mathcal{G})$. For the sake of brevity, when given a strategy profile σ with $\sigma_i = \mathsf{FS}_i \cup \mathsf{VS}_i$ for each player $i \in P$, we denote with $cost_i^V(\sigma) = \sum_{j \in \sigma_i \setminus \mathsf{FS}_i} \frac{c_j}{p_j(\sigma)^\alpha}$ the cost experienced by player i in σ ascribable to resources belonging to her variable strategy only.

Algorithm 1. It takes as input a monotone congestion game with polynomially decreasing latency functions and a parameter γ and outputs a $\gamma(p^{-1} + \rho)$-approximate Nash equilibrium.

1: **for each** $i \in P$ **do**
2: $\quad \sigma_i^0 \leftarrow \mathsf{FS}_i \cup \mathsf{VS}_i^*$
3: **end for**
4: $h \leftarrow 0$
5: **while** there exists a player i s.t. $cost_i^V(\sigma^h) > \gamma \cdot cost_i^V(\sigma_{-i}^h \diamond (\widetilde{br}_i^V(\sigma^h) \cup \mathsf{FS}_i))$ **do**
6: $\quad h \leftarrow h + 1$
7: $\quad \sigma^h \leftarrow \sigma_{-i}^{h-1} \diamond (\mathsf{FS}_i \cup \widetilde{br}_i^V(\sigma^{h-1}))$
8: **end while**
9: **return** σ^h

The following theorem bounds the performance of Algorithm 1 as a function of $\mathcal{G}$, ρ, t and γ.

Theorem 1. *For each $\gamma \in (1, rp)$, Algorithm 1 returns a $\gamma(p^{-1}+\rho)$-approximate Nash equilibrium in time $t \cdot (rp)^{O\left(\frac{\alpha \log p}{\log \gamma}\right)}$.*

Proof. For a strategy profile σ and a player $i \in P$, denote with $\mathsf{VS}_i(\sigma)$ the variable strategy adopted by player i in σ. We start by proving some useful facts. The first one says that, during the execution of the algorithm, a resource of class ℓ may belong to the variable strategies of players of class $\ell - 1$, ℓ or $\ell + 1$ only.

Fact 1. *Fix a resource $j \in cl_\ell^R$ and a strategy profile σ^h generated during the execution of Algorithm 1. Then, for any player i such that $j \in \mathsf{VS}_i(\sigma^h)$, $cl(i) \in \{\ell - 1, \ell, \ell + 1\}$.*

Proof. For a fixed resource $j \in cl_\ell^R$, let us assume, by way of contradiction, that there exist a strategy profile σ^h generated during the execution of Algorithm 1 and a player i such that $j \in \mathsf{VS}_i(\sigma^h)$, for which $cl(i) \notin \{\ell - 1, \ell, \ell + 1\}$.

Consider first the case in which $cl(i) > \ell + 1$. This implies that $cl_\ell^R \subseteq \mathsf{FS}_i$ which contradicts the hypothesis that $j \in \mathsf{VS}_i(\sigma^h)$.

Then consider the case in which $cl(i) < \ell - 1$. This implies that $c(\mathsf{S}_i^*) \leq \Delta^{\ell-2}$ so that, by construction of σ^0, it has to be $h > 0$. Since resource j is of class ℓ, its cost has to be greater than $\Delta^{\ell-1}$. Moreover, since j can be shared by at most p (i.e., all) players, $cost_i^V(\sigma^h) > \frac{\Delta^{\ell-1}}{p^\alpha}$. On the one hand, by line 7 of the algorithm, we have $j \in br_i^V(\sigma^{h-1})$; on the other hand, since $\gamma < rp$, $cost_i^V(\sigma^h) > \frac{\Delta^{\ell-1}}{p^\alpha} > \Delta^{\ell-2} \geq c(\mathsf{S}_i^*)$, thus contradicting the fact that $br_i^V(\sigma^{h-1})$ is the variable strategy for player i of minimum shared cost. □

The second fact says that, during the execution of the algorithm, a resource belonging to any variable strategy of a player of class ℓ may belong to the variable strategies of players of class $\ell - 2$, $\ell - 1$, ℓ, $\ell + 1$ or $\ell + 2$ only.

Fact 2. *Fix a strategy profile σ^h generated during the execution of Algorithm 1, a player $i \in cl_\ell^P$ and a resource $j \in \mathsf{VS}_i(\sigma^h)$. Then, for any player i' such that $j \in \mathsf{VS}_{i'}(\sigma^h)$, $cl(i') \in \{\ell - 2, \ell - 1, \ell, \ell + 1, \ell + 2\}$.*

Proof. By applying twice Fact 1, we first obtain that $cl(j) \in \{\ell - 1, \ell, \ell + 1\}$ and then that $cl(i') \in \{\ell - 2, \ell - 1, \ell, \ell + 1, \ell + 2\}$. □

We now partition the set of player classes into two subsets: light and heavy classes. Both of them will be proved to have different properties that can be suitable exploited in order to show the correctness and the running time of our algorithm.

Definition 2. *A player class cl_ℓ^P is light if $\overline{p}_\ell \geq \frac{\overline{p}_\ell + P_\ell}{\gamma^{1/\alpha}}$, otherwise it is heavy.*

The good property of a light class cl_ℓ^P is that the $\overline{p}_\ell$ players whose fixed strategies contain all the resources that may belong to the variable strategy of any player $i \in cl_\ell^P$ are enough to guarantee that during the execution of the algorithm no player $i \in cl_\ell^P$ ever performs a deviation; so that $\sigma_i^h = \mathsf{FS}_i \cup \mathsf{VS}_i^*$ for each strategy profile σ^h generated during the execution of the algorithm. Let us denote with $\widetilde{P}$ the set of players who are selected at least once at step 5 of Algorithm 1 for updating their strategy.

Fact 3. *For each player $i \in \widetilde{P}$, player class $cl_{cl(i)}^P$ is heavy.*

Proof. Assume, for the sake of contradiction, that there exists a player $i \in \widetilde{P}$ such that $cl_{cl(i)}^P$ is light. Let σ^h be strategy profile at which i performs her first deviation from strategy $\mathsf{FS}_i \cup \mathsf{VS}_i^*$. By line 5 of the algorithm, it must be

$$cost_i^V(\sigma^h) > \gamma \cdot cost_i^V(\sigma_{-i}^h \diamond (br_i^V(\sigma^h) \cup \mathsf{FS}_i)). \tag{1}$$

By $\sigma_i^h = \mathsf{FS}_i \cup \mathsf{VS}_i^*$, we have

$$cost_i^V(\sigma^h) \leq \frac{c(\mathsf{VS}_i^*)}{\overline{p}_{cl(i)}^\alpha}, \tag{2}$$

while, by the minimality of VS_i^* and Fact 2, we get

$$cost_i^V(\sigma_{-i}^h \diamond (br_i^V(\sigma^h) \cup \mathsf{FS}_i)) \geq \frac{c(\mathsf{VS}_i^*)}{(\overline{p}_{cl(i)} + P_{cl(i)})^\alpha}. \tag{3}$$

By using (2) and (3) in (1), we obtain

$$\frac{c(\mathsf{VS}_i^*)}{\overline{p}_{cl(i)}^\alpha} > \frac{\gamma c(\mathsf{VS}_i^*)}{(\overline{p}_{cl(i)} + P_{cl(i)})^\alpha},$$

which, after rearranging, yields $\gamma \overline{p}_{cl(i)}^\alpha < (\overline{p}_{cl(i)} + P_{cl(i)})^\alpha$ thus contradicting the assumption that $cl_{cl(i)}^P$ is light. $\qquad\square$

By Fact 3, we know that only players belonging to heavy classes are involved in the dynamics generated by lines 5–9 of Algorithm 1. Our next task is to bound the number of deviations in this dynamics. To this aim, we first bound the number of heavy classes. The peculiar property of a heavy class cl_ℓ^P is that P_ℓ, that is the number of players belonging to classes going from $cl_{\ell-2}^P$ to $cl_{\ell+2}^P$, is a significant fraction of the $\overline{p}_{\ell+3}$ players belonging to the classes of index at least $\ell + 3$, so that the total number of heavy classes remains bounded as a function of p, γ and α as follows.

Lemma 1. *The number of heavy classes is at most $5 \left\lceil \frac{\alpha \log p}{\log \gamma} \right\rceil$.*

Proof. A *pseudo-sequence of heavy classes* is an ordered sequence $\langle cl_{\ell_1}^P, \ldots, cl_{\ell_k}^P \rangle$ of heavy classes whose mutual distance is at least five, that is, such that $\ell_{i+1} - \ell_i \geq 5$ for each $1 \leq i < k$. Clearly, denoted with k^* the length of the longest pseudo-sequence of heavy classes, it immediately follows that the total number of heavy classes is at most $5k^*$. Thus, in the remaining of the proof, we show that $k^* \leq \left\lceil \frac{\alpha \log p}{\log \gamma} \right\rceil$. Before bounding k^*, we need to introduce some useful facts.

Fact 4. *Fix a pseudo-sequence of heavy classes* $\langle cl_{\ell_1}^P, \ldots, cl_{\ell_k}^P \rangle$. *For each* $1 \leq i < k$, $\overline{p}_{\ell_i} + P_{\ell_i} \leq \overline{p}_{\ell_{i-1}}$.

Proof. By definition, we have $\overline{p}_{\ell_{i-1}} = \sum_{j=\ell_{i-1}+3}^{L} cl_j^P$ and

$$\overline{p}_{\ell_i} + P_{\ell_i} = \sum_{j=\ell_i+3}^{L} cl_j^P + \sum_{j=\ell_i-2}^{\ell_i+2} cl_j^P = \sum_{j=\ell_i-2}^{L} cl_j^P.$$

The claim then follows since $\ell_i - \ell_{i-1} \geq 5$ implies that $\ell_{i-1} + 3 \leq \ell_i - 2$. □

Fact 5. *Fix a pseudo-sequence of heavy classes* $\langle cl_{\ell_1}^P, \ldots, cl_{\ell_k}^P \rangle$. *For each* $1 \leq i \leq k$, $\overline{p}_{\ell_i} < \frac{p}{\gamma^{i/\alpha}}$.

Proof. The proof is by induction on i. As a base case, we have $\overline{p}_{\ell_1} < \frac{\overline{p}_{\ell_1} + P_{\ell_1}}{\gamma^{1/\alpha}} \leq \frac{p}{\gamma^{1/\alpha}}$, where the first inequality follows from the definition of heavy classes. For $i > 1$, we have $\overline{p}_{\ell_i} < \frac{\overline{p}_{\ell_i} + P_{\ell_i}}{\gamma^{1/\alpha}} < \frac{\overline{p}_{\ell_{i-1}}}{\gamma^{1/\alpha}} < \frac{p}{\gamma^{i/\alpha}}$, where the first inequality follows from the definition of heavy classes, the second one follows from Fact 4 and the last one follows from the inductive hypothesis. □

We can now conclude the proof of the lemma. Because of Fact 5, we have that $\overline{p}_{\ell_{k^*}} < \frac{p}{\gamma^{k^*/\alpha}}$. Assume that $k^* > \lceil \log_{\sqrt[\alpha]{\gamma}} p \rceil$. Since k^* is an integer, we get $\overline{p}_{\ell_{k^*-1}} < 1$, i.e., $\overline{p}_{\ell_{k^*-1}} = 0$. This means that all player classes with index at least $\ell_{k^*-1} + 3$ are empty. Since $\ell_{k^*} \geq \ell_{k^*-1} + 5$, it holds that $P_{\ell_{k^*}} = 0$ and therefore, by Definition 2, class ℓ_{k^*} is light, thus rising a contradiction. Hence it must be $k^* \leq \lceil \log_{\sqrt[\alpha]{\gamma}} p \rceil = \left\lceil \frac{\alpha \log p}{\log \gamma} \right\rceil$. □

We now proceed by grouping heavy and light classes into zones according to the following definition.

Definition 3. *A zone* z *is a maximal set of contiguous player classes, starting and ending with a heavy class and such that no two contiguous classes are both light.*

By Lemma 1 and the definition of zones, we have that there exist at most $5 \left\lceil \frac{\alpha \log p}{\log \gamma} \right\rceil$ zones (because each zone has to contain at least a heavy class) and that each zone contains at most $10 \left\lceil \frac{\alpha \log p}{\log \gamma} \right\rceil$ classes (because in the worst-case heavy and light classes are interleaved in a zone, given that two light classes cannot be contiguous). Moreover, again by the definition of zones and by Fact 2, it follows that, for any two players i and i' such that $cl_{cl(i)}^P$ and $cl_{cl(i')}^P$ are two heavy classes belonging to two different zones, $S \cap S' = \emptyset$ for each $S \in \mathcal{VS}_i$ and $S' \in \mathcal{VS}_{i'}$, that is, the congestion of the resources belonging to the variable strategies of the players in a certain zone are not influenced by the choices of the players outside the zone, so that the Nash dynamics inside each zone, generated by Algorithm 1, is an independent process.

Consider a zone z starting at class ℓ and containing x classes (from class ℓ to class $\ell + x - 1$). After line 3 of Algorithm 1, the total contribution to the potential $\Phi(\sigma^0)$ of the resources that may be involved in the dynamics of the players belonging to classes in z is $r\Delta^{\ell+x}p$, because, by Fact 1, any variable strategy of a player of class $\ell + x - 1$ can contain resources of classes $\ell + x - 2$, $\ell + x - 1$ and $\ell + x$. Since each deviation decreases the potential of at least $\frac{\Delta^\ell}{p^\alpha}\left(1 - \frac{1}{\gamma}\right)$, at most $\frac{r\Delta^{\ell+x}p}{\frac{\Delta^\ell}{p^\alpha}\left(1-\frac{1}{\gamma}\right)} = \frac{r\gamma\Delta^x p^{1+\alpha}}{\gamma-1}$ deviations can be performed by the players belonging to classes in z. Thus, since $x \leq 10\left\lceil\frac{\alpha\log p}{\log \gamma}\right\rceil$, $\gamma \in (1, rp)$ and each deviation requires time at most t, we have that Algorithm 1 terminates in time $t \cdot (rp)^{O\left(\frac{\alpha\log p}{\log \gamma}\right)}$.

It remains to show that it returns a $\gamma(p^{-1} + \rho)$-approximate Nash equilibrium. By the condition at line 5, we know that, with respect to the variable strategies, all players are in $\gamma\rho$-equilibrium in the strategy profile σ returned by Algorithm 1. Moreover, for any player i of class ℓ, the cost due to the resources in FS_i can be upper bounded by $r\Delta^{\ell-2}$. Since a best response for player i has to cost at least $\frac{\Delta^{\ell-1}}{p^\alpha}$ (in the most optimistic case the minimum cost strategy S_i^* is shared by all players), by recalling the definition of Δ, we have that

$$r\Delta^{\ell-2} \leq \frac{\gamma}{p} \cdot \frac{\Delta^{\ell-1}}{p^\alpha}. \tag{4}$$

Therefore,

$$
\begin{aligned}
cost_i(\sigma) &= \sum_{j \in \sigma_i \cap \mathsf{FS}_i} \frac{c_j}{p_j(\sigma)^\alpha} + \sum_{j \in \sigma_i \setminus \mathsf{FS}_i} \frac{c_j}{p_j(\sigma)^\alpha} \\
&\leq r\Delta^{\ell-2} + \sum_{j \in \sigma_i \setminus \mathsf{FS}_i} \frac{c_j}{p_j(\sigma)^\alpha} \\
&\leq \frac{\gamma}{p} \cdot \frac{\Delta^{\ell-1}}{p^\alpha} + \gamma\rho \cdot cost_i^V\left(\sigma_{-i} \diamond (br_i^V(\sigma) \cup \mathsf{FS}_i)\right) \tag{5} \\
&\leq \frac{\gamma}{p} \cdot cost_i(\sigma_{-i} \diamond br_i(\sigma)) + \gamma\rho \cdot cost_i(\sigma_{-i} \diamond br_i(\sigma)) \\
&= \gamma(p^{-1} + \rho)cost_i(\sigma_{-i} \diamond br_i(\sigma)),
\end{aligned}
$$

where (5) follows from (4) and from line 5 of Algorithm 1. $\square$

4   Applications to Network Congestion Games

In this section, we discuss the application of our approximation algorithm to network congestion games with polynomially decreasing latency functions. We recall that for these games no positive results have been achieved so far in the literature, except the ones provided by Albers and Lenzner [2] for some Shapley games on undirected graphs. In particular, they show that, for multicast games, a minimum cost network is a H_p-approximate Nash equilibrium. Here, a minimum cost network coincides with a minimum Steiner tree, whose computation is an NP-hard problem and the authors do not discuss whether an approximation of the minimum cost network can still provide an approximate Nash equilibrium with provably good performance guarantee. Thus, if we drift apart from merely existential results to focus on practical (i.e., efficiently computable) ones, we have that an H_p-approximate Nash equilibrium can be computed only in the case of broadcast Shapley games on undirected graphs.

We observe that, given a network congestion game $\mathcal{G}$ defined over a graph G, if there exists an algorithm computing a $\rho(G)$-approximate Nash equilibrium for $\mathcal{G}$ within time $t(G)$, then the same algorithm can be used to approximate $\Pi_{br}(\mathcal{G})$ within a factor of $\rho(G)$ with the same running time. To this aim, fix a graph G and a problem Π on G and define $\mathcal{G}$ in such a way that there is a unique player whose set of strategies is given by all the feasible solutions of problem Π. Clearly, a $\rho(G)$-approximate Nash equilibrium for $\mathcal{G}$ computed in time at most $t(G)$ provides also a $\rho(G)$-approximate solution to Π in time at most $t(G)$. Moreover, it is worth noticing that a similar result can be obtained for games with an arbitrary number p of players, by simply adding $p-1$ dummy players whose available strategies have all cost zero and involve edges not belonging to $E(G)$: such players cannot share any edge with the first player in any ϵ-approximate Nash equilibrium. It follows that, if we denote by $\rho_{LB}^t(\Pi)$ and $\rho_{KN}^t(\Pi)$ the lower bound on the approximability of problem Π and the performance guarantee of the best-known approximation algorithm for Π, respectively, when considering a time constraint t, each algorithm having a time complexity at most t cannot compute an approximate Nash equilibrium for game $\mathcal{G}$ with performance guarantee better than $\rho_{LB}^t(\Pi_{br}(\mathcal{G}))$ and that, given the current state-of-the-art, it is unlikely to design an algorithm of time complexity at most t computing an approximate Nash equilibrium for game $\mathcal{G}$ with performance guarantee better than $\rho_{KN}^t(\Pi_{br}(\mathcal{G}))$.

As an example, consider the family of games $\mathcal{FG}$ in which each player wants to connect a subset of nodes of a directed graph G. Since a fairly easy reduction from the set cover problem shows that it is hard to approximate the directed Steiner tree within a factor better than $O(\log n)$ [9], we obtain that, unless $NP \subseteq DTIME[n^{\log \log n}]$, it is hard to compute an $O(\log n)$-approximate Nash equilibrium for each $\mathcal{G} \in \mathcal{FG}$. Moreover, since the best-known approximation guarantee for the directed Steiner tree problem is $O(\log^2 n)$ in quasi-polynomial time and $O(n^\epsilon)$, for any given $\epsilon > 0$, in polynomial time [5], an algorithm computing an approximate Nash equilibrium for each $\mathcal{G} \in \mathcal{FG}$ achieving the same asymptotical performance guarantees within the same time constraints have to be considered fully satisfactory.

Now consider the application of our algorithm to a network congestion games $\mathcal{G}$. If we want an equilibrium with considerably good performance guarantee, we can run Algorithm 1 with $\gamma = 1 + \epsilon$, for an arbitrarily small $\epsilon > 0$, to obtain a $(1 + \epsilon + o(1))\rho_{KN}^t(\Pi_{br}(\mathcal{G}))$-approximate Nash equilibrium in quasi-polynomial time (provided that also t is a quasi-polynomial function). For instance, if we consider a game $\mathcal{G} \in \mathcal{FG}$, we achieve an $O(\log^2 n)$-approximate Nash equilibrium.

Conversely, if one wants efficiently computable solutions, by setting $\gamma = p^\epsilon$, it is possible to obtain an $O(p^\epsilon)\rho_{KN}^t(\Pi_{br}(\mathcal{G}))$-approximate Nash equilibrium in polynomial time (provided that also t is a polynomial function), for any constant $\epsilon > 0$. Again, for any game $\mathcal{G} \in \mathcal{FG}$, we achieve an $O(p^\epsilon n^\epsilon)$-approximate Nash equilibrium. If n and p are polynomially related, by properly setting ϵ and ϵ', our algorithm provides an $O(p^\epsilon n^{\epsilon'}) = O(n^{\epsilon''})$-approximate Nash equilibrium for any $\epsilon'' > 0$, meaning that in polynomial time it is possible to match the performance of the best known approximation algorithm for the directed Steiner tree problem.

With similar arguments, it is possible to show that, if $\rho_{KN}^t(\Pi_{br}(\mathcal{G})) = O(n^c)$ where $c > 0$ is a constant and t is a polynomial function, our algorithm computes an $O(n^{c+\epsilon})$-approximate Nash equilibrium in polynomial time and, by the above discussion, this means that we are almost optimal (with respect to the best-known approximation ratio). For instance, this situation happens when considering games $\mathcal{G}$ such that $\Pi_{br}(\mathcal{G})$ coincides with the problem of computing the minimum directed steiner forest [10], or the maximum clique.

5 Hardness of Computing a Nash Equilibrium in Multi-Source Shapley Games on Undirected Graphs

The existence of a potential function for congestion games allows us to cast searching for a Nash equilibrium as a local search problem. The states, that is the strategy profiles of the players, are the feasible solutions, and the neighborhood of a state consists of all authorized changes in the strategy of a single player. Then local optima correspond to states where no player can improve individually her cost, that is exactly to Nash equilibria. The potential of a state can be evaluated in polynomial time, and similarly a neighboring state of lower potential can be exhibited, provided that there exists one. This means that the problem of computing a Nash equilibrium in a congestion game belongs to the complexity class PLS (Polynomial Local Search) defined in [11,15].

Syrgkanis [18] proved that computing a Nash equilibrium in multi-source Shapley games played on directed graphs is a PLS-complete problem. Syrgkanis's proof works as follows: starting from an instance $G = (V, E, w)$ of the MAX CUT problem under the flip neighborhood, he constructs a (non-network) Shapley game $\mathcal{G}$ with $n = |V|$ players and exactly two strategies, namely s_i^A and s_i^B, for each player $i \in [n]$ such that, given a Nash equilibrium for $\mathcal{G}$, it is possible to construct in polynomial time a local maximum for the MAX CUT problem defined by G. Then, starting from $\mathcal{G}$, a multi-source Shapley game $\hat{\mathcal{G}}$ played by n players on a directed network is constructed with the property that the set of strategies that each player $i \in [n]$ can eventually adopt in any Nash equilibrium (i.e., non-dominated strategies) is restricted to at most two strategies that simulate strategies s_i^A and s_i^B in $\mathcal{G}$, so that there is a natural one-to-one correspondence between Nash equilibria of $\hat{\mathcal{G}}$ and Nash equilibria of $\mathcal{G}$.

Syrgkanis also shows how to suitably extend $\hat{\mathcal{G}}$ to a multi-source game $\widetilde{\mathcal{G}}$ played on an undirected network guaranteeing again a one-to-one correspondence between Nash equilibria of $\widetilde{\mathcal{G}}$ and Nash equilibria of $\hat{\mathcal{G}}$. Anyway, in order to achieve this last step, he needs to impose latency functions of the form $\frac{c_e(x)}{x}$ (where $c_e(x)$ is a non-decreasing concave function) on some edges of the graph defining $\widetilde{\mathcal{G}}$, so that his reduction does not apply to the Shapley cost sharing protocol.

We show that the problem of computing a Nash equilibrium in multi-source Shapley games played on undirected graphs remains PLS-complete by elaborating on Syrgkanis's proof as follows: starting from $\mathcal{G}$, we directly construct a multi-source Shapley game $\widetilde{\mathcal{G}}$ played on an undirected network, where we add a significant number of dummy players (but still polynomial in the dimensions of the MAX CUT instance G) allowing us to restrict the choice of each of the non-dummy player $i \in [n]$ at any Nash equilibrium to only at most two strategies, so as to simulate again strategies s_i^A and s_i^B in $\mathcal{G}$.

Theorem 2. *The problem of computing a Nash equilibrium in the class of multi-source Shapley games played on undirected networks is* PLS-*complete.*

References

1. Ackermann, H., Röglin, H., Vöcking, B.: On the impact of combinatorial structure on congestion games. J. ACM **55**(6), 25 (2008)
2. Albers, S., Lenzner, P.: On approximate Nash equilibria in network design. Internet Math. **9**(4), 384–405 (2013)
3. Anshelevich, E., Dasgupta, A., Kleinberg, J.M., Tardos, E., Wexler, T., Roughgarden, T.: The price of stability for network design with fair cost allocation. SIAM J. Comput. **38**(4), 1602–1623 (2008)
4. Caragiannis, V., Fanelli, A., Gravin, N., Skopalik, A.: Efficient Computation of Approximate Pure Nash Equilibria in Congestion Games. In: Proceedings of the IEEE 52nd Symposium on Foundations of Computer Science (FOCS), pp. 532–541 (2011)
5. Charikar, M., Chekuri, C., Cheung, T., Dai, Z., Goel, A., Guha, S., Li, M.: Approximation algorithms for directed Steiner problems. J. Algorithms **33**(1), 73–91 (1999)
6. Magniez, F., de Rougemont, M., Santha, M., Zeitoun, X.: The complexity of approximate Nash equilibrium in congestion games with negative delays. In: Chen, N., Elkind, E., Koutsoupias, E. (eds.) Internet and Network Economics. LNCS, vol. 7090, pp. 266–277. Springer, Heidelberg (2011)
7. Chien, S., Sinclair, A.: Convergence to Approximate Nash Equilibria in Congestion Games. In: Proceedings of the 18th Annual ACM-SIAM Symposium on Discrete Algorithms (SODA), pp. 169–178. ACM Press (2007)
8. Fabrikant, A., Papadimitriou, C.H., Talwar, K.: The complexity of pure Nash equilibria. In: Proceedings of the 36th ACM Symposium on Theory of Computing (STOC), pp. 604–612 (2004)
9. Feige, U.: A threshold of $\log n$ for approximating set-cover. In: Proceedings of the 28th ACM Symposium on Theory of Computing (STOC), pp. 314–318 (1996)
10. Feldman, M., Kortsarz, G., Nutov, Z.: Improved approximation algorithms for Directed Steiner Forest. J. Comput. Syst. Sci. **78**(1), 279–292 (2012)
11. Johnson, D., Papadimitriou, C., Yannakakis, M.: How easy is local search? J. Comput. Syst. Sci. **37**(1), 79–100 (1988)

12. Koutsoupias, E., Papadimitriou, C.: Worst-case equilibria. In: Meinel, C., Tison, S. (eds.) STACS 1999. LNCS, vol. 1563, pp. 404–413. Springer, Heidelberg (1999)
13. Monderer, D., Shapley, L.: Potential games. Games Econ. Behav. **14**, 124–143 (1996)
14. Nash, J.F.: Equilibrium points in n-person games. In: Proceedings of the National Academy of Sciences, vol. 36, pp. 48-49 (1950)
15. Papadimitriou, C., Yannakakis, M.: Optimization, approximation, and complexity classes. In: Procedings of ACM Symposium on Theory of Computing (STOC), pp. 229–234 (1988)
16. Rosenthal, R.: A class of games possessing pure-strategy Nash equilibria. Int. J. Game Theory **2**, 65–67 (1973)
17. Skopalik, A., Vöcking, B.: Inapproximability of pure Nash equilibria. In: Proceedings of the 40th ACM Symposium on Theory of Computing (STOC), pp. 355–364 (2008)
18. Syrgkanis, V.: The complexity of equilibria in cost sharing games. In: Saberi, A. (ed.) WINE 2010. LNCS, vol. 6484, pp. 366–377. Springer, Heidelberg (2010)

On Stackelberg Strategies in Affine Congestion Games

Vittorio Bilò[1]([⊠]) and Cosimo Vinci[2]

[1] Department of Mathematics and Physics "Ennio De Giorgi",
University of Salento, Provinciale Lecce-Arnesano,
P.O. Box 193, 73100 Lecce, Italy
vittorio.bilo@unisalento.it
[2] Gran Sasso Science Institute, L'Aquila, Italy
cosimo.vinci@gssi.infn.it

Abstract. We investigate the efficiency of some Stackelberg strategies in congestion games with affine latency functions. A Stackelberg strategy is an algorithm that chooses a subset of players and assigns them a prescribed strategy with the purpose of mitigating the detrimental effect that the selfish behavior of the remaining uncoordinated players may cause to the overall performance of the system. The efficiency of a Stackelberg strategy is measured in terms of the price of anarchy of the pure Nash equilibria they induce. Three Stackelberg strategies, namely Largest Latency First, Cover and Scale, were already considered in the literature and non-tight upper and lower bounds on their price of anarchy were given. We reconsider these strategies and provide the exact bound on the price of anarchy of both Largest Latency First and Cover and a better upper bound on the price of anarchy of Scale.

1 Introduction

Congestion games are, perhaps, the most famous class of non-cooperative games due to their capability to model several interesting competitive scenarios (such as selfish routing, facility location, machine scheduling), while maintaining some nice properties. In these games, there is a set of players sharing a set of *resources*, where each resource has an associated *latency (or cost) function* which depends on the number of players using it (the so-called *congestion*). Each player has an available set of strategies, where each strategy is a non-empty subset of resources, and aims at choosing a strategy minimizing her cost which is defined as the sum of the latencies experienced on all the selected resources.

Congestion games have been introduced by Rosenthal [13]. He proved that each such a game admits a bounded *potential function* whose set of local minima coincides with the set of *pure Nash equilibria* of the game, that is, strategy profiles in which no player can decrease her cost by unilaterally changing

---

This work was partially supported by the PRIN 2010–2011 research project ARS TechnoMedia: "Algorithmics for Social Technological Networks" funded by the Italian Ministry of University.

E. Markakis and G. Schäfer (Eds.): WINE 2015, LNCS 9470, pp. 132–145, 2015.
DOI: 10.1007/978-3-662-48995-6_10

her strategic choice. This existence result makes congestion games particularly appealing especially in all those applications in which pure Nash equilibria are elected as the ideal solution concept.

In these contexts, the study of the performance of pure Nash equilibria, usually measured by the *social welfare*, that is, the sum of the costs experienced by all players, has affirmed as a fervent research direction. To this aim, the notion of *price of anarchy* (Koutsoupias and Papadimitriou [11]) is widely adopted. It compares the social welfare of the worst pure Nash equilibrium with that of an optimal solution, called the *social optimum*, that could be potentially enforced by a dictatorial authority. Several recent works have given bounds on the price of anarchy of different variants of congestion games in which the resource latency functions are polynomially bounded in their congestion [1–6,9,12,15,16].

An interesting intermediate situation, usually referred to as *Stackelberg games*, happens when a central authority, called the *leader*, is granted the power of dictating the strategies of a subset of players. The leader's purpose is to determine a good *Stackelberg strategy*, which is an algorithm that carefully chooses the subset of players (called *coordinated players*) and their assigned strategies, so as to mitigate as much as possible the effects caused by the selfish behavior of the *uncoordinated players*, that is, to lower as much as possible the price of anarchy of the resulting game.

Fotakis [8] considers the application of three Stackelberg strategies, namely Largest Latency First, Cover and Scale, to congestion games with affine latency functions and gives upper and lower bounds on their worst-case price of anarchy. In this work, we improve these results by deriving the exact value of the worst-case price of anarchy of Largest Latency First and Cover and a better upper bound on the worst-case price of anarchy of Scale.

Related Work. Efficient Stackelberg strategies have been intensively studied within the context of *non-atomic* congestion games, that is, the case in which there are infinitely many players so that each player contributes for a negligible fraction to the resources congestion. The works of Roughgarden [14], Swamy [17], Correa and Stier-Moses [7], and Karakostas and Kolliopoulos [10] investigate the price of anarchy achieved in these games by the Stackelberg strategies Largest Latency First and Scale as a function of the fraction of the coordinated players, denoted as α. These two strategies, introduced by Roughgarden in [14], are defined as follows: Largest Latency First assigns the coordinated players to the largest cost strategies in the social optimum, while Scale simply employs the social optimum scaled by α.

The case of our interest, that is (atomic) congestion games with affine latency functions, has been considered before in the literature by Fotakis [8]. For the price of anarchy of the Stackelberg strategy Largest Latency First, he gives an upper bound of $\min\{(20 - 11\alpha)/8, (3 - 2\alpha + \sqrt{5 - 4\alpha})/2\}$ and a lower bound of $5(2-\alpha)/(4+\alpha)$ with the latter holding even for the restricted case of symmetric strategies, i.e., the case in which all players share the same strategic space. He then considers a randomized variant of Scale (since the deterministic one may be infeasible in the realm of atomic games) and shows that the expected price of anarchy is upper bounded by $\max\{(5 - 3\alpha)/2, (5 - 4\alpha)/(3 - 2\alpha)\}$ and lower bounded by $2/(1 + \alpha)$ with the latter holding even for the restricted case of symmetric strategies. He also introduces the Stackelberg strategy λ-Cover which assigns to every resource either at least λ or as many coordinated players as the resource has in the social optimum. For the price of anarchy of this strategy, he proves an upper bound of $(4\lambda - 1)/(3\lambda - 1)$ for affine latency functions and an upper bound of $1 + 1/(2\lambda)$ for linear latency functions. Finally, he also gives upper bounds for strategies obtained by combining λ-Cover with either Largest Latency First or Scale and upper bounds for games played on parallel links.

Our Contribution. We reconsider the three Stackelberg strategies studied by Fotakis in [8] and give either exact or improved bounds on their price of anarchy. In particular, we achieve the following results: for Largest Latency First, we show that the price of anarchy is exactly $(20 - 11\alpha)/8$ for $\alpha \in [0, 4/7]$ and $(4 - 3\alpha + \sqrt{4\alpha - 3\alpha^2})/2$ for $\alpha \in [4/7, 1]$; for λ-Cover, we show that the price of anarchy is exactly $\frac{4\lambda - 1}{3\lambda - 1}$ for affine latency functions and exactly $1 + (4\lambda + 1)/(4\lambda(2\lambda + 1))$ for linear ones; finally, for Scale, we give an improved upper bound of $1 + ((1 - \alpha)(2h + 1))/((1 - \alpha)h^2 + \alpha h + 1)$, where h is the unique integer such that $\alpha \in [(2h^2 - 3)/(2(h^2 - 1)), (2h^2 + 4h - 1)/(2h(h + 2))]$.

Paper Organization. Next section contains all necessary definitions and notation, as well as some preliminary result aimed at simplifying the analysis carried out in Sect. 3: the technical part of the paper in which we prove our bounds of the price of anarchy achieved by the three Stackelberg strategies considered in this work. Due to space limitations, some proofs have been omitted.

2   Model, Definitions, Notation and Preliminaries

For an integer $n \geq 1$, set $[n] := \{1, \ldots, n\}$. We use boldface letters to denote vectors and, given a vector $\boldsymbol{x}$, we denote with x_i its ith component and with $\boldsymbol{x}_{-i} \diamond y$ the vector obtained from $\boldsymbol{x}$ by replacing its ith component with y.

A *congestion game* is a tuple $\mathsf{CG} = (N, R, (\Sigma_i)_{i \in N}, (\ell_j)_{j \in R})$, where N is a set of $n \geq 2$ players, R is a set of resources, $\Sigma_i \subseteq 2^R \setminus \{\emptyset\}$ is the set of strategies available to each player $i \in N$ and $\ell_j : \mathbb{N} \to \mathbb{R}^+$ is the *latency (or cost) function* associated with each resource $j \in R$. The set $\Sigma := \times_{i \in N} \Sigma_i$ denotes the set of all the strategy profiles which can be realized in CG, so that a *strategy profile* $\boldsymbol{\sigma}$ models the state of the game in which each player $i \in N$ is adopting strategy σ_i. Given a strategy profile $\boldsymbol{\sigma}$, the *congestion* of resource $j \in R$, defined as $n_j(\boldsymbol{\sigma}) := |\{i \in N : j \in \sigma_i\}|$, denotes the number of players using resource j in $\boldsymbol{\sigma}$. The cost of player i in $\boldsymbol{\sigma}$ is defined as $c_i(\boldsymbol{\sigma}) = \sum_{j \in \sigma_i} \ell_j(n_j(\boldsymbol{\sigma}))$.

A strategy profile $\boldsymbol{\sigma}$ is a *pure Nash equilibrium* if, for each $i \in N$ and strategy $s \in \Sigma_i$, $c_i(\boldsymbol{\sigma}) \leq c_i(\boldsymbol{\sigma}_{-i} \diamond s)$, that is, no player can lower her cost by unilaterally switching to a different strategy. Each congestion game admits a pure Nash equilibrium by Rosenthal's Theorem [13].

The *social welfare* of a strategy profile is measured by the function $\mathsf{SUM} : \Sigma \to \mathbb{R}^+$ such that $\mathsf{SUM}(\boldsymbol{\sigma}) = \sum_{i \in N} c_i(\boldsymbol{\sigma})$. Let $o(\mathsf{CG})$ be the *social optimum* of

CG, that is, a strategy profile of CG minimizing SUM and denote with NE(CG) the set of pure Nash equilibria of CG. The *price of anarchy* of game CG is defined as the worst-case ratio between the social welfare of a pure Nash equilibrium and that of the social optimum. Formally, $\mathsf{PoA}(\mathsf{CG}) = \max_{\sigma \in \mathsf{NE}(\mathsf{CG})} \frac{\mathsf{SUM}(\sigma)}{\mathsf{SUM}(o(\mathsf{CG}))}$.

Given a congestion game CG and a Stackelberg strategy A, let P_A be the set of coordinated players chosen by A and s_i be the prescribed strategy assigned by A to each player $i \in P_A$. The congestion game CG_A, obtained from CG by coordinating the choices of the players in P_A, is the same as CG with the only difference that, for each player $i \in P_A$, CG_A has $\Sigma_i = \{s_i\}$, that is, each coordinated player becomes a selfish player who has no alternatives except for her prescribed strategy. A Stackelberg strategy is *optimal-restricted* if the coordinated players are assigned the strategy they adopt in a social optimum o, so that $s_i = o_i$ for each coordinated player i. For each optimal-restricted Stackelberg strategy A, since $o(\mathsf{CG}_A) = o(\mathsf{CG})$, the price of anarchy of CG_A becomes $\mathsf{PoA}(\mathsf{CG}_A) = \max_{\sigma \in \mathsf{NE}(\mathsf{CG}_A)} \frac{\mathsf{SUM}(\sigma)}{\mathsf{SUM}(o(\mathsf{CG}))}$.

In this work, we analyze three optimal-restricted Stackelberg strategies which have been previously considered in the literature, see [8]. The first two ones are deterministic, while the latter is randomized, hence, its price of anarchy will be evaluated in expectation. Moreover, in the first and third strategy the number of coordinated players is assumed to be equal to αn for some $\alpha \in [0,1]$ with $\alpha n \in \mathbb{N}$, i.e., the leader controls a fraction α of the players in the game. For a fixed congestion game CG and social optimum $o := o(\mathsf{CG})$, they are defined as follows:

- $\mathsf{LLF}(\alpha)$: for each $\alpha \in [0,1]$ such that $\alpha n \in \mathbb{N}$, the set of coordinated players is chosen equal to the set of the αn players with the highest cost in o, breaking ties arbitrarily. LLF stands for **Largest Latency First**.
- λ-Cover: for each $\lambda \in \mathbb{N} \setminus \{0\}$, the set of coordinated players P is chosen so as to guarantee that $|\{i \in P : j \in o_i\}| \geq \min\{\lambda, n_j(o)\}$ for each $j \in R$. Note that one such a set P might not exist in CG, so that this strategy is not always applicable.
- $\mathsf{Scale}(\alpha)$: for each $\alpha \in [0,1]$ such that $\alpha n \in \mathbb{N}$, the set of coordinated players is randomly chosen with uniform probability among the set of all subsets of N having cardinality αn.

A congestion game is *affine* if the latency function of each resource $j \in R$ can be expressed as $\ell_j(x) = a_j x + b_j$, with $a_j, b_j \geq 0$; moreover, it is *linear* whenever $b_j = 0$ for each $j \in R$. For a given optimal-restricted Stackelberg strategy A, our aim is to focus on the characterization of the price of anarchy of game CG_A, when CG is an affine congestion game.

Let us say that two games CG and CG' defined on the same set of players $[n]$ are *equivalent* if, for each player $i \in [n]$, there exists a bijection π_i between the set of strategies of player i in CG and the set of strategies of the same player in CG' such that, for each strategy profile σ of CG, we have $c_i(\sigma) = c_i(\pi(\sigma))$, where $\pi(\sigma) = (\pi_1(\sigma_1), \ldots, \pi_n(\sigma_n))$. The following lemma, proved in [4], will be of fundamental importance in our analysis.

Lemma 1 ([4]). *For each affine congestion game* CG, *there exists an equivalent linear congestion game* CG'.

Unfortunately, the transformation from affine to linear congestion games used in the proof of the above lemma invalidates the applicability of λ-Cover, so that, for this particular Stackelberg strategy, we will have to consider the cases of affine and linear latency functions separately.

Let $\mathcal{G}$ be any subclass of the class of congestion games. We define the price of anarchy of Stackelberg strategy A in the class of games $\mathcal{G}$ as $\mathsf{PoA}_{\mathcal{G}}(A) = \sup_{CG \in \mathcal{G}} \mathsf{PoA}(CG_A)$. Now let $\mathcal{ACG}$ and $\mathcal{LCG}$ be the class of affine congestion games and that of linear congestion games, respectively. By Lemma 1, it follows that for each affine congestion game CG with $\mathsf{PoA}(CG_A) = x$ there exists a linear congestion game CG' with $\mathsf{PoA}(CG'_A) = x$. Hence, it is possible to conclude that, for any Stackelberg strategy $A \in \{\mathsf{LLF}(\alpha), \mathsf{Scale}(\alpha)\}$,

$$\mathsf{PoA}_{\mathcal{ACG}}(A) = \mathsf{PoA}_{\mathcal{LCG}}(A). \tag{1}$$

As a consequence of (1), we will be allowed to restrict our attention only to the class of linear congestion games when bounding the price of anarchy of these two Stackelberg strategies in the class of affine congestion games. Moreover, we will use the simplified notation $\mathsf{PoA}(A)$ to denote the value $\mathsf{PoA}_{\mathcal{ACG}}(A) = \mathsf{PoA}_{\mathcal{LCG}}(A)$ when considering $A \in \mathsf{LLF}(\alpha), \mathsf{Scale}(\alpha)$.

We conclude this section by observing that, given a linear congestion game CG and a strategy profile $\sigma \in \Sigma$, the social value of σ can be expressed as follows: $\mathsf{SUM}(\sigma) = \sum_{i \in N} c_i(\sigma) = \sum_{i \in N} \sum_{j \in \sigma_i} a_j n_j(\sigma) = \sum_{j \in R} a_j n_j(\sigma)^2$.

3   Bounding the Price of Anarchy

In this section, we provide the exact values of both $\mathsf{PoA}(\mathsf{LLF}(\alpha))$ and $\mathsf{PoA}(\lambda\text{-Cover})$ and give a better upper bound for $\mathsf{PoA}(\mathsf{Scale}(\alpha))$.

3.1   Largest Latency First

Fotakis shows in [8] that $\mathsf{PoA}(\mathsf{LLF}(\alpha)) \leq \min\left\{\frac{20-11\alpha}{8}, \frac{3-2\alpha+\sqrt{5-4\alpha}}{2}\right\}$ and $\mathsf{PoA}(\mathsf{LLF}(\alpha)) \geq \frac{5(2-\alpha)}{4+\alpha}$, where the lower bound holds even for the restricted case of symmetric players. We close this gap for non-symmetric players by proving that $\mathsf{PoA}(\mathsf{LLF}(\alpha)) = \frac{20-11\alpha}{8}$ for $\alpha \in [0, 4/7]$ and $\mathsf{PoA}(\mathsf{LLF}(\alpha)) = \frac{4-3\alpha+\sqrt{4\alpha-3\alpha^2}}{2}$ for $\alpha \in [4/7, 1]$.

For the characterization of the upper bound, we use the primal-dual method introduced by Bilò in [4]. To this aim, fix a linear congestion game CG, a social optimum $o := o(\mathsf{CG})$ and a pure Nash equilibrium σ induced by $\mathsf{LLF}(\alpha)$. Let P be the set of an coordinated players chosen by $\mathsf{LLF}(\alpha)$. For the sake of conciseness, for each $j \in R$, we set $x_j := n_j(\sigma)$, $o_j := n_j(o)$ and $s_j := |\{i \in P : j \in o_i\}|$. Clearly, we have $s_j \leq \min\{x_j, o_j\}$ for each $j \in R$.

We obtain the following primal linear program $PP(\sigma, o)$ defined over the latency functions of the resources in R:

$$\max \sum_{j \in R} a_j x_j^2$$

subject to

$$\sum_{j \in \sigma_i} a_j x_j - \sum_{j \in o_i} a_j (x_j + 1) \leq 0 \quad \forall i \in N \setminus P,$$

$$\sum_{j \in o_i} a_j o_j - \sum_{j \in o_k} a_j o_j \leq 0 \quad \forall i \in N \setminus P, k \in P,$$

$$\sum_{j \in R} a_j o_j^2 = 1$$

$$a_j \geq 0 \quad \forall j \in R,$$

where the first family of constraints is satisfied since each uncoordinated player cannot lower her cost by switching to the strategy she adopts in o and the second one is satisfied because $LLF(\alpha)$ selects the set of $|P| = \alpha n$ players with the highest cost in o.

The dual program $DP(\sigma, o)$, obtained by associating the variables $(y_i)_{i \in N \setminus P}$ with the first family of constraints, the variables $(z_{ik})_{i \in N \setminus P, k \in P}$ with the second family of constraints and the variable γ with the last constraint, is the following:

$$\min \gamma$$

subject to

$$\sum_{i \in N \setminus P: j \in \sigma_i} y_i x_j - \sum_{i \in N \setminus P: j \in o_i} y_i (x_j + 1)$$

$$+ \sum_{i \in N \setminus P, k \in P: j \in o_i} z_{ik} o_j - \sum_{i \in N \setminus P, k \in P: j \in o_k} z_{ik} o_j + \gamma o_j^2 \geq x_j^2 \quad \forall j \in R,$$

$$y_i \geq 0 \quad \forall i \in N \setminus P,$$

$$z_{ik} \geq 0 \quad \forall i \in N \setminus P, k \in P$$

Any feasible solution for $DP(\sigma, o)$ which is independent of the choices of σ and o, i.e., satisfying the dual constraint for each possible pair of integers $(x_j, o_j) \in \{N \cup \{0\}\}^2$, will provide an upper bound on $PoA(CG_{LLF(\alpha)})$ which, by the generality of CG, gives an upper bound on $PoA(LLF(\alpha))$. We get the following result.

Theorem 1. *For each $\alpha \in [0, 1]$,*

$$PoA(LLF(\alpha)) \leq \begin{cases} \frac{20 - 11\alpha}{8} & \text{for } \alpha \in [0, 4/7], \\ \frac{4 - 3\alpha + \sqrt{4\alpha - 3\alpha^2}}{2} & \text{for } \alpha \in [4/7, 1]. \end{cases}$$

Proof. The $\frac{20 - 11\alpha}{8}$ upper bound was already established by Fotakis in [8]. For the second bound, set $\theta = \sqrt{4\alpha - 3\alpha^2}$, $y_i = 1 + \frac{\alpha}{\theta}$ for each $i \in N \setminus P$, $z_{ik} = \frac{1}{2n}\left(3 + \frac{3\alpha - 2}{\theta}\right)$ for each $i \in N \setminus P$ and $k \in P$, and $\gamma = \frac{4 - 3\alpha + \theta}{2}$. With these values, by using $|\{i \in N \setminus P : j \in \sigma_i\}| = x_j - s_j$, $|\{i \in N \setminus P : j \in o_i\}| = o_j - s_j$, $|\{i \in N \setminus P, k \in P : j \in o_i\}| = \alpha n(o_j - s_j)$ and $|\{i \in N \setminus P, k \in P : j \in o_k\}| = (1 - \alpha)n s_j$, the generical dual constraint becomes $f(x_j, o_j, s_j) := 2\alpha x_j^2 - 2(\alpha + \theta)(x_j o_j +$

$o_j - s_j) - (3\theta + 3\alpha - 2)o_j s_j + 2(\alpha + 2\theta)o_j^2 \geq 0$. We show that $f(x_j, o_j, s_j) \geq 0$ for any triple of non-negative integers (x_j, o_j, s_j) such that $s_j \leq \min\{x_j, o_j\}$. Let us first compute the derivative $\frac{\delta f}{\delta x_j}(x_j, o_j, s_j) = 4\alpha x_j - 2(\alpha + \theta)o_j$. This is a linear function on x_j which is negative for $x_j = 0$, hence $f(x_j, o_j, s_j)$ is minimized for $x_j = \frac{(\alpha+\theta)o_j}{2\alpha}$. By substituting, we get that $f(x_j, o_j, s_j) \geq 0$ if $g(o_j, s_j) :=$ $f\left(\frac{(\alpha+\theta)o_j}{2\alpha}, o_j, s_j\right) = 2(3\alpha + 3\theta - 2)o_j^2 - 2((3\alpha + 3\theta - 2)s_j + 2(\alpha + \theta))o_j + 4(\alpha + \theta)s_j \geq$ 0. Again, let us compute the derivative $\frac{\delta g}{\delta s_j}(o_j, s_j) = 4(\alpha + \theta) - 2(3\alpha + 3\theta - 2)o_j$. Since $\alpha + \theta - 1 \geq 0$ for each $\alpha \in [4/7, 1]$, this function is non-increasing for $o_j \geq 2$, so that $g(o_j, s_j)$ is minimized for $s_j = o_j$ when $o_j \geq 2$. By substituting, we get $g(o_j, o_j) \geq 0$ as desired. For the leftover cases of $o_j \leq 1$, we have $f(x_j, 0, 0) = 2\alpha x_j^2 \geq 0$ and $f(x_j, 1, s_j) = 2\alpha x_j^2 - 2(\alpha + \theta)x_j + (2 - \alpha - \theta)s_j + 2\theta$. Since $2 - \alpha - \theta \leq 0$ for each $\alpha \in [4/7, 1]$, we have that this last function is minimized for $s_j = 0$. Hence, we need to show that $f(x_j, 1, 0) = 2\alpha x_j^2 - 2(\alpha + \theta)x_j + 2\theta \geq 0$. By solving for x_j, we get $f(x_j, 1, 0) \geq 0$ when $x_j \leq 1$ or $x_j \geq \theta/\alpha$. Since, $\theta/\alpha \leq 2$ for each $\alpha \in [4/7, 1]$, $f(x_j, 1, 0) \geq 0$ for each non-negative integer x_j and the proof is complete.   □

We now prove matching lower bounds.

Theorem 2. *For each $\alpha \in [0, 4/7]$, $\mathsf{PoA}(\mathsf{LLF}(\alpha)) \geq \frac{20 - 11\alpha}{8}$, while for each $\alpha \in [4/7, 1]$, $\mathsf{PoA}(\mathsf{LLF}(\alpha)) \geq \frac{4 - 3\alpha + \sqrt{4\alpha - 3\alpha^2}}{2}$.*

Proof. For a fixed $\alpha \in [0, 4/7]$, let n be an integer such that $\alpha n \in \mathbb{N}$, $\alpha n \geq 2$ and $(1 - \alpha)n \geq 3$. Consider the linear congestion game CG defined as follows. The set of players is $N = N_1 \cup N_2$, where $|N_1| = (1 - \alpha)n$ and $|N_2| = \alpha n$. The set of resources is $R = U \cup V \cup W$, where $U = \{u_j : j \in [(1 - \alpha)n]\}$, $V = \{v_j : j \in [(1 - \alpha)n]\}$ and $W = \{w_{j,k} : j \in [(1 - \alpha)n], k \in [\alpha n]\}$. Each resource in U has latency function $\ell_U(x) = \frac{(4 - \alpha)n}{2}x$, each resource in V has latency function $\ell_V(x) = \frac{(4 - 7\alpha)n}{2}x$ and each resource in W has latency function $\ell_W(x) = x$. Note that $\alpha \in [0, 4/7]$ guarantees $\frac{(4 - 7\alpha)n}{2} \geq 0$. Each player $i \in N_1$ has two available strategies: the *first strategy* $\{u_i, v_i\}$ and the *second strategy* $\{u_{i+1}, v_{i+1}, v_{i+2}\} \cup \bigcup_{k \in [\alpha n]}\{w_{i,k}\}$, where the sums over the indices have to be interpreted circularly so that, for instance, $u_{(1-\alpha)n+1} = u_1$. Each player $i \in N_2$ has only one available strategy given by $\bigcup_{k \in [(1-\alpha)n]}\{w_{k,i}, w_{k+1,i}\}$, where again the sums over the indices have to be interpreted circularly so that, for instance, $w_{(1-\alpha)n+1,i} = w_{1,i}$. Note that, by $\alpha n \geq 2$, the number of resources in W belonging to the strategy of each player in N_2 is exactly $2(1 - \alpha)n$; moreover, by $(1 - \alpha)n \geq 3$, the first and second strategy of each player in N_1 are disjoint.

First of all, it is not difficult to see that the strategy profile $\boldsymbol{o}$ in which each player in N_1 adopts her first strategy is a social optimum for CG. Now note that, for each player $i \in N_1$, $c_i(\boldsymbol{o}) = \frac{(4 - \alpha)n}{2} + \frac{(4 - 7\alpha)n}{2} = 4(1 - \alpha)n$, whereas, for each player $i \in N_2$, $c_i(\boldsymbol{o}) = 4(1 - \alpha)n$ since each player in N_2 uses $2(1 - \alpha)n$ resources in W each having congestion equal to 2. Hence, it follows that $\mathsf{LLF}(\alpha)$ may choose N_2 as the set of coordinated players. We now show that, under this hypothesis, the strategy profile σ in which each player $i \in N_1$ adopts her second strategy is a pure Nash equilibrium for $CG_{LLF(\alpha)}$. For each player $i \in N_1$, $c_i(\sigma) = \frac{(4-\alpha)n}{2} + 4\frac{(4-7\alpha)n}{2} + 3\alpha n = \frac{(20-23\alpha)n}{2}$ since each player $i \in N_1$ uses 1 resource in U having congestion 1, 2 resources in V both having congestion 2 and αn resources in W each having congestion 3. Let σ' be the strategy profile obtained from σ when player i deviates to her second strategy. In this case, $c_i(\sigma') = 2\frac{(4-\alpha)n}{2} + 3\frac{(4-7\alpha)n}{2} = \frac{(20-23\alpha)n}{2}$ since player i uses 1 resource in U having congestion 2 and 1 resource in V having congestion 3. Hence, it follows that σ is a pure Nash equilibrium for $CG_{LLF(\alpha)}$.

We are now left to compute the ratio $\frac{SUM(\sigma)}{SUM(o)}$. To this aim, note that $n_j(\sigma) = n_j(o) = 1$ for each resource $j \in U$, $n_j(\sigma) = 2$ and $n_j(o) = 1$ for each resource $j \in V$, while, for each resource $j \in W$, $n_j(\sigma) = 3$ and $n_j(o) = 2$. We get

$$\frac{SUM(\sigma)}{SUM(o)} = \frac{(1-\alpha)n\left(\frac{(4-\alpha)n}{2} + 4\frac{(4-7\alpha)n}{2}\right) + 9\alpha(1-\alpha)n^2}{(1-\alpha)n\left(\frac{(4-\alpha)n}{2} + \frac{(4-7\alpha)n}{2}\right) + 4\alpha(1-\alpha)n^2}$$

$$= \frac{20 - 11\alpha}{8}$$

and the claim follows because $PoA(LLF(\alpha)) \geq PoA(CG_{LLF(\alpha)})$.

For a fixed $\alpha \in [4/7, 1]$, set $\theta = \sqrt{4\alpha - 3\alpha^2}$ and let n be an integer such that $\frac{\alpha n}{2} \in \mathbb{N}$, $\frac{(\theta-\alpha)n}{4} \in \mathbb{N}$ and $(1-\alpha)n \geq 2$. Note that, for each $\alpha \in [4/7, 1]$, $\theta - \alpha \geq 0$ so that n is well defined. Consider the linear congestion game CG defined as follows. The set of players is $N = N_1 \cup N_2$, where $|N_1| = (1-\alpha)n$ and $|N_2| = \alpha n$. The set of resources is $R = U \cup W$, where $U = \{u_j : j \in [(1-\alpha)n]\}$ and $W = \{w_{j,k} : j \in [(1-\alpha)n], k \in [\alpha n]\}$. Each resource in U has latency function $\ell_U(x) = \frac{\alpha^2(1-\alpha)n^3}{4}x$ and each resource in W has latency function $\ell_W(x) = x$. Each player $i \in N_1$ has two available strategies: the *first strategy* $\{u_i\}$ and the *second strategy* $\{u_{i+1}\} \cup \bigcup_{k \in [\alpha n], j \in \left[\frac{(\theta-\alpha)n}{4}\right]}\{w_{i+j-1,k}\}$, where again the sums over the indices have to be interpreted circularly. Note that, for each $\alpha \in [4/7, 1]$, $\frac{(\theta-\alpha)n}{4} \leq (1-\alpha)n$ so that the number of resources in W belonging to the second strategy of each player in N_1 is exactly $\frac{\alpha(\theta-\alpha)n^2}{4}$; moreover, by $(1-\alpha)n \geq 2$, the first and second strategy of each player are disjoint. Each player $i \in N_2$ has only one available strategy given by $\bigcup_{k \in [(1-\alpha)n], j \in \left[\frac{\alpha n}{2}\right]}\{w_{k,i+j-1}\}$, where again the sums over the indices have to be interpreted circularly.

First of all, it is not difficult to see that the strategy profile o in which each player in N_1 adopts her first strategy is a social optimum for CG. Now note that, for each player $i \in N_1$, $c_i(o) = \frac{\alpha^2(1-\alpha)n^3}{4}$, whereas, for each player $i \in N_2$, $c_i(o) = \frac{\alpha^2(1-\alpha)n^3}{4}$ since each player in N_2 uses $\frac{\alpha n}{2}(1-\alpha)n$ resources in W each having congestion equal to $\frac{\alpha n}{2}$. Hence, it follows that $LLF(\alpha)$ may choose N_2 as the set of coordinated players. We now show that, under this hypothesis, the strategy profile σ in which each player $i \in N_1$ adopts her second strategy is a pure Nash equilibrium for $CG_{LLF(\alpha)}$. For each player $i \in N_1$, $c_i(\sigma) =$

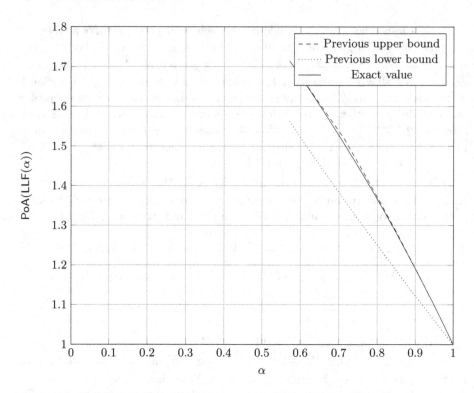

Fig. 1. Comparison between the exact value of $\mathsf{PoA}(\mathsf{LLF}(\alpha))$ and the upper and lower bounds previously known in the literature for the case of $\alpha \in [4/7, 1]$.

$\frac{\alpha^2(1-\alpha)n^3}{4} + \frac{\alpha(\theta-\alpha)n^2}{4}\frac{(\theta+\alpha)n}{4} = \frac{\alpha^2(1-\alpha)n^3}{2}$ since each player $i \in N_1$ uses 1 resource in U having congestion 1 and $\frac{\alpha(\theta-\alpha)n^2}{4}$ resources in W each having congestion $\frac{(\theta+\alpha)n}{4}$. Let σ' be the strategy profile obtained from σ when player i deviates to her second strategy. In this case, $c_i(\sigma') = 2\frac{\alpha^2(1-\alpha)n^3}{4} = \frac{\alpha^2(1-\alpha)n^3}{2}$, since player i uses 2 resources in U both having congestion 2. Hence, it follows that σ is a pure Nash equilibrium for $\mathsf{CG}_{\mathsf{LLF}(\alpha)}$.

We are now left to compute the ratio $\frac{\mathsf{SUM}(\sigma)}{\mathsf{SUM}(o)}$. To this aim, note that $n_j(\sigma) = n_j(o) = 1$ for each resource $j \in U$, while, for each resource $j \in W$, $n_j(\sigma) = \frac{(\theta+\alpha)n}{4}$ and $n_j(o) = \frac{\alpha n}{2}$. We get

$$\frac{\mathsf{SUM}(\sigma)}{\mathsf{SUM}(o)} = \frac{(1-\alpha)n\frac{\alpha^2(1-\alpha)n^3}{4} + \alpha(1-\alpha)n^2\frac{(\alpha+\theta)^2n^2}{16})}{(1-\alpha)n\frac{\alpha^2(1-\alpha)n^3}{4} + \alpha(1-\alpha)n^2\frac{\alpha^2n^2}{4}}$$
$$= \frac{4 - 3\alpha + \sqrt{4\alpha - 3\alpha^2}}{2}$$

and the claim follows because $\mathsf{PoA}(\mathsf{LLF}(\alpha)) \geq \mathsf{PoA}(\mathsf{CG}_{\mathsf{LLF}(\alpha)})$.   □

Our findings, compared with the previous results given by Fotakis in [8], can be summarized as follows. For $\alpha \in [0, 4/7]$, we have shown that the upper bound given by Fotakis is indeed tight by providing a matching lower bound (Theorem 2). For $\alpha \in [4/7, 1]$, we obtain the exact bound (Theorems 1 and 2) by simultaneously improving either the upper and the lower bounds previously given by Fotakis. For a quantitative comparison of the values in this case, see Fig. 1.

3.2 λ-Cover

Fotakis shows in [8] that $\mathsf{PoA}_{\mathcal{ACG}}(\lambda\text{-Cover}) \leq \frac{4\lambda-1}{3\lambda-1}$ and that $\mathsf{PoA}_{\mathcal{LCG}}(\lambda\text{-Cover}) \leq 1 + \frac{1}{2\lambda}$, while no lower bounds are provided. We exactly characterize the performance of λ-Cover by showing that, for each $\lambda \in \mathbb{N}\setminus\{0\}$, $\mathsf{PoA}_{\mathcal{ACG}}(\lambda\text{-Cover}) = \frac{4\lambda-1}{3\lambda-1}$ and $\mathsf{PoA}_{\mathcal{LCG}}(\lambda\text{-Cover}) = 1 + \frac{4\lambda+1}{4\lambda(2\lambda+1)}$.

In order to show the upper bound for linear latency functions, we make use again of the primal-dual method. To this aim, fix a linear congestion game CG, a social optimum o and a pure Nash equilibrium σ induced by λ-Cover applied on a particular choice of the set of coordinated players P. Again, for each $j \in R$, we set $x_j := n_j(\sigma)$, $o_j := n_j(o)$ and $s_j := |\{i \in P : j \in o_i\}|$. In this case, by the definition of λ-Cover, we have $s_j \geq \min\{\lambda, o_j\}$ for each $j \in R$.

We obtain the following primal linear program $\mathsf{PP}(\sigma, o)$:

$$\max \sum_{j \in R} a_j x_j^2$$
$$subject\ to$$
$$\sum_{j \in \sigma_i} a_j x_j - \sum_{j \in o_i} a_j(x_j + 1) \leq 0 \quad \forall i \in N \setminus P,$$
$$\sum_{j \in R} a_j o_j^2 = 1$$
$$a_j \geq 0 \qquad\qquad\qquad \forall j \in R$$

Observe that the proposed program is a subprogram of the one defined for LLF and does not depend on λ. This must not be surprising, since the parameter λ does not affect the players' costs directly, but only gives us an additional property on the resources congestion which can be suitably exploited in the analysis of the dual constraints. The dual program $\mathsf{DP}(\sigma, o)$, obtained by associating the variables $(y_i)_{i \in N \setminus P}$ with the first family of constraints and the variable γ with the last constraint, is the following:

$$\min \gamma$$
$$subject\ to$$
$$\sum_{i \in N \setminus P : j \in \sigma_i} y_i x_j - \sum_{i \in N \setminus P : j \in o_i} y_i(x_j + 1) + \gamma o_j^2 \geq x_j^2 \quad \forall j \in R,$$
$$y_i \geq 0 \qquad\qquad\qquad\qquad\qquad \forall i \in N \setminus P$$

We get the following result.

Theorem 3. *For each* $\lambda \in \mathbb{N} \setminus \{0\}$, $\mathsf{PoA}_{\mathcal{LCG}}(\lambda\text{-Cover}) \leq 1 + \frac{4\lambda+1}{4\lambda(2\lambda+1)}$.

Proof. For each $\lambda \in \mathbb{N}\setminus\{0\}$, set $y_i = \frac{4\lambda+1}{2\lambda+1}$ for each $i \in N\setminus P$, and $\gamma = 1 + \frac{4\lambda+1}{4\lambda(2\lambda+1)}$. With these values, the generical dual constraint becomes

$$\frac{4\lambda+1}{2\lambda+1}\left(\sum_{i\in N\setminus P:j\in\sigma_i} x_j - \sum_{i\in N\setminus P:j\in o_i} x_j + 1\right) + \left(1 + \frac{4\lambda+1}{4\lambda(2\lambda+1)}\right)o_j^2 \geq x_j^2,$$

which is non-negative whenever $f(x_j, o_j, s_j) := 2\lambda x_j^2 - (4\lambda+1)(x_jo_j + o_j - s_j) + \frac{8\lambda^2+8\lambda+1}{4\lambda}o_j^2 \geq 0$. We show that $f(x_j, o_j, s_j) \geq 0$ for any triple of non-negative integers (x_j, o_j, s_j) such that $s_j \geq \min\{\lambda, o_j\}$. First, observe that $f(x_j, o_j, s_j)$ is an increasing function in s_j so that it is minimized for either $s_j = \lambda$ or $s_j = o_j$. For $s_j = o_j$, we get that $f(x_j, o_j, s_j) \geq 0$ if $g(x_j, o_j) := f(x_j, o_j, o_j) = 8\lambda^2 x_j^2 - 4(4\lambda+1)(x_jo_j\lambda) + (8\lambda^2+8\lambda+1)o_j^2 \geq 0$. By solving the associated equality for x_j, we obtain that the discriminant is negative thus showing $g(x_j, o_j) > 0$ in any case. For $s_j = \lambda$, we get that $f(x_j, o_j, s_j) \geq 0$ if $g(x_j, o_j) := f(x_j, o_j, \lambda) = 2\lambda x_j^2 - (4\lambda+1)(x_jo_j + o_j - \lambda) + \frac{8\lambda^2+8\lambda+1}{4\lambda}o_j^2 \geq 0$. The derivative $\frac{\delta g}{\delta x_j}(x_j, o_j) = 4\lambda x_j - (4\lambda+1)o_j$ is a linear function on x_j which is negative for $x_j = 0$, hence $g(x_j, o_j)$ is minimized for $x_j = \frac{(4\lambda+1)o_j}{4\lambda}$. Similarly, $\frac{\delta g}{\delta o_j}(x_j, o_j) = \frac{8\lambda^2+8\lambda+1}{2\lambda}o_j - (4\lambda+1)(x_j+1)$ is a linear function on o_j which is negative for $o_j = 0$, hence $g(x_j, o_j)$ is minimized for $o_j = \frac{2\lambda(4\lambda+1)(x_j+1)}{8\lambda^2+8\lambda+1}$. It follows that $g(x_j, o_j)$ has a unique stationary point $(x^*, o^*) = \left(\frac{(4\lambda+1)^2}{8\lambda+1}; \frac{4\lambda(4\lambda+1)}{8\lambda+1}\right)$ which is a global minimum. Anyway, the values of x^* and o^* are not integral, as it can be easily verified that $x^*, o^* \in (2\lambda, 2\lambda+1)$. Since $g(x_j, o_j)$ is continuous in both x_j and o_j, it follows that the point in $\mathbb{N}^2$ minimizing $g(x_j, o_j)$ belongs to the set $\{(2\lambda, 2\lambda), (2\lambda, 2\lambda+1), (2\lambda+1, 2\lambda), (2\lambda+1, 2\lambda+1)\}$. We get $g(2\lambda, 2\lambda) = 0$, $g(2\lambda, 2\lambda+1) = \frac{16\lambda^2+8\lambda+1}{4\lambda}$, $g(2\lambda+1, 2\lambda) = 0$ and $g(2\lambda+1, 2\lambda+1) = \frac{4\lambda+1}{4\lambda}$ as needed. $\qquad\square$

We now show a matching lower bound.

Theorem 4. *For each* $\lambda \in \mathbb{N} \setminus \{0\}$, $\mathsf{PoA}_{\mathcal{LCG}}(\lambda\text{-Cover}) \geq 1 + \frac{4\lambda+1}{4\lambda(2\lambda+1)}$.

In this case, we obtain the exact value of $\mathsf{PoA}_{\mathcal{LCG}}(\lambda\text{-Cover})$ by improving the upper bound previously given by Fotakis in [8] and providing matching lower bounding instances whereas no trivial lower bounds were previously known. For a quantitative comparison of the values, see Fig. 2.

For affine latency functions, we achieve the exact characterization of the price of anarchy by showing that the upper bound given by Fotakis in [8] is asymptotically tight.

Theorem 5. *For each* $\lambda \in \mathbb{N} \setminus \{0\}$ *and* $\epsilon > 0$, *there exists an affine congestion game* CG *such that* $\mathsf{PoA}(\mathsf{CG}_{\lambda\text{-Cover}}) \geq \frac{4\lambda-1}{3\lambda-1} - \epsilon$.

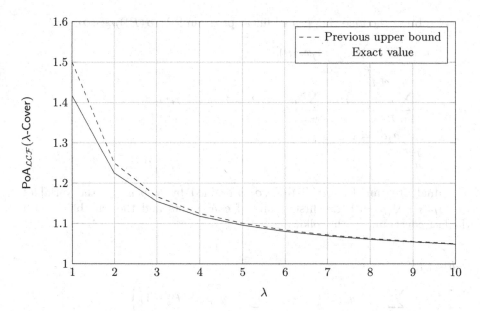

Fig. 2. Comparison between the exact value of PoA$_{\mathcal{LCG}}(\lambda$-Cover$)$ and the upper bound previously known in the literature.

3.3 Scale

Fotakis shows in [8] that $\mathsf{PoA}(\mathsf{Scale}(\alpha)) \leq \max\left\{\frac{5-3\alpha}{2}, \frac{5-4\alpha}{3-2\alpha}\right\}$ and $\mathsf{PoA}(\mathsf{Scale}(\alpha)) \geq \frac{2}{1+\alpha}$, where the lower bound holds for any randomized optimal-restricted Stackelberg strategy even when applied to symmetric players.

For each integer $h \geq 1$, define $r_L(h) = \frac{2h^2-3}{2(h^2-1)}$ and $r_U(h) = \frac{2h^2+4h-1}{2h(h+2)}$, with $r_L(0) := 0$. Note that, since $r_L(0) = 0$, $r_U(\infty) = 1$, and $r_L(h+1) = r_U(h)$ for each $h \geq 1$, it follows that, for each $\alpha \in [0,1]$, there exists a unique integer $h := h(\alpha) \geq 1$ such that $\alpha \in [r_L(h), r_U(h)]$. We show that, for each $\alpha \in [0,1]$, $\mathsf{PoA}(\mathsf{Scale}(\alpha)) \leq 1 + \frac{(1-\alpha)(2h+1)}{(1-\alpha)h^2+\alpha h+1}$. For $h = 1$, which covers the case $\alpha \in [0,5/6]$, and for $h = \infty$, which covers the case $\alpha = 1$, our upper bounds coincide with the ones given by Fotakis, while, in all the other cases, they are better.

Also in this case we make use of the primal-dual method. To this aim, fix a linear congestion game CG and a social optimum o. For each of the $\beta := \binom{n}{\alpha n}$ possible choices of the set of coordinated players P, let $\sigma(P)$ be a pure Nash equilibrium for the induced game. For each $j \in R$, we set $x_j(P) := n_j(\sigma(P))$, $o_j := n_j(o)$ and $s_j(P) := |\{i \in P : j \in o_i\}|$, so that $s_j(P) \leq \min\{x_j(P), o_j\}$ for each $j \in R$ and for each $P \in \mathcal{P}$, where $\mathcal{P}$ denotes the set of all the β subsets of N having cardinality αn.

We obtain the following primal linear program $\mathsf{PP}\left((\boldsymbol{\sigma}(P))_{P\in\mathcal{P}},\boldsymbol{o}\right)$:

$$\max \beta^{-1} \sum_{P\in\mathcal{P}} \sum_{j\in R} a_j x_j(P)^2$$

subject to

$$\sum_{j\in\sigma_i(P)} a_j x_j(P) - \sum_{j\in o_i} a_j(x_j(P)+1) \leq 0 \quad \forall P \in \mathcal{P}, i \in N \setminus P,$$

$$\sum_{j\in R} a_j o_j^2 = 1$$

$$a_j \geq 0 \qquad\qquad\qquad\qquad \forall j \in R$$

The dual program $\mathsf{DP}\left((\boldsymbol{\sigma}(P))_{P\in\mathcal{P}},\boldsymbol{o}\right)$, obtained by associating the variables $(y_{i,P})_{P\in\mathcal{P},i\in N\setminus P}$ with the first family of constraints and the variable γ with the last constraint, is the following:

$$\min \gamma$$

subject to

$$\sum_{P\in\mathcal{P}} \left(\sum_{i\in N\setminus P:j\in\sigma_i(P)} y_{i,P} x_j(P) - \sum_{i\in N\setminus P:j\in o_i} y_{i,P}(x_j(P)+1) \right)$$

$$+\gamma o_j^2 \geq \beta^{-1} \sum_{P\in\mathcal{P}} x_j(P)^2 \qquad\qquad \forall j \in R,$$

$$y_{i,P} \geq 0 \qquad\qquad\qquad\qquad \forall P \in \mathcal{P}, i \in N \setminus P$$

We get the following result.

Theorem 6. *For each $\alpha \in [0,1]$, $\mathsf{PoA}(\mathsf{Scale}(\alpha)) \leq 1 + \frac{(1-\alpha)(2h+1)}{(1-\alpha)h^2+\alpha h+1}$.*

We obtain better upper bounds with respect to those provided by Fotakis for the case of $\alpha \in (5/6, 1)$; for a quantitative comparison of the values, see Fig. 3.

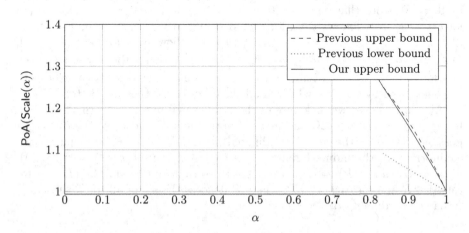

Fig. 3. Comparison between our upper bounds on $\mathsf{PoA}(\mathsf{Scale}(\alpha))$ and the upper and lower bounds previously known in the literature for the case of $\alpha \in [5/6, 1]$.

References

1. Aland, S., Dumrauf, D., Gairing, M., Monien, B., Schoppmann, F.: Exact price of anarchy for polynomial congestion games. SIAM J. Comput. **40**(5), 1211–1233 (2011)
2. Awerbuch, B., Azar, Y., Epstein, L.: The price of routing unsplittable flow. In: Proceedings of the 37th Annual ACM Symposium on Theory of Computing (STOC), pp. 57–66. ACM Press (2005)
3. Bhawalkar, K., Gairing, M., Roughgarden, T.: Weighted congestion games: price of anarchy, universal worst-case examples, and tightness. ACM Trans. Econ. Comput. **2**(4), 14 (2014)
4. Bilò, V.: A unifying tool for bounding the quality of non-cooperative solutions in weighted congestion games. In: Erlebach, T., Persiano, G. (eds.) WAOA 2012. LNCS, vol. 7846, pp. 215–228. Springer, Heidelberg (2013)
5. Caragiannis, I., Flammini, M., Kaklamanis, C., Kanellopoulos, P., Moscardelli, L.: Tight bounds for selfish and greedy load balancing. Algorithmica **61**(3), 606–637 (2011)
6. Christodoulou, G., Koutsoupias, E.: The price of anarchy of finite congestion games. In: Proceedings of the 37th Annual ACM Symposium on Theory of Computing (STOC), pp. 67–73. ACM Press (2005)
7. Correa, J.R., Stier Moses, N.E.: Stackelberg routing in atomic network games. Technical report DRO-2007-03, Columbia Business School (2007)
8. Fotakis, D.: Stackelberg strategies for atomic congestion games. Theor. Comput. Syst. **47**(1), 218–249 (2010)
9. Gairing, M., Lücking, T., Mavronicolas, M., Monien, B., Rode, M.: Nash equilibria in discrete routing games with convex latency functions. J. Comput. Syst. Sci. **74**(7), 1199–1225 (2008)
10. Karakostas, G., Kolliopoulos, S.: Stackelberg strategies for selfish routing in general multicommodity networks. Algorithmica **53**(1), 132–153 (2009)
11. Koutsoupias, E., Papadimitriou, C.: Worst-case equilibria. In: Meinel, C., Tison, S. (eds.) STACS 1999. LNCS, vol. 1563, pp. 404–413. Springer, Heidelberg (1999)
12. Lücking, T., Mavronicolas, M., Monien, B., Rode, M.: A new model for selfish routing. Theoret. Comput. Sci. **406**(3), 187–206 (2008)
13. Rosenthal, R.W.: A class of games possessing pure-strategy Nash equilibria. Int. J. Game Theory **2**, 65–67 (1973)
14. Roughgarden, T.: Stackelberg scheduling strategies. SIAM J. Comput. **33**(2), 332–350 (2004)
15. Roughgarden, T.: Intrinsic robustness of the price of anarchy. Commun. ACM **55**(7), 116–123 (2012)
16. Suri, S., Tóth, C.D., Zhou, Y.: Selfish load balancing and atomic congestion games. Algorithmica **47**(1), 79–96 (2007)
17. Swamy, C.: The effectiveness of Stackelberg strategies and tolls for network congestion games. ACM Trans. Algorithms **8**(4), 36 (2012)

Impartial Selection and the Power of up to Two Choices

Antje Bjelde, Felix Fischer, and Max Klimm$^{(\boxtimes)}$

Institut Für Mathematik, Technische Universität Berlin, Berlin, Germany
{bjelde,fischerf,klimm}@math.tu-berlin.de

Abstract. We study mechanisms that select members of a set of agents based on nominations by other members and that are impartial in the sense that agents cannot influence their own chance of selection. Prior work has shown that deterministic mechanisms for selecting any fixed number of agents are severely limited, whereas randomization allows for the selection of a single agent that in expectation receives at least 1/2 of the maximum number of nominations. The bound of 1/2 is in fact best possible subject to impartiality. We prove here that the same bound can also be achieved *deterministically* by *sometimes but not always* selecting a second agent. We then show a separation between randomized mechanisms that make exactly two or up to two choices, and give upper and lower bounds on the performance of mechanisms allowed more than two choices.

1 Introduction

We consider the setting of impartial selection first studied by Alon et al. [1] and by Holzman and Moulin [6]. The goal in this setting is to select members of a set of agents based on nominations cast by other members of the set, under the assumption that agents will reveal their true opinion about other agents as long as they cannot influence their own chance of selection.

Formally, the impartial selection problem can be modeled by a directed graph with n vertices, one for each agent, in which edges correspond to nominations. A selection mechanism then chooses a set of vertices for any given graph, and it is impartial if the chances of a particular vertex to be chosen do not depend on its outgoing edges. As impartiality may prevent us from simply selecting the vertices with maximum indegree, corresponding to the most highly nominated agents, it is natural to instead approximate this objective. For an integer k, a selection mechanism is called a k-selection mechanism, if it selects at most k vertices of any input graph. A k-selection mechanism is called α-optimal, for $\alpha \leq 1$, if for any input graph the overall indegree of the selected vertices is at least α times the sum of indegree of the k vertices with highest indegrees.

A. Bjelde and M. Klimm—The research was carried out in the framework of MATH-EON supported by the Einstein Foundation Berlin.

F. Fischer—Part of the research was carried out while the author was a member of the Statistical Laboratory at the University of Cambridge.

© Springer-Verlag Berlin Heidelberg 2015
E. Markakis and G. Schäfer (Eds.): WINE 2015, LNCS 9470, pp. 146–158, 2015.
DOI: 10.1007/978-3-662-48995-6_11

In prior work, a striking separation was shown between mechanisms that do not use randomness and those that do. On the one hand, no deterministic exact α-optimal mechanism exists for selecting any fixed number of agents and any $\alpha > 0$ [1]. On the other, a mechanism that considers agents along a random permutation and selects a single agent with a maximum number of nominations from its left achieves a bound of $\alpha = 1/2$ [5]. This bound is in fact best possible subject to impartiality [1].

Our Contribution. We show that randomness can in fact be replaced by the ability to *sometimes but not always* select a second agent: using a fixed permutation instead of a random one but selecting an agent for each direction of that permutation is also 1/2-optimal. The factor of 1/2 is again best possible. A minimal amount of flexibility in the exact number of selected agents is beneficial also in the realm of randomized mechanisms: given a set of three agents, for example, a 3/4-optimal mechanism exists selecting two agents or fewer, whereas the best mechanism selecting exactly two agents is only 2/3-optimal. For an arbitrary number of agents, we construct a randomized exact 7/12-optimal 2-selection mechanism and a randomized (non-exact) 2/3-optimal 2-selection mechanism. Finally, we provide upper and lower bounds on the performance of mechanisms allowed to make more than two choices. A summary of our current state of knowledge is shown in Table 1.

Related Work. The theory of impartial decision making was first considered by de Clippel et al. [4], for the case of a divisible resource to be shared among a set of agents. The difference between divisible and indivisible resources disappears for randomized mechanisms, but the mechanisms of de Clippel et al. allow for fractional nominations and do not have any obvious consequences for our setting. Impartial selection is a rather fundamental problem in social choice theory, with applications ranging from the selection of committees to scientific peer review. The interested reader is referred to the articles of Holzman and Moulin [6] and

Table 1. Bounds on α for α-optimal impartial selection of at most or exactly k agents. For deterministic exact mechanisms (not shown in the table), there is not α-optimal mechanism for any $\alpha > 0$.

k	Deterministic	Randomized exact	Randomized
1	0	$\frac{1}{2}$	$\frac{1}{2}$
2	$\frac{1}{2}$	$\left[\frac{7}{12}, \frac{2}{3}\right]$	$\left[\frac{2}{3}, \frac{3}{4}\right]$
$\vdots$			
k	$\left[\frac{1}{k}, \frac{k-1}{k}\right]$	$\left[\frac{2k-1}{2k}\left(1 - \left(\frac{k-1}{k}\right)^k\right), \frac{k+1}{k+2}\right]$	$\left[\frac{2k-1}{2k}\left(1 - \left(\frac{k-1}{k}\right)^k\right), \frac{k+1}{k+2}\right]$
$\vdots$			
n-2	$\left[\frac{1}{k}, \frac{k-1}{k}\right]$	$\left[\frac{2k-1}{2k}\left(1 - \left(\frac{k-1}{k}\right)^k\right), \frac{7k^3+5k^2-6k+12}{7k^3+13k^2-2k}\right]$	$\left[\frac{2k-1}{2k}\left(1 - \left(\frac{k-1}{k}\right)^k\right), \frac{k+1}{k+2}\right]$
n-1	$\left[\frac{1}{k}, \frac{k-1}{k}\right]$	$\left[\frac{2k-1}{2k}\left(1 - \left(\frac{k-1}{k}\right)^k\right), \frac{k}{k+1}\right]$	$\left[\frac{2k-1}{2k}\left(1 - \left(\frac{k-1}{k}\right)^k\right), \frac{2k}{2k+1}\right]$

Fischer and Klimm [5] for details. Tamura and Ohseto [9] were the first to consider selection mechanisms selecting more than one agent, and showed that these can circumvent some of the impossibility results of Holzman and Moulin. Recently Mackenzie [7] provided a characterization of symmetric randomized selection mechanisms for the special case that each agent nominates exactly one other agent. Inspiration for our title, and indeed for relaxing the requirement to always select the same number of agents, comes from the power of multiple choices in load balancing, where even two choices can lead to dramatically lower average load (e.g., [8]).

Open Problems. With the exception of mechanisms that are asymptotically optimal, when many agents are selected [1] or agents receive many nominations [3], only very little was previously known about the impartial selection of more than one agent. Our understanding of 2-selection is now much better, with some room for improvement in the case of randomized mechanisms. About k-selection for $k > 2$, in particular about deterministic mechanisms for this task, we still know relatively little. This lack of understanding is witnessed by the fact that the optimal deterministic mechanism selecting up to two agents, one for each direction of a permutation, does not generalize in any obvious way to the selection of more than two agents. Meanwhile, the existence of a near-optimal mechanism in the limit of many selected agents suggests that the upper rather than the lower bounds may be correct.

2   Preliminaries

For $n \in \mathbb{N}$, let

$$\mathcal{G}_n = \left\{ (N, E) : N = \{1, \ldots, n\}, E \subseteq (N \times N) \setminus \bigcup_{i \in N} (\{i\} \times \{i\}) \right\}$$

be the set of directed graphs with n vertices and no loops. Let $\mathcal{G} = \bigcup_{n \in \mathbb{N}} \mathcal{G}_n$. For $G = (N, E) \in \mathcal{G}$ and $S, X \subseteq N$ let

$$\delta_S^-(X, G) = |\{(j, i) \in E : G = (N, E), j \in S, i \in X\}|$$

denote the sum of indegrees of vertices in X from vertices in S. We use $\delta^-(X, G)$ as a shorthand for $\delta_N^-(X, G)$ and denote by $\Delta_k(G) = \max_{X \subseteq N, |X| = k} \delta^-(X, G)$. When $X = \{i\}$ for a single vertex i, we write $\delta_S^-(i, G)$ instead of $\delta_S^-(\{i\}, G)$. Most of the time, the graph G will be clear from context. We then write $\delta_S^-(X)$ instead of $\delta_S^-(X, G)$, $\delta^-(X)$ instead of $\delta^-(X, G)$, and Δ_k instead of $\Delta_k(G)$.

For $n, k \in N$, let $\mathcal{X}_n = \{X : X \subseteq \{1, \ldots, n\}\}$ be the set of subsets of the first n natural numbers and let $\mathcal{X}_{n,k} = \{X \in \mathcal{X}_n : |X| = k\}$ be the subset of these sets with cardinality k. A k-selection mechanism for $\mathcal{G}$ is then given by a family of functions $f : \mathcal{G}_n \to [0, 1]^{\bigcup_{\ell=0}^{k} \mathcal{X}_{n,\ell}}$ that maps each graph to a probability distribution on subsets of at most k of its vertices. In a slight abuse of notation, we use f to refer to both the mechanism and individual functions from the family.

We call mechanism f *deterministic* if $f(G) \in \{0, 1\}^{\cup_{\ell=0}^{k} \mathcal{X}_{n,\ell}}$, i.e., if $f(G)$ puts all probability mass on a single set for all $G \in \mathcal{G}$; we call f *exact* if $(f(G))_X = 0$ for every $n \in \mathbb{N}$, $G \in \mathcal{G}_n$, and $X \in \mathcal{X}_n$ with $|X| < k$, i.e., the mechanism never selects a set X of vertices with strictly less than k vertices. Mechanism f is *impartial* on $\mathcal{G}' \subseteq \mathcal{G}$ if on this set of graphs the probability of selecting vertex i does not depend on its outgoing edges, i.e., if for every pair of graphs $G = (N, E)$ and $G' = (N, E')$ in $\mathcal{G}'$ and every $i \in N$, $\sum_{X \in \mathcal{X}_n, i \in X}(f(G))_X = \sum_{X \in \mathcal{X}_n, i \in X}(f(G'))_X$ whenever $E \setminus (\{i\} \times N) = E' \setminus (\{i\} \times N)$. When running an impartial selection mechanism, no agent can modify its probability of being selected by altering its nominations. Note that this does not rule out the possibility of a selected agent to change the number of other agents that are selected as well. All mechanisms we consider are impartial on $\mathcal{G}$, and we simply refer to such mechanisms as impartial mechanisms. Finally, a k-selection mechanism f is *α-optimal* on $\mathcal{G}' \subseteq \mathcal{G}$, for $\alpha \leq 1$, if for every graph in $\mathcal{G}'$ the expected sum of indegrees of the vertices selected by f differs from the maximum sum of indegrees for any k-subset of the vertices by a factor of at most α, i.e., if

$$\inf_{\substack{G \in \mathcal{G} \\ \Delta_k(G) > 0}} \frac{\mathbb{E}_{X \sim f(G)}[\delta^-(X, G)]}{\Delta_k(G)} \geq \alpha.$$

We call a mechanism α-optimal if it is α-optimal on $\mathcal{G}$.

For randomized mechanisms, and as far as impartiality and α-optimality are concerned, we can restrict attention to mechanisms that are *symmetric*, i.e., invariant with respect to renaming of the vertices (e.g., [5]). It may further be convenient to view a k-selection mechanism as assigning probabilities to vertices rather than sets of vertices, with the former summing up to at most k or exactly k for each graph. By the Birkhoff-von Neumann theorem [2], the two views are equivalent.

In the interest of space, we defer most of the proofs to the full version of this paper.

3   Deterministic Mechanisms

Focusing on the exact case, Alon et al. showed that deterministic k-selection mechanisms cannot be α-optimal for any $k \in \{1, \dots, n-1\}$ and any $\alpha > 0$. Although a rather simple observation when $k = 1$, this result is quite surprising for $k > 1$. For $(n-1)$-selection in particular, any deterministic mechanism that is both exact and impartial must sometimes exclude precisely the unique vertex with positive indegree. It is not hard to convince ourselves that relaxing exactness does not help in the case of 1-selection, but we will see momentarily that it is possible to guarantee 1/2-optimality when selecting (up to) two 2 agents without requiring exactness.

Another way to look at this result is that the bound of 1/2 that is best possible for exact randomized mechanisms, can also be achieved by a deterministic mechanism when sometimes but not always one additional vertex is selected.

Input: Graph $G = (N, E)$
Output: Set $\{i_1, i_2\} \subseteq N$ of at most two vertices
1 Let $\pi = (1, \ldots, |N|)$;
2 $i_1 := \Xi_\pi(G)$;
3 $i_2 := \Xi_{\bar{\pi}}(G)$;
4 return $\{i_1, i_2\}$;

Fig. 1. The bidirectional permutation mechanism

Input: Graph $G = (N, E)$, permutation $(\pi_1, \ldots, \pi_{|N|})$ of N
Output: Vertex $i \in N$
1 Set $i := \pi_1$, $d := 0$;
2 **for** $j = 2, \ldots, |N|$ **do**
3 $\quad$ **if** $\delta^-_{\pi_{<j} \setminus \{i\}}(\pi_j) \geq d$ **then**
4 $\quad\quad$ Set $i := \pi_j$, $d := \delta^-_{\pi_{<j}}(\pi_j)$;
5 $\quad$ **end**
6 **end**
7 return i;

Fig. 2. The extraction mechanism Ξ_π

Thus being able to select an additional vertex serves as a perfect substitute for randomness.

Let $N = \{1, \ldots, n\}$. For a graph $G = (N, E)$ and a permutation $\pi = (\pi_1, \ldots, \pi_n)$ of N, denote by

$$E_\pi = \big\{ (u, v) \in E : \pi_i = u, \ \pi_j = v \text{ for some } i, j \text{ with } 1 \leq i < j \leq n \big\}$$

the set of *forward edges* of G with respect to π. Denoting by $\bar{\pi}$ the permutation obtained by reading π backwards, such that $\bar{\pi}_i = \pi_{n+1-i}$ for $i = 1, \ldots, n$. Finally, for a permutation π and $j \in \{1, \ldots, n\}$, let $\pi_{<j} = \{\pi_1, \ldots, \pi_{j-1}\}$.

The first mechanism we consider, which we call the bidirectional permutation mechanism, considers the vertices one by one according to a fixed permutation π and in each step compares the current vertex π_j to a single candidate vertex π_ℓ with $\ell < j$. In determining the indegree of the candidate vertex π_ℓ it takes into account the outgoing edges of vertices $\pi_1, \ldots, \pi_{\ell-1}$. For the indegree of the current vertex π_j it takes into account the outgoing edges of vertices $\pi_1, \ldots, \pi_{j-1}$, with the exception of π_ℓ. If the latter is greater than or equal to the former, π_j becomes the new candidate, and the candidate after the final step is the first vertex selected by the mechanism. The same procedure is then applied with permutation $\bar{\pi}$ to find a second vertex. A formal description of the bidirectional permutation mechanism is given in Fig. 1. It is formulated in terms of the mechanism of Fig. 2, which we call the extraction mechanism and which is identical to a mechanism of Fischer and Klimm except for its use of a given permutation rather than a random one. It is worth noting that the bidirectional permutation mechanism may select only one vertex, namely if the same vertex is chosen for

$$\pi_1 \qquad \pi_2 \qquad \pi_3 \qquad\qquad \pi_1 \qquad \pi_2 \qquad \pi_3$$

(a) (b)

Fig. 3. Graphs for which the bidirectional permutation mechanism returns only one vertex (a) and is only 1/2-optimal (b)

both directions of the permutation. This happens for example in the graph of Fig. 3(a).

To see that the bidirectional permutation mechanism is impartial, we first note that this is true for a single run of the extraction mechanism. Indeed, the outcome of the latter is influenced by the outgoing edges of any given vertex only when that vertex can no longer be selected.

Lemma 1. *The extraction mechanism is impartial.*

Impartiality of the bidirectional permutation mechanism then follows because the union of the results of k impartial 1-selection mechanisms yields an impartial k-selection mechanism.

Lemma 2. *Let $f_1, \ldots, f_k$ be impartial 1-selection mechanisms. Then the mechanism that selects the vertices selected by at least one of the mechanisms $f_1, \ldots, f_k$ is an impartial k-selection mechanism.*

Proof. By impartiality of f_ℓ, $\ell = 1, \ldots, k$, the outgoing edges of a vertex do not influence whether this vertex is selected by f_ℓ. This holds for any ℓ and any vertex, so it also holds for the mechanism that selects the vertices selected by at least one of the mechanisms. □

We now proceed to show that the bidirectional permutation mechanism is 1/2-optimal, starting from the observation that the vertex selected by Ξ_π has a maximum number of incoming forward edges with respect to π.

Lemma 3. *If $i = \Xi_\pi(G)$, then $\delta^-_{\pi_{<i}}(i, G) = \max_{j=1,\ldots,n}\{\delta^-_{\pi_{<j}}(j, G)\}$.*

Proof. It is easy to see that the value of d does not decrease as the mechanism proceeds, so the number of incoming forward edges of the current candidate cannot decrease either. Let $d^* = \max_{j=1,\ldots,n}\{\delta^-_{\pi_{<j}}(j)\}$, and let i^* be a vertex with $\delta^-_{\pi_{<i^*}}(i^*) = d^*$. When i^* is considered by the mechanism, so are at least $d^* - 1$ of its incoming forward edges, one of which may originate from the current candidate i. If $\delta^-_{\pi_{<i}}(i) = d^*$, the selected vertex will have at least d^* incoming forward edges and we are done. If, on the other hand, $\delta^-_{\pi_{<i}}(i) \leq d^* - 1$, then the i^* is made the new candidate. Before that, however, also a possible edge from i to i^* is considered by the mechanism, so that $d = d^*$, regardless whether there is an edge from i to i^*, or not. Again the selected vertex will have at least d^* incoming forward edges, and the claim follows. □

Theorem 1. *The bidirectional permutation mechanism is impartial and 1/2-optimal.*

Proof. Impartiality follows directly from Lemmas 1 and 2.

Now consider a graph $G = (N, E)$, a vertex i^* with $\delta^-(i^*) = \Delta_1$, and let $i_1 = \Xi_\pi(G)$ and $i_2 = \Xi_{\bar\pi}(G)$. By Lemma 3, $\delta^-_{\pi_{<i_1}}(i_1) \geq \delta^-_{\pi_{<i^*}}(i^*)$ and $\delta^-_{\bar\pi_{<i_2}}(i_2) \geq \delta^-_{\bar\pi_{<i^*}}(i^*)$, regardless of whether $i_1 \neq i_2$ or $i_1 = i_2$. Thus

$$\delta^-(\{i_1, i_2\}) \geq \delta^-_{\pi_{<i_1}}(i_1) + \delta^-_{\bar\pi_{<i_2}}(i_2)$$
$$\geq \delta^-_{\pi_{<i^*}}(i^*) + \delta^-_{\bar\pi_{<i^*}}(i^*) = \delta^-(i^*) = \Delta_1 \geq \frac{1}{2}\Delta_2,$$

as claimed.   □

To see that the analysis is tight, consider the graph in Fig. 3(b). For this graph, the mechanism selects vertices π_3 and π_1 with an overall indegree of 1, while the maximum overall indegree of a set of two vertices is 2. We will see later, in Theorem 6, that the bound of 1/2 is in fact best possible.

4   Randomized Mechanisms

In light of the results of the previous section, it is natural to ask whether a relaxation of exactness enables better bounds also for randomized mechanisms. We answer this question in the affirmative and give the first nontrivial bounds for both exact and inexact 2-selection mechanisms, as well as an example that shows a strict separation between the two classes.

We begin by considering an exact mechanism, which we call the 2-partition mechanism with permutation. The mechanism randomly partitions the set of vertices into two sets A_1 and A_2 such that $\mathbb{P}[i \in A_1] = \mathbb{P}[i \in A_2] = 1/2$ for all $i \in N$, $A_1 \cup A_2 = N$, and $A_1 \cap A_2 = \emptyset$. It then selects one vertex from each of the sets by applying the extraction mechanism with a random permutation, while also taking into account incoming edges from the respective other set. Figure 4 shows a formal description of the mechanism. It uses a restricted version of the

---

> **Input:** Graph $G = (N, E)$ with $|N| \geq 2$
> **Output:** Vertices $i_1, i_2 \in N$.
> **1** Assign each $i \in N$ to A_1 or A_2 independently and uniformly at random;
> **2** Choose a permutation $(\pi_1, \ldots, \pi_{|N|})$ of N uniformly at random;
> **3** for $j = 1, 2$ do
> **4** | $i_j := \Xi_{\pi, A_j}(G)$;
> **5** end
> **6** if $A_2 = \emptyset$ then choose i_2 uniformly at random from $A_1 \setminus i_1$;
> **7** if $A_1 = \emptyset$ then choose i_1 uniformly at random from $A_2 \setminus i_2$;
> **8** return $\{i_1, i_2\}$;

Fig. 4. The 2-partition mechanism with permutation

Input: Graph $G = (N, E)$, permutation $(\pi_1, \ldots, \pi_{|N|})$ of N, set $A \subseteq N$
Output: Vertex $i \in N$
1 Set $i := \pi_1$, $d := 0$;
2 **for** $j = 2, \ldots, |N|$ **do**
3     $\quad S := (N \setminus A) \cup (\pi_{<j} \setminus \{i\})$;
4     $\quad$ **if** $\pi_j \in A$ **and** $\delta_S^-(\pi_j) \geq d$ **then**
5     $\quad \quad$ Set $i := \pi_j$, $d := \delta_{S \cup \{i\}}^-(\pi_j)$;
6     $\quad$ **end**
7 **end**
8 **return** i;

Fig. 5. The extraction mechanism $\Xi_{\pi,A}$ restricted to a set $A \subseteq N$

extraction mechanism, given in Fig. 5 and denoted $\Xi_{\pi,A}$ for a set $A \subseteq N$. The properties of the latter can be summarized in terms of the following two results.

Lemma 4. *The restricted extraction mechanism is impartial.*

Lemma 5. *If $i = \Xi_{\pi,A}(G)$, then $\delta^-(i, G) \geq \max_{j \in A}\{\delta_{(N \setminus A) \cup \pi_{<j}}^-(j, G)\}$.*

The proofs of these results follow along the same lines as those of Lemmas 1 and 3. We proceed directly with a result for the 2-partition mechanism with permutation.

Theorem 2. *The 2-partition mechanism with permutation is impartial and 7/12-optimal.*

Proof. Impartiality follows directly from Lemmas 2 and 4.

Now consider a graph $G = (N, E)$, two distinct vertices $i_1^*, i_2^* \in N$ with $\delta^-(i_1^*) + \delta^-(i_2^*) = \Delta_2$, and let i_1 and i_2 be the two vertices selected by the mechanism. We distinguish two cases, depending on whether i_1^* and i_2^* are in the same set or different sets of the partition (A_1, A_2).

First assume that i_1^* and i_2^* are in different sets, and without loss of generality that $i_1^* \in A_1$ and $i_2^* \in A_2$. In the permutation π used by the mechanism and chosen uniformly at random, an arbitrary vertex $i \in N \setminus \{i_1^*, i_2^*\}$ appears before or after each of i_1^* or i_2^* with equal probability, so

$$\mathbb{P}\big[i \in A_1 \cap \pi_{<i_1^*}\big] = \mathbb{P}\big[i \in A_1 \cap \bar{\pi}_{<i_1^*}\big]$$

$$= \mathbb{P}\big[i \in A_2 \cap \pi_{<i_2^*}\big] = \mathbb{P}\big[i \in A_2 \cap \bar{\pi}_{<i_2^*}\big] = \frac{1}{4}.$$

When i_1^* is considered by the mechanism, so are any incoming edges from A_2 and any incoming edges from vertices in A_1 and appearing in π before i_1^*. Thus, by Lemma 5,

$$\mathbb{E}\big[\delta^-(i_1)\big] \geq \mathbb{E}\big[\delta_{A_2 \cup \pi_{<i_1^*}}^-(i_1^*)\big] = \sum_{i \in N} \mathbb{P}\big[i \in A_2 \cup (A_1 \cap \pi_{<i_1^*})\big] \cdot \chi[(i, i_1^*) \in E],$$

where χ denotes the indicator function on Boolean expressions, i.e., $\chi[\phi] = 1$ if expression ϕ holds and $\chi[\phi] = 0$ otherwise. By taking i_2^* out of the sum and using that i_1^* and i_2^* are in different sets of the partition and thus $\mathbb{P}[i_2^* \in A_2] = 1$, we obtain

$$\mathbb{E}[\delta^-(i_1)] = \sum_{i \in N \setminus \{i_2^*\}} \Big(\mathbb{P}[i \in A_2 \cup (A_1 \cap \pi_{<i_1^*})] \cdot \chi[(i, i_1^*) \in E] \Big)$$
$$+ \mathbb{P}[i_2^* \in A_2 \cup (A_1 \cap \pi_{<i_1^*})] \cdot \chi[(i_2^*, i_1^*) \in E]$$
$$\geq \frac{3}{4} \sum_{i \in N} \chi[(i, i_1^*) \in E] = \frac{3}{4} \delta^-(i_1^*).$$

As the same line of reasoning applies to i_2^*, we conclude for this case that

$$\mathbb{E}\left[\frac{\delta^-(i_1, i_2)}{\Delta_2}\right] \geq \frac{3}{4}.$$

Now assume that i_1^* and i_2^* are in the same set of the partition, and without loss of generality that $i_1^*, i_2^* \in A_1$ and $\delta^-(i_1^*) \geq \delta^-(i_2^*)$. In the permutation π used by the mechanism and chosen uniformly at random, an arbitrary vertex $i \in N \setminus \{i_1^*, i_2^*\}$ appears before, between, or after i_1^* and i_2^* with probability $1/3$ each, so

$$\mathbb{P}[i \in A_2] = \frac{1}{2} \quad \text{and}$$
$$\mathbb{P}[i \in A_1 \cap \pi_{<i_1^*} \cap \pi_{<i_2^*}] =$$
$$\mathbb{P}[i \in A_1 \cap ((\pi_{<i_1^*} \cap \bar{\pi}_{<i_2^*}) \cup (\bar{\pi}_{<i_1^*} \cap \pi_{<i_2^*}))] =$$
$$\mathbb{P}[i \in A_1 \cap \bar{\pi}_{<i_1^*} \cap \bar{\pi}_{<i_2^*}] = \frac{1}{6}.$$

If $i_1^* \in \bar{\pi}_{<i_2^*}$, a possible edge from i_2^* to i_1^* would be considered by the mechanism, and by Lemma 5,

$$\mathbb{E}[\delta^-(i_1)] \geq \mathbb{E}[\delta^-_{A_2 \cup \pi_{<i_1^*}}(i_1^*)] = \left(\frac{1}{2} + \frac{2}{6}\right) \delta^-(i_1^*).$$

Analogously, if $i_2^* \in \bar{\pi}_{<i_1^*}$,

$$\mathbb{E}[\delta^-(i_1))] \geq \mathbb{E}[\delta^-_{A_2 \cup \pi_{<i_2^*}}(i_2^*)] = \left(\frac{1}{2} + \frac{2}{6}\right) \delta^-(i_2^*).$$

As each of the two events takes places with probability $1/2$, we conclude for this case that

$$\mathbb{E}\left[\frac{\delta^-(i_1 \cup i_2)}{\Delta_2}\right] \geq \frac{\frac{1}{2}\left(\frac{5}{6}\delta^-(i_1^*) + \frac{5}{6}\delta^-(i_2^*)\right)}{\delta^-(i_1^*) + \delta^-(i_2^*)} = \frac{5}{12}.$$

Averaging over both cases we finally obtain

$$\alpha \geq \frac{1}{2}\left(\frac{3}{4} + \frac{5}{12}\right) = \frac{7}{12}$$

as claimed. $\square$

Input: Graph $G = (N, E)$
Output: Vertex $i \in N$
1 Choose a permutation $(\pi_1, \ldots, \pi_{|N|})$ of N uniformly at random;
2 Invoke the bidirectional permutation mechanism of Figure 1 for G and π

Fig. 6. The randomized bidirectional permutation mechanism

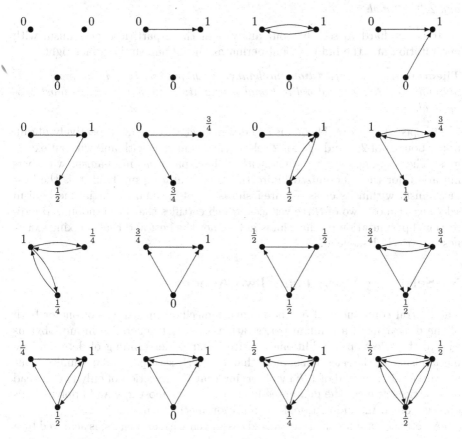

Fig. 7. A 3/4-optimal impartial mechanism for $n = 3$ and $k = 2$ given explicitly by the selection probabilities for all 16 voting graphs. The bound of 3/4 is best possible by Theorem 8.

The 2-partition mechanism with permutation improves on the best deterministic mechanism for 2-selection, and it is natural to ask whether it can be improved upon further by a randomized 2-selection mechanism that is not exact. The answer to this question is not obvious: while the ability to select fewer vertices may make impartiality easier to achieve, actually selecting fewer vertices runs counter to the objective of selecting vertices with a large sum of indegrees. Indeed, in the case of 1-selection, no difference exists between exact and inexact mechanisms. For 2-selection, an obvious approach turns out to be effective:

taking the best deterministic mechanism, which uses both directions of a fixed permutation, and invoking it for a random permutation. The resulting mechanism, which we call the randomized bidirectional permutation mechanism, is shown in Fig. 6.

Theorem 3. *The randomized bidirectional permutation mechanism is impartial and 2/3-optimal.*

It is not hard to see that our analysis of the 2-partition mechanism with permutation and the bidirectional permutation mechanism is in fact tight.

Theorem 4. *The 2-partition mechanism with permutation is at most 7/12-optimal. The randomized bidirectional permutation mechanism is at most 2/3-optimal.*

As special cases of Theorems 7 and 8 in Sect. 6, we will respectively obtain upper bounds of 2/3 and 3/4 for 2-selection mechanisms with and without exactness. These bounds suggest that neither the 2-partition mechanism with permutation nor the randomized bidirectional permutation mechanism is the best mechanism within its class. Figure 7 shows a 3/4-optimal impartial mechanism selecting at most two of three vertices, which certifies that the randomized bidirectional permutation mechanism is indeed not the best and that relaxing exactness is strictly beneficial.

5   Selecting More Than Two Agents

The central component of our best inexact mechanisms, its use of one or both of the directions of a random permutation, does not generalize in any obvious way to the selection of additional vertices. Our understanding of deterministic mechanisms for the selection of more than two vertices is particularly limited, but we can obtain a bound of $1/k$ by observing that the selection of only two instead of k vertices reduces the guarantee by a factor of at most $2/k$ and applying this observation to the bidirectional permutation mechanism.

A better bound for the randomized case, even with exactness, is achieved by a natural generalization of the 2-partition mechanism with permutation that uses a partition into k sets. The general mechanism is described formally in Fig. 8. Its impartiality is easy to see, and we use an argument similar to Lemma 5 to obtain a performance guarantee that approaches $1 - 1/e$ as k grows.

Theorem 5. *The k-partition mechanism with permutation is impartial and α-optimal where $\alpha = \frac{2k-1}{2k}\left(1 - (\frac{k-1}{k})^k\right)$.*

6   Upper Bounds

We conclude by giving upper bounds on the performance of impartial k-selection mechanisms for any value of k, and for both deterministic mechanisms and randomized mechanisms with and without exactness.

Input: Graph $G = (N, E)$ with $|N| \geq 2$
Output: Vertices $i_1, \ldots, i_k \in N$

1 Assign each $i \in N$ independently and uniformly at random to one of k sets
   $A_1, \ldots, A_k$, such that $\mathbb{P}[i \in A_1] = \cdots = \mathbb{P}[i \in A_k] = \frac{1}{k}$ for all $i \in N$,
   $A_1 \cup \cdots \cup A_k = N$, and $A_i \cap A_j = \emptyset$ for all $i \neq j \in N$;
2 Choose a permutation $(\pi_1, \ldots, \pi_{|N|})$ of N uniformly at random;
3 **for** $j = 1, \ldots, k$ **do**
4 $\quad$ $i_j := \Xi_{\pi, A_j}(G)$;
5 **end**
6 **for** $j = 1, \ldots, k$ **do**
7 $\quad$ **if** $A_j = \emptyset$ **then**
8 $\quad\quad$ Choose i_j uniformly at random from $N \setminus \{i_1, \ldots, i_k\}$;
9 $\quad$ **end**
10 **end**
11 Return $(i_1, \ldots, i_k)$;

Fig. 8. The k-partition mechanism with permutation

The first set of bounds applies to deterministic mechanisms and shows that the bidirectional permutation mechanism is the best deterministic mechanism for $k = 2$.

Theorem 6. *Consider a deterministic k-selection mechanism that is α-optimal on $\mathcal{G}_n$, where $k < n$. Then $\alpha \leq (k-1)/k$.*

Proof. Consider a graph $G = (V, E)$ with n vertices where $k + 1$ vertices are arranged in a directed cycle and the remaining vertices do not have any outgoing edges, i.e., $V = \{1, \ldots, n\}$ and $E = \{(i, i+1) : i = 1, \ldots, k\} \cup \{(k+1, 1)\}$. Denote by F the set of vertices selected from G by an arbitrary deterministic k-selection mechanism, and observe that there exists $i \in \{1, \ldots, k+1\} \setminus F$. Let $G' = (V, E \setminus (\{i\} \times V))$, and observe that by impartiality, the mechanism does not select i from G'. The mechanism thus selects at most $k - 1$ out of the k vertices with positive indegree in G' and cannot be more than $(k-1)/k$-optimal. $\square$

The next result concerns randomized mechanisms that are exact. It certifies that the 2-partition mechanism with permutation is best possible within this class when $k = 2$ and $n = 3$, and together with the mechanism of Fig. 7 shows a strict separation between randomized mechanisms with and without exactness. It does not preclude improvements over the 2-partition mechanism with permutation when $n > 3$. The proof requires more effort than that of Theorem 8 below but no additional techniques.

Theorem 7. *Consider a k-selection mechanism that is exact, impartial, and α-optimal on $\mathcal{G}_n$, where $k < n$. Then*

$$\alpha \leq \begin{cases} \frac{1}{2} & \text{if } k = 1 \\ \frac{k}{k+1} & \text{if } 2 \leq k = n - 1 \\ \frac{5}{7} & \text{if } 2 = k = n - 2 \\ \frac{7k^3 + 5k^2 - 6k + 12}{7k^3 + 13k^2 - 2k} & \text{if } 3 \leq k = n - 2 \\ \frac{k+1}{k+2} & \text{otherwise.} \end{cases}$$

Our final result applies to randomized mechanisms without the requirement of exactness, and shows that the mechanism of Fig. 7 for the case when $k = 2$ and $n = 3$ is best possible. A comparison with Theorem 7 further suggests that the influence of the exactness constraint may be limited to cases where almost all vertices are selected.

Theorem 8. *Consider a k-selection mechanism that is impartial and α-optimal on $\mathcal{G}_n$, where $k < n$. Then*

$$\alpha \leq \begin{cases} \frac{1}{2} & \text{if } k = 1 \\ \frac{3}{4} & \text{if } k = 2 \\ \frac{2k}{2k+1} & \text{if } 3 \leq k = n - 1 \\ \frac{k+1}{k+2} & \text{otherwise.} \end{cases}$$

Acknowledgements. We thank the anonymous referees for their suggestions to improve the presentation of the results.

References

1. Alon, N., Fischer, F., Procaccia, A.D., Tennenholtz, M.: Sum of us: Strategyproof selection from the selectors. In: Proceedings of the 13th Conference on Theoretical Aspects of Rationality and Knowledge, pp. 101–110 (2011)
2. Birkhoff, G.: Tres observaciones sobre el algebra lineal. Revista Facultad de Ciencias Exactas, Puras y Aplicadas Universidad Nacional de Tucumán, Serie A **5**, 147–151 (1946)
3. Bousquet, N., Norin, S., Vetta, A.: A near-optimal mechanism for impartial selection. In: Liu, T.-Y., Qi, Q., Ye, Y. (eds.) WINE 2014. LNCS, vol. 8877, pp. 133–146. Springer, Heidelberg (2014)
4. de Clippel, G., Moulin, H., Tideman, N.: Impartial division of a dollar. J. Econ. Theor. **139**(1), 176–191 (2008)
5. Fischer, F., Klimm, M.: Optimal impartial selection. In: Proceedings of the 15th ACM Conference on Economics and Computation, pp. 803–820 (2014)
6. Holzman, R., Moulin, H.: Impartial nominations for a prize. Econometrica **81**(1), 173–196 (2013)
7. Mackenzie, A.: Symmetry and impartial lotteries. Games Econ. Behav. **94**, 15–28 (2015)
8. Mitzenmacher, M., Richa, A.W., Sitaraman, R.: The power of two random choices: a survey of techniques and results. In: Rajasekaran, S., Pardalos, P.M., Reif, J.H., Rolim, J. (eds.) Handbook of Randomized Computing, vol. 1, pp. 255–312. Springer (2001)
9. Tamura, S., Ohseto, S.: Impartial nomination correspondences. Soc. Choice Welfare **43**(1), 47–54 (2014)

Online Allocation and Pricing
with Economies of Scale

Avrim Blum[1], Yishay Mansour[2], and Liu Yang[3]([✉])

[1] Computer Science Department, Carnegie Mellon University, Pittsburgh, USA
avrim@cs.cmu.edu
[2] Computer Science Department, Tel Aviv University, Tel Aviv, Israel
mansour@tau.ac.il
[3] IBM T. J. Watson Research Center, Yorktown Heights, USA
yangli@us.ibm.com

Abstract. Allocating multiple goods to customers in a way that maximizes some desired objective is a fundamental part of Algorithmic Mechanism Design. We consider here the problem of offline and online allocation of goods that have economies of scale, or decreasing marginal cost per item for the seller. In particular, we analyze the case where customers have unit-demand and arrive one at a time with valuations on items, sampled iid from some unknown underlying distribution over valuations. Our strategy operates by using an initial sample to learn enough about the distribution to determine how best to allocate to future customers, together with an analysis of structural properties of optimal solutions that allow for uniform convergence analysis. We show, for instance, if customers have $\{0, 1\}$ valuations over items, and the goal of the allocator is to give each customer an item he or she values, we can efficiently produce such an allocation with cost at most a constant factor greater than the minimum over such allocations in hindsight, so long as the marginal costs do not decrease too rapidly. We also give a bicriteria approximation to social welfare for the case of more general valuation functions when the allocator is budget constrained.

1 Introduction

Imagine it is the Christmas season, and Santa Claus is tasked with allocating toys. There is a sequence of children coming up with their Christmas lists of toys they want. Santa wants to give each child some toy from his or her list (all children have been good this year). But of course, even Santa Claus has to be cost-conscious, so he wants to perform this allocation of toys to children at a near-minimum cost to himself (call this the Thrifty Santa Problem). Now if it was the case that every toy had a fixed price, this would be easy: simply allocate to each child the cheapest toy on his or her list and move on to the next child. But here we are interested in the case where goods have economies of scale. For example, producing a millon toy cars might be cheaper than a million times the cost of producing one toy car. Thus, even if producing a single toy car is more

© Springer-Verlag Berlin Heidelberg 2015
E. Markakis and G. Schäfer (Eds.): WINE 2015, LNCS 9470, pp. 159–172, 2015.
DOI: 10.1007/978-3-662-48995-6_12

expensive than a single Elmo doll, if a much larger number of children want the toy car than the Elmo doll, the minimum-cost allocation might give toy cars to many children, even if some of them also have the Elmo doll on their lists.

The problem faced by Santa (or by any allocator that must satisfy a collection of disjunctive constraints in the presence of economies of scale) makes sense in both offline and online settings. In the offline setting, in the extreme case of goods such as software where all the cost is in the first copy, this is simply weighted set-cover, admitting a $\Theta(\log n)$ approximation to the minimum-cost allocation. We will be interested in the online case where customers are iid from some arbitrary distribution over subsets of item-set $\mathcal{I}$ (i.e., Christmas lists), where the allocator must make allocation decisions online, and where the marginal cost of goods does not decrease so sharply. We show that for a range of cost curves, including the case that the marginal cost of copy t of an item is $t^{-\alpha}$, for some $\alpha \in [0, 1)$, we will be able to get a constant-factor approximation even online so long as the number of customers is sufficiently large compared to the number of items.

A basic structural property we show is that, if the marginal costs are non-increasing, the minimum cost allocation can be compactly described as an ordering of the possible toys, so that as each child comes, Santa simply gives the child the first toy in the ordering that appears on the child's list. We also show that if the marginal costs do not drop too quickly, then if we are given the lists of all the children before determining the allocation, we can efficiently find an allocation that is within a constant factor of the minimum-cost allocation, as opposed to the logarithmic factor required for the set-cover problem. Since, however, the problem we are interested in does not supply the lists before the allocations, but rather requires a decision for each child in sequence, we use ideas from machine learning, as follows: after processing a small initial number of children, we take their wish lists as if they were perfectly representative of the future children, and find an approximately optimal solution based on those, which also will be an ordering over toys. We then take the ordered list of toys from this solution, and use it to allocate to future children (allocating to each child the earliest toy in the ordering that is also on his or her list). We show that, as long as we take a sufficiently large number of initial children, this procedure will find an ordering that will be near-optimal for allocating to the remaining children, using the fact that these compact representations allow for uniform convergence of the cost estimates to the true costs.

More generally, we can imagine the case where, rather than simple lists of items, the lists also provide valuations for each item, and we are interested in the trade-off between maximizing the total of valuations for allocated items while minimizing the total cost of the allocation. In this case, we might think of the allocator as being a large company with many different projects, and each project has some valuations over different resources (e.g., types of laptops for employees involved in that project), where it could use one or another resource but prefers some resources over others. One natural quantity to consider in this context is the social surplus: the difference between the happiness (total of valuations for the allocation) minus the total cost of the allocation. In this case, it turns out the optimal allocation rule can be described by a pricing scheme. In another words, whatever the optimal allocation is, there always exist prices such that if the buyers purchase what they most want at those prices, they will actually produce that allocation. We note that, algorithmically, this is a harder problem than the list-based problem (which corresponds to binary valuations).

Aside from social surplus, it is also interesting to consider a variant in which we have a budget constraint, and are interested in maximizing the total valuation of the allocation, subject to that budget constraint on the total cost of the allocation. It turns out this latter problem can be reduced to a problem known as the weighted budget maximum coverage problem. Technically, this problem is originally formulated for the case in which the marginal cost of a given item drops to zero after the first item of that type is allocated (as in the set cover reduction mentioned above); however, viewed appropriately, we are able to formulate this reduction for arbitrary decreasing marginal cost functions. What we can then do is run an algorithm for the weighted budget maximum coverage problem, and then convert the solution into a pricing. As before, this strategy will be effective for the offline problem, in which all of the valuations are given ahead of time. However, we can extend it to the online setting with iid valuation functions by generating a pricing based on an appropriately-sized initial sample of valuation functions, and then apply that pricing to sequentially generate allocations for the remaining valuations. As long as the marginal costs are not decreasing too rapidly, we can obtain an allocation strategy for which the sum of valuations of the allocated items will be within a constant factor of the maximum possible, minus a small additive term, subject to the budget constraint on the cost.

Our Results and Techniques: We consider this problem under two, related, natural objectives. In the first (the "thrifty Santa" objective) we assume customers have binary $\{0,1\}$ valuations, and the goal of the seller is to give each customer a toy of value 1, but in such a way that minimizes the total cost to the seller. We show that so long as the number of buyers n is large compared to the number of items r, and so long as the marginal costs do not decrease too rapidly (e.g., a rate $1/t^\alpha$ for some $0 \leq \alpha < 1$), we can efficiently perform this allocation task with cost at most a constant factor greater than that of the optimal allocation of items in hindsight. Note that if costs decrease much more rapidly, then even if all customers' valuations were known up front, we would be faced with (roughly) a set-cover problem and so one could not hope to achieve cost $o(\log n)$ times optimal. The second objective we consider, which we apply to customers of arbitrary unit-demand valuation, is that of maximizing total social welfare of customers subject to a cost bound on the seller; for this, we also give a strategy that is constant-competitive with respect to the optimal allocation in hindsight.

Our algorithms operate by using initial buyers to learn enough about the distribution to determine how best to allocate to the future buyers. In fact, there are two main technical parts of our work: the sample complexity and the algorithmic aspects. From the perspective of sample complexity, one key component of this analysis is examining how complicated the allocation rule needs to be in order to achieve good performance, because simpler allocation rules require fewer samples in order to learn. We do this by providing a characterization of what the optimal strategies look like. For example, for the thrifty Santa Claus version, we show that the optimal solution can be assumed wlog to have a simple permutation structure. In particular, so long as the marginal costs are nonincreasing, there is always an optimal strategy in hindsight of this form: order the items according to some permutation and for each bidder, give it the earliest item of its desire in the permutation. This characterization is used inside both our sample complexity results and our algorithmic guarantees. Specifically, we prove that for cost function $\text{cost}(t) = \sum_{\tau=1}^{t} 1/\tau^{\alpha}$, for $\alpha \in [0,1)$, running greedy weighted set cover incurs total cost at most $\frac{1}{1-\alpha}\text{OPT}$. More generally, if the average cost is within some factor of the marginal cost, we have a greedy algorithm that achieves constant approximation ratio. To allocate to new buyers, we simply give it the earliest item of its desire in the learnt permutation. For the case of general valuations, we give a characterization showing that the optimal allocation rule in terms of social welfare can be described by a pricing scheme. That is, there exists a pricing scheme such that if buyers purchased their preferred item at these prices, the optimal allocation would result. Algorithmically, we show that we can reduce to a weighted budgeted maximum coverage problem with single-parameter demand for which there is a known constant-approximation-ratio algorithm [7].

Related Work: In this work we focus on the case of decreasing marginal cost. There have been a large body of research devoted to unlimited supply, which is implicitly constant marginal cost (e.g., [9] Chap. 13), where the goal is to achieve a constant competitive ratio in both offline and online models. The case of increasing marginal cost was studied in [2] where constant competitive ratio was given.

We analyze an online setting where buyers arrive one at a time, sampled iid from some unknown underlying distribution over valuations. Other related online problems with stochastic inputs such as matching problems have been studied in ad auctions [5,8]. Algorithmically, our work is related to the online set cover body of work where [1] gave the first $O(\log m \log n)$ competitive algorithm (here n is the number of elements in the ground set and m is size of a family of subsets of the ground set). The problems we study are also related to online matching problems [3,4,6] in the iid setting; however our problem is more like the "opposite" of online matching in that the cumulative cost curve for us is concave rather than convex.

2   Model, Definitions, and Notation

We have a set $\mathcal{I}$ of r items. We have a set $N = \{1,\ldots,n\}$ indexing n unit demand buyers. Our setting can then generally be formalized in the following terms.

Utility Functions: Each buyer $j \in N$ has a weight $u_{j,i}$ for each item $i \in \mathcal{I}$. We suppose the vectors $u_{j,\cdot}$ are sampled i.i.d. according to a fixed (but arbitrary and unknown) distribution. In the *online* setting we are interested in, the buyers' weight vectors $u_{j,\cdot}$ are observed in sequence, and for each one (before observing the next) we are required to allocate a set of items $T_j \subseteq \mathcal{I}$ to that buyer. The *utility* of buyer j for this allocation is then defined as $u_j(T_j) = \max_{i \in T_j} u_{j,i}$. A few of our results consider a slight variant of this model, in which we are only required to begin allocating goods after some initial $o(n)$ number of customers has been observed (to whom we may allocate items retroactively).

This general setting is referred to as the *weighted unit demand* setting. We will also be interested in certain special cases of this problem. In particular, many of our results are for the *uniform unit demand* setting, in which every $j \in N$ and $i \in \mathcal{I}$ have $u_{j,i} \in \{0,1\}$. In this case, we may refer to the set $S_j = \{i \in \mathcal{I} : u_{j,i} = 1\}$ as the list of items buyer j *wants* (one of).

Production Cost: We suppose there are *cumulative cost functions* $\mathrm{cost}_i : \mathbb{N} \to [0, \infty]$ for each item $i \in \mathcal{I}$, where for $t \in \mathbb{N}$, the value of $\mathrm{cost}_i(t)$ represents the cost of producing t copies of item i. We suppose each $\mathrm{cost}_i(\cdot)$ is nondecreasing.

We would like to consider the case of *decreasing marginal cost*, where $t \mapsto \mathrm{cost}_i(t+1) - \mathrm{cost}_i(t)$ is nonincreasing for each $i \in \mathcal{I}$.

A natural class of decreasing marginal costs we will be especially interested in are of the form $t^{-\alpha}$ for $\alpha \in [0,1)$. That is, $\mathrm{cost}_i(t) = c_0 \sum_{\tau=1}^{t} \tau^{-\alpha}$.

Allocation Problems: After processing the n buyers, we will have allocated some set of items T, consisting of $m_i(T) = \sum_{j \in N} \mathbb{I}_{T_j}(i)$ copies of each item $i \in \mathcal{I}$. We are then interested in two quantities in this setting: the *total (production) cost* $\mathrm{cost}(T) = \sum_{i \in \mathcal{I}} \mathrm{cost}_i(m_i(T))$ and the *social welfare* $SW(T) = \sum_{j \in N} u_j(T_j)$.

We are interested in several different objectives within this setting, each of which is some variant representing the trade-off between reducing total production cost while increasing social welfare.

In the *allocate all* problem, we have to allocate to each buyer $j \in N$ one item $i \in S_j$ (in the uniform demand setting): that is, $SW(T) = n$. The goal is to minimize the total cost $\mathrm{cost}(T)$, subject to this constraint.

The *allocate with budget* problem requires our total cost to never exceed a given limit b (i.e., $\mathrm{cost}(T) \leq b$). Subject to this constraint, our objective is to maximize the social welfare $SW(T)$. For instance, in the uniform demand setting, this corresponds to maximizing the number of satisfied buyers (that get an item from their set S_j).

The objective in the *maximize social surplus* problem is to maximize the difference of the social welfare and the total cost (i.e., $SW(T) - \mathrm{cost}(T)$).

3   Structural Results and Allocation Policies

We now present several results about the structure of optimal (and non-optimal but "reasonable") solutions to allocation problems in the setting of decreasing marginal costs. These will be important in our sample-complexity analysis because they allow us to focus on allocation policies that have inherent complexity that depends only on the number of *items* and not on the number of *customers,* allowing for the use of uniform convergence bounds. That is, a small random sample of customers will be sufficient to uniformly estimate the performance of these policies over the full set of customers.

Permutation and Pricing Policies: A *permutation policy* has a permutation π over $\mathcal{I}$ and is applicable in the case of uniform unit demand. Given buyer j arriving, we allocate to him the minimal (first) demanded item in the permutation, i.e., $\arg\min_{i \in S_j} \pi(i)$. A *pricing policy* assigns a price price_i to each item i and is applicable to general quasilinear utility functions. Given buyer j arriving, we allocate to him whatever he wishes to purchase at those prices, i.e., $\arg\max_{T_j} u_j(T_j) - \sum_{i \in T_j} \text{price}_i$.[1]

We will see below that for uniform unit demand buyers, there always exists a permutation policy that is optimal for the allocate-all task, and for general quasilinear utilities there always exists a pricing policy that is optimal for the task of maximizing social surplus. We will also see that for weighted unit demand buyers, there always exists a pricing policy that is optimal for the allocate-with-budget task; moreover, for any even non-optimal solution (e.g., that might be produced by a polynomial-time algorithm) there exists a pricing policy that sells the same number of copies each item and has social welfare at least as high (and can be computed in polynomial time given the initial solution).

Structural Results:

Theorem 1. *For general quasilinear utilities, any allocation that maximizes social surplus can be produced by a pricing policy. That is, if $\mathcal{T} = \{T_1, \ldots, T_n\}$ is an allocation maximizing $SW(\mathcal{T}) - \text{cost}(\mathcal{T})$ then there exist prices $\text{price}_1, \ldots, \text{price}_r$ such that buyers purchasing their most-demanded bundle recovers $\mathcal{T}$, assuming that the marginal cost function is strictly decreasing.*[2]

Proof. Consider the optimal allocation OPT. Define price_i to be the marginal cost of the next copy of item i under OPT, i.e., $\text{price}_i = \text{cost}_i(\#_i(\text{OPT}) + 1)$. Suppose some buyer j is assigned set T_j in OPT but prefers set T'_j under these prices. Then, $u_j(T'_j) - \sum_{i \in T'_j} \text{price}_i \geq u_j(T_j) - \sum_{i \in T_j} \text{price}_i$, which implies $u_j(T'_j) - u_j(T_j) + \sum_{i \in T_j \setminus T'_j} \text{price}_i - \sum_{i \in T'_j \setminus T_j} \text{price}_i \geq 0$. Now, consider modifying OPT by replacing T_j with T'_j. This increases buyer j's utility by $u_j(T'_j) - u_j(T_j)$, incurs an extra purchase cost *exactly* $\sum_{i \in T'_j \setminus T_j} \text{price}_i$ and a savings of strictly more than $\sum_{i \in T_j \setminus T'_j} \text{price}_i$ (because marginal costs are decreasing). Thus, by the above inequality, this would be a strictly preferable allocation, contradicting the optimality of OPT.   □

Corollary 1. *For uniform unit demand buyers there exists an optimal allocation that is a permutation policy, for the allocate all task.*

---

[1] When more that one subset is applicable, we assume we have the freedom to select any such set. Note that such policies are incentive-compatible.

[2] If the marginal cost function is only non-increasing, we can have the same result, assuming we can select between the utility maximizing bundles.

Proof. Imagine each buyer j had valuation v_{max} on items in S_j where v_{max} is greater than the maximum cost of any single item. The allocation OPT that maximizes social surplus would then minimize cost subject to allocating exactly one item to each buyer and therefore would be optimal for the allocate-all task. Consider the pricing associated to this allocation given by Theorem 1. Since each buyer j is uniform unit demand, he will simply purchase the cheapest item in S_j. Therefore, the permutation π that orders items according to increasing price according to the prices of Theorem 1 will produce the same allocation.   □

We now present a structural statement that will be useful for the allocate-with-budget task.

Theorem 2. *For weighted unit-demand buyers, for any allocation $\mathcal{T}$ there exists a pricing policy that allocates the same multiset of items T (or a subset of T) and has social welfare at least as large as $\mathcal{T}$. Moreover, this pricing can be computed efficiently from $\mathcal{T}$ and the buyers' valuations.*

Proof. Let T be the multiset of items allocated by $\mathcal{T}$. Weighted unit-demand valuations satisfy the gross-substitutes property, so by the Second Welfare Theorem (e.g., see [9] Theorem 11.15) there exists a Walrasian equilibrium: a set of prices for the items in T that clears the market. Moreover, these prices can be computed efficiently from demand queries (e.g., [9], Theorem 11.24), which can be evaluated efficiently for weighted unit-demand buyers. Furthermore, these prices must assign all copies of the *same* item in T the same price (else the pricing would not be an equilibrium) so it corresponds to a legal pricing policy. Thus, we have a legal pricing such that if all buyers were shown only the items represented in T, at these prices, then the market would clear perfectly (breaking any ties in our favor). We can address the fact that there may be items not represented in T (i.e., they had zero copies sold) by simply setting their price to infinity. Finally, by the First Welfare Theorem (e.g., [9] Theorem 11.13), this pricing maximizes social welfare over all allocations of T, and therefore achieves social welfare at least as large as $\mathcal{T}$, as desired.   □

The above structural results will allow us to use the following sketch of an online algorithm. First sample an initial set of ℓ buyers. Then, for the allocate-all problem, compute the best (or approximately best) permutation policy according to the empirical frequencies given by the sample. Or, for the allocate-with budget task, compute the best (or approximately best) allocation according to these empirical frequencies and convert it into a pricing policy. Then run this permutation or pricing policy on the remainder of the customers. Finally, using the fact that these policies have low complexity (they are lists or vectors in a space that depends only on the number of items and not on the number of buyers) compute the size of initial sample needed to ensure that the estimated performance is close to true performance uniformly over all policies in the class.

4   Uniform Unit Demand and the Allocate-All Problem

Here we consider the allocate-all problem for the setting of uniform unit demand. We begin by considering the following natural class of decreasing marginal cost curves such as $1/\sqrt{t}$.

Definition 1. *We say the cost function* $\text{cost}(t)$ *is* α-poly *if the marginal cost of item* t *is* $1/t^\alpha$ *for* $\alpha \in [0,1)$. *That is,* $\text{cost}(t) = \sum_{\tau=1}^{t} 1/\tau^\alpha$.

Theorem 3. *If each cost function is* α-poly, *then there exists an efficient offline algorithm that given a set* X *of buyers produces a permutation policy that incurs total cost at most* $\frac{1}{1-\alpha}$OPT.

Proof. We run the greedy set-cover algorithm. Specifically, we choose the item desired by the most buyers and put it at the top of the permutation π. We then choose the item desired by the most buyers who did not receive the first item and put it next, and so on. For notational convenience assume π is the identity, and let $\mathcal{B}_i$ denote the set of buyers that receive item i, For any set $\mathcal{B} \subseteq X$, let OPT($\mathcal{B}$) denote the cost of the optimal solution to the subproblem $\mathcal{B}$ (i.e., the problem in which we are only required to cover buyers in $\mathcal{B}$). Clearly OPT($\mathcal{B}_r$) $= \text{cost}(|\mathcal{B}_r|) = \sum_{\tau=1}^{|\mathcal{B}_r|} 1/\tau^\alpha \geq \sum_{t=1}^{|\mathcal{B}_r|} \int_1^{|\mathcal{B}_t|} x^{-\alpha}dx = \frac{1}{1-\alpha}|\mathcal{B}_r|^{1-\alpha} - 1$, since any solution using more than one set to cover the elements of $\mathcal{B}_r$ has at least as large a cost.

Now, for the purpose of induction, suppose that some $k \in \{2,\ldots,r\}$ has OPT($\bigcup_{t=k}^{r} \mathcal{B}_t$) $\geq \sum_{t=k}^{r} |\mathcal{B}_t|^{1-\alpha}$. Then, since $\mathcal{B}_{k-1}$ was chosen to be the largest subset of $\bigcup_{t=k-1}^{r} \mathcal{B}_t$ that can be covered by a single item, it must be that the sets used by any allocation for the $\bigcup_{t=k-1}^{r} \mathcal{B}_t$ subproblem achieving OPT($\bigcup_{t=k-1}^{r} \mathcal{B}_t$) have size at most $|\mathcal{B}_{k-1}|$, and thus the marginal costs for each of the elements of $\mathcal{B}_{k-1}$ in the OPT($\bigcup_{t=k-1}^{r} \mathcal{B}_t$) solution is at least $1/|\mathcal{B}_{k-1}|^\alpha$.

This implies OPT($\bigcup_{t=k-1}^{r} \mathcal{B}_t$) $\geq$ OPT($\bigcup_{t=k}^{r} \mathcal{B}_t$) $+ \sum_{x \in \mathcal{B}_{k-1}} 1/|\mathcal{B}_{k-1}|^\alpha =$ OPT($\bigcup_{t=k}^{r} \mathcal{B}_t$) $+ |\mathcal{B}_{k-1}|^{1-\alpha}$. By the inductive hypothesis, this latter expression is at least as large as $\sum_{t=k-1}^{r} |\mathcal{B}_t|^{1-\alpha}$. By induction, this implies OPT(X) $=$ OPT($\bigcup_{t=1}^{r} \mathcal{B}_t$) $\geq \sum_{t=1}^{r} |\mathcal{B}_t|^{1-\alpha}$. On the other hand, the total cost incurred by the greedy algorithm is $\sum_{t=1}^{r} \sum_{\tau=1}^{|\mathcal{B}_r|} 1/\tau^\alpha \leq \sum_{t=1}^{r} \int_0^{|\mathcal{B}_t|} x^{-\alpha}dx = \frac{1}{1-\alpha} \sum_{t=1}^{r} |\mathcal{B}_t|^{1-\alpha}$. By the above argument, this is at most $\frac{1}{1-\alpha}$OPT(X).   □

More General Cost Curves: We can generalize the above result to a broader class of smoothly decreasing cost curves. Due to space constraints, all of the proofs of these results are deferred to the full version of this paper online. Define the average cost of item i given to set $\mathcal{B}_i$ of buyers as $AvgC(i, |\mathcal{B}_i|) = \frac{\text{cost}(|\mathcal{B}_i|)}{|\mathcal{B}_i|}$. Define the marginal cost $MarC(i,t) = \text{cost}_i(t) - \text{cost}_i(t-1)$. Here is a greedy algorithm.

Algorithm: $GreedyGeneralCost(\mathcal{B})$
0. $i = \arg\min AvgC(i, |\mathcal{B}_i|)$, where $\mathcal{B}_i = \{j \in \mathcal{B} : i \in S_j\}$
1.   Call $GreedyGeneralCost(\mathcal{B} - \mathcal{B}_i)$

We make the following assumption:

Assumption 4 $\forall i, t$, $AvgC(i, t) \leq \beta MarC(i, t)$, for some $\beta > 0$.

For example, for the case of an α-poly cost, we have: $MarC(t) = \frac{1}{t^\alpha}$ and $AvgC = \frac{1}{t}\sum_{\tau=1}^{t} \frac{1}{\tau^\alpha} \approx \frac{t^{-\alpha}}{1-\alpha}$; so, therefore we have $\beta = \frac{1}{1-\alpha}$.

Theorem 5. *GreedyGeneralCost achieves approximation ratio β.*

Additionally, a property of β-nice cost functions we will need to use later is the following.

Lemma 1. *For cost satisfying Assumption 4, $\forall x \in \mathbb{N}$, $\forall \epsilon \in (0, 1)$, $\forall i \leq r$,*
$$cost_i(\epsilon x) \leq \epsilon^{\log_2(1+\frac{1}{2\beta})}cost_i(x).$$

Generalization Result: Say n is the total number of customers; ℓ is the size of subsample we do our estimate on; r is the total number of items; $\alpha \in (0, 1]$ is some constant, and the cost is α-poly, so that $cost(t) = \sum_{\tau=1}^{t} 1/\tau^\alpha \simeq \frac{t^{1-\alpha}}{1-\alpha}$. We now show the following uniform convergence over permutation policies, which will justify the use of a near optimal policy for a sample on the larger population.

Theorem 6. *Suppose $n \geq \ell$ and the cost function is α-poly. With probability at least $1 - \delta^{(\ell)}$, for all permutations Π,*

$$cost(\Pi, \ell)(1 + \epsilon)^{-2}\left(\frac{n}{\ell}\right)^{1-\alpha} \leq cost(\Pi, n) \leq cost(\Pi, \ell)(1 + \epsilon)^{2(1-\alpha)}\left(\frac{n}{\ell}\right)^{1-\alpha},$$

where $\delta^{(\ell)} = r2^r(\delta_1 + \delta_2 + \delta_3)$, with $\delta_1 = \exp\left\{-\epsilon^2\left(\frac{\epsilon}{r}\right)^{\frac{1}{1-\alpha}} n/3\right\}$, $\delta_2 = \exp\left\{-\epsilon^2\ell\left(\frac{\epsilon}{r}\right)^{\frac{1}{1-\alpha}}/3\right\}$, and $\delta_3 = \exp\left\{-\left(\frac{\epsilon}{r}\right)^{\frac{1}{1-\alpha}} n\epsilon^2/2\right\}$. Equivalently, for any $\delta \in (0, 1)$, this occurs with probability $1 - \delta$, so long as $\ell \gg (\frac{1}{\epsilon^2})(\frac{r}{\epsilon})^{\frac{1}{1-\alpha}} \ln(2^r/\delta)$.

Proof. Fix a permutation Π. Let π_j denote the event that a customer buys item Π_j and not covered by items Π_1 through Π_{j-1}. Namely, the probability that the customer set of desired items include j and none of the items $1, \ldots, j-1$. Let q_j denote $Pr[\pi_j]$, and let $\hat{q}_j$ denote the fraction of Π_j on the initial ℓ-sample.

Item j to is a "Low probability item" if $q_j < \left(\frac{\epsilon}{r}\right)^{\frac{1}{1-\alpha}}$; and "High probability items" if $q_j \geq \left(\frac{\epsilon}{r}\right)^{\frac{1}{1-\alpha}}$. Let the set "Low" include all "Low probability items"; and the set "High" include all "High probability items".

First we address the case of item j of low probability. The quantity of item j that will sell is at most $\left(\frac{\epsilon}{r}\right)^{\frac{1}{1-\alpha}} n(1 + \epsilon)$ (Chernoff bound) with probability at least $1 - \delta_1$ with $\delta_1 = \exp\{-\epsilon^2\left(\frac{\epsilon}{r}\right)^{\frac{1}{1-\alpha}} n/3\}$. By a union bound, this holds for all low probability item j, with probability at least $1 - |Low|\delta_1$.

Next, we suppose j has high probability. In this case, the quantity of item j will sell is at most $q_j n(1+\epsilon)$, with probability at least $1-\exp\{-\epsilon^2 q_j n/3\} \geq 1-\delta_1$. Again, a union bound implies this holds for all high probability j with probability at least $1 - |\text{High}|\delta_1$.

We have that (by Chernoff bounds), with probability $\geq 1 - \exp\{-\epsilon^2 \ell q_j/3\} \geq 1 - \delta_2$, we have $q_j/\hat{q}_j \leq (1+\epsilon)$. A union bound implies this holds for all high probability j with probability $1 - r\delta_2$.

Furthermore, noting that $q_j n(1+\epsilon) = \hat{q}_j n(1+\epsilon)\frac{q_j}{\hat{q}_j}$, and upper bounding $\frac{q_j}{\hat{q}_j}$ by $1+\epsilon$, we get that $q_j n(1+\epsilon) \leq (1+\epsilon)^2 \hat{q}_j n$, with probability $1 - \delta_2$. Thus, $\text{cost}(\Pi, n) \leq \text{cost}(\text{Low}) + \text{cost}(\text{High}) \leq r\left(\left(\frac{\epsilon}{r}\right)^{\frac{1}{1-\alpha}} n(1+\epsilon)\right)^{1-\alpha} + \sum_{j\in\text{High}}\left((1+\epsilon)^2 \hat{q}_j n\right)^{1-\alpha} \leq \epsilon(1+\epsilon)^{1-\alpha}n^{1-\alpha}+(1+\epsilon)^{2(1-\alpha)}n^{1-\alpha}\sum_{j\in\text{High}}(\hat{q}_j)^{1-\alpha}$. Note that the total cost of all low probability items is at most ϵ-fraction of OPT which is at least $\frac{n^{1-\alpha}}{1-\alpha}$. Also, $(1+\epsilon)^{2(1-\alpha)}n^{1-\alpha}\sum_{j\in\text{High}}(\hat{q}_j)^{1-\alpha} = (1+\epsilon)^{2(1-\alpha)}\left(\frac{n}{\ell}\right)^{1-\alpha}\sum_j(\hat{q}_j\ell)^{1-\alpha} = (1+\epsilon)^{2(1-\alpha)}\left(\frac{n}{\ell}\right)^{1-\alpha}\text{cost}(\Pi,\ell)$, by definition of $\text{cost}(\Pi,\ell)$. Therefore we showed, $\text{cost}(\Pi, n) \leq \epsilon(1+\epsilon)^{1-\alpha}\ell^{1-\alpha}\left(\frac{n}{\ell}\right)^{1-\alpha} + (1+\epsilon)^{2(1-\alpha)}\left(\frac{n}{\ell}\right)^{1-\alpha}\text{cost}(\Pi,\ell) \leq (1+5\epsilon)\left(\frac{n}{\ell}\right)^{1-\alpha}\text{cost}(\Pi,\ell)$.

The lower bound is basically similar. For $j \in Low$, we have $q_j < \left(\frac{\epsilon}{r}\right)^{\frac{1}{1-\alpha}}$ and $\hat{q}_j < \left(\frac{\epsilon}{r}\right)^{\frac{1}{1-\alpha}}(1+\epsilon)$ (by Chernoff bounds). So we have: $\sum_j(\hat{q}_j\ell)^{1-\alpha} \leq \sum_j\left(\left(\frac{\epsilon}{r}\right)^{\frac{1}{1-\alpha}}(1+\epsilon)\ell\right)^{1-\alpha} = r\frac{\epsilon}{r}(1+\epsilon)^{1-\alpha}\ell^{1-\alpha} = \epsilon(1+\epsilon)^{1-\alpha}n^{1-\alpha}\left(\frac{\ell}{n}\right)^{1-\alpha} \leq \epsilon(1+\epsilon)^{1-\alpha}\text{cost}(\Pi,n)\left(\frac{\ell}{n}\right)^{1-\alpha}$. Thus, $\text{cost}(\Pi,\ell) = \sum_{j\in\text{Low}}(\hat{q}_j\ell)^{1-\alpha}+\sum_{j\in\text{High}}(\hat{q}_j\ell)^{1-\alpha} \leq \text{cost}(\Pi,n)\epsilon\left(\frac{\ell}{n}\right)^{1-\alpha}(1+\epsilon)^{1-\alpha} + \sum_{j\in\text{High}}(q_j n)^{1-\alpha}\left(\frac{\ell}{n}\right)^{1-\alpha}\left(\frac{\hat{q}_j}{q_j}\right)^{1-\alpha} \leq \text{cost}(\Pi,n)\epsilon\left(\frac{\ell}{n}\right)^{1-\alpha}(1+\epsilon) + \sum_{j\in\text{High}}(q_j n)^{1-\alpha}\left(\frac{\ell}{n}\right)^{1-\alpha}(1+\epsilon) \leq (1+\epsilon)^2\text{cost}(\Pi,n)\left(\frac{\ell}{n}\right)^{1-\alpha}$ with probability at least $1-\exp\left\{-q_j n\epsilon^2/2\right\} \geq 1-\delta_3$. For low-probability j, the number of item j sold is $\geq \left(\frac{\epsilon}{r}\right)^{\frac{1}{1-\alpha}}n(1-\epsilon)$ with probability $\geq 1-\delta_3$. A union bound extends these to all j with combined probability $1 - r\delta_3$.

Thus we obtain the upper bound $\text{cost}(\Pi,n) \leq \text{cost}(\Pi,\ell)(1+\epsilon)^{2(1-\alpha)}\left(\frac{n}{\ell}\right)^{1-\alpha}$, and the lower bound $\text{cost}(\Pi,n) \geq \text{cost}(\Pi,\ell)(1+\epsilon)^{-2}\left(\frac{n}{\ell}\right)^{1-\alpha}$, with probability at least $1-r2^r(\delta_1+\delta_2+\delta_3)$.

A naive union bound can be done over all the permutations, which will add a factor of $r!$, we can reduce the factor to $r2^r$ by noticing that we are only interested in events of the type π_j, namely a given item (say, j) is in the set of desired items, and another set (say, $\{1,\ldots,j-1\}$) is not in that set. This has only $r2^r$ different events we need to perform the union over. □

Online Performance Guarantees: We define *GreedyGeneralCost*(ℓ, n) as follows. For the first ℓ customers it allocates arbitrary items they desire, and observed their desired sets. Given the sets of the first ℓ customers, it runs *GreedyGeneralCost* and computes a permutation $\hat{\Pi}$ of the items. For the remaining customers it allocates using permutation $\hat{\Pi}$. Namely, each customer is allocated the first item in the permutation $\hat{\Pi}$ that is in its desired set. The following theorem bounds the performance of $GreedyGeneralCost(\ell, n)$ for α-poly cost functions.

Theorem 7. *With probability* $1 - \delta^{(\ell)}$ *(for* $\delta^{(\ell)}$ *as in Theorem 6) The cost of* $GreedyGeneralCost(\ell, n)$ *is at most* $\ell + \frac{(1+\epsilon)^{4-2\alpha}}{1-\alpha}$*OPT.*

Proof. Let $\hat{\Pi}$ be the permutation policy produced by GreedyGeneralCost, after the ℓ first customers. By Theorem 5, $\text{cost}(\hat{\Pi}, \ell) \leq \frac{1}{1-\alpha} \min_{\Pi} \text{cost}(\Pi, \ell)$. By Theorem 6, $\min_{\Pi} \text{cost}(\Pi, \ell) \leq \min_{\Pi} \text{cost}(\Pi, n)(1 + \epsilon)^2 \left(\frac{\ell}{n}\right)^{1-\alpha}$ with probability $1 - \delta^{(\ell)}$. Additionally, on this same event, $\text{cost}(\hat{\Pi}, n) \leq \text{cost}(\hat{\Pi}, \ell)(1 + \epsilon)^{2(1-\alpha)} \left(\frac{n}{\ell}\right)^{1-\alpha}$. Altogether, this implies $\text{cost}(\hat{\Pi}, n) \leq \frac{(1+\epsilon)^{2(1-\alpha)}}{1-\alpha} \left(\frac{n}{\ell}\right)^{1-\alpha} \min_{\Pi} \text{cost}(\Pi, n)(1 + \epsilon)^2 \left(\frac{\ell}{n}\right)^{1-\alpha} = \frac{(1+\epsilon)^{4-2\alpha}}{1-\alpha} \min_{\Pi} \text{cost}(\Pi, n)$. $\quad\square$

Corollary 2. *For any fixed constant* $\delta \in (0, 1)$*, for any* $\ell \geq \frac{3}{\epsilon^2} \left(\frac{r}{\epsilon}\right)^{\frac{1}{1-\alpha}} \ln\left(\frac{3r2^r}{\delta}\right)$*, and* $n \geq \left(\frac{\ell}{\epsilon}\right)^{\frac{1}{1-\alpha}}$*, with probability at least* $1 - \delta$*,* $GreedyGeneralCost(n, \ell)$ *is at most* $\left(\frac{(1+\epsilon)^{4-2\alpha}}{1-\alpha} + \epsilon\right)$ *OPT.*

Generalization for β-nice Costs: We now consider the case of β-nice costs in the online setting. Due to space constraints, the proofs of these results are deferred to the full version of this paper online. We begin with a helper lemma.

Lemma 2. *For any cost* cost *satisfying Assumption 4 with a given* β*, for any* $k \geq 1$*, the cost* cost' *with* $\text{cost}'_i(x) = \text{cost}_i(kx)$ *also satisfies Assumption 4 with the same* β*.*

Now the strategy is to run GreedyGeneralCost with the rescaled cost function $\text{cost}'_i(x) = \text{cost}_i(\frac{n}{\ell}x)$. This provides a β-approximation guarantee for the rescaled problem, which, moreover is a permutation policy. The following shows we have uniform convergence of estimates to true costs for permutation policies.

Theorem 8. *Suppose* $n \geq \ell$ *and the cost function satisfies Assumption 4, and that* $\forall i$*,* $\text{cost}_i(1) \in [1, B]$*, where* $B \geq 1$ *is constant. Let* $\text{cost}'_i(x) = \text{cost}_i(\frac{n}{\ell}x)$*. With probability at least* $1 - \delta^{(\ell)}$*, for any permutations* Π*,*

$$\text{cost}'(\Pi, \ell)\frac{1 - \epsilon}{1 + 2\epsilon - \epsilon^2} \leq \text{cost}(\Pi, n) \leq \text{cost}'(\Pi, \ell)\frac{(1 + \epsilon)^2}{1 - \epsilon},$$

where $\delta^{(\ell)} = r^2 2^{r+1}(\delta_1 + \delta_2)$ *and* $\delta_1 = \exp\{-\frac{\epsilon^3}{3rB(1+\epsilon)}n^{\log_2(1+\frac{1}{2\beta})}\}$*,* $\delta_2 = \exp\{-\ell\frac{\epsilon^3}{rB(1+\epsilon)}n^{\log_2(1+\frac{1}{2\beta})-1}/3\}$*.*

Using the above uniform convergence result, we find that if we run GreedyGeneralCost on the sample of $\ell = o(n)$ initial buyers and apply it to the entire population, we achieve near optimal cost.

Theorem 9. *If* cost *satisfies Assumption 4, and has* $\text{cost}_j(1) \in [1, B]$ *for every* $j \leq r$, *with probability* $\geq 1 - \delta$, *the cost of applying the policy found by* GreedyGeneralCost$(\{1, \ldots, \ell\})$ *to all* n *customers is at most* $\beta \frac{(1+\epsilon)^2(1+2\epsilon-\epsilon^2)}{(1-\epsilon)^2} \text{OPT}(n)$, *where* $\ell = \left\lceil n^{1-\log_2(1+\frac{1}{2\beta})} \frac{3rB(1+\epsilon)}{\epsilon^3} \ln\left(\frac{r^2 2^{r+2}}{\delta}\right) \right\rceil = o(n)$.

Note that Theorem 9 assumes the initial $\ell = o(n)$ buyers can be "previewed" before allocations are made and need not themselves be allocated online.

5   General Unit Demand Utilities

In this section we show how to give a constant approximation for the case of general unit demand buyers in the offline setting in the case when we have a budget B to bound the cost we incur and we would like to maximize the buyers social welfare given this budget constraint. The main tool would be a reduction of our problem to the budgeted maximum coverage problem. Due to space constraints, we defer all proofs in this section to the full version of this paper online.

Definition 2. *An instance of the budgeted maximum coverage problem has a universe* X *of* m *elements where each* $x_i \in X$ *has an associated weight* w_i; *there is a collection of* m *sets* $\mathcal{B}$ *such that each sets* $S_j \in \mathcal{B}$ *has a cost* c_j; *and there is a budget* L. *A feasible solution is a collection of sets* $\mathcal{B}' \subset \mathcal{B}$ *such that* $\sum_{S_j \in \mathcal{B}'} c_j \leq L$. *The goal is to maximize the weight of the elements in* $\mathcal{B}'$, *i.e.,* $w(\mathcal{B}') = \sum_{x_i \in \cup_{S \in \mathcal{B}'} S} w_i$.

While the budgeted maximum coverage problem is NP-complete there is a $(1-1/e)$ approximation algorithm [7]. Their algorithm is a variation of the greedy algorithm, where on the one hand it computes the greedy allocation, where each time a set which maximizes the ratio between weight of the elements covered and the cost of the set is added, as long as the budget constraint is not violated. On the other hand the single best set is computed. The output is the best of the two alternative (either the single best set of the greedy allocation).

Before we show the reduction from a general unit demand utility to the budgeted maximum coverage problem, we show a simpler case where for each buyer j has a value v_j such that of any item i either $v_j = u_{j,i}$ or $u_{j,i} = 0$, which we call *buyer-uniform unit demand*.

Lemma 3. *There is a reduction from the budgeted buyer-uniform unit demand buyers problem to the budgeted maximum coverage problem. In addition the greedy algorithm can be computed in polynomial time on the resulting instance.*

In the above reduction we used very heavily the fact that each buyer j has a single valuation v_j regardless of which desired item it gets. In the following we show a slightly more involved reduction which handles the general unit demand buyers.

Lemma 4. *There is a reduction from the budgeted general unit demand buyers problem to the budgeted maximum coverage problem. In addition the greedy algorithm can be computed in polynomial time on the resulting instance.*

Combining our reduction with approximation algorithm of [7] we have the following theorem.

Theorem 10. *There exists a poly-time algorithm for the budgeted general unit demand buyers problem which achieves social welfare at least $(1 - 1/e)$OPT.*

Generalization: To extend these results to the online setting, we will use Theorem 2 to represent allocations by pricing policies, and then use the results from above to learn a good pricing policy based on an initial sample.

Theorem 11. *Suppose every $u_{j,i} \in [0, C]$. With $\ell = O((1/\epsilon^2)(r^3 \log(rC/\epsilon) + \log(1/\delta)))$ random samples, with probability at least $1 - \delta$, the empirical per-customer social welfare is within $\pm\epsilon$ of the expected per-customer social welfare, uniformly over all price vectors in $[0, C]^r$.*

We also make use of the following result.

Theorem 12. *With $\ell \geq O((1/\epsilon^2)(r^2 + \log(1/\delta)))$ random samples, with probability at least $1 - \delta$, the empirical probability of a customer buying item j is within $\pm\epsilon$ of the actual probability, uniformly over all price vectors and j.*

Consider an algorithm that does not allocate anything to the first $\ell = O((C/\epsilon)^2(r^3 \log(rC/\epsilon) + \log(1/\delta)))$ customers, then finds a $(1-1/e)$-approximate solution to the offline budgeted general unit demand problem on these ℓ customers, with budget B, and cost functions $\text{cost}'_i(x) = \text{cost}_i(x \cdot ((n - \ell)/\ell))$, via the reduction to the budgeted maximum coverage problem. The algorithm then finds a pricing policy price providing at least as good of a social welfare on these ℓ customers, within this budget B. Let ℓ_i denote the number of copies of item i this pricing policy allocates among the ℓ customers. The algorithm then proceeds to allocate to the remaining stream of $n - \ell$ customers using this pricing policy, but if at any time the item i this pricing policy determines should be allocated to the next customer has already had $\ell_i((n - \ell)/\ell)$ copies allocated to customers in the past, then the algorithm does not allocate any item to that customer and simply moves on to the next customer. (As stated, this is not incentive-compatible: we are assuming that if a buyer enters the store and finds his most-desired item is sold-out, he just leaves rather than buying some other item; however, we rectify this in Corollary 3 below.) We have the following result on the performance of this algorithm. The proof is provided in the full version of this paper online.

Theorem 13. *The allocation given by the above algorithm does not exceed the budget B, and if $n \geq O((1/\epsilon)\ell)$, with probability at least $1 - 4\delta$, achieves a social welfare at least $(1 - 1/e)$OPT $- (2(2 - 1/e)(1 + Cr) + C)\epsilon n$.*

To make the above procedure incentive-compatible, if at any time the pricing policy attempts to allocate more than $\ell_i((n-\ell)/\ell)$ copies of item i, then for that customer j we can just allocate the item i' that has the next-highest $u_{j,i'} - \text{price}_{i'}$ among those i' for which the number of copies of item i' this policy has attempted to allocate previously is less than $\ell_{i'}((n - \ell)/\ell)$ (or nothing, if all remaining i' have $u_{j,i'} - \text{price}_{i'} < 0$). A simple modification of the above proof yields the following result on the performance of this algorithm; the proof appears in the full version of this paper online.

Corollary 3. *The allocation given by the above algorithm does not exceed the budget B, and if $n \geq O((1/\epsilon)\ell)$, with probability at least $1 - 4\delta$, the allocation achieves a social welfare at least $(1 - 1/e)\text{OPT} - O(Cr^2\epsilon n)$.*

Acknowledgments. Avrim is supported in part by the National Science Foundation under grants CCF-1101215, CCF-1116892, and IIS-1065251. Yishay is supported in part by a grant from the Israel Science Foundation, a grant from the United States-Israel Binational Science Foundation (BSF), a grant by Israel Ministry of Science and Technology and the Israeli Centers of Research Excellence (I-CORE) program (Center No. 4/11). For part of this work, Liu was supported in part by NSF grant IIS-1065251 and a Google Core AI grant.

References

1. Alon, N., Awerbuch, B., Azar, Y., Buchbinder, N., Naor, J.: The online set cover problem. SIAM J. Comput. **39**(2), 361–370 (2009)
2. Blum, A., Gupta, A., Mansour, Y., Sharma, A.: Welfare and profit maximization with production costs. In: FOCS, pp. 77–86 (2011)
3. Devanur, N.R., Hayes, T.P.: The adwords problem: Online keyword matching with budgeted bidders under random permutations. In: Proc. ACM EC, EC 2009, pp. 71–78 (2009)
4. Devanur, N.R., Jain, K.: Online matching with concave returns. In: Proc. STOC, pp. 137–144 (2012)
5. Goel, G., Mehta, A.: Online budgeted matching in random input models with applications to adwords. In: Proc. SODA, pp. 982–991 (2008)
6. Karp, R.M., Vazirani, U.V., Vazirani, V.V.: An optimal algorithm for on-line bipartite matching. In: Proc. STOC, pp. 352–358 (1990)
7. Khuller, S., Moss, A., Naor, J.: The budgeted maximum coverage problem. Inf. Process. Lett. **70**(1), 39–45 (1999)
8. Mehta, A., Saberi, A., Vazirani, U., Vazirani, V.: Adwords and generalized online matching. J. ACM **54**(5), 22 (2007)
9. Nisan, N., Roughgarden, T., Tardos, E., Vazirani, V.V.: Algorithmic Game Theory. Cambridge University Press, Cambridge (2007)

Multilateral Deferred-Acceptance Mechanisms

Liad Blumrosen[1]([✉]) and Osnat Zohar[2]

[1] Department of Economics, The Hebrew University, Jerusalem, Israel
blumrosen@gmail.com
[2] Bank of Israel, Jerusalem, Israel

Abstract. We study the design of multilateral markets, where agents with several different roles engage in trade. We first observe that the modular approach proposed by Dütting et al. [5] for bilateral markets can also be applied in multilateral markets. This gives a general method to design Deferred Acceptance mechanisms in such settings; these mechanisms, defined by Milgrom and Segal [10], are known to satisfy some highly desired properties.

We then show applications of this framework in the context of *supply chains*. We show how existing mechanisms can be implemented as multilateral Deferred Acceptance mechanisms, and thus exhibit nice practical properties (as group strategy-proofness and equivalence to clock auctions). We use the general framework to design a novel mechanism that improves upon previous mechanisms in terms of social welfare. Our mechanism manages to avoid "trade reduction" in some scenarios, while maintaining the incentive and budget-balance properties.

1 Introduction

Markets are often characterized by multiplicity of participants with diverse characteristics. While much of the mechanism-design literature focuses on bilateral settings that distinguish between two types of agents - buyers and sellers, many real life applications require a more complex description of markets. For instance, dealing with markets in which buyers wish to purchase a bundle of items that are sold separately by different sellers might require a distinction between different types of sellers. Further and more complex distinctions will be needed as markets become more complex and involve trade between diversified agents.

In settings of bilateral trade, and nonetheless in more complex settings where agents engage in several bilateral transactions, it is generally impossible to achieve an efficient allocation while maintaining agents' participation constraints, incentive compatibility and budget balance [12]. Therefore, some goals needs to be sacrificed in order to fully achieve the others. For instance the VCG mechanism maintains agents' incentives and implements the efficient allocation but is generally not budget balanced. Other mechanisms relax incentive compatibility to achieve the efficient allocation and budget balance (see [7] for a survey).

In this paper we devise a family of mechanisms named multilateral deferred-acceptance (MDA) mechanisms. These mechanisms apply the methodology of

© Springer-Verlag Berlin Heidelberg 2015
E. Markakis and G. Schäfer (Eds.): WINE 2015, LNCS 9470, pp. 173–186, 2015.
DOI: 10.1007/978-3-662-48995-6_13

deferred-acceptance (DA) auctions, introduced by Milgrom and Segal [10], to multilateral markets. DA auctions set allocations using an iterative process of rejecting the least attractive bid according to a carefully defined ranking function. Combining this sort of algorithm with threshold payments yields a mechanism with strong incentive properties - other than being truthful, the DA auction is also weakly group-strategy proof (WGSP). This means that no coalition of agents has a joint deviation from truthful bidding that is strictly profitable for all members of the coalition. Another desired feature of DA auctions is that they are equivalent to clock auctions, an auction format which is intuitive for bidders and is thus considered practical. WGSP and equivalence to clock auctions are desired properties of DA auctions that are not generally attained by other mechanisms. For example, the VCG mechanism and greedy mechanisms such as [6] do not possess these properties.

Dütting et al. [5] took a modular approach to adapt DA auctions to two-sided markets. We take the general concept introduced in [5] one step forward and observe that their modular approach can also be applied to multilateral markets with several types of agents. Similar to [5], the mechanism's operation is determined by two elements: separate rankings for each set of agents and a composition rule. In each period the composition rule selects few classes (or groups) of agents. The least desirable agent, according to the corresponding ranking, of each selected class will be rejected. When the mechanism terminates, all unrejected agents are declared winners and threshold payments are set. We generalize the result by [5] and show that any mechanism from this family (MDA mechanisms) is equivalent to a one-sided DA auction; thus, it is strategy-proof, individually rational, WGSP and equivalent to a clock auction.

After introducing the class of MDA mechanisms, we apply them in the context of *supply chains* (see, e.g., [16,17]). Supply chains are collections of markets where each agent engages in at least one bilateral trade, either as a seller of an item he produces, or as a buyer of items (consumption goods or inputs for production). Supply chains are thus composed of several two-sided markets and feature the same impossibilities that exist in bilateral-trade settings of maximizing social welfare while maintaining agents' incentives and budget balance.

We study a model for supply chains that was introduced by Babaioff and Nisan [1] and Babaioff and Walsh [2]. In this model, the supply chain can be viewed as a directed tree-graph with a node per each good. Ingoing edges to a node define the inputs for the production of the relevant good, and producers incur a manufacturing cost which is private information. [1,2] showed a dominant-strategy truthful, budget-balanced mechanism that waives only the least profitable trade (or *"procurement set"*, which is a minimal trade cycle in a supply chain and typically involves multiple agents).

Our first result for supply chains shows that the trade reduction mechanisms of [1,2] can be implemented as MDA mechanisms. Thus, other than being IR, strategy-proof and budget balanced (as proven in [1,2]), these trade reduction mechanisms are also WGSP and equivalent to clock auctions. This is shown under the assumption of homogeneous demand (i.e., all end consumers demand the same bundle).[1]

Our second and main result shows how to use the machinery of MDA mechanisms to construct a novel mechanism that provides an improved outcome in terms of social welfare compared to the above trade reduction mechanisms. It operates by iteratively rejecting procurement sets, but unlike the trade reduction mechanism that always waives one procurement set that engages in trade in an efficient outcome, our mechanism will sometimes result in the efficient allocation (with no reduction of valuable trades). Using the values of agents that have already been declared losers, we bound the payments of active agents and identify situations where the efficient outcome can be implemented while maintaining a balanced budget. In markets where the efficient allocation consists of a small number of procurement sets, this improvement may be substantial. We ran computer simulations of simple supply chain networks. In the simulations, our new mechanism improves upon the trade reduction mechanisms in around 17 % of the instances and saves up to 100 % of the overall efficiency. This provides a good indication that the improvement in efficiency is not a rare phenomenon and can be significant.

The seminal paper by McAfee [8] introduced the trade reduction technique.[2] McAfee's mechanism was given for two-sided markets with unit-demand buyers and unit-supply sellers. This mechanism either implements the efficient allocation or reduces the least valuable profitable trade. [5] proved that the trade reduction mechanism for two-sided markets (a simplified version of [8] that always eliminates one valuable trade) can be implemented via a DA mechanism and it is therefore WGSP and equivalent to a clock auction.[3]

The paper is organized as follows: Sect. 2 defines deferred acceptance mechanisms. Section 3 defines MDA mechanisms and presents their main properties. Section 4 introduces the applications of MDA mechanisms to supply chains.

2   General Deferred-Acceptance Auctions

Consider a set N of single-parameter agents and let $\mathcal{F} \subseteq 2^N$ be the set of feasible sets of agents. An allocation in this setting is represented by a set of winning

---

[1] Our model is a generalization of the linear model in [1]; [2] did not require homogeneous demand.

[2] There is a vast literature on the efficiency of two-sided auctions that followed [8], see, e.g., [3,13,14]. The efficiency of DA auctions was studied in [4].

[3] Our work is inspired by [8] in several ways. First, we sacrifice efficiency in order to satisfy incentive constraints and budget balance and our mechanism loses at most the least valuable procurement set. In addition, McAfee's mechanism computed a price as a function of the "best" losing bids, and if this price cleared the market, no trade reduction would take place. Our mechanism acts in the same spirit and sometimes implements the efficient allocation, but it is not a generalization of McAfee's mechanism. In fact, our mechanism always omits one trade when applied to the degenerate supply chain of a single two-sided market with unit-demand buyers; the benefits of our mechanism stem in more complex markets.

agents $A \in \mathcal{F}$. Every agent $i \in N$ is characterized by a type t_i such that given an allocation A and payments $\{p_i\}_{i \in N} \subseteq \mathbb{R}$, agent i's utility is $t_i + p_i$ if $i \in A$ and p_i if $i \notin A$. An agent's type is assumed to be private information.

In this setting, [10] define DA auctions. Each agent $i \in N$ is required to submit a single bid from a finite set of possible bids $B_i \subseteq \mathbb{R}$. According to submitted bids, an iterative process of rejecting agents is preformed and all agents that are not rejected in the process are declared winners. We now describe this process in detail.

An agent is considered **active** in iteration t if he has not been rejected in any iteration prior to t. Let $A_t \subseteq N$ denote the set of active agents at the beginning of iteration t. Each active agent is assigned a score which is a function of his bid and the bids of all previously rejected agents:

Definition 1. *[10] A **DA scoring function** is a function of the form σ_i^t : $B_i \times B_{N \setminus A_t} \to \mathbb{R}_+$ that is non-decreasing in the first argument.*

The scoring functions form a ranking over the set of active agents in which higher ranked agents are considered less attractive. Following this logic, the DA algorithm iteratively rejects the agents with the highest score, until all agents have a score of zero. Formally:

Definition 2. *[10] Given DA scoring functions, a **DA algorithm** is defined as follows: All bidders are initially active. If all active bidders have a score of zero, the algorithm terminates and the remaining active bidders are declared winners. Otherwise, the algorithm rejects the active bidders with the highest score, removing them from the active set, and iterates.*

A DA auction can now be formally defined as follows. We then define two desired properties of DA auctions.

Definition 3. *[10] A **DA auction** is a sealed-bid auction that computes an allocation using a DA algorithm and makes the corresponding threshold payments to winners.[4] Losing agents are paid zero.*

Definition 4. *A mechanism is **weakly group strategy-proof (WGSP)** if for every profile of truthful reports b, every set of agents $S \subseteq N$ and every strategy profile b_S' of these agents, at least one agent in S has a weakly higher payoff from the profile of truthful reports b than from the strategy profile $(b_{N \setminus S}, b_S')$.*

Namely, a mechanism is WGSP if no coalition of agents can do strictly better by misreporting their values, given that all other agents report their true values.

Definition 5. *A descending **clock auction** is a dynamic mechanism that presents a decreasing sequence of prices to each bidder. Each presentation is followed by a decision period in which each bidder decides whether to exit or continue. When the auction ends, the bidders that have never exited are declared winners and are paid their last (lowest) accepted prices.*

---

[4] Threshold payments will be formally defined in the next section. Informally, these are the highest bids for a winning agent such that he remains a winner.

Theorem 6. *[10] Any DA auction is individually rational (IR), strategy-proof, WGSP and equivalent to a clock auction.*

Milgrom and Segal [10] show that several previously known mechanisms (e.g., [9,11,15]) can be implemented as DA auctions and thus inherit all the properties specified in Theorem 6. In Sect. 4.1 we take a similar approach by showing that the mechanisms in [1,2] can be implemented as MDA mechanisms and thus inherit their properties.

3   Multilateral Deferred-Acceptance Mechanisms

We now turn to settings of multilateral markets in which agents might differ in several aspects other than their types. Consider a setting in which agents have some distinct and known characteristics which allow sorting them into different classes. Let K be the number of agents' classes and denote by N_k the set of agents of class $k \in \{1, ..., K\}$. Thus, the set of all agents N can be described as a union of K disjoint sets $N = N_1 \cup N_2 \cup \cdots \cup N_K$.[5]

3.1   Multilateral Deferred-Acceptance Algorithms

For all $i \in N$ let $B_i \subseteq \mathbb{R}$ be the finite set of possible bids for bidder i, so the input for the MDA algorithm is a vector $b \in \Pi_{i \in N} B_i$.[6] In the spirit of [5], the MDA algorithm is composed of two elements: **scoring functions** and **composition functions**. We now define these two elements and describe how they construct an MDA algorithm. Scoring functions are defined similarly to [10] (Definition 1):

Definition 7. *For each $k = 1, ..., K$ and $i \in N_k \cap A_t$, agent i's **scoring function** $s_{k,i}^t : B_i \times B_{N_k \setminus A_t} \to \mathbb{R}_+$ is non-decreasing in the first argument and assures no ties between agents of the same class.[7]*

The scoring of agent $i \in N_k \cap A_t$ is compared to the scores of all other active agents of class k in period t to form a ranking on the set $N_k \cap A_t$. The "no ties"

---

[5] In a two-sided market with producers and consumers of a homogeneous good, N_1 might be the set of producers and N_2 might be the set of consumers. In that case producers' types will be thought of as production costs, so a producer with a cost c_i will have utility of $-c_i + p_i$ if $i \in A$, and p_i otherwise. Consumers' types will be thought of as their value from possessing one item of the traded good.

[6] As mentioned in Sect. 2, [10] define DA auctions with finite bid spaces. In order to use their results, we do the same. This also requires a more delicate definition of truthfulness and we follow the definition of strategy-proofness in [10] which uses the standard dominant-strategy truthfulness, only with taking care of the finite bid space. We refer the readers to [10] for the exact definition.

[7] In order to keep notation simple, the scoring functions are denoted with superscript t yet they are allowed to depend on the entire history of active agents $(A_1, ..., A_t)$ and not just on the t-period information. In the remainder of the paper, all objects denoted with superscript t are allowed to be history dependent.

requirement does not appear in [10] and it is made here as we sometimes wish
to carefully control the number of agents of each class that are rejected. In order
to simplify the presentation of our mechanisms, from now on, whenever possible
ties occur, we assume the existence of a tie-breaking rule instead of formally
defining scoring functions with no ties.

Composition Functions: In each period a different composition function
is defined. Its inputs are the bids of all previously rejected agents and its output
is a subset of $\{1, ..., K\}$.

Definition 8. *In each period t, a **composition function** C^t is a function of
the form $C^t : B_{N \backslash A_t} \to 2^K$.*

Multilateral Deferred-Acceptance Algorithms: Given a set of scoring
functions and composition functions, an MDA algorithm is defined as follows: In
each period t, the composition function $C^t(b_{N \backslash A_t})$ outputs a subset of classes. For
each class $k \in C^t(b_{N \backslash A_t})$, the algorithm queries the active agents of class k and
rejects the highest scoring one, according to the scoring functions $\{s_{k,i}^t\}_{i \in N_k \cap A_t}$.
This means that in period t the number of agents rejected is $|C^t(b_{N \backslash A_t})|$. If
$C^t(b_{N \backslash A_t}) = \emptyset$, the algorithm terminates and all active agents are declared
winners. The algorithm's operation can be described in the following manner:

1. Initialize the algorithm with $A_1 = N$.
2. For each $t \geq 1$, if $C^t(b_{N \backslash A_t}) = \emptyset$, stop. Accept all currently active agents A_t.
3. If $C^t(b_{N \backslash A_t}) = C^t \neq \emptyset$, define: $A_{t+1} = A_t \setminus \bigcup_{k \in C^t} \underset{i \in N_k \cap A_t}{\arg\max} \, s_{k,i}^t(b_i, b_{N_k \backslash A_t})$
   and return to 2.

3.2   Multilateral Deferred-Acceptance Mechanisms

To complete describing the MDA mechanism we now define the payments.

Definition 9. *Given an MDA algorithm and a vector of bids $b \in \Pi_{i \in N} B_i$, let
$A(b)$ denote the set of winning agents. The **threshold payment** of a winning
agent $i \in A(b)$ is defined as $\sup\{b_i' \in B_i | i \in A(b_i', b_{-i})\}$*

Definition 10. *An **MDA mechanism** computes allocation using an MDA
algorithm and makes the corresponding threshold payments to winning agents.
Losing agents pay zero.*

In the full version of the paper we show that every MDA mechanism can be
implemented as a DA auction. The method we use is similar to the one used
in [5] for the generalization of DA auctions to two-sided markets.

Proposition 11. *For every MDA mechanism there is an equivalent DA auction.*

Proposition 11 establishes that MDA mechanisms inherit all the properties
of DA auctions. Together with Theorem 6, we conclude that:

Corollary 12. *Any MDA mechanism is IR, strategy-proof, WGSP and equivalent to a clock auction.*

4   Applications to Supply Chains

We now present a model of supply chains which follows [2] and generalizes the linear supply-chain model in [1]. Consider an economy with K types of items, denoted $1, .., K$. We begin by describing the production of these items, assuming that all items of a specific type are manufactured in the same manner, using the same inputs.[8]

Assumption 1 *Each product is manufactured with a* **unique manufacturing technology***.*

Production in this economy can be described as a directed a-cyclical graph with K nodes representing the K different types of items. In this graph an edge (j, k) indicates that the production of item k uses item j as an input and the weight of the edge is the number of items of type j needed for production. Since any directed a-cyclical graph has a topological ordering,[9] assume WLOG that this ordering is given by the numbering of items' types. This means that the manufacturing of a type-k item makes use only of items of types $1, ..., k - 1$.

This production structure allows us to characterize the production of a type-k item with a **production vector** $q^k = (q_{1,k}, ..., q_{K,k})' \in \mathbb{Z}_{\leq 0}^{k-1} \times \{1\} \times \{0\}^{K-k}$. Arguments $1, ..., k - 1$ of the production vector are non-positive integers representing the quantities of inputs required for production (i.e., $-q_{j,k}$ for $j = 1, ..., k - 1$ is the weight of the edge (j, k)).[10] $q_{k,k} = 1$ indicates that one unit of item k is being produced in the process. Items $k + 1, ..., K$ are not involved in the production of item k, so $q_{k+1,k} = \cdots = q_{K,k} = 0$.

We further assume that each producer in the economy can manufacture a single item. Thus, all the producers that manufacture an item of type $k \in \{1, ..., K\}$ are substitutes and will be regarded as agents of class k. Let $c_i \in [0, \bar{c}_k]$ denote the production cost of producer i of class k. A producer's class and production vector are assumed to be common knowledge but his cost is private information.

We now turn to describe end consumers in the economy. These agents lack any production abilities but they benefit from consumption. All consumers are single minded, meaning that each consumer values only one particular bundle of items and will gain zero utility from consuming any smaller bundle.

Up until now all our assumptions follow the ones made in [2]; We now add an additional assumption, which was also assumed in [1]:

Assumption 2 **Homogeneous demand** - *All end consumers demand the same bundle* $d = (d_1, ..., d_K)' \in \mathbb{N}^K$ *where* d_k *is the demand from item* k.

---

[8] In the full paper, we relax the assumption of unique manufacturing technology and design an MDA mechanism for scenarios where a certain good can be produced from different types of inputs.

[9] A topological ordering of a directed a-cyclical graph is an ordering of the nodes such that for every edge (j, k), the node j comes before k in that ordering.

[10] The assumption that $q_{j,k}$ for $j < k$ is an integer, rather than a real number, is without loss of generality since any amount of items can be regarded as one unit. For example, if item j is flour and all items $k > j$ are produced using amounts of flour in multiples of 0.5 kg, set one unit of item j to be 0.5 kg of flour.

The demanded bundle d is commonly known but consumer i's valuation for this bundle, $v_i \in [0, \bar{v}]$, is private information. We refer to consumers as agents of class $K + 1$.

Define the **production matrix** as $Q = (q^1, ..., q^K)$. If there are μ_k producers of class k and $\mu = (\mu_1, ..., \mu_K)'$ then the supply of items is given by the vector $Q \cdot \mu$ where the k'th argument is the supply of item k. Denote by $\tilde{\mu}_k$ the number of class-k producers needed to meet the demand d of one consumer, i.e., the vector $\tilde{\mu} = (\tilde{\mu}_1, ..., \tilde{\mu}_K)$ solves $Q \cdot \tilde{\mu} = d$.[11]

Assume WLOG that initially there is no excess demand of any item, i.e., that $Q \cdot \mu \geq |N_{K+1}| \cdot d$, where $\mu = (|N_1|, ..., |N_K|)'$. An equivalent requirement is that $\mu \geq |N_{K+1}| \cdot \tilde{\mu}$. If this condition is not met, reject the highest bidding consumers until there is no excess demand.

A *procurement set* ([2]) is a set of agents that contains one consumer and the minimal amount of producers needed to meet his demand. Formally:

Definition 13. *A **procurement set** is a set of agents containing one consumer and $\tilde{\mu}_k$ producers of class k for every $1 \leq k \leq K$.*

Example 14. Figure 1 depicts a simple supply chain; fabric will be referred to as item 1, hats as item 2 and shirts as item 3. The production structure is such that producing either one hat or one shirt requires one roll of fabric. This implies that the production vectors are $q^1 = (1, 0, 0)'$, $q^2 = (-1, 1, 0)'$, $q^3 = (-1, 0, 1)'$ and the production matrix is $Q = (q^1, q^2, q^3)$.

Each consumer in this example demands a bundle of one hat and two shirts, i.e., $d = (0, 1, 2)'$. This implies that $\tilde{\mu} = Q^{-1} \cdot d = (3, 1, 2)'$ which means that a procurement set in this example contains one consumer, three producers of fabric, one producer of hats and two producers of shirts.

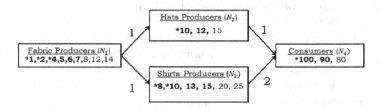

Fig. 1. A simple supply chain. The optimal allocation is marked in bold and the Trade Reduction allocation is marked with asterisks.

Definition 15. *[2] Given a set of bids, the **Trade Reduction allocation** is obtained from the optimal allocation by reducing one procurement set. The procurement set reduced is the one with the lowest value consumer and the highest*

---

[11] Q is a unitriangular matrix with negative integers on the entries above the main diagonal. Since $d \in \mathbb{N}^K$, it can be shown that $\tilde{\mu} = Q^{-1} \cdot d$ is a vector of non-negative integers and thus appropriately represents numbers of agents.

cost producers out of all the procurement sets in the optimal allocation. Together with threshold payments, determined by submitted bids, this allocation rule establishes the **Trade Reduction mechanism.**

Example 16. In Fig. 1 the optimal allocation is marked in bold and the Trade Reduction allocation is marked with asterisks. The threshold payments for winning agents in the examples: fabric producers are paid 5 each, the hat producer is paid 12, shirts producers are paid 13 and the consumer pays 90.

Proposition 17. *[2] The Trade Reduction mechanism is IR, strategy-proof, budget balanced and incurs the loss of the least valuable procurement set in the optimal allocation.*

4.1   The MDA Trade Reduction Mechanism

The MDA mechanism we present in this section implements the Trade Reduction allocation in a process that is equivalent to iterative rejection of procurement sets. In each iteration, the MDA algorithm examines the most valuable procurement set out of the ones that were rejected so far. The algorithm calculates the net cost of this procurement set, i.e., the sum of costs of all the producers in the set minus the value of the consumer in the set. While this net cost is strictly positive, the algorithm keeps on rejecting more procurement sets. Immediately after the mechanism rejects one procurement set with a non-negative net cost, it terminates and accepts all currently active agents. This way only one efficient procurement set is rejected, but all the rest are accepted.

We now turn to the formal definition of the mechanism. For each $k = 1, ..., K$ let B_k be a finite set of possible bids for all producers of class k, such that $B_k \subseteq [0, \bar{c}_k]$ and $\max B_k > \bar{c}_k$. Let $B_{K+1} \subseteq [-\bar{v}, 0]$ be a finite set possible bids for all consumers, such that $\max B_{K+1} > 0$.[12]

Scoring Functions: Producers of each class are ranked in an ascending order of costs and consumers are ranked in a descending order of values, i.e., all agents are ranked in a descending order of attractiveness. Formally:

$$\begin{aligned} \forall t, k \in \{1, ..., K\}, i \in A_t \cap N_k, \qquad & s^t_{k,i} = c_i \\ \forall t, i \in A_t \cap N_{K+1}, \qquad & s^t_{K+1,i} = \bar{v} - v_i \end{aligned} \tag{1}$$

The composition rule is based on two auxiliary functions:

---

[12] Consumers' bid spaces are defined as subsets of $[-\bar{v}, 0]$ so we can treat all agents, producers and consumers, in a similar manner such that higher bidding agents are less attractive. Thus, the mechanism will determine negative monetary transfers for consumers and positive transfers for producers. The maximal (minimal) possible bid of a producer (consumer) is set to be higher (lower) than his highest possible cost (lowest possible value) to insure that participation is strictly preferable to non-participation (see [10]).

1. **The Net Cost Function** NC^t: For each period t, denote by $NC^t(b_{N\setminus A_t})$ the net cost of the most valuable procurement set rejected so far. Formally:

$$NC^t(b_{N\setminus A_t}) = \sum_{k=1}^{K}\sum_{l=1}^{\tilde{\mu}_k} c^t_{k,(l)} - v^t_{max} \qquad (2)$$

where $c^t_{k,(l)}$ is the l'th lowest cost reported in $b_{N_k\setminus A_t}$ and v^t_{max} is the highest value reported by a rejected consumer (i.e., $-v^t_{max}$ is the minimal bid in $b_{N_{K+1}\setminus A_t}$).[13]

2. **The Excess Supply Function** ES^t: Let μ^t_k denote the number of active agents of class k in period t, i.e., $\mu^t_k = |N_k \cap A_t|$. The aggregate demand in period t is equal to $\mu^t_{K+1}{\cdot}d$ and the number of producers of class $k \in \{1, ..., K\}$ needed to meet it is $\mu^t_{K+1} \cdot \tilde{\mu}_k$. If there are more producers of class k than that, regard the class-k producers as being in excess. According to this logic, the excess supply function indicates the classes of producers that are in excess in period t. Formally, $ES^t(A_t) = \{k|1 \le k \le K, \mu^t_k > \mu^t_{K+1} \cdot \tilde{\mu}_k\}$.

Composition Functions: Now we can define the composition functions, using the auxiliary functions NC^t and ES^t. For every period t, define:

$$C^t(b_{N\setminus A_t}) = \begin{cases} ES^t(A_t) & \text{if } ES^t(A_t) \ne \emptyset \\ \{1, ..., K+1\} & \text{if } ES^t(A_t) = \emptyset \text{ and } NC^t(b_{N\setminus A_t}) > 0 \\ \emptyset & \text{if } ES^t(A_t) = \emptyset \text{ and } NC^t(b_{N\setminus A_t}) \le 0 \end{cases} \qquad (3)$$

In words, the algorithm first rejects excess producers, as determined by the first line in (3). This is repeated until there are no excess producers, i.e., until $ES^t = \emptyset$, which means that supply equals demand (recall that we assumed that initially there is no excess demand). From this point, the algorithm's operation can be described as an iteration of three steps:

1. If there is no excess supply ($ES^t = \emptyset$), examine the net cost of the most valuable procurement set rejected so far, NC^t. If NC^t is non-positive - terminate (third line in (3)). Otherwise, continue to step 2.
2. Reject one agent of each class $1, ..., K+1$ (second line in (3)).
3. As long as there is excess supply ($ES^t \ne \emptyset$), reject one agent of each class $k \in ES^t$ (first line in (3)). Once $ES^t = \emptyset$, return to step 1.

It is worth noting that each time steps 1–3 are completed, one procurement set is rejected. The rejection begins with eliminating the highest bidding agent of each class and continues with eliminating excess supply. Since each procurement set includes only one consumer, this process is equivalent to rejecting one procurement set, and that is the highest costing active procurement set.

As we show in the full version of the paper, the MDA trade reduction mechanism is in fact equivalent to the Trade Reduction mechanism:

---

[13] For any $k = 1, .., K$ such that $|N_k \setminus A_t| < \tilde{\mu}_k$, set $c^t_{k,(|N_k\setminus A_t|+1)} = c^t_{k,(|N_k\setminus A_t|+2)} = ... = c^t_{k,(\tilde{\mu}_k)} = \max B_k$ and if no consumer was rejected prior to period t, set $v^t_{max} = 0$. Specifically, for $t = 1$ set $NC^1(\emptyset) = \sum_{k=1}^K \tilde{\mu}_k \max B_k$.

Proposition 18. *Consider an MDA mechanism that is defined by the scoring functions (1) and the composition functions (3). This mechanism is equivalent to the Trade Reduction mechanism (Definition 15).*

We can now use Proposition 18 (together with Proposition 11) to infer additional properties of the Trade Reduction mechanism of [1,2].

Corollary 19. *The Trade Reduction mechanism is WGSP and equivalent to a clock auction.*

4.2   The Modified Trade Reduction Mechanism

The class of MDA mechanisms allows considerable freedom in design while maintaining the incentive properties common to all MDA mechanisms. We use this feature to modify the Trade Reduction mechanism and improve its social welfare. The improvement is possible since the MDA trade reduction mechanism (Sect. 4.1) uses an inaccurate measure of the deficit - the net cost of the last rejected procurement set. This causes the mechanism to reject more trades than is actually needs in order to keep the budget balanced. The new mechanism uses a different measure of deficit and thus waives efficient trades less frequently. For this, we need the following definition:

Definition 20. *[10] For each agent $i \in N_k \cap A_t$, the **t-period threshold payment** $p_{k,i}^t(b_{N \setminus A_t})$, is the maximal bid that would have kept i active until iteration t, holding all other bids fixed.*

First note that the mechanism's final threshold payment for a winning agent is equal to his T-period threshold payment, where T is the final period. This follows directly from the definition of threshold payments (Definition 9).

Second, note that for an active agent $i \in N_k \cap A_t$, the t-period threshold payment is determined only by bids of rejected agents of class k. This is true since agent i's bid can not affect $C^{t'}(b_{N \setminus A_t})$ for $t' < t$ and thus can not affect which classes of agents are chosen for rejection prior to period t. The only effect agent i's bid has is on his ranking relative to other agents of class k. Furthermore, in cases where agents are ranked solely by their bids (as will be the case here), the t-period threshold payment of a class-k agent is equal to the bid of the last rejected agent of his class. This means that all active agents of class k have the same t-period threshold payments.

Definition 21. *Consider an MDA mechanism in which agents are ranked only by their bids. For all t and k, the **t-period threshold payments for active agents of class k** is:*

$$p_k^t = \begin{cases} \min\limits_{j \in N_k \setminus A_{t-1}} b_j & \text{if } N_k \setminus A_{t-1} \neq \emptyset \\ \max B_k & \text{if } N_k \setminus A_{t-1} = \emptyset \end{cases}$$

We now turn to the formal definition of the Modified Trade Reduction mechanism by defining the scoring and composition functions.

Scoring Functions: Similar to Sect. 4.1, producers are ranked in an ascending order of costs and consumers are ranked in a descending order of values (see (1) for the formal definition).

The composition rule is similar to the one presented in Sect. 4.1 with the slight difference that it rejects procurement sets according to a *lower bound* on net costs, instead of the net costs themselves. The lower bound, or **minimal net cost**, is a function of the t-period threshold payments:

$$MNC^t(b_{N \setminus A_t}) = \sum_{k=1}^{K} \tilde{\mu}_k p_k^t + p_{K+1}^t \tag{4}$$

Let the excess supply function ES^t be defined as in Sect. 4.1. Now define the **composition function** for each period t as follows:

$$C^t(b_{N \setminus A_t}) = \begin{cases} ES^t(A_t) & \text{if } ES^t(A_t) \neq \emptyset \\ \{1, ..., K+1\} & \text{if } ES^t(A_t) = \emptyset \text{ and } MNC^t(b_{N \setminus A_t}) > 0 \\ \emptyset & \text{if } ES^t(A_t) = \emptyset \text{ and } MNC^t(b_{N \setminus A_t}) \leq 0 \end{cases} \tag{5}$$

Definition 22. *The **Modified Trade Reduction mechanism** is an MDA mechanism with the scoring functions (1) and the composition functions (5).*

The Modified Trade Reduction mechanism operates as follows. First it rejects the least valuable procurement set. Fixing this allocation, the t-period threshold payments are calculated together with the implied deficit, which is proportional to MNC^t. If the deficit is non-positive, the mechanism terminates. Otherwise, the least valuable active procurement set is rejected, and so on. The critical stage of the mechanism comes after it removes enough procurement sets and reaches an efficient allocation. Then, it computes a "within-class" threshold payments for each class of agents: the value of the most valuable agent that does not win in the efficient allocation. It then checks what would happen if all active agents paid their within-group threshold: if there is no deficit, then the mechanism outputs the efficient allocation. Otherwise, a trade reduction is performed.[14]

Theorem 23. *The Modified Trade Reduction mechanism satisfies:*

1. *It is IR, strategy-proof, WGSP and equivalent to a clock auction.*
2. *It is weakly budget balanced.*
3. *For every realization of values and costs, the mechanism either sets the optimal allocation or incurs the loss of the least valuable procurement set.*

---

[14] Note that this procedure is not equivalent to the following mechanism: run VCG if it is budget balanced, otherwise run a trade reduction. This mechanism is not truthful, as the VCG payment of an agent can be determined by agents of other classes (who therefore can manipulate the outcome). The Modified Trade Reduction mechanism uses bounds on the payments that are determined only by the agents of each class, and therefore it is strategy-proof.

Item 1 holds since the modified mechanism is an MDA mechanism. For the proof of item 2 see the full version of the paper. Next, we prove item 3.

Note that in each period, $p^t_{K+1} = -v^t_{max}$ and $p^t_k = c^t_{k,(1)}$ for all $k = 1, ..., K$. Now use the definitions of $NC^t(b_{N \setminus A_t})$ and $MNC^t(b_{N \setminus A_t})$ ((2) and (4) respectively) to get that for all t:

$$NC^t(b_{N \setminus A_t}) = \sum_{k=1}^{K} \sum_{l=1}^{\tilde{\mu}_k} c^t_{k,(l)} - v^t_{max} \geq$$

$$\sum_{k=1}^{K} \tilde{\mu}_k c^t_{k,(1)} - v^t_{max} = \sum_{k=1}^{K} \tilde{\mu}_k p^t_k + p^t_{K+1} = MNC^t(b_{N \setminus A_t}) \quad (6)$$

Since the modified mechanism terminates once $MNC^t \leq 0$ and the MDA trade reduction mechanism terminates once $NC^t \leq 0$, the former mechanism terminates (weakly) prior to the latter. Since the procurement sets are ordered in the same manner in both mechanisms, the allocation determined by the modified mechanism contains the Trade Reduction allocation but is possibly larger.

According to equation (6), MNC^t is a lower bound on the net costs of all previously rejected procurement sets. While MNC^t is positive, all rejected procurement sets have positive net costs and the mechanism rejects more procurement sets. The mechanism terminates once a procurement set with a non-positive lower bound on its net cost is rejected. By that time all the procurement sets with positive net costs were rejected. The proof of 3 follows.

The Modified mechanism improves upon the Trade Reduction mechanism in scenarios where there is variance in the values of agents, such that the "within-class" threshold is sufficiently far from the values of the next losing agent. It follows that we need consumers to demand more than one unit of some item for having a different outcome than the Trade Reduction mechanism. (Indeed, when applying the modified mechanism to the two-sided market in [8], it identifies with the Trade Reduction Mechanism and eliminates one efficient trade.)

Simulations: We ran some computer simulations of simple supply chains where the advantages of the modified mechanism can kick in. Our simulations considered economies with $n = 2, ..., 10$ buyers and $2n$ sellers of a homogeneous good, where each buyer is interested in a bundle of two units and values were drawn from the uniform distribution on $[0, 1]$. The modified mechanism improves upon the Trade Reduction mechanism in around 17 % of the instances, even when the market grows substantially. In small economies the improvement can be up to a 100 % of the overall efficiency. More details appear in the full version.

Diverse Manufacturing Technologies: The MDA framework allows us to explore more general settings of supply chains. Specifically, we are able to relax the assumption of unique manufacturing technologies (Assumption 1) and examine a setting in which part of the production process may be conducted with different types of inputs. The details appear in the full version of the paper.

Acknowledgments. We thank Moshe Babaioff for helpful discussions. Liad Blumrosen was supported by the Israel Science Foundation (grant No. 230/10).

References

1. Babaioff, M., Nisan, N.: Concurrent auctions across the supply chain. J. Artif. Intell. Res. **21**, 595–629 (2004)
2. Babaioff, M., Walsh, W.E.: Incentive-compatible, budget-balanced, yet highly efficient auctions for supply chain formation. Decision Sup. Sys. **39**(1), 123–149 (2005)
3. Cripps, M.W., Swinkels, J.M.: Efficiency of large double auctions. Econometrica **74**(1), 47–92 (2006)
4. Dütting, P., Gkatzelis, V., Roughgarden, T.: The performance of deferred-acceptance auctions. In: Proceedings of the Fifteenth ACM Conference on Economics and Computation, pp. 187–204. ACM (2014)
5. Dütting, P., Roughgarden, T., Talgam-Cohen, I.: Modularity and greed in double auctions. In: Proceedings of the Fifteenth ACM Conference on Economics and Computation, pp. 241–258. ACM (2014)
6. Lehmann, D., Oćallaghan, L.I., Shoham, Y.: Truth revelation in approximately efficient combinatorial auctions. J. ACM (JACM) **49**(5), 577–602 (2002)
7. Lubin, B., Parkes, D.C.: Approximate strategyproofness. Curr. Sci. **103**(9), 1021–1032 (2012)
8. McAfee, R.P.: A dominant strategy double auction. J. Econ. Theor. **56**(2), 434–450 (1992)
9. Mehta, A., Roughgarden, T., Sundararajan, M.: Beyond moulin mechanisms. In: 8th ACM Conference on Electronic Commerce, EC 2007, pp. 1–10 (2007)
10. Milgrom, P., Segal, I.: Deferred-acceptance auctions and radio spectrum reallocation. In: Proceedings of the Fifteenth ACM Conference on Economics and Computation, pp. 185–186. ACM (2014)
11. Moulin, H.: Incremental cost sharing: characterization by coalition strategy-proofness. Soc. Choice Welfare **16**(2), 279–320 (1999)
12. Myerson, R.B., Satterthwaite, M.A.: Efficient mechanisms for bilateral trading. J. Econ. Theor. **29**(2), 265–281 (1983)
13. Rustichini, A., Satterthwaite, M.A., Williams, S.R.: Convergence to efficiency in a simple market with incomplete information. Econometrica **62**(5), 1041–63 (1994)
14. Satterthwaite, M.A., Williams, S.R.: The optimality of a simple market mechanism. Econometrica **70**(5), 1841–1863 (2002)
15. Singer, Y.: Budget feasible mechanisms. In: The 51st Annual Symposium on Foundations of Computer Science, FOCS 2010, pp. 765–774 (2010)
16. Vorobeychik, Y., Kiekintveld, C., Wellman, M.P.: Empirical mechanism design: methods, with application to a supply-chain scenario. In: ACM Conference on Electronic Commerce, pp. 306–315 (2006)
17. Walsh, W., Wellman, M., Ygge, F.: Combinatorial auctions for supply chain formation. In: Proceedings of the ACM Conference on Electronic Commerce, pp. 260–269 (2000)

Testing Consumer Rationality Using Perfect Graphs and Oriented Discs

Shant Boodaghians[1]([⊠]) and Adrian Vetta[2]

[1] Department of Mathematics and Statistics, McGill University, Montreal, Canada
shant.boodaghians@mail.mcgill.ca
[2] Department of Mathematics and Statistics, and School of Computer Science,
McGill University, Montreal, Canada
vetta@math.mcgill.ca

Abstract. Given a consumer data-set, the axioms of revealed preference proffer a binary test for rational behaviour. A natural (non-binary) measure of the degree of rationality exhibited by the consumer is the minimum number of data points whose removal induces a rationalisable data-set. We study the computational complexity of the resultant CONSUMER RATIONALITY problem in this paper. This problem is, in the worst case, equivalent (in terms of approximation) to the *directed feedback vertex set* problem. Our main result is to obtain an exact threshold on the number of commodities that separates easy cases and hard cases. Specifically, for two-commodity markets the CONSUMER RATIONALITY problem is polynomial time solvable; we prove this via a reduction to the *vertex cover* problem on perfect graphs. For three-commodity markets, however, the problem is NP-complete; we prove this using a reduction from PLANAR 3-SAT that is based upon oriented-disc drawings.

1 Introduction

The theory of revealed preference, introduced by Samuelson [24,25], has long been used in economics to test for rational behaviour. Specifically, given a set of m commodities with price vector p, we wish to determine whether the consumer always demands an affordable bundle x of maximum utility. To test this question, assume we are given a collection of *consumer data* $\{(p_1, x_1), (p_2, x_2), \ldots, (p_m, x_m)\}$. Each pair (p_i, x_i) denotes the fact that the consumer purchased the bundle of goods $x_i \in \mathbf{R}^n$ when the prices were $p_i \in \mathbf{R}^n$. (Here $\mathbf{R} = \mathbb{R}_{\geq 0}$ denotes the set of non-negative real numbers.) Now, assuming the consumer is rational, the selection of x_i reveals information about the consumer's preferences; in particular, suppose that $p_i \cdot x_i \geq p_i \cdot x_j$ for some $j \neq i$. This means that the bundle x_j was affordable, and available for selection, when x_i was chosen. In this case, we say x_i is *directly revealed preferred* to x_j and denote this $x_i \succeq x_j$. Furthermore, suppose we observe that $x_i \succeq x_j$ and that $x_j \succeq x_k$. Then, by transitivity of preference, we say x_i is *indirectly revealed preferred* to x_k.

© Springer-Verlag Berlin Heidelberg 2015
E. Markakis and G. Schäfer (Eds.): WINE 2015, LNCS 9470, pp. 187–200, 2015.
DOI: 10.1007/978-3-662-48995-6_14

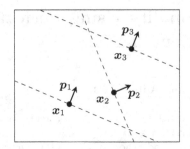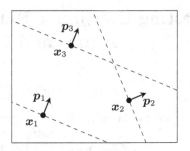

Fig. 1. A rational consumer and an irrational consumer.

For clarity of presentation, we will assume that all the chosen bundles are distinct and that all revealed preferences are strict (no ties). For a rational consumer, the data-set should then have the following property:

The Generalized Axiom of Revealed Preference.[1]
If $x_1 \succeq x_2$, $x_2 \succeq x_3, \ldots, x_{k-2} \succeq x_{k-1}$ and $x_{k-1} \succeq x_k$ then $x_k \not\succeq x_1$.

Moreover, Afriat [1] showed that the Generalized Axiom of Revealed Preference (GARP) is also sufficient for the construction of a utility function which *rationalises* the data-set. That is, Afriat showed that if the consumer data satisfies GARP then one can construct a utility function $v : \mathbf{R}^n \to \mathbf{R}$ such that v is maximised at x_i among the set of affordable bundles at prices p_i. Hence, GARP is a necessary and sufficient condition for consumer rationality.

We can represent the preferences revealed by the consumer data via a directed graph, $D_{\succeq} = (V, A)$. This directed *revealed preference graph* contains a vertex $x_i \in V$ for each data-pair (p_i, x_i), and an arc from x_i to x_j if and only if $x_i \succeq x_j$. Observe that GARP holds *if and only if* the revealed preference graph is acyclic. Consequently, Afriat's theorem implies that the consumer is rational if and only if $D_{\succeq}$ contains no directed cycles.

For example, Fig. 1 displays visually two sets of consumer data. Each bundle x_i is paired with its price vector p_i, and a dotted line is drawn through x_i perpendicular to p_i. Note that $p_i x_i \geq p_i y$ if and only if y lies on the opposite side of the dotted line to the drawing of p_i. Hence, for the first consumer (left), we have $x_3 \succeq x_2$, $x_3 \succeq x_1$ and $x_2 \succeq x_1$. This produces an acyclic revealed preference graph $D_{\succeq}$ and, therefore, her behaviour can be rationalized. On the otherhand, the second consumer (right) reveals $x_3 \succeq x_2 \succeq x_3$. This produces a directed 2-cycle in $D_{\succeq}$ and, so, her behaviour cannot be rationalised.

1.1   A Measure of Consumer Rationality

We have seen that graph acyclicity can be used to provide a test for consumer rationality. However such a test is binary and, in practice, leads to the imme-

---

[1] When ties are possible, this formulation is called the *strong axiom of revealed preference*; see Houthakker [17]. We refer the reader to the survey by Varian [29] for details concerning the assorted axioms of revealed preference.

diate conclusion of irrationality, as observed data sets typically induce cycles in the revealed preference graph. Consequently, there has been a large body of experimental and theoretical work designed to measure how close to rational the behaviour of a consumer is. Examples include measurements based upon best-fit perturbation errors (*e.g.* Afriat [2] and Varian [30]), measurements based upon counting the number of rationality violations present in the data (*e.g.* Swofford and Whitney [28] and Famulari [15]), and measurements based upon the maximum size of a *rational* subset of the data (*e.g.* Koo [21] and Houtman and Maks [18]). Gross [11] provides a review and analysis of some of these measures. Recently new measures have been designed by Echenique et al. [10], Apesteguia and Ballester [3], and Dean and Martin [6].

Combinatorially, perhaps the most natural measure is simply to count the number of "irrational" purchases. That is, what is the minimum number of data-points whose removal induces a rational set of data? The associated decision problem is called the CONSUMER RATIONALITY problem.

CONSUMER RATIONALITY
Instance: Consumer data $(p_1, x_1), \ldots, (p_m, x_m) \in \mathbf{R}^n \times \mathbf{R}^n$, and an integer k.
Problem: Is there a sub-collection of at most k data points whose removal produces a data set satisfying GARP?

We note that this CONSUMER RATIONALITY problem is dual to the measure of Houtman and Maks [18]. Using the graphical representation, it can be seen that the consumer rationality problem is a special case of the DIRECTED FEEDBACK VERTEX SET problem. In fact, as we explain in Sect. 2, when there are many goods, the two problems are equivalent. However, the consumer rationality problem becomes easier to approximate as the number of commodities falls. Indeed, the main contribution of this paper is to obtain an exact threshold on the number of commodities that separates easy cases (polynomial) and hard cases (NP-complete). In particular, we prove the problem is polytime solvable for a two-commodity market (Sect. 3), but that it is NP-complete for a three-commodity market (Sect. 4).

2 The General Case: Many Commodities

In this section we show that the CONSUMER RATIONALITY problem in full generality is computationally equivalent to the DIRECTED FEEDBACK VERTEX SET (DFVS) problem.

DIRECTED FEEDBACK VERTEX SET
Instance: A directed graph $D = (V, A)$, and an integer k.
Problem: Is there a set S of at most k vertices such that the induced subgraph $D[V \setminus S]$ is acyclic? (Such a set S is called a *feedback vertex set*.)

First, observe that the CONSUMER RATIONALITY PROBLEM is a special case of the DIRECTED FEEDBACK VERTEX SET PROBLEM: we have seen that the dataset is rationalizable if and only if the preference graph is acyclic. Thus, the minimum feedback vertex set in the preference graph $D_\succeq$ clearly corresponds to the minimum number of data points that must be removed to create a rationalizable data-set.

On the other hand, provided the number of commodities is large, DFVS is a special case of the CONSUMER RATIONALITY PROBLEM. Specifically, Deb and Pai [7] show that for any directed graph D there is a data-set on $m = n$ commodities whose preference graph is $D_\succeq = D$; for completeness, we include the short proof of this result.

Lemma 2.1. *[7] Given sufficiently many commodities, we may construct any digraph as a preference graph.*

Proof. Let D be any digraph on n nodes. We will construct n pairs in $\mathbf{R}^n \times \mathbf{R}^n$ such that $D_\succeq \cong D$. Denote $\boldsymbol{p}^i = (p_1^i, \ldots, p_n^i)$, and set $p_i^i = 1$, $p_j^i = 0$ for $j \neq i$. Similarly, denote $\boldsymbol{x}^i = (x_1^i, \ldots, x_n^i)$, and set $x_j^i = 1$ if $i = j$, 0 if $(i, j) \in D$, and 2 if $(i, j) \notin D$. We then have, $\boldsymbol{p}_i \cdot \boldsymbol{x}_i = 1$, $\boldsymbol{p}_i \cdot \boldsymbol{x}_j = 0$ if we want an arc from i to j, and $\boldsymbol{p}_i \cdot \boldsymbol{x}_j = 2$ if we do not want an arc, as desired. $\square$

It follows that any lower and upper bounds on approximation for (the optimization version of) DFVS immediately apply to (the optimization version of) the CONSUMER RATIONALITY problem. The exact hardness of approximation for DFVS is not known. The best upper bound is due to Seymour [26] who gave an $O(\log n \log \log n)$ approximation algorithm. With respect to lower bounds, the DIRECTED FEEDBACK VERTEX SET problem is NP-complete [19]. Furthermore, as we will see in Sect. 3, the CONSUMER RATIONALITY problem is at least as hard to approximate as VERTEX COVER. It follows that DFVS problem cannot be approximated to within a factor 1.36 [8] unless $P = NP$. Also, assuming the Unique Games Conjecture [20], the minimum directed feedback vertex set cannot be approximated to within any constant factor [14, 27].

Lemma 2.1 shows the equivalence with DIRECTED FEEDBACK VERTEX SET applies when the number of commodities is at least the size of the data-set. However, Deb and Pai [7] also show that for an m-commodity market, there exists a directed graph on $O(2^m)$ vertices that cannot be realised as a preference graph. This suggests that the hardness of the consumer rationality problem may vary with the quantity of goods. Indeed, we now prove that this is the case.

3   The Case of Two Commodities

We begin by outlining the basic approach to proving polynomial solvability for two goods. As described, the CONSUMER RATIONALITY problem is a special case of DVFS. For two goods, however, rather than considering all directed cycles, it is sufficient to find a vertex hitting set for the set of *digons* (directed cycles consisting of two arcs). The resulting problem can be solved by finding a minimum vertex cover in a corresponding auxiliary undirected graph. The VERTEX COVER problem is, of course, itself hard [8]. But we prove that the auxiliary undirected graph is perfect, and VERTEX COVER is polytime solvable in perfect graphs.

3.1   Two-Commodity Markets and the Vertex Cover Problem

So, our first step is to show that it suffices to hit only digons. Specifically, we prove that every vertex-minimal cycle in the revealed preference graph $D_\succeq$ is a digon. This fact corresponds to the result that for two goods the *Weak Axiom of Revealed Preference* is equivalent to the *Generalised Axiom of Revealed Preference*. This equivalence was noted by Samuelson [25] and formally proven by Rose [23] in 1958; for a recent structurally motivated proof see [16].

Lemma 3.1. *[23] For two commodities, every minimal cycle is a digon.*

A direct graphical proof is presented in the full version of the paper. Lemma 3.1 implies that a vertex set that intersects every digon will also intersect each directed cycle of any length. Hence, to solve the CONSUMER RATIONALITY problem for two goods, it suffices to find a minimum cardinality hitting vertex set for the digons of $D_\succeq$. We can do this by transforming the problem into one of finding a minimum vertex cover in an undirected graph. Recall the VERTEX COVER problem is:

VERTEX COVER

INSTANCE: Given an undirected graph $G = (V, E)$ and an integer k.
PROBLEM: Is there a set S of at most k vertices such that every edge has an
endpoint in S?

The transformation is then as follows: given the directed revealed preference graph $D_\succeq$ we create an *auxiliary undirected graph* $G_\succeq$. The vertex set $V(G_\succeq) = V(D_\succeq)$ so the undirected graph also has a vertex for each bundle x_i. There is an edge (x_i, x_j) in $G_\succeq$ if and only if x_i and x_j induce a digon in $D_\succeq$. It is easy to verify that a vertex cover in $G_\succeq$ corresponds to a hitting set for digons of $D_\succeq$.

Let's see some simple examples for the auxiliary graph $G_\succeq$. First consider Fig. 2(a), where bundles are placed on a concave curve. Now every pair of vertices x_i and x_j induce a digon in $D_\succeq$. Thus $G_\succeq$ is an undirected clique. Now consider Fig. 2(b). The vertices on the left induce a directed path in $D_\succeq$; the vertices along the bottom also induce a directed path in $D_\succeq$. However each pair consisting of one vertex on the left and one vertex on the bottom induce a digon in $D_\succeq$. Thus $G_\succeq$ is a complete bipartite graph.

3.2   Perfect Graphs

An undirected graph G is perfect if the chromatic number of any induced subgraph is equal to the cardinality of the maximum clique in the subgraph. In 1961, Berge [4] made the famous conjecture that an undirected graph is perfect if and only if it contains neither an odd length hole nor an odd length antihole. Here a *hole* is a chordless cycle with at least four vertices. An *antihole* is the complement of a chordless cycle with at least four vertices. Berge's conjecture was finally proven by Chudnovsky, Robertson, Seymour and Thomas [5] in 2006.

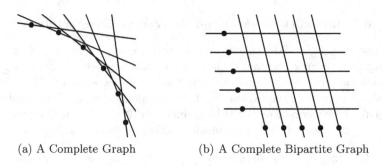

(a) A Complete Graph   (b) A Complete Bipartite Graph

Fig. 2. Examples of the auxiliary undirected graph.

Theorem 3.1 (The Strong Perfect Graph Theorem [5]**).** *An undirected graph is perfect if and only if it contains no odd holes and no odd antiholes.*

There are many important classes of perfect graphs, for example, cliques, bipartitie graphs, chordal graphs, line graphs of bipartite graphs, and comparability graphs.[2] Interestingly, we now show that the class of 2D auxiliary revealed preference graphs are also perfect. To prove this, we will need the following geometric lemma, but first, we introduce the required notation.

Let $x = (x_1, x_2) \in \mathbf{R}^2$, and define

$$x^{\searrow} := \{(y_1, y_2) \in \mathbf{R}^2 : y_1 \geq x_1, y_2 \leq x_2\} ,$$

i.e. the points which lie "below and to the right" of x in the plane. Define $x^{\nearrow}$, $x^{\nwarrow}$ and $x^{\swarrow}$ similarly. In addition, define $x^{\searrow\!\!\searrow}$ $x^{\nwarrow\!\!\nwarrow}$, $x^{\nearrow\!\!\nearrow}$ and $x^{\swarrow\!\!\swarrow}$ by replacing the inequalities with strict inequalities. Furthermore, if ℓ is a line in the plane of non-positive slope which intersects the positive quadrant, we say a point *lies below* ℓ if it lies in the same closed half-plane as the origin. For each data pair $(\boldsymbol{p}_i, \boldsymbol{x}_i)$, we define ℓ_i to be the line through $\boldsymbol{x}_i$ perpendicular to $\boldsymbol{p}_i$. Hence, in our setting $\boldsymbol{x}_i \succeq \boldsymbol{x}_j$ if and only if $\boldsymbol{x}_j$ lies below ℓ_i. Note that, if $\boldsymbol{x}_i \succeq \boldsymbol{x}_j$, then we may not have $\boldsymbol{x}_j \in \boldsymbol{x}_i^{\nearrow\!\!\nearrow}$ since $\boldsymbol{p}_i$ is non-negative.

Lemma 3.2. *Let $\{x_i, x_j, x_k\}$, listed in order, be an induced path in the 2D auxiliary revealed preference graph $G_{\succeq}$. If $\boldsymbol{x}_i \in \boldsymbol{x}_j^{\searrow}$ then $\boldsymbol{x}_k \in \boldsymbol{x}_j^{\searrow}$. (Similarly, if $\boldsymbol{x}_i \in \boldsymbol{x}_j^{\nwarrow}$ then $\boldsymbol{x}_k \in \boldsymbol{x}_j^{\nwarrow}$.)*

Proof. Recall the assumption that the bundles distinct, that is, $\boldsymbol{x}_i \neq \boldsymbol{x}_j$ for all $i \neq j$. Because $\{x_i, x_j\}$ is an edge in the auxiliary undirected graph $G_{\succeq}$, we know that $\boldsymbol{x}_i \succeq \boldsymbol{x}_j$ and $\boldsymbol{x}_j \succeq \boldsymbol{x}_i$. Therefore it cannot be the case that $\boldsymbol{x}_i \in \boldsymbol{x}_j^{\nearrow\!\!\nearrow}$ or $\boldsymbol{x}_j \in \boldsymbol{x}_i^{\nearrow\!\!\nearrow}$. Thus, either $\boldsymbol{x}_j \in \boldsymbol{x}_i^{\searrow}$ or $\boldsymbol{x}_j \in \boldsymbol{x}_i^{\nwarrow}$, but not both. Similarly, because $\{x_j, x_k\}$ is an edge in $G_{\succeq}$, either $\boldsymbol{x}_k \in \boldsymbol{x}_j^{\searrow}$ or $\boldsymbol{x}_k \in \boldsymbol{x}_j^{\nwarrow}$.

---

[2] By the (Weak) Perfect Graph Theorem [22], the complements of these classes of graphs are also perfect.

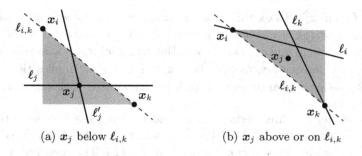

(a) x_j below $\ell_{i,k}$ (b) x_j above or on $\ell_{i,k}$

Fig. 3. Induced path on three vertices.

Now, without loss of generality, let $x_i \in x_j^{\searrow}$. For a contradiction, assume that $x_k \in x_j^{\searrow}$. Hence, we have $x_j \in x_i^{\searrow} \cap x_k^{\searrow}$. Suppose x_j lies strictly below the line $\ell_{i,k}$ through x_i and x_k. But then we cannot have both $x_j \succeq x_i$ and $x_j \succeq x_k$. This is because the line ℓ_j must cross the segment of $\ell_{i,k}$ between x_i and x_k if it is to induce either of the two preferences. Thus, the line ℓ_j separates x_i and x_k and, so, at most one of bundles can lie below the line. This is illustrated in Fig. 3(a).

On the other hand, suppose x_j lies on or above the line $\ell_{i,k}$ through x_i and x_k. Now we know that $x_i \succeq x_j$. This implies that $x_i \succeq x_k$, as illustrated in Fig. 3(b). Furthermore, we know that $x_k \succeq x_j$ which implies that $x_k \succeq x_i$. Thus $\{x_i, x_k\}$ is an edge in $G_\succeq$. This contradicts the fact that $\{x_i, x_j, x_k\}$ is an induced path. $\qquad\square$

Lemma 3.3. *The 2D auxiliary revealed preference graph $G_\succeq$ contains no odd holes on at least 5 vertices.*

Proof. Take a hole $C_k = \{x_0, x_1, \ldots, x_{k-1}\}$, listed in order, where $k \geq 5$ is odd. For any $0 \leq i \leq k - 1$, the three vertices $\{x_{i-1}, x_i, x_{i+1}\}$ induce a path in $G_\succeq$. Consequently, by Lemma 3.2, either both x_{i-1} and x_{i+1} are in $x_i^{\searrow}$ or both x_{i-1} and x_{i+1} are in $x_i^{\searrow}$. In the former case, colour x_i yellow. In the latter case, colour x_i red. Thus we obtain a 2-coloring of C_k. Since k is odd, there must be two adjacent vertices, x_i and x_{i+1}, with the same colour. Without loss of generality, let both vertices be yellow. Thus, x_{i+1} is $x_i^{\searrow}$ and x_i is in $x_{i+1}^{\searrow}$. This contradicts the distinctness of x_i and x_{i+1}. $\qquad\square$

Lemma 3.4. *The 2D auxiliary revealed preference graph $G_\succeq$ contains no anti-holes on at least 5 vertices.*

Proof. Note that the complement of an odd hole on five vertices is also an odd hole. Thus, by Lemma 3.3, the graph $G_\succeq$ may not contain an antihole on five vertices.

Next consider an antihole $\bar{C}_k = \{x_0, x_1, \ldots, x_{k-1}\}$, listed in order, with $k \geq 6$. The neighbours in $\bar{C}_k$ of x_i, for any $0 \leq i \leq k - 1$, are $\Gamma_i = \{x_{i+2}, x_{i+3}, \ldots, x_{i-2}\}$. We claim that either every vertex of Γ_i is in $x_i^{\searrow}$ or every vertex of Γ_i is in $x_i^{\searrow}$. To see this note that (x_{i+2}, x_{i+3}) is not an edge, and therefore $\{x_{i+2}, x_i, x_{i+3}\}$ is an induced path in $G_{\succeq}$. By Lemma 3.2, without loss of generality, both x_{i+2} and x_{i+3} are in $x_i^{\searrow}$. But $\{x_{i+3}, x_i, x_{i+4}\}$ is also an induced path in $G_{\succeq}$. Consequently, as x_{i+3} is in $x_i^{\searrow}$, Lemma 3.2 implies that x_{i+4} is in $x_i^{\searrow}$. Repeating this argument through to the induced path $\{x_{i-3}, x_i, x_{i-2}\}$ gives the claim.

Now consider the three vertices x_0, x_2 and x_4. Since $k \geq 6$ these vertices are pairwise adjacent in $\bar{C}_k$. Without loss of generality, by the claim, Γ_0 is in $x_0^{\searrow}$. Thus, x_2 and x_4 are in $x_0^{\searrow}$. However x_0 is in $\Gamma_2 \cap \Gamma_4$. Thus every vertex in Γ_2 is in $x_2^{\searrow}$ and every vertex in Γ_2 is in $x_4^{\searrow}$. Hence, x_4 is in $x_2^{\searrow}$ and x_2 is in $x_4^{\searrow}$, a contradiction. $\qquad\square$

Lemmas 3.3 and 3.4 together show, by applying the Strong Perfect Graph Theorem, that the auxiliary undirected graph is perfect.

Theorem 3.2. *The 2D auxiliary revealed preference graph $G_{\succeq}$ is perfect.* $\qquad\square$

3.3   A Polynomial Time Algorithm

In classical work, Grötschel, Lovász and Schrijver [12,13] show that the VERTEX COVER problem in a perfect graph can be solved in polynomial time via the ellipsoid method.

Theorem 3.3. *[12] The VERTEX COVER problem is solvable in polynomial time on a perfect graph.* $\qquad\square$

But by Theorem 3.2, the auxiliary undirected graph is perfect. Since the consumer rationality problem for two commodities corresponds to a vertex cover problem on this auxiliary undirected graph, we have:

Theorem 3.4. *In a two-commodity market, the CONSUMER RATIONALITY problem is solvable in polynomial time.* $\qquad\square$

4   The Case of Three Commodities

We have shown that for two commodities, the consumer rationality problem can be solved in polynomial time. We now prove the problem is NP-complete if there are three (or more) commodities by presenting a reduction from PLANAR 3-SAT. The proof has three parts: first we transform an instance of PLANAR 3-SAT to an instance of VERTEX COVER in an associated undirected *gadget graph*. Second, we show that a vertex cover in the gadget graph corresponds to a directed feedback vertex set in a directed *oriented disc graph*. Finally, we prove that every oriented disc graph corresponds to a preference graph in a three-commodity market. Consequently, we can solve PLANAR 3-SAT using an algorithm for the three-commodity case of the CONSUMER RATIONALITY problem.

We begin by defining the class of oriented-disc graphs. Let $\{x_1, \ldots, x_n\}$ be points in the plane and let $\{B_1, \ldots, B_n\}$ be closed discs of varying radii such

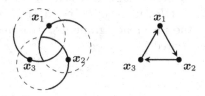

Fig. 4. An oriented disc drawing and its corresponding oriented disc graph.

that B_i contains x_i on its boundary. We call this collection of points and discs an *oriented-disc drawing*. Given a drawing, we construct a directed graph $D = (V, A)$ on the vertex set $V = \{x_1, \ldots, x_n\}$. There is an arc from x_i to x_j in D if x_j, $j \neq i$, is contained in the disc B_i. A directed graph that can be built in this manner is called an *oriented-disc graph*.

An example is given in Fig. 4. The oriented-disc drawing is shown on the left and the the resulting oriented disc graph, a directed cycle on 3 vertices, is shown on the right. (We remark that, for enhanced clarity in the larger figures that follow, the boundary circles are drawn half-dotted.) Note that, even if the discs have uniform radii, the resulting oriented-disc graphs need not be symmetric – that is, (x_i, x_j) can be an arc even if (x_j, x_i) is not. This is due to the fact that x_i lies on the boundary, not at the centre, of its disc B_i. We now start by proving the third part of the reduction: every oriented disc graph corresponds to a preference graph in a three-commodity market.

Lemma 4.1. *Every oriented-disc graph corresponds to a preference graph induced by consumer data in a three-commodity market.*

Proof. Let D be any oriented-disc graph. We wish to build a three-commodity data set whose preference graph is D. Recall that the plane is homomorphic to the 2-dimensional sphere minus a point. Moreover, the inverse of the *stereographic projection* is a map from the plane to a sphere which preserves the shape of circles; see, for example, [9]. This motivates us to attempt to draw the points and discs on the unit sphere centered at $(1, 1, 1) \in \mathbf{R}^3$. To do this, we scale the oriented-disc drawing appropriately and embed it in a small region on the "underside" of the sphere, that is, around the point where the inwards normal vector is $(1, 1, 1)$. An example of this, where the oriented-disc graph is the directed 3-cycle, is shown in Fig. 5(a).

We now need to create the corresponding collection of consumer data. Let $\{x_1, \ldots, x_n\}$ be the n points of some oriented-disc drawing of D embedded onto the underside of the sphere. Note that the intersection of a sphere and a plane is a circle. Furthermore, a plane through a point on the sphere will create a circle containing that point. Thus we may select the x_i to be the bundles chosen by the market and we may choose p_i such that the plane with normal p_i that passes through x_i intersects the sphere exactly along the boundary of the embedding of the disc B_i. An example is shown in Fig. 5(b). Because p_i is non-negative it points into the sphere. Therefore, x_i is revealed preferred to every point on the inside of the embedding of B_i; it is not revealed preferred to any other point on the sphere. Hence, the preference graph $D_\succeq$ is isomorphic to the original oriented-disc graph, as desired.   □

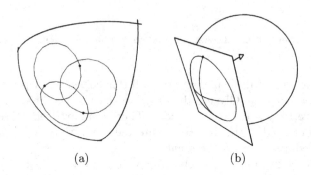

(a)   (b)

Fig. 5. A 3-cycle embedded on a sphere section, and a disc on a sphere.

Now, recall the first part of the reduction: we wish to transform an instance of PLANAR 3-SAT to an instance of VERTEX COVER in an associated undirected *gadget graph*. Our gadget graph is based upon a network used by Wang and Kuo [31] to prove the hardness of MAXIMUM INDEPENDENT SET in undirected unit-disc graphs. However, we are able to simplify their non-planar network by using an instance of PLANAR 3-SAT rather than the general 3-SAT. This simplification will be useful when implementing the second part of the reduction.

Let φ be an instance of PLANAR 3-SAT with variables $u_1, \ldots, u_n$ and clauses $C_1, \ldots, C_m$. Recall that φ is *planar* if the bipartite graph H_φ consisting of a vertex for each variable, a vertex for each clause, and edges connecting each clause to its three variables, is planar. The associated, undirected, *gadget graph* G_φ is constructed as shown in Fig. 6. For each clause $C = (u_i \vee u_j \vee u_k)$, add a 3-cycle to the graph whose vertices are labelled by the appropriate literals for the variables u_i, u_j and u_k. We call these the *clause gadgets*. For each variable u_i, add a large cycle of even length whose vertices are alternatingly labelled as the literals u_i and $\bar{u}_i$. We call these the *variable gadgets*. Finally, add an edge from each variable in the clause gadgets to some vertex on the corresponding variable gadget with the opposite label – we choose a different variable vertex for each clause it is contained in.

The next lemma is equivalent to the result shown by Wang and Kuo [31]. We provide a direct proof in the full version of the paper.

Lemma 4.2. *[31] The PLANAR 3-SAT instance φ is satisfiable if and only if G_φ has vertex cover set of size at most $2m + \frac{1}{2}\sum_{i=1}^n r_i$, where r_i is the number of vertices in the variable gadget's cycle for u_i.*

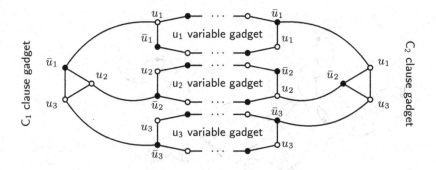

Fig. 6. The gadget graph G_φ for $\varphi = (\bar{u}_1 \vee u_2 \vee u_3) \wedge (u_1 \vee \bar{u}_2 \vee u_3)$.

Hence, to solve for the satisfiability of φ, it suffices to test whether G_φ admits a vertex cover with at most $2m + \frac{1}{2} \sum_{i=1}^n r_i$ vertices. It remains to show the second of the three parts of the reduction. That is, we need to show that this VERTEX COVER problem in the undirected gadget graph can be solved by finding a minimum directed feedback vertex set in an oriented-disc graph D_φ. The basic idea is straightforward (albeit that the implementation is intricate). The oriented-disc graph D_φ will contain a digon for each edge in some G_φ. However, it will also contain a collection of additional arcs. The key fact will be that these additional arcs form an acyclic subgraph of D_φ. Thus every cycle in D_φ must induce a digon. Consequently, a minimum directed feedback vertex set need only intersect each digon to ensure that every cycle is hit. As argued previously, hitting the underlying graph formed by the digons of D_φ corresponds to selecting a vertex cover in G_φ, as desired. We now formalise this argument.

Lemma 4.3. *For every instance φ of* PLANAR 3-SAT, *there exists an oriented-disc graph D_φ on which the* DIRECTED FEEDBACK VERTEX SET *problem is equivalent to the* VERTEX COVER *problem on G_φ.*

Proof. We prove this by explicitly constructing the oriented-disc drawing. Recall the disc graph D_φ should contain a digon for each edge in G_φ. To do this, we begin with sufficiently a large planar drawing of H_φ, the planar bipartite network associated with φ. At each clause vertex, we place an oriented-disc construction for the clause gadget. This construction, along with its resulting graph, is shown in Fig. 7. The figure shows a clause gadget and a section of each of the neighbouring three variable gadgets to which it is attached. Observe from the figure that, as claimed, the set of arcs created in D_φ which are not in a digon, form an acyclic subgraph of D_φ.

It remains to construct the large cycles for the variable gadgets, and connect them to the clause gadgets. However, parts of these cycles are already included in the clause gadgets. Thus, it suffices to join these cycle segments together via paths of digons. This can be done via the oriented disc constructions shown in Fig. 8. To draw the cycle for some variable, say u_i, we note that u_i's vertex in the planar network H_φ shares and edge with every clause gadget which connects

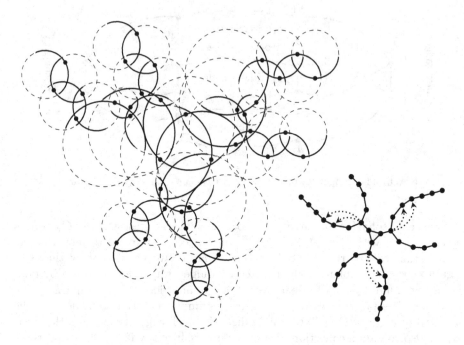

Fig. 7. Oriented-disc construction of the clause gadget, and its resulting graph.

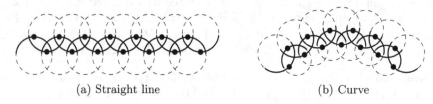

(a) Straight line   (b) Curve

Fig. 8. Paths of bidirected edges as oriented-disc drawings.

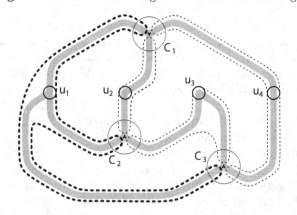

Fig. 9. G_φ as an oriented-disc graph.

to u_i's gadget. Hence, as illustrated in Fig. 9, we may follow along the edges of H_φ to construct the cycle. For example, in the figure, the variable cycle for u_1 (highlighted) follows the topology of the edges incident to u_1's vertex, and joins the clause gadgets (circled) to one another.

Observe that constructions in Fig. 8 produce paths of digons in D_φ, where every arc produced is contained in a digon. It follows that the only arcs in D_φ that are not in digons are in the neighbourhoods of the clause gadgets and, as we have seen, these are acyclic. But then, to hit all the cycles in D_φ, it suffices to hit all the digons, which, in turn, corresponds to a vertex cover in G_φ, completing the proof.   □

This completes all the steps in the reduction and we obtain:

Theorem 4.1. *The* CONSUMER RATIONALITY *problem is NP-complete for a market with at least 3 commodities.*   □

References

1. Afriat, S.: The construction of a utility function from expenditure data. Int. Econ. Rev. **8**, 67–77 (1967)
2. Afriat, S.: On a system of inequalities in demand analysis: an extension of the classical method. Int. Econ. Rev. **14**, 460–472 (1967)
3. Apesteguia, J., Ballester, M.: A measure of rationality and welfare. Journal of Political Economy (2015, to appear)
4. Berge, C.: Färbung von Graphen deren sämtliche beziehungsweise deren ungerade Kreise starr sind (Zusammenfassung). Wiss. Z. Martin-Luther-Univ. Halle-Wittenberg Math.-Natur. Reihe **10**, 114–115 (1961)
5. Chudnovsky, M., Robertson, N., Seymour, P., Thomas, R.: The strong perfect graph theorem. Ann. Math. **164**, 51–229 (2006)
6. Dean, M., Martin, D.: Measuring rationality with the minimum cost of revealed preference violations. Review of Economics and Statistics (2015, to appear)
7. Deb, R., Pai, M.: The geometry of revealed preference. J. Math. Econ. **50**, 203–207 (2014)
8. Dinur, I., Safra, S.: On the hardness of approximating minimum vertex cover. Ann. Math. **162**(1), 439–485 (2005)
9. Earl, R.: Geometry II: 3.1 Stereographic Projection and the Riemann Sphere (2007). https://people.maths.ox.ac.uk/earl/G2-lecture5.pdf
10. Echenique, F., Lee, S., Shum, M.: The money pump as a measure of revealed preference violations. J. Polit. Econ. **119**(6), 1201–1223 (2011)
11. Gross, J.: Testing data for consistency with revealed preference. Rev. Econ. Stat. **77**(4), 701–710 (1995)
12. Grötschel, M., Lovász, L., Schrijver, A.: Polynomial algorithms for perfect graphs. Ann. Discret. Math. **21**, 325–356 (1984)
13. Grötschel, M., Lovász, L., Schrijver, A.: Geometric Algorithms and Combinatorial Optimisation. Springer-Verlag, Berlin (1988)
14. Guruswami, V., Hastad, J., Manokaran, R., Raghavendra, P., Charikar, M.: Beating the random ordering is hard: every ordering CSP is approximation resistant. SIAM J. Comput. **40**(3), 878–914 (2011)

15. Famulari, M.: A household-based, nonparametric test of demand theory. Rev. Econ. Stat. **77**, 372–383 (1995)
16. Heufer, J.: A geometric approach to revealed preference via Hamiltonian cycles. Theor. Decis. **76**(3), 329–341 (2014)
17. Houthakker, H.: Revealed preference and the utility function. Economica New Ser. **17**(66), 159–174 (1950)
18. Houtman, M., Maks, J.: Determining all maximal data subsets consistent with revealed preference. Kwantitatieve Methoden **19**, 89–104 (1950)
19. Karp, R.: Reducibility among combinatorial problems. In: Miller, R.E., Thatcher, J.W., Bohlinger, J.D. (eds.) Complexity of Computer Computations, pp. 85–103. Plenum, New York (1972)
20. Khot, S.: On the power of unique 2-prover 1-round games. In: Proceedings of STOC, pp 767–775 (2002)
21. Koo, A.: An emphirical test of revealed preference theory. Econometrica **31**(4), 646–664 (1963)
22. Lovász, L.: Normal hypergraphs and the perfect graph conjecture. Discret. Math. **2**(3), 253–267 (1972)
23. Rose, H.: Consistency of preference: the two-commodity case. Rev. Econ. Stud. **25**, 124–125 (1958)
24. Samuelson, P.: A note on the pure theory of consumer's behavior. Economica **5**(17), 61–71 (1938)
25. Samuelson, P.: Consumption theory in terms of revealed preference. Economica **15**(60), 243–253 (1948)
26. Seymour, P.: Packing directed circuits fractionally. Combinatorica **15**(2), 281–288 (1995)
27. Svensson, O.: Hardness of vertex deletion and project scheduling. In: Gupta, A., Jansen, K., Rolim, J., Servedio, R. (eds.) APPROX 2012 and RANDOM 2012. LNCS, vol. 7408, pp. 301–312. Springer, Heidelberg (2012)
28. Swafford, J., Whitney, G.: Nonparametric test of utility maximization and weak separability for consumption, leisure and money. Rev. Econ. Stat. **69**, 458–464 (1987)
29. Varian, H.: Revealed preference. In: Szenberg, M., et al. (eds.) Samulesonian Economics and the 21st Century, pp. 99–115. Oxford University Press, New York (2005)
30. Varian, H.: Goodness-of-fit in optimizing models. J. Econometrics **46**, 125–140 (1990)
31. Wang, D., Kuo, Y.: A study on two geometric location problems. Inf. Process. Lett. **28**(6), 281–286 (1988)

Computation of Stackelberg Equilibria
of Finite Sequential Games

Branislav Bošanský[1,2]([✉]), Simina Brânzei[1], Kristoffer Arnsfelt Hansen[1],
Peter Bro Miltersen[1], and Troels Bjerre Sørensen[3]

[1] Department of Computer Science, Aarhus University, Aarhus, Denmark
simina.branzei@gmail.com,
{arnsfelt,bromille}@cs.au.dk
[2] Department of Computer Science, Faculty of Electrical Engineering,
Czech Technical University in Prague, Prague, Czech Republic
branislav.bosansky@agents.fel.cvut.cz
[3] IT-University of Copenhagen, Copenhagen, Denmark
trbj@itu.dk

Abstract. The Stackelberg equilibrium is a solution concept that
describes optimal strategies to commit to: Player 1 (*the leader*) first
commits to a strategy that is publicly announced, then Player 2 (*the follower*) plays a best response to the leader's choice. We study Stackelberg
equilibria in finite sequential (i.e., extensive-form) games and provide
new exact algorithms, approximate algorithms, and hardness results for
finding equilibria for several classes of such two-player games.

1 Introduction

The Stackelberg competition is a game theoretic model introduced by
von Stackelberg [10] for studying market structures. The original formulation
of a Stackelberg duopoly captures the scenario of two firms that compete by
selling homogeneous products. One firm (*the leader*) first decides the quantity
to sell and announces it publicly, while the second firm (*the follower*) decides
its own production only after observing the announcement of the first firm. The
leader firm must have a commitment power (e.g., it is the monopoly in an industry) and cannot undo its publicly announced strategy, while the follower firm
(e.g., a new competitor) plays a best response to the leader's chosen strategy.
The Stackelberg competition has been an important model in economics ever
since and the solution concept of a *Stackelberg equilibrium* has been studied
in a rich body of literature in computer science with a number of real-world
applications [12].

In this paper, we are interested in the problem of *efficient computation of the
optimal strategy that the leader should commit to in an extensive-form game, given
a description of the game*. We study this problem for several classes of extensive-
form games (EFGs) and variants of the Stackelberg solution concept (i.e., kinds
of strategies to commit to) giving both efficient algorithms and computational
hardness results. Our results can be classified by the following parameters:

© Springer-Verlag Berlin Heidelberg 2015
E. Markakis and G. Schäfer (Eds.): WINE 2015, LNCS 9470, pp. 201–215, 2015.
DOI: 10.1007/978-3-662-48995-6_15

- *Information.* Information captures how much a player knows about the opponent's moves (past and present). We study *turn-based games* (TB), where for each state there is a unique player that can perform an action, and *concurrent-move games* (CM), where the players act simultaneously in at least one state.
- *Chance.* A game with chance nodes allows stochastic transitions between states; otherwise, the transitions are deterministic (made through actions of the players).
- *Graph.* We focus on *trees* and *directed acyclic graphs* (DAGs) as the main representations. Given such a graph, each node represents a different state in the game, while the edges represent the transitions between states.
- *Strategies.* We study several major types of strategies that the leader can commit to, namely *pure* (P), *behavioral* (B), and *correlated behavioral* (C).

All of our results are summarized in Table 1; (1) We design a more efficient algorithm for computing optimal strategies for turn-based games on DAGs. Compared to the previous state of the art (due to Letchford and Conitzer [6,7]), we reduce the complexity by a factor proportional to the number of terminal states (see row 1 in Table 1). (2) We show that correlation often reduces the computational complexity of finding optimal strategies. In particular, we design several new polynomial time algorithms for computing the optimal correlated strategy to commit to (see rows 3, 9, 11). (3) We study approximation algorithms for the NP-hard problems in this framework and provide fully polynomial time approximation schemes for finding pure and behavioral Stackelberg equilibria for turn-based games on trees with chance nodes (see rows 7, 8). We leave open

Table 1. Overview of the computational complexity results containing both existing and new results provided by this paper (marked with *). Information column: TB stands for *turn-based* and CM for *concurrent moves*. Strategies: P stands for *pure*, B for *behavioral*, and C for *correlated*. Finally, $|\mathcal{S}|$ denotes the number of decision points in the game and $|\mathcal{Z}|$ the number of terminal states.

	INFORMATION	CHANCE	GRAPH	STRATEGIES	COMPLEXITY	SOURCE
1.*	TB	✗	DAG	P	$O\left(|\mathcal{S}| \cdot (|\mathcal{S}| + |\mathcal{Z}|)\right)$	Theorem 5
2.	TB	✗	Tree	B	$O\left(|\mathcal{S}| \cdot |\mathcal{Z}|^2\right)$	[7]
3.*	TB	✗	Tree	C	$O\left(|\mathcal{S}| \cdot |\mathcal{Z}|\right)$	Theorem 6
4.	TB	✗	DAG	B	NP-hard	[6]
5.*	TB	✗	DAG	C	NP-hard	Theorem 8
6.	TB	✓	Tree	B	NP-hard	[7]
7.*	TB	✓	Tree	P	FPTAS	Theorem 12
8.*	TB	✓	Tree	B	FPTAS	Theorem 11
9.*	TB	✓	Tree	C	$O\left(|\mathcal{S}| \cdot |\mathcal{Z}|\right)$	Theorem 7
10.*	CM	✗	Tree	B	NP-hard	Theorem 9
11.*	CM	✓	Tree	C	polynomial	Theorem 10

the question of finding an approximation for concurrent-move games on trees without chance nodes (see row 10).

Due to the space constraints we focus on the positive results in this version of the paper; the proofs for all the results can be found in the full version.

1.1 Related Work

There is a rich body of literature studying the problem of computing Stackelberg equilibria. The computational complexity of the problem is known for one-shot games [3], Bayesian games [3], and selected subclasses of extensive-form games [7] and infinite stochastic games [8]. Similarly, many practical algorithms are also known and typically based on solving multiple linear programs [3], or mixed-integer linear programs for Bayesian [9] and extensive-form games [1].

The extension of the Stackelberg notion to correlated strategies appeared in several works [2,8,13]. Conitzer and Korzhyk analyzed correlated strategies in one-shot games providing a single linear program for their computation [2]. Letchford et al. showed that the problem of finding optimal correlated strategies to commit to is NP-hard in infinite discounted stochastic games[1] [8]. Xu et al. focused on using correlated strategies in a real-world security based scenario [13].

2 Preliminaries

We consider finite two-player sequential games. Note that for every finite set K, $\Delta(K)$ denotes probability distributions over K and $\mathcal{P}(K)$ denotes the set of all subsets of K.

Definition 1. A two-player sequential game is given by a tuple $G = (\mathcal{N}, \mathcal{S}, \mathcal{Z}, \rho, \mathcal{A}, u, \mathcal{T}, \mathcal{C})$, where:

- $\mathcal{N} = \{1, 2\}$ is a set of two players;
- $\mathcal{S}$ is a set of non-terminal states;
- $\mathcal{Z}$ is a set of terminal states;
- $\rho : \mathcal{S} \to \mathcal{P}(\mathcal{N}) \cup \{c\}$ is a function that defines which player(s) act in a given state, or whether the node is a chance node (case in which $\rho(s) = c$);
- $\mathcal{A}$ is a set of actions; we overload the notation to restrict the actions only for a single player as $\mathcal{A}_i$ and for a single state as $\mathcal{A}(s)$;
- $\mathcal{T} : \mathcal{S} \times \prod_{i \in \rho(s)} \mathcal{A}_i \to \{\mathcal{S} \cup \mathcal{Z}\}$ is a transition function between states depending on the actions taken by all the players that act in this state. Overloading notation, $\mathcal{T}(s)$ also denotes the children of a state s: $\mathcal{T}(s) = \{s' \in \mathcal{S} \cup \mathcal{Z} \mid \exists a \in \mathcal{A}(s); \mathcal{T}(s,a) = s'\}$;
- $\mathcal{C} : \mathcal{A}_c \to [0,1]$ are the chance probabilities on the edges outgoing from each chance node $s \in \mathcal{S}$, such that $\sum_{a \in \mathcal{A}_c(s)} \mathcal{C}(a) = 1$;
- Finally, $u_i : \mathcal{Z} \to \mathbb{R}$ is the utility function for player $i \in \mathcal{N}$.

---

[1] To be precise they assumed that the correlated strategies can use a finite history.

In this paper we study Stackelberg equilibria, thus player 1 will be referred to as the *leader* and player 2 as the *follower*.

We say that a game is *turn-based* if there is a unique player acting in each state (formally, $|\rho(s)| = 1 \; \forall s \in \mathcal{S}$) and with *concurrent moves* if both players can act simultaneously in some state. Moreover, the game is said to have *no chance* if there exist no chance nodes; otherwise the game is *with chance*.

A *pure strategy* $\pi_i \in \Pi_i$ of a player $i \in \mathcal{N}$ is an assignment of an action to play in each state of the game ($\pi_i : \mathcal{S} \to \mathcal{A}_i$). A *behavioral strategy* $\sigma_i \in \Sigma_i$ is a probability distribution over actions in each state $\sigma_i : \mathcal{A} \to [0, 1]$ such that $\forall s \in S, \forall i \in \rho(s) \sum_{a \in \mathcal{A}_i(s)} \sigma_i(a) = 1$.

The expected utility of player i given a pair of strategies (σ_1, σ_2) is defined as follows:

$$u_i(\sigma_1, \sigma_2) = \sum_{z \in \mathcal{Z}} u_i(z) p_\sigma(z),$$

where $p_\sigma(z)$ denotes the probability that leaf z will be reached if both players follow the strategy from σ and due to stochastic transitions corresponding to $\mathcal{C}$.

A strategy σ_i of player i is said to represent a *best response* to the opponent's strategy σ_{-i} if $u_i(\sigma_i, \sigma_{-i}) \geq u_i(\sigma_i', \sigma_{-i}) \; \forall \sigma_i' \in \Sigma_i$. Let $\mathcal{BR}(\sigma_{-i}) \subseteq \Pi_i$ denote the set of all the pure best responses of player i to strategy σ_{-i}.

Definition 2. *A strategy profile $\sigma = (\sigma_1, \sigma_2)$ is a Stackelberg Equilibrium if σ_1 is an optimal strategy of the leader given that the follower best-responds to its choice. Formally, a Stackelberg equilibrium in* pure *strategies is defined as*

$$(\sigma_1, \sigma_2) = \underset{\sigma_1' \in \Pi_1, \sigma_2' \in \mathcal{BR}(\sigma_1')}{\arg \max} u_1(\sigma_1', \sigma_2')$$

while a Stackelberg equilibrium in behavioral *strategies is defined as*

$$(\sigma_1, \sigma_2) = \underset{\sigma_1' \in \Sigma_1, \sigma_2' \in \mathcal{BR}(\sigma_1')}{\arg \max} u_1(\sigma_1', \sigma_2')$$

Next, we describe the notion of a Stackelberg equilibrium where the leader can commit to a correlated strategy in a sequential game. The concept was suggested and investigated by Letchford *et al.* [8], but no formal definition exists. Formalizing such a definition below, we observe that the definition is essentially the "Stackelberg analogue" of the notion of *Extensive-Form Correlated Equilibria (EFCE)*, introduced by von Stengel and Forges [11]. This parallel turns out to be technically relevant as well.

Definition 3. *A probability distribution ϕ on pure strategy profiles Π is called a Stackelberg Extensive-Form Correlated Equilibrium (SEFCE) if it maximizes the leader's utility (that is, $\phi = \arg \max_{\phi' \in \Delta(\Pi)} u_1(\phi')$) subject to the constraint that whenever the play reaches a state s where the follower can act, the follower is recommended an action a according to ϕ such that the follower cannot gain by unilaterally deviating from a in state s (and possibly in all succeeding states), given the posterior on the probability distribution of the strategy of the leader, defined by the actions taken by the leader so far.*

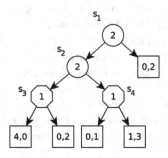

Fig. 1. An example game with different outcomes depending on whether the leader commits to behavioral or to correlated strategies. The leader acts in nodes s_3 and s_4, the follower acts in nodes s_1 and s_2. Utility values are shown in the terminal states, first value is the utility for the leader, second value is the utility of the follower.

Example. We give an example to illustrate both variants of the Stackelberg solution concept. Consider the game in Fig. 1, where the follower moves first (in states s_1, s_2) and the leader second (in states s_3, s_4). By committing to a behavioral strategy, the leader can gain utility 1 in the optimal case – leader commits to play left in state s_3 and right in s_4. The follower will then prefer playing right in s_2 and left in s_1, reaching the leaf with utilities $(1, 3)$. Note that the leader cannot gain by committing to strictly mixed behavioral strategies.

Now, consider the case when the leader commits to correlated strategies. We interpret the probability distribution over strategy profiles ϕ as signals send to the follower in each node where the follower acts, while the leader is committing to play with respect to ϕ and the signals sent to the follower. This can be shown in node s_2, where the leader sends one of two signals to the follower, each with probability 0.5. In the first case, the follower receives the signal to move left, while the leader commits to play uniform strategy in s_3 and left in s_4. In the second case, the follower receives the signal to move right, while the leader commits to play right in s_4 and left in s_3. By using this correlation, the leader is able to get the utility of 1.5, while ensuring the utility of 2 for the follower; hence, the follower will follow the only recommendation in node s_1 to play left.

The example gives an intuition about the structure of the probability distribution ϕ in SEFCE. In each state of the follower, the leader sends a signal to the follower and commits to follow the correlated strategy if the follower admits the recommendation, while simultaneously committing to punish the follower for each deviation. This punishment is simply a strategy that minimizes the follower's utility and will be useful in many proofs; next we introduce some notation for it.

Let σ^m denote a behavioral strategy profile, where in each sub-game the leader plays a minmax behavior strategy based on the utilities of the follower and the follower plays a best response. Moreover, for each state $s \in \mathcal{S}$, we denote by $\mu(s)$ the expected utility of the follower in the sub-game rooted in state s if both players play according to σ^m (i.e., the value of the corresponding zero-sum sub-game defined by the utilities of the follower).

Note that being a probability distribution over pure strategy profiles, an SEFCE is, a priori, an object of exponential size in the size of the description of the game, when it is described as a tree. This has to be dealt with before we can consider computing it. The following lemma gives a compact representation of the correlated strategies in an SEFCE and the proof yields an algorithm for constructing the probability distribution ϕ from the compact representation. It is this compact representation that we seek to compute.

Lemma 4. *For any turn-based or concurrent-move game in tree form, there exists an SEFCE $\phi \in \Delta(\Pi)$ that can be compactly represented as a behavioral strategy profile $\sigma = (\sigma_1, \sigma_2)$ such that $\forall z \in \mathcal{Z}\ p_\phi(z) = p_\sigma(z)$ and ϕ corresponds to the following behavior:*

- *the follower receives signals in each state s according to $\sigma_2(a)$ for each action $a \in \mathcal{A}_2(s)$*
- *the leader chooses the action in each state s according to $\sigma_1(a)$ for each action $a \in \mathcal{A}_1(s)$ if the state s was reached by following the recommendations*
- *both players switch to the minmax strategy σ^m after a deviation by the follower.*

Proof. Let ϕ' be an SEFCE. We construct the behavioral strategy profile σ from ϕ' and then show how an optimal strategy ϕ can be constructed from σ and σ^m.

To construct σ, it is sufficient to specify a probability $\sigma(a)$ for each action $a \in \mathcal{A}(s)$ in each state s. We use the probability of state s being reached (denoted $\phi'(s)$) that corresponds to the sum of pure strategy profiles $\phi'(\pi)$ such that the actions in strategy profile π allow state s to be reached. Formally, there exists a sequence of states and actions starting at the root $s_0, a_0, \ldots, a_{k-1}, s_k$ such that for every $j = 0, \ldots, k-1$ it holds that $a_j = \pi(s_j)$, $s_{j+1} = \mathcal{T}(s_j, a_j)$ (or s_{j+1} is the next decision node of some player if $\mathcal{T}(s_j, a_j)$ is a chance node), $s_0 = s_{root}$, and $s_k = s$. Let us denote $\Pi(s)$ to be such a set of pure strategy profiles, for which such a sequence exists for state s. Moreover, we denote $\Pi(s,a) \subseteq \Pi(s)$ to be strategy profiles that not only reach s but prescribe action a to be played in state s. Now, $\sigma(a) = \frac{\sum_{\pi' \in \Pi(s,a)} \phi'(\pi')}{\phi'(s)}$, where $\phi'(s) = \sum_{\pi' \in \Pi(s)} \phi'(\pi')$. In case $\phi'(s) = 0$, we set behavior strategy in σ arbitrarily.

Now, we construct a strategy ϕ that corresponds to the desired behavior and show that it is indeed an optimal SEFCE strategy. We need to specify a probability for every pure strategy profile $\pi = (\pi_1, \pi_2)$. Consider the sequence of states and actions that corresponds to executing the actions from the strategy profile π. Let $s_0^l, a_0^l, \ldots, a_{k_l-1}^l, s_{k_l}$ be one of q possible sequences of states and actions (there can be multiple such sequences due to chance nodes), such that $j = 0, \ldots, k_l - 1$, $a_j^l = \pi(s_j^l)$, $s_{j+1}^l = \mathcal{T}(s_j^l, a_j^l)$ (or s_{j+1}^l is one of the next decision nodes of some player immediately following the chance node(s) $\mathcal{T}(s_j^l, a_j^l)$), $s_0^l = s_{root}$, and $s_{k_l}^l \in \mathcal{Z}$. The probability for the strategy profile π corresponds to the probability of executing the sequences of actions multiplied by the probability that the remaining actions prescribe minmax strategy σ^m in case the follower deviates:

$$\phi(\pi) = \left(\prod_{l=1}^{q} \prod_{j=0}^{k_l-1} \sigma(a_j^l) \right) \cdot \prod_{a'=\pi(s')|s'\in\mathcal{S}\setminus\{s_0^1,\ldots,s_{k_0-1}^1,s_0^2,\ldots,s_{k_q-1}^q\}} \sigma^m(a').$$

Correctness. By construction of σ and ϕ it holds that probability distribution over leafs remains the same as in ϕ'; hence, $\forall z \in \mathcal{Z}\ p_{\phi'}(z) = p_\sigma(z) = p_\phi(z)$ and thus the expected utility of ϕ for the players is the same as in ϕ'. Second, we need to show that the follower has no incentive to deviate from the recommendations in ϕ. By deviating to some action a' in state s the follower gains $\mu(\mathcal{T}(s,a'))$ since both players play according to σ^m after a deviation. In ϕ', the follower can get for the same deviation at best some utility value $v_2(\mathcal{T}(s,a'))$, which by the definition of the minmax strategies σ^m is greater or equal than $\mu(\mathcal{T}(s,a'))$. Now since the expected utility value of the follower for following the recommendations is the same in ϕ as in ϕ' and the follower has no incentive to deviate in ϕ' because of the optimality, she has no incentive to deviate in ϕ either. ☐

3   Computing Exact Strategies in Turn-Based Games

Theorem 5. *There is an algorithm that takes as input a turn-based game in DAG form with no chance nodes and outputs a Stackelberg equilibrium in pure strategies. The algorithm runs in time $O(|\mathcal{S}|(|\mathcal{S}| + |\mathcal{Z}|))$.*

Proof. Our algorithm performs three passes through all the nodes in the graph.

First, the algorithm computes the minmax values $\mu(s)$ of the follower for each node in the game by backward induction. Second, the algorithm computes a *capacity* for each state in order to determine which states of the game are reachable (i.e., there exists a commitment of the leader and a best response of the follower such that the state can be reached by following their strategies). The capacity of state s, denoted $\gamma(s)$, is defined as the minimum utility of the follower that needs to be guaranteed by the outcome of the sub-game starting in state s in order to make this state reachable. By convention $\gamma(s_{root}) = -\infty$ and we initially set $\gamma(\mathcal{S} \cup \mathcal{Z} \setminus \{s_{root}\}) = \infty$ and mark them as open. Next, the algorithm evaluates each open state s, parents of which have all been marked as closed. We distinguish whether the leader, or the follower makes the decision: (1) in case s is a leader's node, the algorithm sets $\gamma(s') = \min(\gamma(s'), \gamma(s))$ for all children $s' \in \mathcal{T}(s)$; (2) in case s is a follower's node, the algorithm sets $\gamma(s') = \min(\gamma(s'), \max(\gamma(s), \max_{s''\in\mathcal{T}(s)\setminus\{s'\}} \mu(s'')))$ for all children $s' \in \mathcal{T}(s)$. Finally, we mark state s as closed.

We say that leaf $z \in \mathcal{Z}$ is a *possible outcome*, if $\mu(z) = u_2(z) \geq \gamma(z)$. Now, the solution is such a possible outcome that maximizes the utility of the leader, i.e. $\arg\max_{z\in\mathcal{Z}\ u_2(z)\geq\gamma(z)} u_1(z)$. The strategy is now constructed by following nodes from leaf z back to the root while using nodes s' with capacities $\gamma(s') \leq \mu(z)$. Thanks to the construction of capacities, such a path exists and forms a part of the Stackelberg strategy. The leader commits to the strategy leading to max min utility for the follower in the remaining states that are not part of this path.

Complexity Analysis. Computing the max min values can be done in $O(|\mathcal{S}|(|\mathcal{S}| + |\mathcal{Z}|))$ by backward induction due to the fact the graph is a DAG. In the second pass, the algorithm solves the widest-path problem from a single source to all leafs. In each node, the algorithm calculates capacities for every child. In nodes where the leader acts, there is a constant-time operation performed for each child. However, we need to be more careful in nodes where the follower acts. For each child $s' \in \mathcal{T}(s)$ the algorithm computes a maximum value $\mu(s')$ of all of the siblings. We can do this efficiently by computing two maximal values of $\mu(s')$ for all $s' \in \mathcal{T}(s)$ (say s^1, s^2) and for each child then the term $\max_{s'' \in \mathcal{T}(s) \setminus \{s'\}} \mu(s'')$ equals either to s^1 if $s' \neq s^1$, or to s^2 if $s' = s^1$. Therefore, the second pass can again be done in $O(|\mathcal{S}|(|\mathcal{S}| + |\mathcal{Z}|))$. Finally, finding the optimal outcome and constructing the optimal strategy is again at most linear in the size of the graph. Therefore the algorithm takes at most $O(|\mathcal{S}|(|\mathcal{S}| + |\mathcal{Z}|))$ steps. $\square$

Theorem 6. *There is an algorithm that takes as input a turn-based game in tree form with no chance nodes and outputs an SEFCE in the compact representation. The algorithm runs in time $O(|\mathcal{S}||\mathcal{Z}|)$.*

Proof. We improve the algorithm from the proof of Theorem 4 in [7]. The algorithm contains two steps: (1) a bottom-up dynamic program that for each node s computes the set of possible outcomes, (2) a downward pass constructing the optimal correlated strategy in the compact representation.

For each node s we keep set of points H_s in two-dimensional space, where the x-dimension represents the utility of the follower and the y-dimension represents the utility of the leader. These points define the convex set of all possible outcomes of the sub-game rooted in node s (we assume that H_s contains only the points on the boundary of the convex hull). We keep each set H_s sorted by polar angle.

Upward Pass. In leaf $z \in \mathcal{Z}$, we set $H_z = \{z\}$. In nodes s where the leader acts, the set of points H_s is equal to the convex hull of the corresponding sets of the children H_w. That is, $H_s = \text{Conv}(\cup_{w \in \mathcal{T}(s)} H_w)$.

In nodes s where the follower acts, the algorithm performs two steps. First, the algorithm removes from each set H_w of child w the outcomes from which the follower has an incentive to deviate. To do this, the algorithm uses the maxmin u_2 values of all other children of s except w and creates a new set $\hat{H}_w$ that we term the *restricted set*. Formally, the restricted set $\hat{H}_w$ is defined as an intersection of the convex set representing all possible outcomes H_w and all outcomes defined by the halfspace restricting the utility x of the follower by the inequality $x \geq \max_{w' \in \mathcal{T}(s); w' \neq w} \min_{p' \in H_{w'}} u_2(p')$. Now, H_s is equal to the convex hull of the corresponding restricted sets $\hat{H}_w$ of the children w. That is, $H_s = \text{Conv}(\cup_{w \in \mathcal{T}(s)} \hat{H}_w)$.

Finally, in the root of the game tree, the outcome of the Stackelberg Extensive-Form Correlated Equilibrium is the point with maximal payoff of player 1: $p_{SE} = \arg\max_{p \in H_{s_{root}}} u_1(p)$.

Downward Pass. We now construct the compact representation of commitment to correlated strategies that ensures the outcome p_{SE} calculated in the upward pass. The method for determining the optimal strategy in each node is similar to the method strategy(s, p'') used in the proof of Theorem 4 in [7]. This method, given a node s and a point p'' that lies on the boundary of H_s, specifies how to commit to correlated strategies in the sub-tree rooted in node s. Moreover, the previous proof in [7] also showed that it is sufficient to consider mixtures of at most two actions in each node and allowing correlated strategies does violate their proof.

In nodes s where the leader acts, the algorithm needs to find two points p, p' in the boundaries of children H_w and $H_{w'}$, such that the desired point p'' is a convex combination of $p \in H_w$ and $p' \in H_{w'}$. If $w = w'$, then the strategy in node s is to commit to pure strategy leading to node w. If $w \neq w'$, then the strategy to commit to in node s is a mixture: with probability α to play action leading to w and with probability $(1 - \alpha)$ to play action leading to w', where $\alpha \in [0,1]$ such that $p'' = \alpha p + (1 - \alpha)p'$. Finally, for every child $s' \in T(s)$ we call the method strategy with appropriate p (or p') in case $s' = w$ (or w'), and with the threat value corresponding to $\mu(s')$ for every other child.

In nodes s where the follower acts, the algorithm again needs to find two points p, p' in the *restricted* boundaries of children $\hat{H}_w$ and $\hat{H}_{w'}$, such that the desired point p'' is a convex combination of $p \in \hat{H}_w$ and $p' \in \hat{H}_{w'}$. The reason for using the restricted sets is because the follower must not have an incentive to deviate from the recommendation. Similarly to the previous case, if $w = w'$, then the correlated strategy in node s is to send the follower signal leading to node w while committing further to play strategy strategy(w, p) in sub-tree rooted in node w, and to play the minmax strategy in every other child s' corresponding to value $\mu(s')$. If $w \neq w'$, then there is a mixture of possible signals: with probability α the follower receives a signal to play the action leading to w and with probability $(1 - \alpha)$ signal to play the action leading to w', where $\alpha \in [0,1]$ is again such that $p'' = \alpha p + (1 - \alpha)p'$. As before, by sending the signal to play certain action, the leader commits to play strategy strategy(w, p) (or strategy(w', p')) in sub-tree rooted in node w (or w') and committing to play the minmax strategy leading to value $\mu(s')$ for every other child s'.

Correctness. Due to the construction of the set of points H_s that are maintained for each node s, these points correspond to the convex hull of all possible outcomes in the sub-game rooted in node s. In leafs, the algorithm adds the point corresponding to the leaf. In the leader's nodes, the algorithm creates a convex combinations of all possible outcomes in the children of the node. The only places where the algorithm removes some outcomes from these sets are nodes of the follower. If a point is removed from H_w in node s, there exists an action of the follower in s that guarantees the follower a strictly better expected payoff than the expected payoff of the outcome that correspond to the removed point. Therefore, such an outcome is not possible as the follower will have an incentive to deviate. The outcome selected in the root node is the possible outcome that maximizes the payoff of the leader of all possible outcomes; hence, it is optimal for the leader. Finally, the downward pass constructs the compact representation of the optimal correlated strategy to commit to that reaches the optimal outcome.

Complexity Analysis. Computing boundary of the convex hull H_s takes $O(|\mathcal{Z}|)$ time in each level of the game tree since the children sets H_w are already sorted [4, p. 6]. Moreover, since we keep only nodes on the boundary of the convex hull, $\sum_{s \in \mathcal{S}} |H_s| \leq |\mathcal{Z}|$ for all nodes in a single level of the game tree also bounds the number of lines that need to be checked in the downward pass. Therefore, each pass takes at most $O(|\mathcal{S}||\mathcal{Z}|)$ time.   □

Theorem 7. *There is an algorithm that takes as input a turn-based game in tree form with chance nodes and outputs the compact form of an SEFCE for the game. The algorithm runs in time $O(|\mathcal{S}||\mathcal{Z}|)$.*

Proof. We can use the proof from Theorem 6, but need to analyze what happens in chance nodes in the upward pass. The algorithm computes in chance nodes the Minkowski sum of all convex sets in child nodes and since all sets are sorted and this is a planar case, this operation can be again performed in linear time [4, p. 279]. The size of set H_s is again bounded by the number of all leafs [5].   □

Theorem 8. *Given a turn-based game in DAG form and a number α, it is NP-hard to decide if the leader gets payoff at least α in an SEFCE.*

For DAGs, the algorithm described in the previous proofs for trees fails, since it can prescribe different strategies in a state depending on the path with which the state is reached. We prove the hardness by following the proof of Theorem 13 from [6] for behavioral strategies and verifying that it holds also for correlated strategies. The proof can be found in the full version of the paper.

4   Computing Exact Strategies in Concurrent-Move Games

Theorem 9. *Given a concurrent-move games in tree form with no chance nodes and a number α, it is NP-hard to decide if the leader achieves payoff at least α in a Stackelberg equilibrium in behavior strategies.*

We prove this theorem by a reduction from KNAPSACK; the proof can be found in the full version of the paper.

Theorem 10. *For a concurrent-move games in tree form, the compact form of an SEFCE for the game can be found in polynomial time by solving a single linear program.*

Proof. We construct a linear program (LP) based on the LP for computing Extensive-Form Correlated Equilibria (EFCE) [11]. We use the compact representation of SEFCE strategies (described by Lemma 4) represented by variables $\delta(s)$ that denote a joint probability that state s is reached when both players, and chance, play according to SEFCE strategies.

The size of the original EFCE LP (both, the number of variables and constraints) is quadratic in the number of sequences of players. However, the LP for EFCE is defined for a more general class of imperfect-information games without chance. In our case, we can exploit the specific structure of a concurrent-move game and together with the Stackelberg assumption reduce the number of constraints and variables. First, the deviation from a recommended strategy causes the game to reach a different sub-game in which the strategy of the leader can be chosen (almost) independently to the sub-game that follows the recommendation. Second, the strategy that the leader should play after the deviations is a minmax strategy, with which the leader punishes the follower by minimizing the utility of the follower as much as possible. Thus, by deviating to action a' in state s, the follower can get at best the minmax value of the sub-game starting in node $T(s, a')$ that we denote as $\mu(T(s, a'))$. The values $\mu(s)$ for each state $s \in S$ can be computed beforehand using backward induction.

$$\max_{\delta, v_2} \sum_{z \in \mathcal{Z}} \delta(z) u_1(z) \qquad \text{s.t.} \tag{1}$$

$$\delta(s_{root}) = 1 \tag{2}$$

$$0 \geq \delta(s) \geq 1 \qquad \forall s \in S \tag{3}$$

$$\delta(s) = \sum_{s' \in T(s)} \delta(s') \qquad \forall s \in S; \rho(s) = \{1, 2\} \tag{4}$$

$$\delta(T(s, a_c)) = \delta(s) \mathcal{C}(s, a_c) \qquad \forall s \in S \, \forall a \in \mathcal{A}_c(s); \rho(s) = \{c\} \tag{5}$$

$$v_2(z) = u_2(z) \delta(z) \qquad \forall z \in \mathcal{Z} \tag{6}$$

$$v_2(s) = \sum_{s' \in T(s)} v_2(s') \qquad \forall s \in S \tag{7}$$

$$\sum_{a_1 \in \mathcal{A}_1(s)} v_2(T(s, a_1 \times a_2)) \geq \sum_{a_1 \in \mathcal{A}_1(s)} \delta(T(s, a_1 \times a_2)) \mu(T(s, a_1 \times a_2'))$$

$$\forall s \in S \, \forall a_2, a_2' \in \mathcal{A}_2(s) \tag{8}$$

The LP works as follows: Variables δ represent the compact form of the correlated strategies. The probability of reaching the root state is 1 (Eq. (2)) and δ must be between 0 and 1 (Eq. (3)). Next, so called network-flow constraints must hold: the probability of reaching a state equals the sum of probabilities of reaching all possible children (Eq. (4)) and it must correspond with the probability of actions in chance nodes (Eq. (5)). The objective ensures finding a correlated strategy that maximizes the leader's utility.

Next, we need to guarantee that the follower has no incentive to deviate from the recommendations given by δ. We use variables $v_2(s)$ to represent the expected payoff for the follower in a sub-game rooted in node $s \in S$ when played according to δ; defined by Eqs. (6 and 7). Now, each action that is recommended by δ must guarantee the follower at least the utility she gets by deviating from the recommendation. This is ensured by Eq. (8), where the expected utility for recommended action a_2 is expressed by the left side of the constraint, while the expected utility for deviating is expressed by the right side of the constraint. Note that the expected utility on the right side is calculated by considering the posterior probability after receiving the recommendation a_2 and the minmax values of children states after playing a_2'; $\mu(\mathcal{T}(s, a_1 \times a_2'))$.

Therefore, variables δ found by solving this linear program correspond to the compact representation of the optimal SEFCE strategy. $\square$

5   Approximating Optimal Strategies

In this section, we describe fully polynomial time approximation schemes for finding a Stackelberg equilibrium in behavioral strategies as well as in pure strategies for turn based games on trees with chance nodes. We first prove:

Theorem 11. *There is an algorithm that takes as input a turn-based game on a tree with chance nodes and a parameter ϵ, and computes a behavioral strategy for the leader. That strategy, combined with some best response of the follower, achieves a payoff that differs by at most ϵ from the payoff of the leader in a Stackelberg equilibrium in behavioral strategies. The algorithm runs in time $O(\epsilon^{-3}(UH_T)^3 T)$, where $U = \max_{\sigma, \sigma'} u_1(\sigma) - u_1(\sigma')$, T is the size of the game tree and H_T is its height.*

The exact version of this problem was shown to be NP-hard by Letchford and Conitzer [7]. Their hardness proof was a reduction from KNAPSACK and our algorithm is closely related to the classical approximation scheme for this problem. Due to space constraints, we only present the algorithm, while the detailed proof of correctness is relegated to the full version of the paper.

Our scheme uses dynamic programming to construct a table of values for each node in the tree. Each table contains a discretized representation of the possible tradeoffs between the utility that the leader can get and the utility that can at the same time be offered to the follower. In the full version of the paper, we show that the cumulative error in the leaders utility is bounded additively by the height of the tree. This error only depends on the height of the tree and not the utility. By an initial scaling of the leader utility by a factor D, the error can be made arbitrarily small, at the cost of extra computation time. This scaling is equivalent to discretizing the leaders payoff to multiples of some small $\delta = 1/D$. For simplicity, we only describe the scheme for binary trees, since nodes with higher branching factor can be replaced by small equivalent binary trees.

An important property is that only the leader's utility is discretized, since we need to be able to reason correctly about the follower's actions. The tables are indexed by the leader's utility and contains values that are the follower's utility. More formally, for each sub-tree T we will compute a table A_T with the following guarantee for each index k in each table:

(a) the leader has a strategy for the game tree T that offers the follower utility $A_T[k]$ while securing utility *at least k* to the leader.
(b) no strategy of the leader can (starting from sub-tree T) offer the follower utility *strictly more* than $A_T[k]$, while securing utility *at least* $k + H_T$ to the leader, where H_T is the height of the tree T.

This also serves as our induction hypothesis for proving correctness. For commitment to pure strategies, a similar table is used with the same guarantee, except quantifying over pure strategies instead.

We will now examine each type of node, and for each show how the table is constructed. For each node T, we let L and R denote the two successors (if any), and we let A_T, A_L, and A_R denote their respective tables. Each table will have $n = H_T U/\epsilon$ entries.

If T is a leaf with utility (u_1, u_2), the table can be filled directly from the definition:

$$A_T[k] := \begin{cases} u_2 & \text{, if } k \leq u_1 \\ -\infty & \text{, otherwise} \end{cases}$$

Both parts of the induction hypothesis are trivially satisfied by this.

If T is a leader node, and the leader plays L with probability p, followed up by the strategies that gave the guarantees for $A_L[i]$ and $A_R[j]$, then the leader would get an expected $pi + (1-p)j$, while being able to offer $pA_L[i] + (1-p)A_R[j]$ to the follower. For a given k, the optimal combination of the computed tradeoffs becomes: $A_T[k] := \max_{i,j,p}\{pA_L[i] + (1-p)A_R[j] \mid pi + (1-p)j \geq k\}$. This table can be computed in time $O(n^3)$ by looping over all $0 \leq i, j, k < n$, and taking the maximum with the extremal feasible values of p.

If T is a chance node, where the probability of L is p, and the leader combines the strategies that gave the guarantees for $A_L[i]$ and $A_R[j]$, then the leader would get an expected $pi + (1-p)j$ while being able to offer $pA_L[i] + (1-p)A_R[j]$ to the follower. For a given k, the optimal combination of the computed tradeoffs becomes: $A_T[k] := \max_{i,j}\{pA_L[i] + (1-p)A_R[j] \mid pi + (1-p)j \geq k\}$. The table A_T can thus be filled in time $O(n^3)$ by looping over all $0 \leq i, j, k < n$, and this can even be improved to $O(n^2)$ by a simple optimization.

If T is a follower node, then if the leader combines the strategy for $A_L[i]$ in L with the minmax strategy for R, then the followers best response is L iff $A_L[i] \geq \mu(R)$, and similarly it is R if $A_R[j] \geq \mu(L)$. Thus, the optimal combination becomes

$$A_T[k] := \max(A_L[k] \downarrow_{\mu(R)}, A_R[k] \downarrow_{\mu(L)}) \qquad x \downarrow_\mu := \begin{cases} x, & \text{if } x \geq \mu \\ -\infty, & \text{otherwise} \end{cases}$$

The table A_T can be filled in time $O(n)$.

Putting it all together, each table can be computed in time $O(n^3)$, and there is one table for each node in the tree, which gives the desired running time. Let A_T be the table for the root node, and let $i' = \max\{i \mid A_T[i] > -\infty\}$. The strategy associated with $A_T[i']$ guarantees utility that is at most H_T from the best possible guarantee in the scaled game, and therefore at most ϵ from the best possible guarantee in the original game.

This completes the proof of Theorem 11. Next, we prove the analogous statement for the case of pure strategies (again, the exact problem was shown to be NP-hard by Conitzer and Letchford).

Theorem 12. *There is an algorithm that takes as input a turn-based game on a tree with chance nodes and a parameter ϵ, and computes a pure strategy for the leader. That strategy, combined with some best response of the follower, achieves a payoff that differs by at most ϵ from the payoff of the leader in a Stackelberg equilibrium in pure strategies. The algorithm runs in time $O(\epsilon^{-2}(UH_T)^2 T)$, where $U = \max_{\sigma,\sigma'} u_1(\sigma) - u_1(\sigma')$, T is the size of the game tree and H_T is its height.*

Proof. (Sketch) In essence, the algorithm is the same as in the case of behavioral strategies, except that leader nodes only have $p \in \{0, 1\}$. The induction hypothesis is the same, except the quantifications are over pure strategies instead. For a given k, the optimal combination of the computed tradeoffs becomes: $A_T[k] := \max\{A_c[i] \mid i \geq k \wedge c \in \{L, R\}\}$. The table A_T can be computed in time $O(n)$. The performance of the algorithm is slightly better than in the behavioral case, since the most expensive type of node in the behavioral case can now be handled in linear time. Thus, computing each table now takes at most $O(n^2)$ time, which gives the desired running time. $\square$

Acknowledgements. We would like to thank the reviewers for useful feedback. This work was supported by the Danish National Research Foundation and The National Science Foundation of China (under the grant 61361136003) for the Sino-Danish Center for the Theory of Interactive Computation, by the Center for Research in Foundations of Electronic Markets (CFEM) supported by the Danish Strategic Research Council, and by the Czech Science Foundation (grant no. 15-23235S).

References

1. Bosansky, B., Cermak, J.: Sequence-form algorithm for computing stackelberg equilibria in extensive-form games. In: Proceedings of AAAI (2015)
2. Conitzer, V., Korzhyk, D.: Commitment to correlated strategies. In: Proceedings of AAAI, pp. 632–637 (2011)
3. Conitzer, V., Sandholm, T.: Computing the optimal strategy to commit to. In: Proceedings of ACM-EC, pp. 82–90 (2006)
4. De Berg, M., Van Kreveld, M., Overmars, M., Schwarzkopf, O.C.: Computational Geometry, 2nd edn. Springer, Heidelberg (2000)
5. Gritzmann, P., Sturmfels, B.: Minkowski addition of polytopes: computational complexity and applications to Gröbner bases. SIAM J. Discrete Math. **6**(2), 246–269 (1993)
6. Letchford, J.: Computational aspects of Stackelberg games. Ph.D. thesis, Duke University (2013)
7. Letchford, J., Conitzer, V.: Computing optimal strategies to commit to in extensive-form games. In: Proceedings of ACM-EC, pp. 83–92. ACM (2010)

8. Letchford, J., MacDermed, L., Conitzer, V., Parr, R., Isbell, C.L.: Computing optimal strategies to commit to in stochastic games. In: Proceedings of AAAI, pp. 1380–1386 (2012)
9. Paruchuri, P., Pearce, J., Marecki, J., Tambe, M., Ordonez, F., Kraus, S.: Playing games for security: an efficient exact algorithm for solving bayesian stackelberg games. In: Proceedings of AAMAS, pp. 895–902 (2008)
10. von Stackelberg, H.: Marktform und gleichgewicht. Springer-Verlag (1934)
11. von Stengel, B., Forges, F.: Extensive-form correlated equilibrium: definition and computational complexity. Math. Oper. Res. **33**(4), 1002–1022 (2008)
12. Tambe, M.: Security and Game Theory: Algorithms, Deployed Systems, Lessons Learned. Cambridge University Press, Cambridge (2011)
13. Xu, H., Rabinovich, Z., Dughmi, S., Tambe, M.: Exploring information asymmetry in two-stage security games. In: Proceedings of AAAI (2015)

Welfare and Rationality Guarantees for the Simultaneous Multiple-Round Ascending Auction

Nicolas Bousquet[1,2]($\boxtimes$), Yang Cai[3], and Adrian Vetta[4]

[1] Department of Mathematics and Statistics, McGill University, Montreal, Canada
[2] LIRIS, Ecole Centrale Lyon, Écully, France
nicolas.bousquet@ec-lyon.fr
[3] School of Computer Science, McGill University, Montreal, Canada
cai@cs.mcgill.ca
[4] Department of Mathematics and Statistics, and School of Computer Science,
McGill University, Montreal, Canada
vetta@math.mcgill.ca

Abstract. The simultaneous multiple-round auction (SMRA) and the combinatorial clock auction (CCA) are the two primary mechanisms used to sell bandwidth. Recently, it was shown that the CCA provides good welfare guarantees for general classes of valuation functions [7]. This motivates the question of whether similar welfare guarantees hold for the SMRA in the case of general valuation functions.

We show the answer is no. But we prove that good welfare guarantees still arise if the degree of complementarities in the bidder valuations are bounded. In particular, if bidder valuations functions are α-near-submodular then, under *truthful bidding*, the SMRA has a welfare ratio (the worst case ratio between the social welfare of the optimal allocation and the auction allocation) of at most $(1 + \alpha)$. However, for $\alpha > 1$, this is a bicriteria guarantee, to obtain good welfare under truthful bidding requires relaxing individual rationality. We prove this bicriteria guarantee is asymptotically (almost) tight.

Finally, we examine what strategies are required to ensure individual rationality in the SMRA with general valuation functions. First, we provide a weak characterization, namely *secure bidding*, for individual rationality. We then show that if the bidders use a profit-maximizing secure bidding strategy the welfare ratio is at most $1 + \alpha$. Consequently, by bidding securely, it is possible to obtain the same welfare guarantees as truthful bidding without the loss of individual rationality.

Keywords: Ascending auctions · SMRA · Welfare guarantee · Individual rationality · Near-submodular

1 Introduction

The question of how best to allocate spectrum dates back over a century, and the case in favour of selling bandwidth was first formalized in the academic

© Springer-Verlag Berlin Heidelberg 2015
E. Markakis and G. Schäfer (Eds.): WINE 2015, LNCS 9470, pp. 216–229, 2015.
DOI: 10.1007/978-3-662-48995-6_16

literature as far back as 1959 by Ronald Coase [9]. Over the past twenty years there have been large number spectrum auctions world-wide and, amongst these, the Simultaneous Multi-Round Auction (SMRA) and the Combinatorial Clock Auction (CCA) have proved to be extremely successful.

Both of these multiple-item auctions are based upon the same underlying mechanism. At time t, each item j has a price p_j^t. Given the current prices, each bidder i then selects her preferred set S_i^t of items. The price of any item that has excess demand then rises in the next time period and the process is repeated. There are important differences between the two auctions however. The SMRA uses *item bidding*, that is, the auctioneer views the selection of S_i^t as a collection of bids, one bid for every item of S_i^t. It also utilizes the concept of a *standing high bid* [12]. Any item (with a positive price) has a *provisional winner*. That bidder will win the item unless a higher bid is received in a later round. If such a bid is received then the standing high bid is increased and a new provisional winner assigned (chosen at random in the case of a tie). Item bidding and standing high bids lead to a major drawback, the *exposure problem*. Namely, a large set may be desired but such a bid may result in being allocated only a smaller undesirable subset. If the bidder valuation functions satisfy the gross-substitutes property then this problem does not arise. Indeed, given truthful bidding, Milgrom [21] showed that the SMRA will terminate in a Walrasian Equilibrium that maximizes social welfare; see also [14,16] who studied a similar auction mechanism. The exposure problem is also absent when the bidder valuation functions are submodular. In that case, Fu et al. [13] show that the final allocation, whilst not necessarily a Walrasiam Equilibrium, does provide at least half of the optimal social welfare.[1] For these classes of valuation function, the SMRA is *individually rational* in every time period. That is, if bidder i is provisionally allocated set S at round t then the value of S to i is at least the price of S.

For broader classes of valuation function that permit *complementarities*, though, the exposure problem does arise under the SMRA. This is a practical issue because in spectrum auctions bidder valuation functions typically do exhibit complementarities. The CCA [23] was designed to deal with such complementarities. Specifically, the CCA uses *package bidding* rather than item bidding. A package bid is an all-or-nothing bid. Consequently, a bidder cannot be allocated a subset of her bid; in particular, a bidder cannot be allocated an undesirable subset. Unfortunately, the basic CCA mechanism cannot provide for non-trivial approximate welfare guarantees, even for auctions with additive valuation functions and a small number of bidders and items [7]. It is perhaps surprising, then, that a minor adjustment to the CCA mechanism leads to good welfare guarantees for *any* class of valuation function. Specifically, if bid increments are made proportional to excess demand the welfare of the CCA is within an $O(k^2 \cdot \log n \log^2 m)$ factor of the optimal welfare [7]. Here n is the number of items, m is the number of bidders and k is the maximum cardinality of a set desired by the bidders. The fact that the CCA can generate high welfare for general valuation functions motivates the work in this paper. Is it possible that the SMRA also performs well with general valuation functions?

---

[1] Their proof is not for the SMRA, but it can be adapted to apply there.

Our Results. The short answer to the question posed above is NO, the SMRA cannot unconditionally guarantee high social welfare for valuation functions that exhibit complementarities (see for instance [8]).

It turns out however that we can quantify precisely the welfare guarantee in terms of the magnitude of the complementarities exhibited by the valuation function. To explain this we require a few definitions. Each bidder $i \in B$ has value $v_i(S)$ for any set of items $S \subseteq \Omega$. The valuation function $v_i(\cdot)$ is monotonically non-decreasing (free-disposal). Each bidder has a quasi-linear utility, that is, its *utility* for a set S is $v_i(S) - p(S)$, where $p(S)$ is the price of S. The social welfare of an allocation $S = \{S_1, \ldots, S_n\}$, where the S_i are pairwise-disjoint subsets of the items, is $\omega(S) = \sum_i v_i(S_i)$. Next, to quantify the extent of complementarities, let the *degree of submodularity* [1] of a function f be

$$\mathcal{D}(f) = \min_{x \in \Omega \setminus B} \min_{A,B:A \subset B} \frac{f(A \cup x) - f(A)}{f(B \cup x) - f(B)}$$

Note that f is submodular if and only if $\mathcal{D}(f) \geq 1$. We say that f is α *near-submodular* if $\mathcal{D}(f) \geq \frac{1}{\alpha}$. A similar concept to near-submodularity, called *bounded complementarity*, is introduced by Lehman et al. [18].

The parameter α turns out to be the key in explaining the performance of the SMRA. To explain this we require one more concept. We say that a bidder $i \in B$ is λ-*individually rational* if $\lambda \cdot v_i(S_i^t) \geq p(S_i^t)$ in each round t. Note that if $\lambda = 1$ then we have individual rationality. We say that an auction mechanism is λ-*individually rational* if every bidder is λ-individually rational. We then prove in Sect. 3:

Theorem 1. *If bidders have α near-submodular valuations then, under (conditional) truthful bidding[2],*
(i) The SMRA outputs an allocation S with $\omega(S) \geq \frac{1}{1+\alpha} \cdot \omega(S^*)$ where S^* is the optimal allocation.
(ii) The auction is α-individually rational.

The bi-criteria guarantees in Theorem 1 are (almost) tight. There are examples with α near-submodular valuations where the SMRA is only α-individually rational and the welfare guarantee tends to $\frac{1}{1+\alpha}$. Despite the fact that SMRA has arbitrarily poor welfare guarantees, it seems to perform very well in practice. Theorem 1 provides an explanation for this, and confirms empirical results, since complementarities exist but are typically bounded in magnitude in most spectrum auctions. Indeed, the SMRA has been proposed for auctions where valuation functions have weak complementarities [3].

There are, however, two major drawbacks inherent in Theorem 1. The first drawback is that it relies upon *truthful bidding*, that is, in each round the bidder selects the feasible set that maximizes utility. But, as we explain in Sect. 2, there are many reasons why a bidder will not bid truthfully in the SMRA. One of

---

[2] A detailed discussion on truthful and conditional truthful biddings will follow in Sect. 2.

these reasons is that, in a spectrum auction, a bidder may not even know its own valuation function [11]. Bidders typically can however make comparisons between similar sets. Thus, a natural method by which a bidder can select a bid is via local improvement.

We show, in Sect. 4, that local improvement leads to similar guarantees as truthful bidding (albeit with an additional α factor in the denominator for the welfare guarantee).

Theorem 2. *If bidders have α near-submodular valuations then, under (conditional) local improvement bidding,*
(i) The SMRA outputs an allocation S with $\omega(S) \geq \frac{1}{1+\alpha^2} \cdot \omega(S^*)$ where S^* is the optimal allocation.
(ii) The auction is α-individually rational.

Again, the bounds in Theorem 2 are (almost) tight.

The second drawback is that Theorem 1 shows that the SMRA is not individually rational. That is, it may produce outcomes which give negative utility to some bidders. Consequently, in Sect. 5 we provide a detailed study of what bidding strategies are required to ensure the individual rationality of the SMRA, and what are the consequences for welfare when such strategies are used. Towards this end, we characterize the individual rationality of the SMRA in terms of *secure bidding*. We then prove, in Sect. 5.2, that secure bidding has a good welfare guarantees, provided the bidders make profit maximizing secure bids.

Theorem 3. *If bidders have α near-submodular valuations then, under (conditional) profit maximizing secure bidding, the SMRA outputs an allocation S with $\omega(S) \geq \frac{1}{1+\alpha} \cdot \omega(S^*)$ where S^* is the optimal allocation. Moreover, the auction is individually rational.*

Consequently, by bidding securely, it is possible to obtain the same welfare guarantees as truthful bidding without the loss of individual rationality!

2   The Simultaneous Multiple-Round Ascending Auction

The SMRA was first proposed by Milgrom, Wilson and McAfee for the 1994 FCC spectrum auction. It is an ascending price auction that simultaneously sells many items. Let B be a set of n bidders and let Ω be a collection of m items. For each item $j \in \Omega$ the auction posits an item-price p_j^t at the start of round t. Moreover, the SMRA has a unique *standing high bidder* for each item with a positive price. Specifically, at the start of round t, bidder i is the standing high bidder for a set of items S_i^t; we call S_i^t the *provisional (winning) set* for bidder i.

<u>The SMRA Mechanism:</u> Initially $p_j^0 = 0$ for each item $j \in \Omega$, and $S_i^0 = \emptyset$ for each bidder $i \in B$ and $t = 0$. The auction then iterates over rounds as follows. In round t, bidder i bids for a set $T_i^t \subseteq \Omega \setminus S_i^t$ under the assumption that the price of each item $j \in \Omega \setminus S_i^t$ is incremented to $p_j^t + \epsilon$. We call T_i^t the *conditional bid* for i. The term conditional is used as the auction mechanism automatically

assumes that bidder i also makes a bid of price p_j^t for every item $j \in S_i^t$ (recall, bidder i is the provisional winner of the items S_i^t).

The item-prices and provisional sets are then updated. Take an item j and suppose that j is in exactly k of the conditional bids. If $k = 0$ then no bidder has placed a bid on item j at the incremented price $p_j^t + \epsilon$. Thus we set $p_j^{t+1} = p_j^t$ and the standing high bidder for j remains the same, i.e. if $j \in S_i^t$ then $j \in S_i^{t+1}$. If $k > 0$ (we say that j is in *excess demand*) then at least one bidder has accepted the incremented price $p_j^t + \epsilon$. Thus we set $p_j^{t+1} = p_j^t + \epsilon$. The mechanism then randomly selects a bidder i amongst these k bidders and places $j \in S_i^{t+1}$. Note that, in this case, the standing high bidder must change as the previous standing high bidder was only assumed to bid the non-increment price p_j^t.

The mechanism then proceeds to the next round. The auction terminates when the conditional bids T_i^t of all bidders are empty, at which point each bidder i is permanently allocated her provisional set S_i^t for a price $\sum_{j \in S_i^t} p_i^t$.

An extremely important property of the SMRA is that the use of standing high bidders implies that every item with a positive price is sold.

Observation 4. *In an SMRA auction, every item with a positive price is sold.* □

Truthful Bidding in the SMRA. A key factor in determining the practical success of the auction is accurate price discovery (see, for example Cramton [10,11]). This, in turn, relies upon bidding that is truthful or, at least, approximately truthful. There are two pertinent issues here. Firstly, is the SMRA mechanism compatible with truthful bidding? Specifically, the use of conditional bidding implicitly implies that bidders are forced to rebid on their provisional sets. However, suppose that T_i^t is the optimal conditional bid, that is

$$T_i^t \in \mathrm{argmax}_{T \subseteq \Omega \setminus S_i^t} \left(v_i(T \cup S_i^t) - v_i(S_i^t) - \sum_{j \in T}(p_j^t + \epsilon) \right)$$

It need not be the case that the implicit bid $S_i^t \cup T_i^t$ is truthful. In particular, we may have $S_i^t \cup T_i^t \notin \mathrm{argmax}_{T \subseteq \Omega} \left(v_i(T) - \sum_{j \in T \cap S_i^t} p_j^t - \sum_{j \in T \setminus S_i^t}(p_j^t + \epsilon) \right)$.

Recall, here, that bidder i has a personalized set of prices: $((\mathbf{p})_{S_i^t}, (\mathbf{p} + \epsilon \cdot \mathbf{1})_{\Omega \setminus S_i^t})$. Indeed, at round t, bidder i has an ϵ discount on the prices of S_i^t.

Interestingly, truthful bidding is compatible with the SMRA (for any price trajectories) precisely if the valuation function satisfies the gross substitutes property [21]. The gross substitutes property[3] was defined by Kelso and Crawford [16] and used by them to prove the existence of Walrasian equilibrium. Moreover, with gross substitutes, the SMRA will converge to a Walrasian equilibrium; furthermore such an equilibrium will maximize social welfare (given negligible price increments) – see Milgrom [21,22].

---

[3] A valuation function satisfies the *gross substitutes property* if, given any set of prices, increasing the price of some goods does not decrease demand for another good.

Secondly, even if truthful bidding is compatible with the SMRA, it is unlikely that the bidders will actually bid truthfully. For example, in bandwidth auctions, firms typically have ranked bandwidth targets and budget constraints that are more important than profit maximization. Moreover, the valuation function is often not known in advance, rather it is "learned" as the auction proceeds. Regardless, the SMRA and the CCA do both incorporate a set of bidding activity rules to encourage truthful bidding. In the CCA these include revealed preference bidding rules that are difficult to game [4,6]. However, the bidding rules in the SMRA are weaker and strategic bidding is common – examples include demand reduction, parking, and hold-up strategies [11].

Consequently, as well as examining truthful (optimal conditional) bidding, we will examine the natural strategy of local improvement bidding that consists of attempting to add one item, delete one item, or replace one item in the current proposed solution. Gul and Stacchetti [14] prove that this local improvement method finds an optimal demand set, given any set of prices, if the valuation function has the gross substitutes property. We examine the quality of outcomes, for more general valuation functions, when this local search method is used in the SMRA in Sect. 4.

From now on, we will just consider conditional bidding and then will omit the term conditional when we refer to conditional truthful bidding or conditional profit maximizing bidding.

3   Bi-criteria Guarantees for the SMRA Under Truthful Bidding

We now prove Theorem 1 and show that, under truthful bidding, the worst case welfare and rationality guarantees are dependent upon the degree of submodularity in the bidder valuation functions.

Theorem 5. *Given α-near-submodular truthful bidders, the SMRA outputs an α-individually rational allocation.*

Proof. In order to show α-individual rationality upon termination, let us prove a stronger result. Specifically, we will show that for any time t and any bidder i, every set $S' \subseteq S_i^t$ satisfies $\alpha \cdot v_i(S') \geq p(S')$. We proceed by induction on t. The statement trivially holds for $t = 0$. For the induction hypothesis, assume that bidder i is allocated the set S_i^t in round t where

$$\alpha \cdot v_i(S') \geq p^t(S') \qquad \forall S' \subseteq S_i^t \tag{1}$$

We now require the following claim:

Claim. Let $X \subseteq S_i^t \cup T_i^t$ be such that $\alpha \cdot v_i(X) \geq p^t(X \cap S_i^t) + p^{t+1}(X \setminus S_i^t)$. Then, for every $x \in T_i^t \setminus X$, we have

$$\alpha \cdot v_i(X \cup x) \geq p^t(X \cap S_i^t) + p^{t+1}(X \cup x \setminus S_i^t)$$

Proof. Take any $x \in T_i^t \setminus X$. By α near-submodularity, we have

$$\frac{v_i(X \cup x) - v_i(X)}{v_i(S_i^t \cup T_i^t) - v_i(S_i^t \cup T_i^t \setminus x)} \geq \frac{1}{\alpha}$$

Consequently,

$$\alpha \cdot v_i(X \cup x) - \alpha \cdot v_i(X) \geq v_i(S_i^t \cup T_i^t) - v_i(S_i^t \cup T_i^t \setminus x) \geq p^{t+1}(x) = p^t(x) + \epsilon$$

Here the second inequality follows from truthful bidding. Otherwise, $T_i^t \setminus x$ is a more profitable bid than T_i^t. The equality arises as $x \notin S_i$.

By the condition in the statement of the claim, we have $\alpha \cdot v_i(X) \geq p^t(X \cap S_i^t) + p^{t+1}(X \setminus S_i^t)$. Therefore

$$\alpha \cdot v_i(X \cup x) \geq p^t(X \cap S_i^t) + p^{t+1}(X \setminus S_i^t) + p^{t+1}(x)$$
$$= p^t(X \cap S_i^t) + p^{t+1}(X \cup x \setminus S_i^t)$$

Again, the equality arises as $x \notin S_i$. □

By iteratively applying the previous claim over items in a set $\hat{X} \subseteq T_i^t \setminus X$, we obtain

Claim. Let $X \subseteq S_i^t$ be such that $\alpha \cdot v_i(X) \geq p^t(X)$. Then, for every $\hat{X} \subseteq T_i^t \setminus X$, we have $\alpha \cdot v_i(X \cup \hat{X}) \geq p^t(X \cap S_i^t) + p^{t+1}(X \cup \hat{X} \setminus S_i^t)$. □

Now take any $\hat{S} \subseteq S_i^{t+1}$. To complete the proof of Theorem 5, we must show that $\alpha \cdot v_i(\hat{S}) \geq p^{t+1}(\hat{S})$. For this purpose, set $S' = \hat{S} \cap S_i^t$ and set $T' = \hat{S} \setminus S_i^t$. By the induction hypothesis, we have that $\alpha \cdot v_i(S') \geq p^t(S')$.

Thus we may apply the second claim with $X = S'$ and $\hat{X} = T'$ to obtain

$$\alpha \cdot v_i(\hat{S}) = \alpha \cdot v_i(S' \cup T') \geq p^t(S' \cap S_i^t) + p^{t+1}((S' \cup T') \setminus S_i^t) = p^t(S') + p^{t+1}(T')$$

Furthermore, note that $S' \subseteq S_i^t \cap S_i^{t+1}$. In order to be the provisional winner of an item j in both rounds t and $t+1$, it must be the case that no other bidder bid for item j at the price $p^{t+1}(j)$. Thus the price of j at time $t+1$ remains $p^t(j)$. Hence $p^{t+1}(S') = p^t(S')$, and so

$$\alpha \cdot v_i(\hat{S}) \geq p^t(S') + p^{t+1}(T') = p^{t+1}(S') + p^{t+1}(T') = p^{t+1}(\hat{S})$$

Theorem 5 follows by induction. □

We remark that this proof implies a stronger conclusion: if bidder i is truthful then she is α-individually rational *regardless* of the strategies of other bidders. To conclude the proof of Theorem 1, we just have to prove the following:

Theorem 6. *Given α-near-submodular truthful bidders, the SMRA outputs an allocation $\mathcal{S} = (S_1, \ldots, S_n)$ with social welfare $\omega(\mathcal{S}) \geq \frac{1}{1+\alpha} \cdot \omega(S_i^*)$ where $S^* = (S_1^*, \ldots, S_n^*)$ is an allocation of maximum welfare.*

Proof. Assume the auction terminates in round t with a set of prices $\mathbf{p}^t$. Thus $T_i^t = \emptyset$ for each bidder i. In particular, by truthfulness, we have that

$$v_i(S_i \cup (S_i^* \setminus S_i)) - p^t(S_i^* \setminus S_i) \leq v_i(S_i \cup \emptyset) - p^t(\emptyset) = v_i(S_i)$$

Thus

$$v_i(S_i^*) \leq v_i(S_i \cup S_i^*) \leq v_i(S_i) + p^t(S_i^* \setminus S_i)$$

We now obtain a $(1 + \alpha)$ factor welfare guarantee.

$$\sum_{i=1}^{n} v_i(S_i^*) \leq \sum_{i=1}^{n} \left(v_i(S_i) + p^t(S_i^* \setminus S_i) \right) \leq \sum_{i=1}^{n} v_i(S_i) + \sum_{i=1}^{n} p^t(S_i^*)$$

$$\leq \sum_{i=1}^{n} v_i(S_i) + \sum_{i=1}^{n} p^t(S_i) \leq \sum_{i=1}^{n} v_i(S_i) + \sum_{i=1}^{n} \alpha \cdot v_i(S_i)$$

$$= (1 + \alpha) \cdot \sum_{i=1}^{n} v_i(S_i)$$

Here the third inequality follows because the SMRA mechanism utilizes provisional winners. This implies that every item with a positive price is sold at the end of the auction. Consequently, $\sum_{i=1}^{n} p^t(S_i) \geq \sum_{i=1}^{n} p^t(S_i^*)$. The fourth inequality follows as the auction allocation is α-individual rational, as shown in Theorem 5.   □

By combining Theorems 5 and 6 we obtain Theorem 1.

Tightness of the Bi-criteria Guarantees. The bounds in Theorem 1 are almost tight. To see this, consider the following example. There are k items $X = \{x_1, x_2, \ldots, x_k\}$. Let there be a large number L of identical bidders. For any $S \subset X$, each bidder i has a valuation 1 for set S if $|S| = 1$ and $(|S|-1)\cdot\alpha+1$ otherwise. It is easy to verify that this function is α near-submodular.

The optimal welfare is obtained by allocating the entire set X to a single bidder achieving social welfare $(k - 1) \cdot \alpha + 1$. Now let us examine the allocation produced by the SMRA. Initially, all prices are 0 and the truthful bid for each bidder is to demand the entire set X. Indeed, every bidder keeps bidding on the entire set (except for the items that she is the standing high bidder) until every item has price greater than $\frac{1}{k}((k - 1) \cdot \alpha + 1)$. At this point, no profitable bids can be made and all bidders drop out.

In each round, the randomly chosen standing high bidders are all distinct with probability at least $(1 - \frac{k-1}{L-1})^k$. For $L >> k$, this probability tends to 1. So by the end of the auction, the k items are allocated to k different bidders with probability almost 1. Since the social welfare of this allocation is only k, the expected social welfare of the SMRA is around k. When k goes to infinity, the welfare ratio tends to α.

Next consider the rationality of this allocation. Each winner was allocated exactly one item with probability almost 1, and the final price of that item is $\frac{1}{k}((k - 1) \cdot \alpha + 1)$. The bidder has only value 1 for the item. When k goes to infinity, this tends to α-rationality for the winners. We remark that even for $k = 2$ items, the previous example ensures that the welfare guarantee cannot be improved beyond $\frac{\alpha}{2}$ since the optimal welfare is $(\alpha+1)$ and the expected welfare of the SMRA is 2.

4   Bi-criteria Guarantees Under Locally Optimal Bidding

As discussed in Sect. 2, the assumption of truthful bidding is unrealistic in the SMRA. Consequently, here we examine an alternate natural bidding method. Given S_i^{t-1}, a bid $T_i^t \subseteq \Omega \setminus S_i^{t-1}$ is *locally optimal* if $v_i(S_i^{t-1} \cup T_i^t) - p^t(T_i^t) \geq v_i(S_i^{t-1} \cup X) - p^t(X)$ for all $X \subseteq \Omega \setminus S_i^{t-1}$, where $|X \setminus T_i^t| \leq 1$ and $|T_i^t \setminus X| \leq 1$. Observe that a locally optimal bid can be obtained via a local improvement algorithm that, given the current solution, seeks to add one item, delete one item, or replace one item. Analysing this local improvement method is useful because local comparison is a key tool used by bidders in real bandwidth auctions. Thus, there are practical reasons to suspect that bidders will not make bids that are clearly not locally optimal. From the theoretical viewpoint, this specific local improvement method is interesting because it is guaranteed, given any set of prices, to output an optimal set if the valuations satisfy the gross substitute property [14]. Due to space restrictions, the reader is referred to the full version [8] for a proof of the following result and tight examples.

Theorem 2. *If bidders have α-near-submodular valuations and make locally optimal bids, then the SMRA has welfare ratio $\frac{1}{1+\alpha^2}$ and is α-individually rational.*

5   Individually Rational Bidding

As shown in Theorems 2 and 5, truthful and locally-optimal bidding can only ensure approximate individual rationality in the SMRA. Consequently, such bidding strategies are highly risky. In this section, we investigate what bidding strategies are risk-free and what are the welfare implications of such strategies.

We call a risk-free strategy *conservative*, and show in Sect. 5.1 that conservative bidding is (weakly) characterized by *secure bidding*. Specifically, secure bidding always produces individually rational outcomes. Conversely, if the other bidders use secure bids then the only way a bidder can ensure an individually rational solution is by also bidding securely. This result holds even with stronger assumptions on the bidding strategies of the other bidders, for example, that they make profit-maximizing secure bids.

We then examine the welfare consequences of secure bidding. Our main result, in Sect. 5.2, is that then the welfare ratio is at most $1 + \alpha$ provided the bidders make profit maximizing secure bids. This result is surprising in that we are able to match the welfare guarantee of truthful bidding without having to lose individual rationality.

5.1   Secure Bidding

We say that a bidding strategy is *conservative* if it cannot lead to a bidder having negative utility. Thus, conservative strategies are individually rational. To understand what strategies are conservative, we first need to understand what constitutes a bidding strategy. In the SMRA, a bidder can select a bid based upon the auction history she observed, for example, the sequence of price vectors, her sequence of conditional bids, and on her sequence of provisional sets of items. Thus, we consider a bidding strategy to be a function of these three factors.[4]

We say that a conditional bid T_i^t is *secure* for bidder i (given the provisional winning set S_i^t) if $v_i(S') \geq p(S')$ for every $S' \subseteq S_i^t \cup T_i^t$. A bidding strategy is *secure* if every conditional bid it makes is secure. It is easy to verify that any secure bidding strategy is individually rational. We now show that bidding securely in every round is essentially the only individually rational strategy.

Lemma 1. *Let t be an integer and $T_i^{\hat{t}}, S_i^{\hat{t}}, \mathbf{p}^{\hat{t}}$ be the conditional bid of bidder i, the provisional winning set of bidder i and the price vector at round $\hat{t}$ for any $\hat{t} \leq t$. If bidder i makes a non-secure bid in round $t + 1$, then there exist secure bidders who can bid consistent with the history and ensure that bidder i has negative utility in the final allocation.*

Proof. Assume that the conditional bid T_i^{t+1} of bidder i at some round t is not secure, then there exists $S' \subseteq S_i^{t+1} \cup T_i^{t+1}$ such that S' satisfies $v_i(S') < p_i^{t+1}(S')$. Let us prove that there exists an auction such that, with high probability, (i) the set allocated to i is S', (ii) at any time $\hat{t} \leq t$, the provisional winning set of i is $S_i^{\hat{t}}$ and (iii) the price vector at round $\hat{t}$ is $\mathbf{p}^{\hat{t}}$.

The auction is as follows: there are many copies of the same bidder 1 whose valuation function is v_1. Let M be an integer larger than the maximum of the prices at any round $\hat{t} \leq t$ and the maximum valuation of any subset of items for bidder i. The valuation function v_1 of all the copies of bidder 1 is additive[5] and the value of each item is the following:

$$v_1(s) = \begin{cases} M + 2 \cdot \epsilon & \text{if } s \in \Omega \setminus S' \\ p^t(s) & \text{if } s \in S' \setminus S_i^t \\ p^t(s) - \epsilon & \text{if } s \in S' \cap S_i^t \end{cases}$$

Claim. Assume that i bids on $T_i^{\hat{t}}$ at any round $\hat{t} \leq t$. There is a sequence of secure bids such that, for every $\hat{t} \leq t$, with high probability

(i) the price vector is exactly $\mathbf{p}^{\hat{t}}$ at the end of round $\hat{t}$,
(ii) bidder i is the standing high bidder of the set $S_i^{\hat{t}+1}$.[6]

---

[4] In some SMRA mechanisms, bidders also know the excess demand of each item.
[5] A valuation function v is additive if $v(S) = \sum_{s \in S} v(s)$.
[6] Recall that i is the standing high bidder of $S_i^{\hat{t}}$ at the beginning of round $\hat{t}$, which explains the index difference.

Proof. By induction on t, let us prove that if the copies of bidder 1 use the following strategy, the conclusion holds. If the price of item s does not increase from round $\hat{t}$ to round $\hat{t}+1$ then no copy of bidder 1 bids on it at round $\hat{t}$; if the price of item s increases and $s \in S_i^{\hat{t}}$ then no copy of bidder 1 bids on it at round $\hat{t}$; if the price of s increases and $s \notin S_i^{\hat{t}}$ then all the copies of bidder 1 bid on it at round $\hat{t}$. By construction of the valuation function v_1, at any time $\hat{t} \leq t$, the value of any item $s \in \Omega \setminus (S' \cap S_i^{t+1})$ for copies of bidder 1 is at least its price. Moreover, if $s \in S_i^t$ then copies of bidders 1 do not bid on s at price $p^t(s)$ by construction. It is easy to verify that $v_1(s)$ is larger than the price of s at any round where copies of 1 bid on it. As $v_1(\cdot)$ is additive, all bids by copies of bidder 1 up to round t are secure.

Let us show that items in excess demand are those whose prices increase between rounds $\hat{t}-1$ and $\hat{t}$. If the price of an item in $\Omega \setminus S_i^{\hat{t}}$ is distinct in $\mathbf{p}^{\hat{t}-1}$ and $\mathbf{p}^{\hat{t}}$, then all the copies of bidder 1 bid on it, and it is in excess demand. Now assume $s \in S_i^{\hat{t}}$, if the price of s increases, then $s \in S_i^{\hat{t}} \setminus S_i^{\hat{t}-1}$ (the provisional winner must change when there is a price increment). Thus bidder i bids on s and then s is in excess demand.

Now let us show that with high probability, bidder i is the standing high bidder of the items in $S_i^{\hat{t}}$. Since the prices of any item s in $S_i^{\hat{t}-1} \cap S_i^{\hat{t}}$ do not increase, copies of bidder 1 do not bid on s at round $\hat{t}$. Thus s is still in $S_i^{\hat{t}}$. Moreover, bidder i is the unique bidder in excess demand for the items in $S_i^{\hat{t}} \setminus S_i^{\hat{t}-1}$. So the provisional set of bidder i contains $S_i^{\hat{t}}$. Let us prove that it does not contain any other item s with high probability. First assume that $s \in S_i^{\hat{t}-1} \setminus S_i^{\hat{t}}$. Thus the price of s increases. And since i was the standing high bidders of these items at round $\hat{t}-1$, she cannot be the standing high bidder anymore at round $\hat{t}$. Assume now that i bids on $s \notin S_i^{\hat{t}-1} \cup S_i^{\hat{t}}$. Then by construction, all the other copies of 1 also bid on s and then, with high probability (since there are many copies of bidder 1), s is not allocated to bidder i, which completes the proof of the claim. □

Now assume that at round $t+1$, bidder i decides to bid on T_i^{t+1}. Starting from round $t+1$, copies of bidder 1 securely bid on subsets in the complement of S' until the prices of all items in $\Omega \setminus S$ reach $M+2\epsilon$. Note that since no copy of 1 bid on any item in S', all the items in S' are in the provisional set of i at the end of round $t+1$. Copies of 1 continue to perform the same bids until they drop out. On the other hand, bidder i can perform any bid.

Let us first show that the set allocated to i contains S'. At the end of round $t+1$, the price of item s in S' is $p^t(s)$ if $s \in S' \cap S_i^t$ and $p^t(s) + \epsilon$ if $s \in S' \setminus S_i^t$. Thus the price of s is above $v_1(s)$ and then copies of 1 cannot bid anymore on s since they make secure bids. Since $S' \subseteq S_i^{t+1}$, the set of items allocated to i by the SMRA contains the set S'.

Assume now that $s \notin S'$ is allocated to i at the end of the procedure. Since copies of 1 continue to bid on it until its price is at least $M+\epsilon$. This implies that bidder i bids on it at price at least $M+\epsilon$. Thus the price of the set allocated to i is at least $M+\epsilon$, which is above the value of any set for bidder i by definition of M. So i is not individually rational. Otherwise, bidder i is allocated the set S' and by definition of S', we have $p^t(S') > v_i(S')$ and then bidder i receives negative utility.   □

So, if the bids of the other bidders are secure, then performing a non-secure bid may lead to negative utility. One may ask if a similar statement still holds if stronger assumptions are made concerning the strategies of the other bidders. This is indeed the case. The following lemma states that even if we know the other bidders are truthful (or if they make profit-maximizing secure bids), making any non-secure bid is not individually rational (the proof is deferred to the full version [8]).

Lemma 2. *Let t be an integer and $T_i^{\hat{t}}, S_i^{\hat{t}}, \mathbf{p}^{\hat{t}}$ be the conditional bid of bidder i, the provisional winning set of bidder i and the price vector at round $\hat{t}$ for any $\hat{t} \leq t$. Assume that there is an item of value 0 for i with price $\epsilon \cdot \hat{t}$ at any round $\hat{t} \leq t$. If bidder i makes a non-secure bid in round $t + 1$, then there exist truthful (or profit-maximizing secure) bidders who can bid consistent with the history and ensure that bidder i has negative utility in the final allocation.*

5.2   Social Welfare Under Secure Bidding

The previous results ensure that secure bidding strategies are essentially the only way to guarantee individual rationality. In this section, we will assume that bidders strategies are secure. The following simple lemma ensures that any allocation where each bidder is allocated at least one item can be obtained in the SMRA with bidders only making secure bids. Due to space restriction, we refer the reader to the full version [8] for the proofs of the next two lemmas.

Lemma 3. *Any allocation where each bidder is allocated at least one item can be obtained via the SMRA with secure bidders. In particular if each bidder is allocated at least one item then the optimal allocation can be obtained if bidders are secure.*

Lemma 3 is unsatisfactory in two ways. First, if there are bidders that are allocated nothing then the situation can be far more complex. Specifically, it may then be the case, see Lemma 4, that secure bidding cannot provide a guarantee on welfare.

Lemma 4. *If super-additive bidders only make secure bids, then there is no guarantee on the social welfare of the SMRA.*

The second unsatisfactory aspect of Lemma 3 is that the structure of the bids used there is extremely artificial, since the bidders need to know all the valuation functions in order to calculate the secure bids. Theorem 3 shows we can circumvent both of these problems if the bidders' valuation functions are α near-submodular. Then a good welfare guarantee can be obtained if the bidders make profit-maximizing secure bids. Namely, in every round every bidder i bids on a secure set T_i such that the utility of $S_i \cup T_i$ is maximized over all possible secure bids.

Theorem 3. *If bidders have α near-submodular valuations then, under (conditional) profit maximizing secure bidding, the SMRA outputs an allocation $\mathcal{S}$ with $\omega(\mathcal{S}) \geq \frac{1}{1+\alpha} \cdot \omega(\mathcal{S}^*)$ where $\mathcal{S}^*$ is the optimal allocation. Moreover, the auction is individually rational.*

Proof. Let $\mathcal{S}^* = \{S_1^*, \ldots, S_n^*\}$ be the optimal allocation, and let $\mathcal{S} = \{S_1, \ldots, S_n\}$ be the assignment output by the SMRA. By assumption, S_i is the most profitable secure set in the final round, and the conditional bid T_i is empty in the final round and then S_i was the provisional set for bidder i in the penultimate round. Let $S_i^* \setminus S_i = \{x_1, x_2, \ldots, x_k\}$ and let $X^j = \{x_1, x_2, \ldots, x_j\}$, for each $j \leq k$.

First, as all bidders use secure bidding strategies, the auction is individually rational. Now we bound the welfare ratio. For every item $x_j \in S_i$, since the conditional bid T_i is empty, there are two possibilities.

- **Case 1:** $\{x_j\}$ is a secure conditional set but not as profitable as $\emptyset$. Then $v_i(S_i \cup \{x_j\}) - p(x_j) < v_i(S_i)$. Let Q_{ij} be S_i, and we have $p(x_j) \geq v_i(Q_{ij} \cup \{x_j\}) - v_i(Q_{ij})$.
- **Case 2:** $\{x_j\}$ is an insecure conditional set. Then there exist a set $Q \subseteq S_i$ such that $v_i(Q \cup \{x_j\}) < p(Q \cup \{x_j\})$. On the other hand, since S_i is a secure set, $v_i(Q) \geq p(Q)$. Let Q_{ij} be Q, and we have $p(x_j) \geq v_i(Q_{ij} \cup \{x_j\}) - v_i(Q_{ij})$.

In both cases, we have $p(x_j) \geq v_i(Q_{ij} \cup \{x_j\}) - v_i(Q_{ij})$. Using these inequalities, we can bound $v_i(S_i^*)$.

$$v_i(S_i^*) \leq v_i(S_i \cup X^k) = v_i(S_i) + \sum_{j=1}^{k} \left(v_i(S_i \cup X^j) - v_i(S_i \cup X^{j-1}) \right)$$

$$\leq v_i(S_i) + \sum_{j=1}^{k} \alpha \cdot \left(v_i(Q_{ij} \cup \{x_j\}) - v_i(Q_{ij}) \right) \leq v_i(S_i) + \alpha \cdot \sum_{j=1}^{k} p(x_j)$$

$$\leq v_i(S_i) + \alpha \cdot p(S_i^*)$$

The second inequality is because $Q_{ij} \subseteq S_i$ and $v_i(\cdot)$ is α-near-submodular. The third inequality is derived from the case analysis above. Finally, we are ready to bound the welfare ratio.

$$\sum_{i=1}^{n} v_i(S_i^*) \leq \sum_{i=1}^{n} v_i(S_i) + \alpha \cdot \sum_{i=1}^{n} p(S_i^*) \leq \sum_{i=1}^{n} v_i(S_i) + \alpha \cdot \sum_{i=1}^{n} p(S_i) \leq (1 + \alpha) \cdot \sum_{i=1}^{n} v_i(S_i)$$

The second inequality is a consequence of Observation 4. The last inequality holds because the auction is individually rational. □

The bound in Theorem 3 is almost tight. This can be seen by adapting the tight example in Sect. 3.

Thus, under secure bidding we are able to match the welfare guarantee of truthful bidding without having to lose individual rationality.

References

1. Alkalay-Houlihan, C., Vetta, A.: False-name bidding and economic efficiency in combinatorial auctions. In: Proceedings of the Twenty-Eighth AAAI Conference on Artificial Intelligence (AAAI), pp. 2339–2347 (2014)
2. Ausubel, L., Cramton, P.: Auctioning many divisible goods. J. Eur. Econ. Assoc. **2**, 480–493 (2004)
3. Ausubel, L., Cramton, P.: Auction Design for Wind Rights. Report to Bureau of Ocean Energy Management, Regulation and Enforcement (2011)
4. Ausubel, L., Cramton, P., Milgrom, P.: The clock-proxy auction: a practical combinatorial auction design. In: Cramton, P., Shoham, Y., Steinberg, R. (eds.) Combinatorial Auctions, pp. 115–138. MIT Press, Cambridge (2006)
5. Bichler, M., Georee, J., Mayer, S., Shabalin, P.: Spectrum auction design: simple auctions for complex sales. Telecommun. Policy **38**(7), 613–622 (2014)
6. Boodaghians, S., Vetta, A.: The combinatorial world (of auctions) according to GARP. To appear in Proceedings of Symposium in Algorithmic Game Theory (SAGT) (2015)
7. Bousquet, N., Cai, Y., Hunkenschröder, C., Vetta, A.: On the economic efficiency of the combinatorial clock auction. arxiv, #1507.06495 (2015)
8. Bousquet, N., Cai, Y., Vetta, A.: Welfare and rationality guarantees for the simultaneous multiple-round ascending auction. arxiv, #1510.00295 (2015)
9. Coase, R.: The Federal Communications Commission. J. Law Econ. **2**, 1–40 (1959)
10. Cramton, P.: Simultaneous ascending auctions. In: Cramton, P., Shoham, Y., Steinberg, R. (eds.) Combinatorial Auctions, pp. 99–114. MIT Press, Cambridge (2006)
11. Cramton, P.: Spectrum auction design. Rev. Ind. Organ. **42**(2), 161–190 (2013)
12. Crawford, P., Knoer, E.: Job matching with heterogenous firms and workers. Econometrica **49**(2), 437–450 (1981)
13. Fu, H., Kleinberg, R., Lavi, R.: Conditional equilibrium outcomes via ascending processes with applications to combinatorial auctions with item bidding. In: Proceedings of the Thirteenth Conference on Electronic Commerce (EC), p. 586 (2012)
14. Gul, F., Stacchetti, E.: Walrasian equilibrium with gross substitutes. J. Econ. Theory **87**, 95–124 (1999)
15. Gul, F., Stacchetti, E.: The English auction with differentiated commodities. J. Econ. Theory **92**, 66–95 (2000)
16. Kelso, A., Crawford, P.: Job matching, coalition formation, and gross substitutes. Econometrica **50**(6), 1483–1504 (1982)
17. Klemperer, P.: Auctions: Theory and Practice. Princeton University Press, Princeton (2004)
18. Lehmann, B., Lehmann, D., Nisan, N.: Combinatorial auctions with decreasing marginal utilities. Games Econ. Behav. **55**(2), 270–296 (2006)
19. McMillan, J.: Selling spectrum rights. J. Econ. Perspect. **8**(3), 145–162 (1994)
20. McMillan, J.: Why auction the spectrum? Telecommun. Policy **3**, 191–199 (1994)
21. Milgrom, P.: Putting auction theory to work: the simultaneous ascending auction. J. Polit. Econ. **108**, 245–272 (2000)
22. Milgrom, P.: Putting Auction Theory to Work. Cambridge University Press, Cambridge (2004)
23. Porter, D., Rassenti, S., Roopnarine, A., Smith, V.: Combinatorial auction design. Proc. Natl. Acad. Sci. U.S.A. **100**(19), 11153–11157 (2003)

Combinatorial Auctions with Conflict-Based Externalities

Yun Kuen Cheung[1](✉), Monika Henzinger[1], Martin Hoefer[2],
and Martin Starnberger[1]

[1] Faculty of Computer Science, University of Vienna, Vienna, Austria
yun.kuen.cheung@univie.ac.at
[2] Max-Planck-Institut für Informatik, Saarland University, Saarbrücken, Germany

Abstract. Combinatorial auctions (CA) are a well-studied area in algorithmic mechanism design. However, contrary to the standard model, empirical studies suggest that a bidder's valuation often does not depend solely on the goods assigned to him. For instance, in adwords auctions an advertiser might not want his ads to be displayed next to his competitors' ads. In this paper, we propose and analyze several natural graph-theoretic models that incorporate such negative externalities, in which bidders form a directed conflict graph with maximum out-degree Δ. We design algorithms and truthful mechanisms for social welfare maximization that attain approximation ratios depending on Δ.

For CA, our results are twofold: (1) A *lottery* that eliminates conflicts by discarding bidders/items independent of the bids. It allows to apply any truthful α-approximation mechanism for conflict-free valuations and yields an $\mathcal{O}(\alpha\Delta)$-approximation mechanism. (2) For fractionally sub-additive valuations, we design a rounding algorithm via a novel combination of a semi-definite program and a linear program, resulting in a *cone program*; the approximation ratio is $\mathcal{O}((\Delta \log \log \Delta)/\log \Delta)$. The ratios are almost optimal given existing hardness results.

For adwords auctions, we present several algorithms for the most relevant scenario when the number of items is small. In particular, we design a truthful mechanism with approximation ratio $o(\Delta)$ when the number of items is only logarithmic in the number of bidders.

1 Introduction

Combinatorial auctions (CA) are an important area in algorithmic mechanism design due to wide-spread applications in resource allocation and e-commerce, e.g., spectrum or adwords auctions [5]. In the standard CA, a set of items is assigned to a set of bidders in order to maximize social welfare, which is given by the total valuations of bidders for their assigned items. This assumes that each bidder values exclusively the set of items assigned to him — his

This work was funded by the Vienna Science and Technology Fund (WWTF) through project ICT10-002. This work is supported by DFG through Cluster of Excellence MMCI.

© Springer-Verlag Berlin Heidelberg 2015
E. Markakis and G. Schäfer (Eds.): WINE 2015, LNCS 9470, pp. 230–243, 2015.
DOI: 10.1007/978-3-662-48995-6_17

valuation is *independent of the assignment of other items to other bidders*. In many applications (see [20,21] for examples), however, such an assumption is not justified since bidder preferences have a significant dependence on how items are assigned to other bidders. Such a dependence is called *externality*.

Mechanism design for CA with externalities in the most general form is difficult, primarily due to the huge complexity of bidders' preferences with externalities, which then also leads to the computational complexity issue for (approximately) maximizing social welfare. Prior work has studied more restricted scenarios, e.g., when there is only one item on sale, or when the bidders' preferences are simple (e.g., unit-demand). In this work, we focus on a simple type of externalities called *conflict-based externality*, which is readily motivated by sponsored search auctions (SSA); in our model, there are multiple items on sale, and the bidders' preferences might be more complex than unit-demand ones.

SSA are one of the most popular special cases of CA, where ad slots on a search result page are assigned to advertisers. *Negative externality* arises when, for example, a car-rental company has much smaller value for an ad slot if an ad of another prominent rental company is shown right next to it. More generally, for an advertiser there might be a number of competitors, and an assignment yields value to the bidder only if the ads of competitors are not displayed simultaneously. The existence of negative externalities in sponsored search has been confirmed empirically [11]. Moreover, similar negative externalities also arise in other prominent applications of CA, e.g., in secondary spectrum auctions where interferences induce negative externalities; or when selling luxury goods, where the value of a buyer for items from an exclusive brand drops when other buyers also obtain items from the same brand. These examples give rise to a natural and simple graph-based model of externalities: each bidder is a node in a directed graph, and a directed edge indicates that a bidder sees another bidder as a competitor; assigning an item to a bidder yields value only if none of the competitors receives any item (or just any "similar" or "better" item).

Negative externalities in auctions have recently received attention, but — perhaps surprisingly — the natural and simple idea sketched above has not been analyzed in a rigorous and general fashion. We propose three graph-theoretic models that incorporate these conflict-based externalities. We study approximation algorithms and truthful mechanisms under the models. Formally, we assume there is a directed conflict graph on the set of bidders. Each edge (i, j) indicates a conflict: i has no value for any assignment in which j receives an item. More generally, we also consider cases where conflicts arise only among certain pairs of items, or different values for assignments that include or avoid certain conflicts. Our algorithms cope with externalities via new extensions of algorithmic techniques for independent set problems *in combination with* algorithms for conflict-free CA. We also provide additional results for the prominent special case of SSA. Before we state our results, we proceed with a formal introduction and discussion of the models on conflict-based externalities treated in this paper.

Auctions with Conflict-Based Externalities. In all models, we have a bidder set B of n bidders and an item set I of m items. Each item can be given to at most one bidder. For each $i \in B$, there is a valuation function $v_i : 2^I \to \mathbb{R}^+$, where $v_i(S_i)$ represents the value for receiving item set $S_i \subseteq I$. In the SSA case, the items in I are ad slots. Each slot k has a click-through rate $\alpha_k \geq 0$. Each bidder i has a valuation per click of $v_i \geq 0$ in one slot. Then $v_i(S_i) = \max_{k \in S_i} v_i \cdot \alpha_k$, a unit demand valuation function with free disposal.

The valuation $v_i(S_i)$ will be extended, due to externalities, to $v_i^c(S)$, a valuation that depends on the complete allocation $S = (S_1, \cdots, S_n)$. The goal is to find an allocation S that maximizes social welfare $SW(S) = \sum_{i \in B} v_i^c(S)$.

CA with Bidder Conflicts. The set of bidders B is the vertex set of a *(bidder) conflict graph* $G = (B, E)$, which is a directed graph. Each bidder i has a valuation function $v_i : 2^I \to \mathbb{R}^+$. Given a complete allocation $S = (S_1, \cdots, S_n)$,

$$v_i^c(S) = \begin{cases} v_i(S_i) & \text{if } \bigcup_{j:(i,j)\in E} S_j = \emptyset \\ 0 & \text{otherwise.} \end{cases}$$

This models the situation that advertiser i is not interested in showing its ad together with an ad from a competitor j, represented by an edge $(i, j) \in E$.

The introduction of conflicts turns social welfare maximization NP-hard; in the special case SSA with all $v_i = 1$, all $\alpha_k = 1$, and $m = n$, it reduces to the maximum independent set problem.

CA with Bidder and Item Conflicts. There are two conflict structures in this model, each represented by a directed graph. The bidder set B is the vertex set of a *bidder conflict graph* $G = (B, E)$. The item set I is the vertex set of an *item conflict graph* $G_I = (I, E_I)$. Both graphs are directed. Intuitively speaking, if $(i, j) \in E$ and $(k, \ell) \in E_I$, then bidder i has no use for item k if j receives item ℓ. Formally, for any allocation S, bidder i has a set D_i of *useless items*, defined as $D_i := \{k \in S_i \mid \exists \ell \in S_j : (i, j) \in E \text{ and } (k, \ell) \in E_I\}$, and $v_i^c(S) := v_i(S_i \setminus D_i)$.

An intuitive example is *ordered conflicts*, where ad slots are ordered on a page top-down, and a bidder has a conflict only if a competitor receives a slot *above* him. This can be modelled by numbering slots top-down and $E_I = \{(k, \ell) \mid k, \ell \in I, \ell < k\}$. Another intuitive example is *neighbor conflicts*, where ad slots are arranged horizontally, and a bidder has conflict only if a competitor receives a slot *right next to* him. This can be modelled by numbering slots from left to right and $E_I = \{(k, \ell) \mid k, \ell \in I, |k - \ell| = 1\}$.

Note that CA with bidder conflicts is a sub-case of this model, when G_I is the complete digraph.

The results in this paper depend on two parameters Δ and Δ_I of the conflict graphs, which are the maximum out-degrees of the graphs G and G_I respectively.

CA with Bidder Conflicts and Conflict Value. In CA with bidder conflicts, we assume that the valuation of a bidder drops to $v_i(S) = 0$ as soon as a competitor receives any item. We can generalize this assumption to a second valuation function $w_i(S_i)$: if $\bigcup_{j:(i,j)\in E} S_j = \emptyset$, then $v_i^c(S) = v_i(S_i)$; otherwise, $v_i^c(S) = w_i(S_i)$.

This model can be reduced to the model with bidder-conflicts only. Given an instance of CA with bidder conflicts and conflict value, we build an instance without conflict value as follows: for each bidder i, we add an auxiliary bidder i_c, where $v_{i_c}(S_i) = w_i(S_i)$. In the bidder conflict graph, we add the edges (i, i_c) and (i_c, i). This increases Δ by exactly 1. Now if bidder i is conflicted, we can take all items assigned to it and assign them to bidder i_c instead. In this way, we can transform any allocation into the instance without conflict value and obtain the same social welfare. It is straightforward to observe that social welfare maximization in both instances is equivalent. This, however, does not directly apply to truthfulness.

There are numerous further ways to extend our models, e.g., to combinations of item conflicts and conflict values, weighted conflicts, etc. Studying their properties are interesting avenues for future work.

Our Contribution. For CA with conflict-based externalities, we design and analyze poly-time approximation algorithms and truthful mechanisms which provide almost best possible approximation guarantees of maximizing social welfare. To state our results, we first define *fractionally sub-additive* valuations (see [8]). a valuation function $v : 2^I \to \mathbb{R}$ is fractionally sub-additive if it satisfies the following property for any $S, T_1, T_2, \cdots, T_k \in 2^I$ and $0 \leq \alpha_1, \alpha_2, \cdots, \alpha_k \leq 1$: if for all $j \in S$, $\sum_{\ell:\ j \in T_\ell} \alpha_\ell \geq 1$, then $v(S) \leq \sum_{\ell=1}^k \alpha_\ell \cdot v(T_\ell)$. The class of fractionally sub-additive valuations is known to strictly contain the more well-known unit-demand valuations, linear valuations, gross substitute valuations and submodular valuations.

For CA with bidder conflicts, we use well-known techniques for independent set problem to eliminate conflicts, which is in a spirit similar to *lottery*, to give a reduction to conflict-free CA. Given any α-approximation algorithm for the unconflicted problem, we obtain an $\mathcal{O}(\alpha\Delta)$-approximation algorithm for CA with bidder conflicts (Theorem 1). If the original algorithm is a truthful mechanism, our reduction preserves the truthfulness. Moreover, our reduction preserves the use of randomization (deterministic, universally truthful, truthful in expectation). If the bidders have fractionally sub-additive valuations, our results extend to CA with bidder and item conflicts (Theorem 2).

The next natural question to ask is whether one can improve the approximation ratio to $o(\Delta)$. Since our problem generalizes the weighted independent set (WIS) problem, the ratio must be $\Omega(\Delta/\log^4 \Delta)$ [2], even for unit-demand valuations. We answer the question positively: if the bidders have fractionally sub-additive valuations and if there is a *demand oracle* for each bidder, we design an $\mathcal{O}((\Delta \log \log \Delta)/\log \Delta)$-approximation algorithm (Theorem 3). This implies, for example, ratios of $\mathcal{O}((\Delta \log \log \Delta)/\log \Delta)$ for sponsored search, unit-demand, or more general gross-substitute valuations. The dependence on Δ mirrors the best-known approximation ratio for WIS. Our algorithm combines an approach for WIS based on semi-definite programming (SDP) with the standard approach for CA based on linear programming (LP) to design a cone program relaxation and a rounding scheme. To the best of our knowledge, we are the first to combine an SDP with an LP in this fashion, and to show how to analyze it. We believe this technique might be of independent interest in other applications.

It is an interesting open problem if this approach can be turned into a truthful mechanism, or be generalized to CA with bidder and item conflicts.

We then focus on SSA with bidder conflicts. Even in this special case, the hardness bound of $\Omega(\Delta/\log^4 \Delta)$ applies. We consider a restriction to a small number of slots that is natural in the context of sponsored search. For the case of $m = \mathcal{O}(\log n)$ slots, we present a truthful mechanism based on SDP that obtains an $\mathcal{O}(\Delta \cdot \sqrt{(\log \log \Delta)/(\log \Delta)})$-approximation (Theorem 4). To obtain the desired truthfulness property, the first step of our mechanism is to gather a statistic from a sampling of bidders who will not be allocated any item, which is similar the first two steps in the framework of Dobzinski et al. [6] for designing truthful mechanisms. However, the subsequent steps of our algorithm will be different from theirs. Also, we get an $\mathcal{O}(\log m)$-approximation algorithm based on partial enumeration that runs in time $\mathcal{O}((m\Delta)^m)$ (Theorem 5); the algorithm can be turned into truthful-in-expectation mechanisms with the same approximation guarantee, and it extends to CA with bidder and item conflicts.

Related Work. The study of auctions with externalities was initiated by seminal work of Jehiel et al. [20,21] in the single-item setting. The externality in this work is *identity-dependent*, i.e., each bidder can have a different valuation when different bidders obtain the item. The preference of each bidder can thus be represented by a low dimensional $\mathbb{R}^{n+1}$ vector, which reflects the bidder's valuation on the $(n+1)$ possible outcomes. In our model, a bidder is indifferent between the bidders who he conflicts with, but our model allows multiple items in an auction.

Gomes et al. [11] gave empirical evidence that externalities exist in real-life SSA. Externalities in online advertising were investigated by [9] using a probabilistic model. CA with externalities were presented in [4,12,23], and maximizing social welfare was shown to be NP-hard. In [10] a sponsored search setting was treated where each advertiser has two valuations, one if his ad is shown exclusively and one if it is shown together with other ads. This is a special case of our model for CA with bidder conflicts and conflict values. A different line of work considered bidder-independent externalities in the click-through rates of SSA [1,22,27]. All this work considered only the unit-demand setting.

Our model of SSA with bidder conflicts has been proposed and studied before by Papadimitriou and Garcia-Molina [26]. They consider an approach based on exact optimization algorithms using ILP, implement truthfulness using VCG, and experimentally evaluate their approach with respect to running time and revenue on a dataset from the Yahoo! Webscope. However, they do not consider polynomial-time algorithms, provable approximation ratios, or extensions to CA with more general valuations.

Our work is related to approximation algorithms for weighted independent set problem, a central problem in the study of approximation algorithms and computational hardness over the past four decades. For a survey on some of the work on approximation algorithms, see, e.g., [13]; here, we just mention a number of directly related results. The problem is known to be NP-hard to approximate within a ratio of $n^{1-\epsilon}$ [16], and even in undirected Δ-regular graphs it remains hard for a ratio of $\mathcal{O}(\Delta/\log^4 \Delta)$ [2]. A trivial greedy algorithm obtains an approximation ratio of $(\Delta + 1)$ in undirected graph with maximum degree Δ. For directed graphs, which arise in our application, a simple randomized (4Δ)-approximation algorithm exists. The best-known approximation algorithms for undirected graphs with maximum degree Δ attain ratios of $\mathcal{O}((\Delta \log \log \Delta)/\log \Delta)$ [14,15]. They are based on rounding suitable SDP relaxations, and below we build on these techniques and their analysis to provide algorithms for our cases, which involve directed graphs.

More recently, the study of asymmetric and edge-weighted versions of independent set has found interest, especially in the context of secondary spectrum auctions [17–19,28], where bidders are wireless devices that strive to obtain channel access under interference constraints. In these scenarios, bidders become vertices in a conflict graph. Each channel is an item that can be given to any subset of bidders representing an independent set in the graph.

2 CA with Bidder and Item Conflicts via Lottery

In this section, we present results for CA with bidder and item conflicts. We assume that either (i) Δ is bounded and G_I is arbitrary, or (ii) Δ_I is bounded, G is arbitrary and bidders have fractionally sub-additive valuations.

Theorem 1. *Given a (maximal-in-range) deterministic α-approximation algorithm f for CA without conflicts, there exists a (truthful maximal-in-range) deterministic $(16\Delta\alpha/3)$-approximation algorithm f^c for CA with bidder and item conflicts satisfying condition (i).*

The main idea of Theorem 1 is to first generate a "good" conflict-free bidder set B^c, and then apply the blackbox algorithm f w.r.t. the bidders in B^c. Initially, each bidder is in B^c with probability $1/(2\Delta)$. Then, if there are still bidder conflicts within B^c, we remove those bidders having conflicts. Overall, each bidder is in B^c with probability of at least $1/(4\Delta)$, and this will translate into a randomized $(4\Delta\alpha)$-approximation algorithm f^c.

We derandomize the above algorithm using the standard technique of pairwise independent distributions [24]; this will lead to an increase of the approximation ratio from $4\Delta\alpha$ to $16\Delta\alpha/3$. Since no bidder can alter B^c by changing his valuation, truthfulness is preserved from f to f^c. The details of derandomization is given in the full paper [3].

For case (ii), the ideas are similar, but here we generate a conflict-free item set I^c using the same technique as in (i). We require fractionally sub-additive valuations to avoid the existence of complementary goods, which should not be deleted independently. We obtain the following result; its proof is given in the full paper [3].

Theorem 2. *For CA with bidder and item conflicts satisfying conditions (i) and (ii), given a (maximal-in-range) deterministic α-approximation algorithm f for CA without conflicts, there exists a (truthful maximal-in-range) deterministic $(16\alpha/3) \cdot \min\{\Delta, \Delta_I\}$-approximation algorithm f^c.*

3   CA with Bidder Conflicts via Cone Program Relaxation

We design an approximation algorithm via a combination of (i) an SDP for WIS problem and (ii) an LP for conflict-free CA. This yields a *cone program*, which we round its solution to yield a good allocation. Cone program (CP) is a generalization of the more well-known LP and SDP. Briefly speaking, a CP is a program that optimizes a linear function in the intersection of some hyperspaces and a *proper cone*. More about CP will be given in the full paper [3].

To the best of our knowledge, we are the first to combine this SDP and a LP in this fashion. Also, our analysis for the cone program rounding algorithm is novel, combining of the analyses for (i) and (ii) so as to get the benefits of both. We prove the following result:

Theorem 3. *For CA with bidder conflicts, suppose that the bidders have fractionally sub-additive (FSA) valuations. If there is a demand oracle for each bidder (we shall define this soon), then there exists an $\mathcal{O}\left((\Delta \log \log \Delta)/\log \Delta\right)$-approximation algorithm of social welfare that runs in $\mathsf{poly}(m, n)$-time.*

In the rest of this section, we first include some standard facts about CA without conflicts, followed by our algorithm and its analysis. We slightly abuse the notation and use S to denote a subset of I.

CA with No Conflicts. The optimal social welfare of a CA without conflicts can be represented by the program ILP-NC: maximizing $\sum_{i \in B} \sum_{S \neq \emptyset} v_i(S) \cdot x_{i,S}$ subject to three sets of constraints: (i) $\forall i \in B$, $\sum_{S \neq \emptyset} x_{i,S} \leq 1$; (ii) $\forall k \in I$, $\sum_{S \ni k} \sum_{i \in B} x_{i,S} \leq 1$; (iii) $\forall i \in B$ and $\forall S \subseteq I$, $x_{i,S} \in \{0,1\}$.

In general, solving ILP-NC is NP-hard. The usual remedy is to solve its LP relaxation LPR-NC, i.e., relaxing (iii) from $x_{i,S} \in \{0,1\}$ to $x_{i,S} \in [0,1]$, to obtain a *fractional* solution, and *round* it to an integral solution.

There are $\Omega\left(2^m n\right)$ variables in LPR-NC, but it can be solved in $\mathsf{poly}(m, n)$-time if there is a *demand oracle* for each bidder: given the prices of the items $p_1, p_2, \cdots, p_m$, the demand oracle of bidder i returns a set $S \subseteq I$ that maximizes $v_i(S) - \sum_{k \in S} p_k$. The demand oracles serve as *separation oracles* for the dual of LPR-NC, thus allow solving LPR-NC efficiently using the ellipsoid algorithm [25].

For CA without conflicts where bidders have FSA valuations, a rounding algorithm called *fair contention resolution algorithm* (FCRA) [8, Sect. 1.2] attains approximation ratio $1 - \frac{1}{e}$. In Lemma 1 below, we state the precise result on FCRA, which will be useful for our conflict setting; we need the following notation: $\forall B' \subseteq B$, let LPR(B') denote the program LPR-NC with the item set I and the bidder set restricted to B'. Given any feasible point $\{x_{i,S}\}_{i \in B', S \subseteq I}$ of LPR(B'), $\forall i \in B'$, let $L_i\left(\{x_{i,S}\}_{i \in B', S \subseteq I}\right) := \sum_{S \neq \emptyset} v_i(S) \cdot x_{i,S}$.

Lemma 1 ([8]). *Suppose the bidders in B have FSA valuations. Given any feasible point $\{x_{i,S}\}_{i \in B, S \subseteq I}$ of LPR(B), FCRA outputs a randomized allocation in which each $i \in B$ obtains expected welfare of at least $\left(1 - \frac{1}{e}\right) \cdot L_i\left(\{x_{i,S}\}_{i \in B, S \subseteq I}\right)$.*

Let FCRA $(B, \{x_{i,S}\}_{i \in B, S \subseteq I})$ denote the randomized allocation in Lemma 1. For any $B' \subseteq B$, let $\hat{x}(B')$ denote the optimal solution to LPR(B').

Algorithm. Halperin [15] designed an SDP and a rounding scheme for WIS with approximation guarantee $\mathcal{O}\left((\Delta \log \log \Delta)/\log \Delta\right)$. We conglomerate his SDP with LPR-NC for our problem, which is equivalent to solving the discrete program ICP-C below.

As the constraint (1) involves a product of variables, an LP relaxation is not admissible. As LP is a subclass of SDP, one might think that an SDP relaxation suffices. However, this is not true. If we use a "fully" SDP relaxation, each constraint $x_{i,S} \leq 1$ will be converted to a non-conic constraint in the SDP relaxation. This will introduce exponentially many dual variables, prohibiting an ellipsoid algorithm on its dual to run in poly-time.

Thus, we relax to CPR-C, a "mixture" of LP and SDP; note that in CPR-C, $w_0, w_i \in \mathbb{R}^{n+1}$. CPR-C is a CP. In the full paper [3], we show that strong duality holds between CPR-C and its dual, and we can solve the dual in $\mathsf{poly}(m, n)$-time using the ellipsoid algorithm, assuming that we have a demand oracle for each bidder. We then round the fractional solution of CPR-C as in Algorithm 1.

(ICP-C)

$$\max \sum_{i \in B} \sum_{S \neq \emptyset} v_i(S) \cdot x_{i,S}$$

subject to

$$\sum_{S \neq \emptyset} x_{i,S} \leq 1, \qquad \forall i \in B$$

$$\sum_{S \ni k} \sum_{i \in B} x_{i,S} \leq 1, \qquad \forall k \in I$$

$$\frac{1 + w_i}{2} = \sum_{S \neq \emptyset} x_{i,S}, \qquad \forall i \in B$$

$$(1 + w_i)(1 + w_j) = 0, \qquad \forall (i,j) \in E \quad (1)$$

$$w_i \in \pm 1, \qquad \forall i \in B$$

$$x_{i,S} \in \{0, 1\}, \qquad \forall i \in B, S \subseteq I.$$

(CPR-C)

$$\max \; Z := \sum_{i \in B} \sum_{S \neq \emptyset} v_i(S) \cdot x_{i,S}$$

subject to

$$\sum_{S \neq \emptyset} x_{i,S} \leq 1$$

$$\sum_{S \ni k} \sum_{i \in B} x_{i,S} \leq 1$$

$$\frac{1 + w_0 \cdot w_i}{2} = \sum_{S \neq \emptyset} x_{i,S} \qquad (2)$$

$$(w_0 + w_i) \cdot (w_0 + w_j) = 0 \qquad (3)$$

$$\|w_0\| = \|w_i\| = 1. \qquad (4)$$

$$x_{i,S} \geq 0.$$

Intuitions of the Algorithm and its Analysis. Let $(Z^*, \{x^*\}, \{w^*\})$ be the solution to CPR-C. Fo any $B' \subseteq B$, let $Z^*(B') := \sum_{i \in B'} \sum_{S \neq \emptyset} v_i(S) \cdot x_{i,S}^*$.

We partition the bidders according to the values of $1 + w_0^* \cdot w_i^*$ into three sets B_0, B_1 and B_2. Items are allocated to one of the sets; the best one is chosen. The methods of allocating items to B_2 and B_1 (Steps 3 and 4) are well motivated by Halperin's algorithm – first selecting a "good" independent subset of bidders from them, and then apply FCRA for conflict-free CA; we call this "IS-then-FCRA". In the full paper [3], we prove that $\mathcal{A}_2$ and $\mathcal{A}_1$ attains expected social welfares of at least $\left(1 - \frac{1}{e}\right) Z^*(B_2)$ and $\Omega\left(\frac{\log \Delta}{\Delta \log \log \Delta}\right) \cdot Z^*(B_1)$ respectively.

For B_0, we face two difficulties which force us to use an approach quite different from Halperin's. Firstly, Halperin's algorithm is for undirected graph while in our application the graph is directed. Secondly, we notice that the "IS-then-FCRA" approach will not work for B_0, and we ought to do the opposite – first apply FCRA by ignoring conflicts (see the next paragraph), and then resolve

---

Algorithm 1. Approximation Algorithm via Cone Program Relaxation.

1   Solve CPR-C to obtain the solution $(Z^*, \{x^*\}, \{w^*\})$.

2   Set $\tau \leftarrow \frac{3\log\log\Delta}{4\log\Delta}$, which is less than $1/2$. Partition the bidders into three sets B_0, B_1, B_2: $B_0 = \{i \mid 0 \le 1 + w_0^* \cdot w_i^* \le 2\tau\}$, $B_1 = \{i \mid 2\tau < 1 + w_0^* \cdot w_i^* \le 1\}$, $B_2 = \{i \mid 1 < 1 + w_0^* \cdot w_i^* \le 2\}$.

3   Let $J_2 = B_2$. $\mathcal{A}_2 \leftarrow \mathsf{FCRA}(J_2, \hat{x}(J_2))$.

4   For the bidders in B_1, do as follows:

   – Project all vectors in $\{w_i^* \mid i \in B_1\}$ to $(w_0^*)^\perp$, the space orthogonal to w_0^*, then normalize them. Let $\{w_i'\}$ denote the projected normal vectors. Note that $(w_0^*)^\perp$ has dimension n, so we can treat each w_i' as an n-dimensional vector.

   – Choose a random n dimensional vector $r = (r_1, r_2, \cdots, r_n)$, where each r_i follows the standard normal distribution with density function $\phi(x) = \frac{1}{\sqrt{2\pi}}e^{-x^2/2}$.

   – Let $\gamma := (1 - 2\tau)/(2 - 2\tau)$. Let $B_1' := \left\{ i \in B_1 \mid w_i' \cdot r \ge \sqrt{\frac{2\gamma}{1-\gamma}} \log\Delta \right\}$.

   – Let $J_1 := B_1' \setminus \{i \in B_1' \mid \exists j \in B_1' \text{ such that } (i,j) \in E\}$. $\mathcal{A}_1 \leftarrow \mathsf{FCRA}(J_1, \hat{x}(J_1))$.

5   For the bidders in B_0, do as follows:

   – Let $\{q_{i,S}\}_{i \in B_0, S \subseteq I}$ denote the following distribution: $\forall S \ne \emptyset$, $q_{i,S} = \frac{x_{i,S}^*}{2\tau\Delta}$, and $q_{i,\emptyset} = 1 - \sum_{S \ne \emptyset} q_{i,S}$.

   – $\{T_i\}_{i \in B_0} \leftarrow \mathsf{FCRA}\left(B_0, \{q_{i,S}\}_{i \in B_0, S \subseteq I}\right)$.

   – (Conflict handling.) Let $\mathcal{A}_0$ denote the following allocation: for each bidder $i \in B_0$, if there exists another bidder j such that $(i,j) \in E$ and $T_j \ne \emptyset$, bidder i gets nothing in $\mathcal{A}_0$; otherwise bidder i gets T_i in $\mathcal{A}_0$.

6   Return the best allocation among $\mathcal{A}_0, \mathcal{A}_1, \mathcal{A}_2$.

---

any remaining conflicts. These force us to have an analysis for B_0 quite different from Halperin's one.

We provide more intuitions for B_0. The bidders in B_0 have low values of $\frac{1+w_0^* \cdot w_i^*}{2} = \sum_{S \ne \emptyset} x_{i,S}^*$. The values $x_{i,S}^*$ are typically viewed as probability densities. Low values of $\frac{1+w_0^* \cdot w_i^*}{2}$ allow room to "expand" these densities by a factor of $1/\tau$, where $\tau < \frac{1}{2}$. However, to handle conflicts, we ought to "dwell" these densities by a factor of $1/(2\Delta)$ afterwards. Then we apply FCRA with the "expanded then dwelled" densities to obtain a sufficiently good allocation to B_0.

We show that $\mathcal{A}_0$ attains an expected social welfare of $\Omega\left(\frac{\log\Delta}{\Delta\log\log\Delta}\right) \cdot Z^*(B_0)$; a proof sketch is given below, and the complete proof is given in the full paper [3]. Finally, note that $Z^*(B_0) + Z^*(B_1) + Z^*(B_2) = Z^*$, so the best among $\mathcal{A}_0, \mathcal{A}_1, \mathcal{A}_2$ attains an expected social welfare of $\Omega\left(\frac{\log\Delta}{\Delta\log\log\Delta}\right) \cdot Z^*$.

Analysis sketch on Step 5. Observe that $2\tau\Delta > 1$ for sufficiently large Δ, so the vector q, which collects $\{q_{i,S}\}_{i \in B_0, S \subseteq I}$, is a feasible point of $\mathsf{LPR}(B_0)$.

For the analysis of this step, we need to unwind FCRA. Taking the feasible point q as input, the algorithm first selects a random set S_i for each bidder i as follows: a non-empty set S is selected with probability $q_{i,S}$, and the empty set is selected with probability $1 - \sum_{S \ne \emptyset} q_{i,S}$. Note that $S_1, S_2, \cdots, S_n$ may not be disjoint, so a *resolution scheme* is needed to generate disjoint sets $T_1, T_2, \cdots, T_n$, which are the sets stated in Step 5, while $\forall i$, $T_i \subseteq S_i$. By Lemma 1, $\mathbf{E}\left[v_i(T_i)\right]$, the expected welfare of bidder i (modulo conflicts), is at least $\left(1 - \frac{1}{e}\right)\frac{Z^*(\{i\})}{2\tau\Delta}$.

To handle conflicts, the algorithm resets the allocation of some bidders to the empty set. We will show that for each bidder $i \in B_0$, at least half of his expected welfare (modulo conflicts) is retained after conflict handling.

For every $i \in B_0$, let F_i be the event: $\forall j$ with $(i,j) \in E$, $S_j = \emptyset$, and let $\overline{F_i}$ be the complement of F_i. We will prove in the full paper [3] that for all $i \in B_0$, $\mathbf{E}\left[v_i(T_i) \mid F_i\right] \geq \mathbf{E}\left[v_i(T_i) \mid \overline{F_i}\right]$. Intuitively this inequality depicts that bidder i gets more when facing less competition from bidders he conflicts with.

Since $\mathbf{E}\left[v_i(T_i)\right] = \mathbf{E}\left[v_i(T_i) \mid F_i\right] \cdot \mathbf{Pr}\left[F_i\right] + \mathbf{E}\left[v_i(T_i) \mid \overline{F_i}\right] \cdot \mathbf{Pr}\left[\overline{F_i}\right]$, and $\mathbf{E}\left[v_i(T_i) \mid F_i\right] \geq \mathbf{E}\left[v_i(T_i) \mid \overline{F_i}\right]$, it follows that $\mathbf{E}\left[v_i(T_i) \mid F_i\right] \geq \mathbf{E}\left[v_i(T_i)\right]$.

Next, it is easy to prove that $\mathbf{Pr}\left[F_i\right] \geq 1/2 \geq \mathbf{Pr}\left[\overline{F_i}\right]$. Observe that bidder i's allocation is reset during conflict handling only if $\overline{F_i}$ holds. Hence, the expected welfare of bidder i after conflict handling is at least

$$\mathbf{E}\left[v_i(T_i) \mid F_i\right] \cdot \mathbf{Pr}\left[F_i\right] \geq \left(1 - \frac{1}{e}\right)\frac{Z^*(\{i\})}{2\tau\Delta} \cdot \frac{1}{2} = \Omega\left(\frac{\log\Delta}{\Delta\log\log\Delta}\right) \cdot Z^*(\{i\}).$$

Then the expected social welfare is at least $\sum_{i \in B_0} \Omega\left(\frac{\log\Delta}{\Delta\log\log\Delta}\right) \cdot Z^*(\{i\}) = \Omega\left(\frac{\log\Delta}{\Delta\log\log\Delta}\right) \cdot Z^*(B_0)$.

We obtain the following proposition as a notable special case.

Proposition 1. *There is a poly-time $\mathcal{O}\left((\Delta\log\log\Delta)/\log\Delta\right)$-approximation algorithm for the WIS problem in a directed graph G with out-degree at most Δ.*

By Theorem 2, we have an $\mathcal{O}\left(\min\{\Delta, \Delta_I\}\right)$-approximation algorithm for CA with bidder and item conflicts, in which bidders have FSA valuations. An interesting open problem is whether we can improve the approximation guarantee to $o\left(\min\{\Delta, \Delta_I\}\right)$. We note that if each bidder has linear valuation, the problem reduces to the WIS problem in the *tensor product* of the graphs G and G_I, which might be of independent interest. We describe a CP approach for this problem in the full paper [3], however, we do not succeed to analyze.

4   Sponsored Search with Limited Number of Slots

In this section we consider sponsored search with bidder conflicts. Some of our results extend to ordered conflicts and more general graph-based slot conflicts. In light of the application, we concentrate on the case with a small number m of slots. Note that a trivial enumeration solves the problem in time $\mathcal{O}(n^m)$. Moreover, it is unlikely that significantly faster algorithms exist that solve the problem exactly, even for $m \leq \log n$; it is W[1]-hard to decide Log-Independent-Set, i.e., given $k \leq \log n$, deciding if G has an independent set of size at least k cannot be done in time $f(k) \cdot n^c$ for constant c unless FPT = W[1] [7]. Thus, we present two approximation algorithms. The first one uses semi-definite programming and has polynomial running time for $m \in \mathcal{O}(\log n)$.

Algorithm 2. Sponsored search auction with conflicts

---

1 Assign all bidders in B independently with probability $1/2$ to set B_1 and set
   $B_2 \leftarrow B \setminus B_1$. With $v_1 \geq v_2 \geq \cdots \geq v_n$ and $h = |B_1|$ define the functions
   $\phi : [h] \rightarrow [n]$ and $\chi : [n - h] \rightarrow [n]$ such that $B_1 = \{\phi(1), \ldots, \phi(h)\}$ and
   $\phi(j) < \phi(j + 1)$ for $j \in [h - 1]$ and $B_2 = \{\chi(1), \ldots, \chi(n - h)\}$ and
   $\chi(j) < \chi(j + 1)$ for $j \in [n - h - 1]$;
2 Set $q \leftarrow 1$ with probability $\frac{1}{2}$ and set $q \leftarrow 2$ otherwise;
3 if $n - h \geq \lceil \frac{m}{4} \rceil + 1$ then $t \leftarrow \chi(\lceil \frac{m}{4} \rceil + 1)$ else $t \leftarrow \infty$ and $v_t \leftarrow 0$;
4 if $q = 1$ then
5   Set $r_1 \leftarrow v_t$; Set $B_1^1 \leftarrow \{\phi(j) | j \in [h]$ and $\phi(j) < t\}$;
6   if $t \leq m + 1$ then set $\mathcal{A}$ to the set of all subsets of B_1^1 else $\mathcal{A} \leftarrow \emptyset$
7 else
8   Set $r_2 \leftarrow v_t \cdot \frac{1}{8} R(\Delta)$; Set $B_1^2 \leftarrow \{\phi(j) | j \in [h]$ and $v_{\phi(j)} \geq r_2\}$;
9   Set $J \leftarrow$ (unweighted) independent set in B_1^2 computed by using the WIS
       algorithm (Proposition 1) giving bidders in B_1^2 in random order and with
       equal weights; $\mathcal{A} \leftarrow \{J\}$;
10 Add m bidders without conflicts and with valuation r_q to B and each set in $\mathcal{A}$;
11 For each set $A \in \mathcal{A}$ let $\mathcal{M}(A)$ define all the conflict-free matchings of bidders in
   A to slots; define $\mathcal{M} = \bigcup_{A \in \mathcal{A}} \mathcal{M}(A)$;
12 Select allocation $M' \in \arg\max_{M \in \mathcal{M}} \sum_{i \in B} v_i(M)$;
13 Every real-bidder a in B pays $p_a \leftarrow \max_{M \in \mathcal{M}} \sum_{i \in B \setminus \{a\}} (v_i(M) - v_i(M'))$;

The second one is a partial enumeration approach and runs in polynomial time
if $m \in \mathcal{O}((\log n)/(\log \max(\Delta + 1, \log n)))$.

Sponsored Search via Semidefinite Programming. We study sponsored
search with bidder conflicts and $m \in \mathcal{O}(\log n)$. We assume for simplicity that
$n \geq m \geq 6$. If $m > n$, we could add $(m - n)$ dummy bidders with valuation
zero. We assume consistent tie-breaking among bidders with the same valuation.
Recall that in this setting bidders are unit demanded, and thus we can represent
an allocation S of slots to bidders by a matching M_S in a bipartite bidder-
slot-graph. We define $v_i(M_S) = v_i(S)$ for all $i \in B$. We call a matching M_S
conflict-free if $D_i \cap S_i = \emptyset$ for all $i \in B$. Note that for every matching there
exists a conflict-free matching with the same social welfare; we simply unassign
all the slots in $\bigcup_{i \in B} D_i \cap S_i$. Furthermore, we define the expected social welfare
$SW(M) := \mathbf{E}[\sum_{i \in B} v_i(M)]$ for a (randomized) matching M. The mechanism
is presented in Algorithm 2. We will provide a proof sketch for analyzing its
approximation guarantee; the complete proof is deferred to the full paper [3].
We use the notation $R(\Delta) := \sqrt{\log \log \Delta / \log \Delta}$.

Let t, r_1, r_2, B_1^1, and B_1^2 be defined as in Algorithm 2. We show that if
the optimal conflict-free assignment of bidders to slots OPT was restricted to
a random subset OPT'' of the $t - 1$ most valuable edges, where each of those
edges is picked with probability $1/2$, then $SW(\text{OPT}'') \geq SW(\text{OPT})/16$. Thus,
it suffices to compare the performance of a mechanism with OPT''. We run two
different mechanisms, ALG_1 and ALG_2, each with probability $1/2$, and receive at
least $1/2$ of the maximum of their social welfares SW_1 and SW_2.

If ALG_1 performs very well, i.e., if $SW_1 > SW(\mathrm{OPT}'')/(\Delta R(\Delta))$, we achieve the result promised in Lemma 2. ALG_1 tries out all possibilities to find the best non-conflicting matching for bidders in B_1^1. If ALG_1 does not perform very well, we can show that OPT'' must get at least a quarter of its social welfare from bidders in $B_1^2 \setminus B_1^1$. In this case, we build an (unweighted) independent set J of all bidders in B_1^2 using the WIS algorithm described in Proposition 1, which guarantees that the number of bidders in J is at least an $\mathcal{O}(1/(\Delta R(\Delta)^2))$-fraction of the optimal number for bidders in $B_1^2 \setminus B_1^1$. As in OPT'' every bidder in $B_1^2 \setminus B_1^1$ contributes at most with valuation r_1 to $SW(\mathrm{OPT}'')$ and in ALG_2 every bidder in J contributes at least with valuation r_2 to SW_2, the overall approximation ratio of ALG_2 is $\mathcal{O}(\Delta R(\Delta)^2 \cdot r_1/r_2) = \mathcal{O}(\Delta R(\Delta))$.

Lemma 2. *The matching M' computed in Algorithm 2 is in expectation an $\mathcal{O}(\Delta \sqrt{\log \log \Delta / \log \Delta})$-approximation of the optimal social-welfare.*

We show that the mechanism runs in $\mathsf{poly}(n, \Delta)$ time for certain restrictions on the number of slots m, and it is universally truthful. The crucial idea for showing truthfulness is to prove that no bidder has an incentive to alter the set of matchings $\mathcal{M}$. Thus, even though the range of allocations $\mathcal{M}$ depends on the valuations of the bidders, no bidder has an incentive to change it.

Theorem 4. *For sponsored search with bidder conflicts and $m \in \mathcal{O}(\log n)$, Algorithm 2 is a universally-truthful mechanism that attains approximation guarantee of $\mathcal{O}(\Delta \sqrt{\log \log \Delta / \log \Delta})$. It runs in time $\mathsf{poly}(n, \Delta)$.*

Sponsored Search via Partial Enumeration. We treat a slightly more general *small-supply* case with $m \leq n/\Delta$. For this case we observe that the problem can be solved optimally in linear time when all bidders i have uniform values $v_i = v$. For non-uniform values v_i, we will strive for a truthful mechanism that solves the problem approximately but much faster than the trivial enumeration that solves the problem exactly in $\mathcal{O}(n^m)$ time. Note that there is an m-approximation algorithm that assigns slot 1 to the highest bidder, obtains value $\max_{k,i} \alpha_k \cdot v_i$, and runs in time $\mathcal{O}(n)$. Thus, we obtain the following trade-off.

Theorem 5. *In sponsored search with bidder and slot conflicts, there is a universally-truthful mechanism that yields an $\mathcal{O}(\log m)$-approximation of social welfare and runs in time $\mathcal{O}(n + (m(\Delta + 1))^m)$.*

Lemma 3. *In sponsored search with bidder conflicts, there is an $\mathcal{O}(\log m)$-approximation algorithm that runs in time $\mathcal{O}(n + (m(\Delta + 1))^m)$.*

Proof. The algorithm is extremely simple for uniform values $v_i = v$ for all $i \in B$ if $m \leq n/(\Delta + 1)$. Initially, every bidder is active. We assign slot 1 to the bidder i with smallest out-degree, label i and its all out-neighbors to be inactive. We repeat this procedure with slots $2, 3, \ldots, m$. Since $m \leq n/(\Delta + 1)$, we will be able to assign all slots in this way. This yields an optimum solution and takes time $\mathcal{O}(n)$. If the v_i are different, we apply logarithmic scaling. Let $v_{\max} = \max_{i \in B} v_i$. We consider $\lceil \log_2(2m) \rceil$ classes, where class k contains bidders i with value $v_i \in (v_{\max}/2^k, v_{\max}/2^{k-1}]$. The unclassified bidders have a value which is at most $v_i \leq v_{\max}/(2m)$. Thus, by discarding this set of bidders, we discard at most $1/2$ of the optimum value.

For the remaining bidders, we pick $k \in \{1, 2, \ldots, \lceil \log_2(2m) \rceil\}$ uniformly at random and consider $V_k = \{i \in B \mid v_i > v_{\max}/2^k\}$, the union of all bidders in classes $1, \ldots, k$. Let $n_k = |V_k|$. If $n_k/(\Delta + 1) \geq m$, then we can apply the above algorithm for identical values to V_k. Otherwise, if $n_k/(\Delta + 1) \leq m$, then $n_k \leq (\Delta+1)m$, and a complete enumeration takes time at most $\mathcal{O}((m(\Delta+1))^m)$. In either case, we obtain the optimum for V_k under the assumption that every bidder has value $v_{\max}/2^k$, and hence at least half of the value that the optimum gets from bidders in class k. In expectation over the random choice of k, this shows that we recover an $\mathcal{O}(\log m)$-fraction of the optimum.

The highest valuation can be found in time $\mathcal{O}(n)$. Computing the threshold and reducing the set of considered bidders can be done in time $\mathcal{O}(n)$. Applying the previous algorithm can be done in time $\mathcal{O}(n)$, enumeration takes time $\mathcal{O}((m(\Delta + 1))^m)$. □

Note that for a particular choice of k, the algorithm described in the proof of Lemma 3 is applied in the induced subgraph of V_k and produces an optimum solution under the assumption that all nodes have the same valuation. If this results from the greedy algorithm for the independent set of bidders, it also remains an optimum solution with arbitrary additional slot conflicts. If this results from enumeration, we can apply the enumeration also for additional slot conflicts within the same running time. Thus, we obtain the same running time and approximation ratio also for sponsored search with bidder and slot conflicts.

By the sampling arguments in [6,18] we can turn the algorithm into a universally truthful mechanism with the same asymptotic running time and approximation ratio. The idea is as follows. First, choose a random bit q. If $q = 0$, partition B into B_1 and B_2 randomly and set $v_{\max}$ be the highest valuation in B_1. However, we run the algorithm in Proposition 3 on B_2 only; if bidder $i \in B_2$ gets assigned slot ℓ he has to pay $\alpha_\ell \cdot v_{\max}/2^k$. If $q = 1$, we keep the best slot and remove all others, and run a second price auction among all bidders in B. This ensures that the claimed approximation ratio even if there is a dominant bidder, i.e., a bidder who contributes at least a constant fraction of the optimal social welfare. This completes the proof of Theorem 5.

References

1. Aggarwal, G., Feldman, J., Muthukrishnan, S.M., Pál, M.: Sponsored search auctions with Markovian users. In: Papadimitriou, C., Zhang, S. (eds.) WINE 2008. LNCS, vol. 5385, pp. 621–628. Springer, Heidelberg (2008)
2. Chan, S.O.: Approximation resistance from pairwise independent subgroups. In: 45th STOC, pp. 447–456 (2013)
3. Cheung, Y.K., Henzinger, M., Hoefer, M., Starnberger, M.: Combinatorial auctions with conflict-based externalities. CoRR abs/1509.09147 (2015). http://arxiv.org/abs/1509.09147
4. Conitzer, V., Sandholm, T.: Computing optimal outcomes under an expressive representation of settings with externalities. J. Comput. Syst. Sci. **78**(1), 2–14 (2012)

5. Cramton, P., Shoham, Y., Steinberg, R. (eds.): Combinatorial Auctions. MIT Press, Cambridge (2006)
6. Dobzinski, S., Nisan, N., Schapira, M.: Truthful randomized mechanisms for combinatorial auctions. J. Comput. Syst. Sci. **78**(1), 15–25 (2012)
7. Downey, R., Fellows, M.: Parametrized Complexity. Springer, New York (1999)
8. Feige, U., Vondrák, J.: The submodular welfare problem with demand queries. Theory Comput. **6**(1), 247–290 (2010)
9. Ghosh, A., Mahdian, M.: Externalities in online advertising. In: 17th WWW, pp. 161–168 (2008)
10. Ghosh, A., Sayedi, A.: Expressive auctions for externalities in online advertising. In: 19th WWW, pp. 371–380 (2010)
11. Gomes, R., Immorlica, N., Markakis, E.: Externalities in keyword auctions: an empirical and theoretical assessment. In: Leonardi, S. (ed.) WINE 2009. LNCS, vol. 5929, pp. 172–183. Springer, Heidelberg (2009)
12. Haghpanah, N., Immorlica, N., Mirrokni, V., Munagala, K.: Optimal auctions with positive network externalities. ACM Trans. Econ. Comput. **1**(2), 13:1–13:24 (2013)
13. Halldórsson, M.: A survey on independent set approximations. In: 1st APPROX, pp. 1–14 (1998)
14. Halldórsson, M.M.: Approximations of weighted independent set and hereditary subset problems. J. Graph Alg. Appl. **4**(1), 1–16 (2000)
15. Halperin, E.: Improved approximation algorithms for the vertex cover problem in graphs and hypergraphs. SIAM J. Comput. **31**(5), 1608–1623 (2002)
16. Håstad, J.: Clique is hard to approximate within $n^{1-\varepsilon}$. Acta Math. **182**(1), 105–142 (1999)
17. Hoefer, M., Kesselheim, T.: Secondary spectrum auctions for symmetric and submodular bidders. In: 13th EC, pp. 657–671 (2012)
18. Hoefer, M., Kesselheim, T.: Brief announcement: universally truthful secondary spectrum auctions. In: 25th SPAA, pp. 99–101 (2013)
19. Hoefer, M., Kesselheim, T., Vöcking, B.: Approximation algorithms for secondary spectrum auctions. ACM Trans. Internet Techn. **14**(2–3), 16 (2014)
20. Jehiel, P., Moldovanu, B., Stacchetti, E.: How (not) to sell nuclear weapons. Am. Econ. Rev. **86**, 814–829 (1996)
21. Jehiel, P., Moldovanu, B., Stacchetti, E.: Multidimensional mechanism design for auctions with externalities. J. Econ. Theory **85**, 258–293 (1999)
22. Kempe, D., Mahdian, M.: A cascade model for externalities in sponsored search. In: Papadimitriou, C., Zhang, S. (eds.) WINE 2008. LNCS, vol. 5385, pp. 585–596. Springer, Heidelberg (2008)
23. Krysta, P., Michalak, T.P., Sandholm, T., Wooldridge, M.: Combinatorial auctions with externalities. In: 9th AAMAS, pp. 1471–1472 (2010)
24. Luby, M., Wigderson, A.: Pairwise independence and derandomization. Found. Trends Theor. Comput. Sci. **1**(4), 237–301 (2005)
25. Nisan, N., Segal, I.: The communication requirements of efficient allocations and supporting prices. J. Econ. Theory **129**(1), 192–224 (2006)
26. Papadimitriou, P., Garcia-Molina, H.: Sponsored search auctions with conflict constraints. In: 5th WSDM, pp. 283–292 (2012)
27. Roughgarden, T., Tardos, É.: Do externalities degrade gsps efficiency. In: 8th Ad-Auctions Workshop (2012)
28. Zhou, X., Gandhi, S., Suri, S., Zheng, H.: eBay in the Sky: strategy-proof wireless spectrum auctions. In: 14th MobiCom, pp. 2–13 (2008)

Applications of α-Strongly Regular Distributions to Bayesian Auctions

Richard Cole[⊠] and Shravas Rao

Computer Science Department, Courant Institute, NYU, New York, USA
cole@cs.nyu.edu

Abstract. Two classes of distributions that are widely used in the analysis of Bayesian auctions are the Monotone Hazard Rate (MHR) and Regular distributions. They can both be characterized in terms of the rate of change of the associated virtual value functions: for MHR distributions the condition is that for values $v < v'$, $\phi(v') - \phi(v) \geq v' - v$, and for regular distributions, $\phi(v') - \phi(v) \geq 0$. Cole and Roughgarden introduced the interpolating class of α-Strongly Regular distributions (α-SR distributions for short), for which $\phi(v') - \phi(v) \geq \alpha(v' - v)$, for $0 \leq \alpha \leq 1$. In this paper, we investigate five distinct auction settings for which good expected revenue bounds are known when the bidders' valuations are given by MHR distributions. In every case, we show that these bounds degrade gracefully when extended to α-SR distributions. For four of these settings, the auction mechanism requires knowledge of these distribution(s) (in the other setting, the distributions are needed only to ensure good bounds on the expected revenue). In these cases we also investigate what happens when the distributions are known only approximately via samples, specifically how to modify the mechanisms so that they remain effective and how the expected revenue depends on the number of samples.

1 Introduction

Much of the recent computer science research on revenue-maximizing auctions uses Bayesian analysis to measure auction performance (see [5] for an overview), although there is also a considerable body of work on worst-case revenue maximization (see [6]). Typically the analyses seek to compare the revenue for the given mechanism to a measure of the optimal revenue, expressing this as an approximation factor.

In Bayesian analyses the bidders valuations are assumed to be drawn from one or more distributions, either one common distribution for all the bidders, or separate distributions for distinct groups of bidders, possibly with each bidder being in a distinct group. Almost all previous Bayesian analyses have been

R. Cole—The work of Richard Cole work was supported in part by NSF Grants CCF-1217989 and CCF-1527568.

S. Rao—The work of Shravas Rao was supported in part by the National Science Foundation Graduate Research Fellowship Program under Grant No. DGE-1342536.

E. Markakis and G. Schäfer (Eds.): WINE 2015, LNCS 9470, pp. 244–257, 2015.
DOI: 10.1007/978-3-662-48995-6_18

for one of three settings: all distributions, regular distributions, and Monotone Hazard Rate (MHR) distributions, with MHR being the more restrictive. For example, Myerson's analysis [8] of the expected revenue of the optimal auction for the sale of a single item is most natural when the buyer values are given by regular value distributions (different buyers may have values drawn from distinct distributions). Many other results, including those we will consider in this paper, are currently known only for MHR distributions, and for the most part do not extend to regular distributions.

Recently, Cole and Roughgarden [3] introduced the notion of α-Strongly Regular distributions, α-SR distributions for short; these interpolate between MHR and Regular distributions. They gave two examples of settings for which results previously shown for MHR distributions extended smoothly to α-SR distributions. However, the main focus of their work was to investigate what happens in auctions, and in Myerson's auction in particular, when distributions are known only approximately, rather than exactly, and how to analyze the resulting expected revenue as a function of the number of samples.

In this paper we carry out a more thorough investigation of α-SR distributions, and specifically to what extent known results for MHR distributions extend to α-SR distributions. We consider five auction settings, listed in Table 1. For each problem, we show that the prior result extends smoothly. In addition, for four of these problems, the auction uses knowledge of the distribution in its decision making. For these two settings, we propose variants of the auctions which allow efficiency in terms of revenue to be maintained, and we also determine how the expected revenue varies as a function of the number of samples.

The technical challenges in this work were two-fold. First, we had to extend a variety of results concerning properties of MHR distributions to α-SR distributions. While some of these results are straightforward extensions of analogous results for MHR distributions, in other cases new proofs were needed, as the previous arguments depended on convexity properties that need not hold outside the MHR domain. For the most part, once these new results were obtained, analyzing the auction revenue was simply a matter of replacing an MHR bound with the corresponding α-SR bound, as illustrated in Sect. 3.

Second, in working with samples we had to adjust some of the mechanisms to take account of the fact that they were using an approximation of the actual distributions. For example, for the result in Theorem 9, we take the apparently optimal solution based on the approximate distributions, and adjust it in a non-uniform manner; as can be seen in the full version of the paper, the resulting solution achieves an approximation factor similar to what is obtained given exact distributions.

In sum, this work strongly suggests that results that hold w.r.t. MHR distributions will often degrade gracefully when extended to α-SR distributions. The one result we did not succeed in extending was Theorem 3.14 in [4]. It would be interesting to know if there are problems for which this is provably not the case. This work also suggests that the optimal mechanism given full knowledge of the distributions may need non-trivial modifications to achieve good performance when faced with sample-based empirical distributions.

In Sect. 2 we review some standard definitions and results. In Sect. 3 we explain the approach taken to prove Theorems 1–5 and give the analysis for Theorems 1 and 2. In the full version of the paper we explore what happens when the distributions are known approximately via samples, proving Theorems 6–9

Table 1. Results for the mechanisms we analyze.

Mechanism	MHR	α-SR	with samples
VCG for downward closed revenue approx	3 [7]	$\frac{2+\alpha}{\alpha}$ (Theorem 1)	n/a
VCG-L mechanism revenue vs. VCG welfare	e [4]	$\frac{1}{\alpha^{1/(1-\alpha)}}$ (Theorem 2)	result below
k-bidders		$\frac{1}{\alpha^{1/(1-\alpha)}} \cdot \frac{(1+\gamma)^4}{1-\xi(1+\gamma)^2(1-k\delta)}$ (Theorem 6)	
Downward closed, known budgets social welfare approx	4(1 + e) [2]	$\frac{4}{\alpha} + 2\frac{\alpha+1}{\alpha^{(2-\alpha)/(1-\alpha)}}$ (Theorem 3)	result below
		$\frac{4}{\alpha} + 2\frac{\alpha+1}{\alpha^{(2-\alpha)/(1-\alpha)}(1-\epsilon)(1-k\delta)}$ (Theorem 7)	
Downward closed, private budgets, single parameter revenue approx	3(1 + e) [2]	$3\left(1 + \frac{1}{\alpha^{1/(1-\alpha)}}\right)$ (Theorem 4)	result below
k-bidders		$\frac{3}{1-k\delta}\left(1 + \frac{1}{\alpha^{1/(1-\alpha)}(1-\max\{\sqrt{8\gamma/\alpha}, 4\gamma+\xi\})}\right)$ (Theorem 8)	
Public budget, universally IC revenue approx	$192e^2$ [1]	$\frac{192}{\alpha}\left(\frac{2-\alpha}{\alpha}\right)^{1/(1-\alpha)}$ (Theorem 5)	see Theorem 9

In Table 1, Column 2 gives known results for MHR distributions expressed as an approximation factor; Column 3 gives the corresponding results for α-SR distributions, and Column 4, where applicable, the results under sampling of the distributions. δ, ξ, and γ are parameters used to specify the number m of samples and which need to satisfy $\gamma\xi m \geq 4$, $(1+\gamma)^2 \leq 3/2$, and $m \geq \frac{6(1+\gamma)}{\gamma^2\xi}\max\{\ln 3, \ln\frac{3}{\delta}\}$. Reasonable choices are $\xi = \delta$, $\delta = \gamma/k$, and $\gamma \leq 1/5$ as small as needed to give the desired approximation factor. All the sampling results assume there are k classes of bidders each with their own distribution. Note that when α tends to 1, the limit values for all the bounds in column 3, are the prior known bounds for MHR distributions.

Our goal with this work is two fold. First, we aim to show that results for MHR distributions can often be extended to α-SR distributions. Second, by providing a tool-kit of results about α-SR distributions we hope to encourage other authors to attempt to extend their MHR results to α-SR distributions.

2   Preliminaries

Recall that for a distribution F, the virtual valuation $\phi(v)$ is given by

$$\phi(v) = v - \frac{1 - F(v)}{f(v)},$$

where f is the derivative of F. Sometimes, we might define F on a discrete set $\{1, \ldots, L\}$, for some L, in which case we define the virtual valuation as

$$\phi(v) = v - \frac{1 - F(v)}{F(v) - F(v - 1)},$$

where $F(0) = 0$. Unless otherwise stated, we will assume that F is a continuous distribution. It is often useful to use the hazard rate, $h(v) = f(v)/(1 - F(v))$ (or $h(v) = (F(v) - F(v-1))/(1 - F(v))$ in the case of a discrete distribution); then $\phi(v) = v - 1/h(v)$. Note that f, F, and h are always non-negative.

Given a value v, it can be useful to refer to the quantile, $q(v) = 1 - F(v)$. Additionally, we let $v(q)$ be the value at quantile q.

Also recall that the monopoly price is the least price r such that $\phi(r) \geq 0$.

The following definition of α-SR distributions was introduced in [3].

Definition 1. *A distribution F is α-SR if for all $x < y$,*

$$\phi(y) - \phi(x) \geq \alpha(y - x).$$

Note that monotone-hazard (MHR) rate distributions are 1-SR, and regular distributions are 0-SR. If F is a continuous distribution, then Definition 1 is equivalent to stating that $\frac{d\phi}{dv} \geq \alpha$.

The following worst-case α-SR distributions, first given in [3], will be used to show that several of our results are tight:

$$F^{\alpha}(v) = 1 - \left(1 + \frac{1 - \alpha}{\alpha} v\right)^{-\frac{1}{1-\alpha}} , \quad f^{\alpha}(v) = \frac{1}{\alpha}\left(1 + \frac{1 - \alpha}{\alpha} v\right)^{-\frac{2 - \alpha}{1 - \alpha}} .$$

These distributions have power-law tails with parameter $c = 2 + \frac{\alpha}{1 - \alpha}$, i.e. $f^{\alpha}(v) = \theta(v^{-c})$ for large v.

3   Approximation Algorithms for α-SR Distributions

The versions of all of Theorems 1–5 for MHR distributions rely on various quantitative properties of MHR distributions. The new results depend on generalizing these properties to α-SR distributions; some of these extensions are quite nontrivial. We illustrate by proving Theorems 1 and 2, and their associated lemmas. The formal statements and proofs of the remaining results can be found in the full version of the paper.

3.1   Revenue of VCG with Duplicates

Theorem 1 bounds the expected revenue of Vickrey-Clarke-Groves (VCG) with duplicates as described in [7]. Recall that the VCG mechanism chooses the feasible set of bidders with the maximum total value to be the winners, and charges each bidder appropriately. With duplicates, VCG is run on the set of bidders, along with a single additional copy of each bidder, so that each bidder and its copy have independent and identical distributions on their valuations, are interchangable, and cannot both be part of the winning set of bidders. In Theorem 1, as α tends to 1, our bound on the approximation factor tends to 3, the tight bound previously achieved for MHR distributions in [7].

Theorem 1. *Let $0 < \alpha < 1$. For every downward-closed environment with valuations drawn independently from distributions that are α-SR, the expected revenue of VCG with duplicates is a $\left(\frac{2+\alpha}{\alpha}\right)$-approximation to the expected revenue of the optimal mechanism without duplicates.*

Proof. Lemma 1 below replaces Lemma 4.1 in the proof of Theorem 4.2 in [7]. The rest of the proof is unchanged.

Lemma 1. *Let $0 < \alpha < 1$ and let F be an α-SR distribution, and ϕ be its virtual valuation function. Then, for all t,*

$$\mathbb{E}_{v_1, v_2 \sim F}[\max\{v_1, v_2\} | \max\{v_1, v_2\} \geq t] \leq$$
$$\left(\frac{2+\alpha}{\alpha}\right) \mathbb{E}_{v_1, v_2 \sim F}[\max(\phi(v_1), \phi(v_2)) | \max\{v_1, v_2\} \geq t].$$

In Lemma 5 below we prove Lemma 1 for the case $t = 0$, which it turns out is when the bound is tightest, as we later show in concluding the proof of Lemma 1.

To prove Lemma 5 we will use the following structural properties of α-SR distributions F and their density functions f.

1. A lower bound on $f(q)$ (Lemma 2): for $q \leq q_0 \leq 1$, $f(q) \geq f(q_0) \left(\frac{q}{q_0}\right)^{2-\alpha}$.
2. The *single crossing property* (Lemma 3): if for some v_0, $F(v_0) > F^\alpha(v_0)$, then $F(v) \geq F^\alpha(v)$ for all $v \geq v_0$, where F^α is a tight distribution: $f(q) = f(1) \cdot q^{2-\alpha}$.
3. Lemma 4: $\frac{\alpha}{1+\alpha} \int_0^\infty (1 - F(v)) dv \leq \int_0^\infty (1 - F(v))^2 dv$.

Lemma 2. *Let F be an α-SR distribution, and let f be the density function for v. Let $q_0 \in [0, 1]$. Then for $q \leq q_0$,*

$$f(q) \geq f(q_0) \left(\frac{q}{q_0}\right)^{2-\alpha}.$$

Proof. Recall that $\phi(v) = v - q(v)/f(v)$ and $\frac{dv}{dq} = -1/f(v)$. Hence $\frac{d\phi}{dq} = \frac{-2}{f(v)} + \frac{q(v)}{f(v)^2} \frac{df}{dq}$. The condition $\frac{d\phi}{dv} \geq \alpha$ yields $\frac{d\phi}{dq} = \frac{d\phi}{dv} \frac{dv}{dq} \leq \frac{-\alpha}{f(v)}$. Thus $\frac{q(v)}{f(v)^2} \frac{df}{dq} \leq \frac{2-\alpha}{f(v)}$ or $\frac{d}{dq} \ln f \leq \frac{1}{q}(2 - \alpha)$. For $q \leq q_0$, this yields $f(q) \geq f(q_0) \left(\frac{q}{q_0}\right)^{2-\alpha}$ as desired.

Lemma 3. *Let F be an α-SR distribution, and let F^α be an α-SR distribution such that $f(q) = f(q_0)\left(\frac{q}{q_0}\right)^{2-\alpha}$ for all $q_0 \in [0,1]$ and $q \leq q_0$. If for some v_0, $F(v_0) > F^\alpha(v_0)$, then $F(v) \geq F^\alpha(v)$ for all $v \geq v_0$.*

Proof. Assume for a contradiction that the statement of the lemma does not hold. In particular, assume that there are v_2 and v_4 with $v_4 > v_2$, $F(v_2) > F^\alpha(v_2)$, but $F(v_4) < F^\alpha(v_4)$. Then there must exist v_3, with $v_2 < v_3 < v_4$, such that the function $F(v) - F^\alpha(v)$ crosses the x-axis from above at $v = v_3$. It follows that $f(v_3) - f^\alpha(v_3) < 0$, where f and f^α are the density functions, or derivatives, of F and F^α respectively.

Suppose that the function $F(v) - F^\alpha(v)$ crosses the x-axis from below at $v_1 < v_2$. If no such v_1 exists, let $v_1 = 0$. Then it follows that $f(v_1) - f^\alpha(v_1) \geq 0$. This is true even if $v_1 = 0$, as in this case, for all v in the interval $[v_1, v_2]$, $F(v) - F^\alpha(v) \geq 0$.

Let $q(v)$ be the quantile of v in F, and let $q^\alpha(v)$ be the quantile of v in F^α. Note that $q(v_1) = q^\alpha(v_1)$ and $q(v_3) = q^\alpha(v_3)$. By Lemma 2, for all $v \geq v_1$,

$$f(v) \geq f(v_1)\left(\frac{q(v)}{q(v_1)}\right)^{2-\alpha}.$$

Because $f(v_1) \geq f^\alpha(v_1)$ and $q(v_1) = q^\alpha(v_1)$, the above is bounded below by

$$f^\alpha(v_1)\left(\frac{q(v)}{q^\alpha(v_1)}\right)^{2-\alpha}.$$

On setting $v = v_3$, as $q(v_3) = q^\alpha(v_3)$, we obtain the bound

$$f(v_3) \geq f^\alpha(v_1)\left(\frac{q^\alpha(v_3)}{q^\alpha(v_1)}\right)^{2-\alpha}.$$

However, the right-hand side is equal to $f^\alpha(v_3)$, which contradicts the statement that $f(v_3) - f^\alpha(v_3) < 0$.

Lemma 4. *Let $0 < \alpha < 1$ and let F be an α-SR distribution. Then*

$$\frac{\alpha}{1+\alpha}\int_0^\infty (1 - F(v))dv \leq \int_0^\infty (1 - F(v))^2 dv.$$

Proof. We start by defining the distribution G by rescaling F's argument so that $\int_0^\infty (1 - G(v))dv = 1$. Let

$$G(v) = F\left(v \cdot \left(\int_0^\infty (1 - F(w))dw\right)^{-1}\right) = F(v \cdot \lambda)$$

where $\lambda = \left(\int_0^\infty (1 - F(w))dw\right)^{-1}$. Note that $\int_0^\infty (1 - G(v))dv = \int_0^\infty (1 - F(\lambda v))dv = \int_0^\infty (1 - F(w))\lambda dw = 1$. As G is obtained by rescaling F's argument, it is easy to see that G is also α-SR, and that

$$\frac{\int_0^\infty (1 - G(v))dv}{\int_0^\infty (1 - G(v))^2 dv} = \frac{\int_0^\infty (1 - F(v))dv}{\int_0^\infty (1 - F(v))^2 dv}.$$

Therefore proving the lemma for G implies the lemma for F.

Let G^α be defined analogously with respect to the worst case distribution F^α. A straightforward calculation shows that the distribution G^α satisfies the inequality in the lemma. Therefore it is enough to prove that

$$\frac{\int_0^\infty (1 - G(v))dv}{\int_0^\infty (1 - G(v))^2 dv} \le \frac{\int_0^\infty (1 - G^\alpha(v))dv}{\int_0^\infty (1 - G^\alpha(v))^2 dv}. \tag{1}$$

As both G and G^α are normalized so that $\int_0^\infty (1 - G(v))dv = \int_0^\infty (1 - G^\alpha(v))dv = 1$, we can show (1), and consequently the lemma, by showing that

$$\int_0^\infty (1 - G(v))^2 dv \ge \int_0^\infty (1 - G^\alpha(v))^2 dv,$$

i.e. that

$$\int_0^\infty (1 - G(v))^2 dv - \int_0^\infty (1 - G^\alpha(v))^2 dv \ge 0$$

or equivalently that

$$\int_0^\infty [(1 - G(v)) - (1 - G^\alpha(v))] \cdot [(1 - G(v)) + (1 - G^\alpha(v))]dv \ge 0.$$

We apply Lemma 3 to G and G^α. Because G^α is the normalized version of the worst case distribution, the conditions of Lemma 3 hold. It follows that there exists a v_0 such that $G(v) \ge G^\alpha(v)$ when $v \ge v_0$, and $G(v) \le G^\alpha(v)$ when $v < v_0$. (Possibly $v_0 = \infty$.)

Both $1 - G$ and $1 - G^\alpha$ are decreasing functions and hence so is $(1 - G) + (1 - G^\alpha)$. Thus,

$$\int_0^\infty [(1 - G(v)) - (1 - G^\alpha(v))] \cdot [(1 - G(v)) + (1 - G^\alpha(v))]dv$$

$$= \int_{v_0}^\infty [(1 - G(v)) - (1 - G^\alpha(v))] \cdot [(1 - G(v)) + (1 - G^\alpha(v))]dv$$

$$+ \int_0^{v_0} [(1 - G(v)) - (1 - G^\alpha(v))] \cdot [(1 - G(v)) + (1 - G^\alpha(v))]dv$$

$$\ge [(1 - G(v_0)) + (1 - G^\alpha(v_0))] \int_{v_0}^\infty [(1 - G(v)) - (1 - G^\alpha(v))]dv$$

$$\text{as } (1 - G(v)) - (1 - G^\alpha(v)) \le 0 \text{ when } v \ge v_0$$

$$+ [(1 - G(v_0)) + (1 - G^\alpha(v_0))] \int_0^{v_0} [(1 - G(v)) - (1 - G^\alpha(v))]dv$$

$$\text{as } (1 - G(v)) - (1 - G^\alpha(v)) \ge 0 \text{ when } v < v_0$$

$$= [(1 - G(v_0)) + (1 - G^\alpha(v_0))] \int_0^\infty [(1 - G(v)) - (1 - G^\alpha(v))]dv$$

$$= 0 \qquad \qquad \text{as } \int_0^\infty [(1 - G(v)) - (1 - G^\alpha(v))] = 0.$$

Lemma 5. *Let $0 < \alpha < 1$ and let F be an α-SR distribution. Then,*

$$\mathbb{E}_{v_1,v_2 \sim F}[\max\{v_1, v_2\}] \le \left(\frac{2+\alpha}{\alpha}\right) \mathbb{E}_{v_1,v_2 \sim F}[\min\{v_1, v_2\}]$$

$$= \left(\frac{2+\alpha}{\alpha}\right) \mathbb{E}_{v_1,v_2 \sim F}[\max\{\phi(v_1), \phi(v_2)\}].$$

Proof. We first note that the equality follows by Myerson's Lemma [8]. We now prove the inequality. Note that $\Pr[\max\{v_1, v_2\} \ge x] = 1 - F(x)F(x)$. Then

$$\mathbb{E}_{v_1,v_2 \sim F}[\max\{v_1, v_2\}] = \int_0^\infty x \frac{d}{dx}[F(x)^2]dx$$

$$= \int_0^\infty 1 - F(x)F(x)dx \quad \text{(on integrating by parts).}$$

Similarly, $\Pr[\min\{v_1, v_2\} \ge x] = (1 - F(x))(1 - F(x))$. Thus

$$\mathbb{E}_{v_1,v_2 \sim F}[\min\{v_1, v_2\}] = \int_0^\infty (1 - F(x))(1 - F(x))dx.$$

Therefore,

$$\frac{\mathbb{E}_{v_1,v_2 \sim F}[\max\{v_1, v_2\}]}{\mathbb{E}_{v_1,v_2 \sim F}[\min\{v_1, v_2\}]} = \frac{\int_0^\infty 1 - F(x)F(x)dx}{\int_0^\infty (1 - F(x))(1 - F(x))dx} =$$

$$\frac{\int_0^\infty 2(1 - F(x)) - (1 - F(x))^2 dx}{\int_0^\infty (1 - F(x))^2 dx} = \frac{2\int_0^\infty (1 - F(x))dx}{\int_0^\infty (1 - F(x))^2 dx} - 1.$$

By applying Lemma 4, we see this is bounded above by

$$\left(\frac{2(1+\alpha)}{\alpha} - 1\right) = \frac{2+\alpha}{\alpha} \quad \text{as desired.}$$

Proof (Lemma 1). By Lemma 5, Lemma 1 holds when $t = 0$. We now prove the result for $t > 0$.

Let $C(\alpha) = \left(\frac{2+\alpha}{\alpha}\right)$, and note that as F is regular, ϕ is increasing, and hence $\max(\phi(v_1), \phi(v_2)) = \phi(\max\{v_1, v_2\})$. Then, by substituting $\max\{v_1, v_2\} - \frac{1}{h(\max\{v_1,v_2\})}$ for $\phi(\max\{v_1, v_2\})$ in the statement of Lemma 1, we see that it is equivalent to

$$\mathbb{E}_{v_1,v_2 \sim F}\left[\left.(C(\alpha) - 1)\max\{v_1, v_2\} - \frac{C(\alpha)}{h(\max\{v_1, v_2\})}\,\right|\, \max\{v_1, v_2\} \ge t\right] \ge 0.$$

We rewrite this as

$$C(\alpha) \cdot \mathbb{E}_{v_1,v_2 \sim F}\left[(1 - \alpha)\max\{v_1, v_2\} - \frac{1}{h(\max\{v_1, v_2\})} + \right.$$

$$\left.\left.\left(\alpha - \frac{1}{C(\alpha)}\right)\max\{v_1, v_2\}\,\right|\, \max\{v_1, v_2\} \ge t\right] \ge 0.$$

As $\frac{d\phi}{dv} \ge \alpha$, $\frac{d(\phi - \alpha v)}{dv} \ge 0$, and consequently, $(1 - \alpha)\max\{v_1, v_2\} - \frac{1}{h(\max\{v_1,v_2\})}$ is always non-decreasing as a function of $\max\{v_1, v_2\}$. Additionally, we note that $1/C(\alpha) \le \alpha$. Therefore, conditioning on the event that $\max\{v_1, v_2\} \ge t$ only increases the expected value.

3.2   Revenue of the VCG-L Mechanism

The VCG-L mechanism, as defined in [4], is used in settings in which each bidder has an attribute (a classification) and for each attribute there is a corresponding known distribution from which the bidder's valuation is drawn. The VCG-L mechanism uses the reserve prices, one per bidder, as defined in Sect. 2, as follows. First, the VCG mechanism is run. Second, all bidders whose valuation is less than their reserve price are removed. Finally, each winning bidder is charged the larger of its reserve price and its VCG payment from the first step.

In [4] the expected revenue of the VCG-L mechanism was shown to achieve a $1/e$ approximation of the welfare, or efficiency, of the VCG mechanism, for MHR distributions, which is tight. In Theorem 2 we extend the analysis to α-distributions; the bound is again tight, as shown by the case of a single bidder drawn from the worst-case distribution F^α. We note that the mechanism does not achieve a constant factor approximation in the case of regular distributions [4].

Theorem 2. *For every downward-closed environment with valuations drawn independently from α-SR distributions where $0 < \alpha < 1$, the expected efficiency of the VCG mechanism is at most a $\frac{1}{\alpha^{1/(1-\alpha)}}$ fraction of the expected revenue of the VCG-L mechanism with monopoly reserves.*

Proof. Lemma 9 below replaces Lemma 3.10 in the proof of Theorem 3.11 in [4]. The rest of the proof is unchanged.

The proof of Lemma 9 uses the fact that $(\alpha + 1)/\alpha \leq \alpha^{-1/(1-\alpha)}$, shown in Lemma 6, and lower and upper bounds on the hazard rate $h(v)$, given in Lemmas 7 and 8, respectively.

Lemma 6. *For $0 < \alpha < 1$, $(\alpha + 1)/\alpha \leq \alpha^{-1/(1-\alpha)}$.*

Proof. By rearranging the terms, we see that proving the lemma is equivalent to proving that $(\alpha + 1)^{1-\alpha} \leq (1/\alpha)^\alpha$. We replace α with $1/x$, and therefore it is enough to prove that for $x > 1$,

$$\left(\frac{1}{x} + 1\right)^{1-1/x} \leq x^{1/x}.$$

Again, by rearranging terms, it is enough to show that

$$\left(\frac{x+1}{x}\right)^x = \left(1 + \frac{1}{x}\right)^x \leq 1 + x.$$

The left-hand side is at most e, and therefore the inequality is true when $x \geq e-1$.

When $x < e - 1$, using the power series expansion for the left-hand side, we can bound it by $1 + 1 + (x - 1)/(2x) = 5/2 - 1/(2x)$. The right-hand side is bounded above by $1 + x$ if and only if $3x - 1 \leq 2x^2$, which holds when $x > 1$, as desired.

In the proof of Lemma 9, and other lemmas, we often refer to the cumulative hazard rate, $H(v) = \int_0^v h(x)dx$. We can relate F and H by the following identity, which follows by differentiating $\ln(1 - F(v))$.

$$1 - F(v) = e^{-H(v)}. \tag{2}$$

The following lemma gives a lower bound on $h(v)$ which will be used in Lemma 9.

Lemma 7. *Let $0 \leq \alpha \leq 1$ and let F be an α-SR distribution with virtual valuation function ϕ. Then for all $v_1 \leq v_2$,*

$$\frac{1}{(1 - \alpha)(v_2 - v_1) + 1/h(v_1)} \leq h(v_2).$$

Proof. When $\alpha = 1$, this states that $h(v_2) \geq h(v_1)$ (for $\phi(v_2) - \phi(v_1) = v_2 - 1/h(v_2) - (v_1 - 1/h(v_1)) \geq v_2 - v_1$ in this case).

By definition, as ϕ is α-SR, $\phi(v_2) - \phi(v_1) \geq \alpha(v_2 - v_1)$. Substituting $\phi(v) = v - 1/h(v)$ yields

$$\left(v_2 - \frac{1}{h(v_2)}\right) - \left(v_1 - \frac{1}{h(v_1)}\right) \geq \alpha(v_2 - v_1)$$

i.e.

$$(1 - \alpha)(v_2 - v_1) + \frac{1}{h(v_1)} \geq \frac{1}{h(v_2)},$$

from which the desired inequality follows.

Using almost the same proof as above, we obtain the following upper bound on $h(v)$, also used in Lemma 9. This was also given in [3] for the special case of $v_2 = r$.

Lemma 8. *Let $0 \leq \alpha \leq 1$ and let F be an α-SR distribution with virtual valuation function ϕ. Then for all $v_1 \leq v_2$ such that $1/h(v_2) - (1 - \alpha)(v_2 - v_1) > 0$,*

$$h(v_1) \leq \frac{1}{1/h(v_2) - (1 - \alpha)(v_2 - v_1)}.$$

Proof. Again, when $\alpha = 1$, this states that $h(v_1) \leq h(v_2)$.

As in the proof of Lemma 7,

$$\frac{1}{h(v_1)} \geq \frac{1}{h(v_2)} - (1 - \alpha)(v_2 - v_1).$$

If $1/h(v_2) - (1 - \alpha)(v_2 - v_1) > 0$, then taking the reciprocal of both sides yields the desired inequality.

Note that for continuous distributions, the condition $1/h(v_2) - (1 - \alpha)(v_2 - v_1) > 0$ holds when $v_2 = r$, where r is the reserve price, as $1/h(r) = r$. Also note that Lemmas 7 and 8 hold in the case that F is defined on a discrete set.

We now state and prove Lemma 9

Lemma 9. *Let $0 < \alpha < 1$ and let F be an α-SR distribution, with monopoly price r and revenue function $\hat{R}$. Let $V(t)$ denote the expected welfare of a single-item auction with a posted price of t and a single bidder with valuation drawn from F. For every non-negative number $t \geq 0$,*

$$\hat{R}(\max\{t, r\}) \geq \alpha^{1/(1-\alpha)} V(t).$$

Proof. As in the proof in [4] of the corresponding lemma for MHR distributions, we split this into two cases, $t \leq r$ and $t \geq r$. In both cases, we can write the left-hand side as $s \cdot (1 - F(s)) = s \cdot e^{-H(s)}$, where $H(v) = \int_0^v h(v)$, and $s = \max\{t, r\}$.

Case 1: $t \leq r$.

We start from the fact that $V(t) \leq \int_0^\infty e^{-H(v)} dv,$ $\qquad\qquad$ (3)

as shown in Lemma 3.10 in [4] if h is non-negative, which therefore still applies for the case of α-SR distributions.

In order to upper bound $V(t)$, we start by lower bounding $H(v)$. Because $h(v)$ is always non-negative, $H(v)$ is always non-negative. When $v \leq r$, this will be the only lower bound we use. Otherwise, we lower bound $H(v)$ using the lower bound for $h(v)$ from Lemma 7 when $v \geq r$. In particular, if $v \geq r$, then

$$H(v) = \int_0^v h(v)dv = \int_0^r h(v)dv + \int_r^v h(v)dv = H(r) + \int_r^v h(v)dv$$

$$\geq H(r) + \int_r^v \frac{1}{(1-\alpha)(v-r)+r} dv \qquad\qquad \text{(by Lemma 7)}$$

$$= H(r) + \frac{1}{1-\alpha} \ln\left((1-\alpha)v + \alpha r\right)\Big|_r^v$$

$$= H(r) + \frac{1}{1-\alpha} \ln\left((1-\alpha)\frac{v}{r} + \alpha\right).$$

Therefore,

$$V(t) \leq \int_0^\infty e^{-H(v)} dv = \int_0^r e^{-H(v)} dv + \int_r^\infty e^{-H(v)} dv$$

$$\leq \int_0^r e^{-H(v)} dv + e^{-H(r)} \int_r^\infty e^{-\frac{1}{1-\alpha} \ln((1-\alpha)\frac{v}{r}+\alpha))} dv$$

$$= \int_0^r e^{-H(v)} dv + e^{-H(r)} \int_r^\infty [(1-\alpha)\frac{v}{r} + \alpha]^{-1/(1-\alpha)} dv$$

$$= \int_0^r e^{-H(v)} dv - e^{-H(r)} \frac{1-\alpha}{\alpha} \cdot \frac{r}{1-\alpha} \cdot \left(\frac{1-\alpha}{r} v + \alpha\right)^{-\alpha/(1-\alpha)}\Big|_r^\infty$$

$$= \int_0^r e^{-H(v)} dv - e^{-H(r)} \frac{r}{\alpha} \left(\frac{1-\alpha}{r} v + \alpha\right)^{-\alpha/(1-\alpha)}\Big|_r^\infty$$

$$= \int_0^r e^{-H(v)} dv + e^{-H(r)} \frac{r}{\alpha} \qquad\qquad\qquad (4)$$

We rewrite this as $\left(e^{H(r)} \int_0^r e^{-H(v)} dv + \dfrac{r}{\alpha} \right) e^{-H(r)}.$

In order to upper bound this, we consider $e^{H(r)} \int_0^r e^{-H(v)} dv$ by itself. Note that if $v \leq r$, on applying Lemma 8 with $v_1 = v$ and $v_2 = r$,

$$h(v) \leq \frac{1}{r - (1 - \alpha)(r - v)} = \frac{1}{(1 - \alpha)v + \alpha r}.$$

It follows that $H(r) - H(v) = \displaystyle\int_v^r h(v) dv \leq \int_v^r \frac{1}{(1 - \alpha)v + \alpha r} dv$

$$= \frac{1}{1 - \alpha} \ln \left((1 - \alpha)v + \alpha r \right) \Big|_v^r$$

$$= \frac{1}{1 - \alpha} \ln \left(\frac{r}{(1 - \alpha)v + \alpha r} \right).$$

Therefore, $e^{H(r)} \displaystyle\int_0^r e^{-H(v)} dv \leq \int_0^r e^{\frac{1}{1-\alpha} \ln \left(\frac{r}{(1-\alpha)v + \alpha r} \right)} dv$

$$= \int_0^r \left(\frac{(1 - \alpha)v}{r} + \alpha \right)^{-1/(1-\alpha)} dv$$

$$= -\frac{r}{\alpha} \left(\frac{1 - \alpha}{r} v + \alpha \right)^{-\alpha/(1-\alpha)} \Big|_0^r$$

$$= \frac{r}{\alpha} \left(\alpha^{-\alpha/(1-\alpha)} - 1 \right).$$

Plugging this into our bound for $V(t)$ yields

$$V(t) \leq \left(\frac{r}{\alpha} \left(\alpha^{-\alpha/(1-\alpha)} - 1 \right) + \frac{r}{\alpha} \right) e^{-H(r)}$$
$$= \alpha^{-1/(1-\alpha)} \cdot r \cdot e^{-H(r)}$$
$$= \hat{R}(r) \alpha^{-1/(1-\alpha)}, \qquad\qquad \text{(as } \hat{R}(r) = r e^{-H(r)} \text{ by (2))}$$

and as $\max\{t, r\} = r$ in this case, $\hat{R}(\max\{t, r\}) \geq \alpha^{1/(1-\alpha)} V(t)$ as desired.

Case 2: $t \geq r$.

From the proof of Lemma 3.10 in [4],

$$V(t) = e^{-H(t)} \cdot \left[t + \int_t^\infty e^{-(H(v) - H(t))} dv \right].$$

For $v \geq t$, $H(v) - H(t) = \int_t^v h(v)dv \geq \int_t^v \frac{1}{(1-\alpha)(v-r)+r} dv$

$$\text{(by Lemma 7 with } v_2 = v, v_1 = r)$$

$$= \frac{1}{1-\alpha} \ln \left((1-\alpha)v + \alpha r\right) \Big|_t^v$$

$$= \frac{1}{1-\alpha} \ln \left(\frac{(1-\alpha)v + \alpha r}{(1-\alpha)t + \alpha r}\right).$$

Therefore,

$$\int_t^\infty e^{-(H(v)-H(t))} dv \leq \int_t^\infty e^{\frac{-1}{1-\alpha} \ln\left(\frac{(1-\alpha)v+\alpha r}{(1-\alpha)t+\alpha r}\right)} dv$$

$$= \int_t^\infty \left(\frac{(1-\alpha)v + \alpha r}{(1-\alpha)t + \alpha r}\right)^{-1/(1-\alpha)} dv$$

$$= \frac{-(1-\alpha)}{\alpha} \cdot \frac{(1-\alpha)t + \alpha r}{1-\alpha} \cdot \left(\frac{(1-\alpha)v + \alpha r}{(1-\alpha)t + \alpha r}\right)^{-\alpha/(1-\alpha)} \Big|_t^\infty$$

$$= \frac{(1-\alpha)t + \alpha r}{\alpha} \leq \frac{t}{\alpha}. \tag{5}$$

It follows that $V(t) = e^{-H(t)} \cdot \left[t + \int_t^\infty e^{-(H(v)-H(t))} dv\right]$

$$\leq e^{-H(t)} \cdot \left[t + \frac{t}{\alpha}\right] = e^{-H(t)} \cdot t \cdot \left(\frac{\alpha+1}{\alpha}\right) \tag{6}$$

$$\leq e^{-H(t)} \cdot t \cdot \alpha^{-1/(1-\alpha)} \qquad \text{(by Lemma 6)}$$

$$= \hat{R}(t)\alpha^{-1/(1-\alpha)}.$$

Acknowledgements. We thank the referees for their thoughtful comments.

References

1. Bhattacharya, S., Goel, G., Gollapudi, S., Munagala, K.: Budget constrained auctions with heterogeneous items. In: STOC, pp. 379–388 (2010)
2. Chawla, S., Malec, D.L., Malekian, A.: Bayesian mechanism design for budget-constrained agents. In: EC, pp. 253–262 (2011)
3. Cole, R., Roughgarden, T.: The sample complexity of revenue maximization. In: Symposium on Theory of Computing, STOC 2014, New York, NY, USA, pp. 243–252, 31 May–03 June 2014
4. Dhangwatnotai, P., Roughgarden, T., Yan, Q.: Revenue maximization with a single sample. In: EC, pp. 129–138 (2010)
5. Hartline, J.D.: Mechanism design and approximation. Book draft (2014)

6. Hartline, J.D., Karlin, A.: Profit maximization in mechanism design. In: Nisan, N., Roughgarden, T., Tardos, É., Vazirani, V.V. (eds.) Algorithmic Game Theory, chapt. 13, pp. 331–362. Cambridge University Press, Cambridge (2007)
7. Hartline, J.D., Roughgarden, T.: Simple versus optimal mechanisms. In: EC, pp. 225–234 (2009)
8. Myerson, R.B.: Optimal auction design. Math. Oper. Res. 6(1), 58–73 (1981)

The Curse of Sequentiality in Routing Games

José Correa[1], Jasper de Jong[2], Bart de Keijzer[3],
and Marc Uetz[2]([⊠])

[1] Universidad de Chile, Santiago, Chile
correa@uchile.cl
[2] University of Twente, Enschede, The Netherlands
{j.dejong-3,m.uetz}@utwente.nl
[3] Sapienza University of Rome, Rome, Italy
dekeijzer@dis.uniroma1.it

Abstract. In the "The curse of simultaneity", Paes Leme et al. show that there are interesting classes of games for which sequential decision making and corresponding subgame perfect equilibria avoid worst case Nash equilibria, resulting in substantial improvements for the price of anarchy. This is called the sequential price of anarchy. A handful of papers have lately analysed it for various problems, yet one of the most interesting open problems was to pin down its value for linear atomic routing (also: *network congestion*) games, where the price of anarchy equals 5/2. The main contribution of this paper is the surprising result that the sequential price of anarchy is unbounded even for linear symmetric routing games, thereby showing that sequentiality can be arbitrarily worse than simultaneity for this class of games. Complementing this result we solve an open problem in the area by establishing that the (regular) price of anarchy for linear symmetric routing games equals 5/2. Additionally, we prove that in these games, even with two players, computing the outcome of a subgame perfect equilibrium is NP-hard.

1 Introduction

The concept of the price of anarchy, introduced by Koutsoupias and Papadimitriou [10], has spurred a lot of research over the past 15 years that has contributed significantly to establish the area of algorithmic game theory. Not only Nash equilibria, but also alternative equilibrium concepts have been addressed. One recent and interesting example of the latter is the *sequential price of anarchy* (*SPoA*), recently introduced by Leme et al. [14], that aims at understanding the quality of subgame perfect equilibrium outcomes of a game.

This work is partially supported by: 3TU.AMI (http://www.3tu.nl/), Dutch Mathematics Cluster DIAMANT, EU FET project MULTIPLEX no. 317532, ERC StG Project PAAl 259515, Google Research Award for Economics and Market Algorithms, EU-IRSES project EUSACOU, Millennium Nucleus Information and Coordination in Networks ICM/FIC RC130003.

E. Markakis and G. Schäfer (Eds.): WINE 2015, LNCS 9470, pp. 258–271, 2015.
DOI: 10.1007/978-3-662-48995-6_19

Similar to the price of anarchy (*PoA*) [10], the sequential price of anarchy measures the cost of decentralization. However, while the price of anarchy compares the quality of a worst case Nash equilibrium to the quality of an optimal solution, the sequential price of anarchy considers the possible outcomes of a game where players choose their strategies sequentially in some arbitrary order. It then compares the quality of the outcome of the worst possible subgame perfect equilibrium [16] to the quality of an optimal solution. Note that for games with perfect information, subgame perfect equilibria coincide with sequential equilibria as introduced by Kreps and Wilson [11]. In that sense, subgame perfect equilibria are indeed the "right" equilibrium concept for sequential routing games. It turns out that there are interesting examples of games where this notion leads to improved worst case guarantees, and in this sense avoid the "curse of simultaneity" [14] inherent in some simultaneous move games. Indeed, for a handful of games, the *SPoA* has indeed been proven to be lower than the *PoA* [8,9,14], while for others, this is not the case [1,5].

In this paper we consider one of the most basic types of congestion games, namely atomic network routing games with linear latencies. Here, the *PoA* has long been known to be equal to 5/2 [3,6], while de Jong and Uetz [9] recently showed that the *SPoA* is less than the *PoA* for a small number of players leading them to conjecture that the *SPoA* is at most 5/2. Our main result is to disprove this conjecture. We thereby establish a sharp contrast between the *PoA* and the *SPoA* in network routing games. Indeed, we prove that even in the symmetric case, i.e. when all players share the same origin and destination, the *SPoA* is not bounded by any constant and can be as large as $\Omega(\sqrt{n})$, with n being the number of players.

The crucial part of our proof is a "contingency plan of actions" for every player and every possible move of all previous players that leads to a subgame perfect outcome. This is generally very difficult, since the strategies of the players are of exponential size. We are however able to design a plan leading to an unbounded *SPoA* that can be described in a succinct manner: The core idea, that we believe may be of independent interest, is to design a master plan of actions that all players are supposed to follow, together with a punishing action that players only apply when some previous player deviates from the master plan. The main technical difficulty is to design a construction such that the punishing actions do not lead to a higher cost for the player applying it, so that subgame perfection is achieved.

To complement the previous result, we resolve an open problem posed by Bhawalkar et al. [4] about the *PoA* for symmetric atomic network routing games with linear latencies. Indeed, we prove that this equals 5/2, as it is the case for the nonsymmetric network case [3] and the symmetric case for general congestion games [6] (not necessarily networked).

Finally, we prove a number of additional results for the symmetric two player case. We start by observing that even for just two players subgame perfect equilibria are more complex than Nash equilibria. In particular, the corresponding outcome is generally not a Nash equilibrium of the simultaneous game, as opposed to the *crowding games* studied by Milchtaich [13]. Furthermore, we show that computing the outcome of a subgame perfect equilibrium is in general NP-hard. Although we know from [14] that computing subgame perfect equilibria is PSPACE-complete in general congestion games, that reduction requires a non-constant number of players. Our result shows that the problem remains at least NP-hard even when the number of players is two. To conclude, we pin down the exact sequential price of anarchy for the symmetric two player case, showing that it equals 7/5. This constitutes an improvement over the 3/2 upper bound in the more general non-symmetric case [9], but is higher than the straightforward 4/3 bound for the *PoA*.

2   Model and Notation

Throughout we consider a special case of *atomic congestion games*, namely, symmetric atomic network routing games with linear latency functions. The input of an instance $I \in \mathcal{I}$ consists of a directed graph $G = (V, E)$, with designated source and target nodes $s, t \in V$, and for each arc $e \in E$ a linear latency function with coefficient d_e. There are n players that all want to travel from s to t, so that the possible actions of all players consist of all directed (s,t)-paths in G. Note that all players have the same set of actions at their disposal, hence the term *symmetric*. We will denote by m the number of arcs $|E|$. We refer to the possible paths a player can choose the *actions* and to a vector of paths, one for each player, $A = (A_1, \ldots, A_n)$ as an *outcome* or *action profile*.

The cost of a player i for choosing a specific (s,t)-path A_i depends on the number of players on each arc on that path. Specifically, for an outcome $A = (A_1, \ldots, A_n)$, let $n_e(A) := \sum_{i=1}^{n} |A_i \cap \{e\}|$ denote the number of players using arc e, then the cost of that arc for each player using it equals $n_e(A)d_e$, and therefore the cost for player i, choosing path A_i, is defined as[1]

$$c_i(A) = \sum_{e \in A_i} d_e \cdot n_e(A) \, .$$

This induces the *social cost* $C(A) = \sum_{i=1}^{n} c_i(A)$, i.e., the sum of the costs of the players.[2]

A pure Nash equilibrium is an outcome A in which no player can decrease her costs by unilaterally deviating, i.e. switching to an action that is different from A_i. The price of anarchy *PoA* [10] measures the quality of any Nash equilibrium relative to the quality of a globally optimal allocation, *OPT*. Here *OPT* is an outcome minimizing $C(\cdot)$. More specifically, for an instance I,

$$PoA(I) = \max_{NE \in NE(I)} \frac{C(NE)}{C(OPT)}, \tag{1}$$

---

[1] Our upper bound on the *SPoA* for two players also holds with affine functions.
[2] Note that we consider a *utilitarian* social cost function. This is one of the standard models, yet different than the *egalitarian* makespan objective as studied, e.g., in [10].

where $NE(I)$ denotes the set of all Nash equilibria for instance I. The price of anarchy of a class of instances $\mathcal{I}$ is defined by $PoA(\mathcal{I}) = \sup_{I \in \mathcal{I}} PoA(I)$.

Our goal is to evaluate the quality of *subgame perfect equilibria* of an induced extensive form game that we call the *sequential* version of the game [12,16]. In the sequential game, players choose an action from the set of (s,t)-paths, but instead of doing so simultaneously, they choose their actions in an arbitrary predefined order $1, 2, \ldots, n$, so that the i-th player must choose action A_i, observing the actions of players preceding i, but of course not observing the actions of the players succeeding her.[3] A strategy S_i then specifies for player i the full contingency plan of actions she would choose for each potential choice of actions $A_{<i} := (A_1, \ldots, A_{i-1})$ chosen by her predecessors. We use $S_i(A_{<i})$ to denote the action that i plays under strategy S_i when $A_{<i}$ is the vector of actions chosen by players $1, \ldots, i - 1$. We refer to a choice of strategies $S = (S_1, \ldots, S_n)$ by each of the players as a *strategy profile*. Note the explicit distinction between action (profile) and strategy (profile). The *outcome resulting from* S is then the set of actions chosen by the players when they play according to the strategy profile S.

Subgame perfect equilibria, defined by Selten [16], are defined as strategy profiles S that induce pure Nash equilibria in any subgame of the extensive form game. In other words, a strategy profile S is a subgame perfect equilibrium if for all i and for any choice of actions $A_{<i}$ of players $1, \ldots, i - 1$, player i cannot decrease her cost by switching to an action different from $S_i(A_{<i})$, in the *subgame* where the actions of $1, \ldots, i - 1$ are fixed to $A_{<i}$, and $i + 1, \ldots, n$ play strategies $(S_{i+1}, \ldots, S_n)$. Subgame perfect equilibria reflect farsighted strategic behaviour of players that observe the state of the game and reason strategically about choices of subsequent players. Analogous to (1), the sequential price of anarchy of an instance I is defined by

$$SPoA(I) = \max_{SPE \in SPE(I)} \frac{C(SPE)}{C(OPT)}, \tag{2}$$

where $SPE(I)$ denotes the set of all outcomes of subgame perfect equilibria of instance I. The sequential price of anarchy of a class of instances $\mathcal{I}$ is defined as in [14] by $SPoA(\mathcal{I}) = \sup_{I \in \mathcal{I}} SPoA(I)$. Throughout the paper, when the class of instances is clear from the context, we write PoA and $SPoA$.

Extensive form games can be represented in a *game tree* (see Fig. 1 for an example), with the nodes on one level representing the possible states of the game that a single player can encounter, and the arcs emanating from any node representing the possible actions of that player in the given state. The nodes of the game tree are called *information sets* or *states*. We will refer to a state by a pair $(A_{<i}, i)$ where $A_{<i}$ is the choice of actions of the players $1, \ldots, i - 1$ in that state, and i is the next player who has to choose her action. Since we deal with a game with perfect information, subgame perfect equilibria can be computed by backward induction. In particular, it is known that subgame perfect

---

[3] However, since players are fully rational and fully informed, at equilibrium they anticipate the others' behavior and therefore make optimal choices anticipating the followers actions.

equilibria always exist; see e.g. [15]. Note however that, if S is a subgame perfect equilibrium, the resulting outcome A need not be a Nash equilibrium of the corresponding strategic form game, as will also be witnessed in the next section.

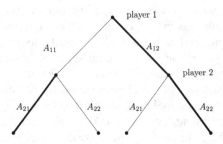

Fig. 1. Game tree for a symmetric sequential game with two players. The nodes are the states. Note that A_{11} and A_{21} are actions of players one and two respectively, but denote the same action (recall that we have a symmetric game). The same holds for A_{12} and A_{22}. Fat lines denote a subgame perfect strategy $S = (S_1, S_2)$ where $S_1 = A_{12}$, $S_2(A_{11}) = A_{21}$ and $S_2(A_{12}) = A_{22}$. The outcome resulting from S would be (A_{12}, A_{22}), i.e., the rightmost path of the game tree.

3 Warm-Up: The Two-Player Case

As a way to illustrate the difficulties behind subgame perfect equilibria in general we focus for the moment on the two player case and point out two phenomena that showcase the fundamental difference between the concept of subgame perfect equilibrium and that of Nash equilibrium.

First we give an instance in which the resulting actions of a subgame perfect equilibrium do not correspond to a Nash equilibrium. This contrasts with the case of parallel links [9], and with the so-called crowding games [13]. Based on this particular instance we additionally prove that the sequential price of anarchy for the two player case equals 7/5. This exceeds the price of anarchy (which equals 4/3), but it is smaller than the sequential price of anarchy for the asymmetric case (which equals 3/2 [9]).[4] Secondly, we show that even in the two-player case, computing the outcome of a subgame perfect equilibrium is NP-hard.

3.1 The Sequential Price of Anarchy

Consider the two-player instance depicted in Fig. 2, with five vertices and eight arcs. The vertices $1, 2, \ldots, 5$ are numbered from left to right and from top to bottom so that $s = 1$ and $t = 5$. The linear latency functions are given by the numbers next to the respective arcs.

It can be easily verified that the following is a subgame perfect equilibrium:

---

[4] In [9] a lower bound example is given for general congestion games which can be easily transformed to network routing games.

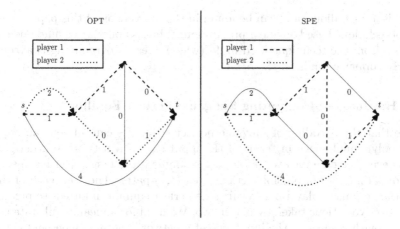

Fig. 2. Lower bound example for 2 players. Numbers are arc latencies.

- Player 1 chooses path $(1, 2, 3, 4, 5)$.
- Player 2 chooses:
    - $(1, 5)$ if player 1 chooses $(1, 2, 3, 4, 5)$,
    - $(1, 2, 4, 5)$ if player 1 chooses $(1, 2, 3, 5)$,
    - $(1, 2, 3, 5)$ if player 1 chooses $(1, 2, 4, 5)$,
    - Any (best response) path for all remaining choices of player 1.

In this equilibrium outcome player 1 chooses the dashed path on the right, that is vertices $(1, 2, 3, 4, 5)$, while player 2 chooses the dotted path on the right, which is simply the straight arc going from 1 to 5. One may think that player 1 has an incentive to deviate to the path $(1, 2, 3, 5)$ since the cost of going straight from 3 to 5 is 0. However, if player one does this, player two would pick path $(1, 2, 4, 5)$ and therefore player 1's cost would still be 3. This implies that indeed the outcome of the subgame perfect equilibrium is not a Nash equilibrium. Note furthermore that player 1's cost is 3 and player 2's is 4, for a total social cost of 7, while in the socially optimal situation, depicted to the left of the figure, the social cost is 5. So in particular this shows that the *SPoA* is at least $7/5$.

In the above, the subgame perfect equilibrium is not unique. However, the latencies can be slightly perturbed so uniqueness is achieved, while the cost of the equilibrium remains arbitrarily close to 7 and that of the optimum remains arbitrarily close to 5. To this end consider the same instance but changing the latency of the $(1, 2)$ arc of latency 2 to $2 + \epsilon$, that of the $(1, 5)$ arc from 4 to $4 + \epsilon$, and those of arcs $(2, 4)$ and $(3, 5)$ from 0 to ϵ.

With the latter observation not only the sequential price of anarchy but also the *sequential price of stability*[5] equals $7/5$ in the two-player case. This is because it is possible to prove a matching upper bound, even for the more general class of symmetric affine congestion games. The proof of this upper

---

[5] Just like the price of stability as defined in [2], the sequential price of stability is the ratio of the outcome of the best subgame perfect equilibrium over the optimum.

bound is a bit tedious, and can be found in the full version of this paper. It uses a proof technique based on linear programming, but is nonetheless fundamentally different from the technique used in [9] (where linear programming is also used to derive upper bounds on the *SPoA*).

3.2   Hardness of Computing Subgame Perfect Equilibria

Notice that the encoding of subgame perfect strategies can, in general, require super-polynomial space in terms of the input size of a network routing game. This is even the case for two players, for example if the first player has a super-polynomial number of possible actions, i.e., (s, t)-paths. Then, for each of these potential actions of player one, a subgame perfect equilibrium needs to prescribe the respective actions taken by player two. We head for a meaningful statement, however, with respect to the input size of a network routing game, and not the output. Therefore we consider the computational problem to only compute the *outcome* resulting from a subgame perfect equilibrium. This exactly corresponds to a single path in the game tree, which for two players has depth two. This has polynomial size, as it is just one path per player. The problem to compute such an equilibrium path in the game tree, however, turns out to be hard.

Theorem 1. *Computing an action profile resulting from a subgame perfect equilibrium of symmetric linear network routing games or symmetric affine congestion games is (strongly)* NP-*hard for any number of players $n \geq 2$.*

The proof is by a reduction from the Hamiltonian path problem, and is deferred to the full version of the paper. Moreover, we can also show NP-completeness of the decision problem that asks if in a two-player game, the cost of the first player is below some threshold k in a subgame perfect equilibrium.

4   The n-Player Case

Our main result is as follows.

Theorem 2. *The sequential price of anarchy of symmetric linear network routing games is not bounded by any constant.*

We prove the theorem by constructing a sequence of lower bound instances where the sequential price of anarchy gets arbitrarily large. Intuitively, the construction of these instances works as follows. To obtain a *SPoA* of x, an instance consists of x segments. In *OPT*, the majority of players chooses only a single free resource per segment, while in the worst case subgame perfect equilibrium, the majority of players form groups of $\sqrt{x}$ players who choose the same sets of $\sqrt{x}$ resources per segment. Any player who deviates from this strategy is punished by some of her successors. The tricky part of the construction is to make sure that all punishing strategies are credible. This is achieved in the following way: There are slightly more players than disjoint paths. As an effect, the last player has to necessarily share every arc in her chosen path with (at least) one other player. That will result in the situation that this player can credibly "threaten" any other player j by choosing the arcs that player j chooses, if player j does not stick to a certain action. More generally, we extend this idea so that a whole group of players can force a common predecessor into a certain action. This is achieved in such a way that the "concerted" threatening is not too expensive for every single threatener, but very expensive for the common predecessor.

Definition of Instance Γ_x. Formally, in order to obtain a sequential price of anarchy of x, where $x \geq 4$ is a square number, we construct the following instance Γ_x: Let p be a sufficiently large integer. There are $n = p\sqrt{x} + 5x^2$ players. The network consists of x segments $R_i, i \in [x]$. Segment R_i consists of $2(1 + p\sqrt{x} + 4x^2)$ nodes $\{i, (2i, 1), (2i, 2), \ldots, (2i, p\sqrt{x} + 4x^2), (2i + 1, 1), (2i + 1, 2), \ldots, (2i + 1, p\sqrt{x} + 4x^2), i + 1\}$. Note that node $i + 1$ is in both segments R_i and R_{i+1}. There is an arc with latency 0 from node i to node $(2i, j)$ for all $j \in \{1, \ldots, (p\sqrt{x} + 4x^2)\}$. There is an arc with latency $1/x$, from $(2i, j)$ to $(2i + 1, j)$ for all $j \in \{1, \ldots, (p\sqrt{x} + 4x^2)\}$. There is an arc with latency 0 from $(2i + 1, j)$ to $i + 1$ for all $j \in \{1, \ldots, (p\sqrt{x} + 4x^2)\}$. There is an arc with latency 0 from $(2i + 1, j)$ to $(2i, k)$ for all $j \in \{1, \ldots, (p\sqrt{x} + 4x^2)\}$ and for all $k \in \{j, \ldots, (p\sqrt{x} + 4x^2)\}$. Note that between any nodes $i, i + 1$, there exist $2^{p\sqrt{x}+4x^2}$ different paths: one for every subset of arcs with latency $1/x$ of segment R_i. For brevity, when we refer from now on to *arcs*, we mean the arcs of which the latency function is not identically zero, i.e., arcs with latency $1/x$.

Node 1 is the source s, and node $x+1$ is the sink t. Now any feasible action of a player consists of at least one arc from each segment $R_i, i \in [x]$. This example is shown in Fig. 3. In the remainder of the section, we say that in a state $(A_{<i}, i)$, an arc e is *free* if no player $1, \ldots, i - 1$ has chosen e in her path.

Optimal Social Cost of Γ_x. In the optimal outcome A^*, each player chooses exactly one arc from each segment, and players share arcs as little as possible. Straightforward counting based on the above definitions yields that the optimal social cost is $C(A^*) = p\sqrt{x} + 3x^2 + (2x^2)2 = p\sqrt{x} + 7x^2$.

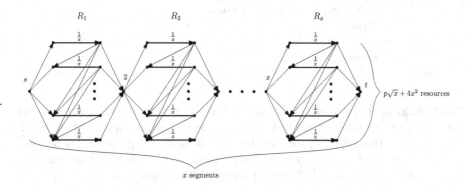

Fig. 3. A lower bound instance of a network routing game. Players travel from s to t.

Definition of Strategy Profile S **for** Γ_x. In order to describe our worst-case subgame perfect equilibrium strategy, we first define the following actions, relative to the state in which a player must choose her action:

- *Greedy*: In each segment, choose the single arc chosen by the fewest number of players. In case of ties, the tie-breaking rule as described below is used.
- *Punish(j)*: Denote by R a segment where all arcs chosen by player j are chosen by less than x players from $[j]$. Denote by e an arc from R that is chosen by the largest number of players among the arcs chosen by j (breaking ties in a consistent way). The action *Punish(j)* is then defined as choosing e in R, and any free arc in each other segment.
- *Fill*: Choose $\sqrt{x}$ free arcs in each segment.
- *Copy*: Choose exactly the same arcs as the previous player.

Note that the above actions are defined relative to a given state in the game. The actions *Greedy* and *Copy* are well-defined for each state, while the actions *Punish(j)* and *Fill* only exist for a subset of the states.

Using these actions, we now define a bad subgame perfect equilibrium $S = (S_1, \ldots, S_n)$ for Γ_x. For each state $(A_{<i}, i)$, strategy S_i prescribes to play an action $S_i(A_{<i})$, which is determined as follows.

1: **if** every player $j \in [i-1]$ plays according to S_j **then**
2:     **if** i has at least $5x^2$ successors **then**
3:         **if** i is the first player, or if the previous $\sqrt{x}-1$ players chose *Copy* **then**
4:             *Fill*
5:         **else**
6:             *Copy*
7:     **else**
8:         *Greedy*
9: **else**
10:     **if** exactly 1 player $j \in [i-1]$ does not play according to S_j **then**
11:         **if** j has chosen less than x^2 arcs in each segment **then**
12:             **if** S_j prescribed j to choose *Fill* or *Copy* **then**
13:                 **if** there exists a segment such that all arcs e chosen by j contain less than x players in total **then**
14:                     *Punish(j)*
15:                 **else**
16:                     *Greedy*
17:         **else**
18:             *Greedy*

Tie-Breaking Rule: When the strategy S_i prescribes that a player i chooses an arc chosen by the smallest number of players, and there is a set E' of arcs with this property, the following tie-breaking rule is used: All predecessors of i are ordered. The set of all players that deviate from S comes first in this ordering. After that comes the set of all other players. Within these two sets, the players are ordered by index from high to low. Now the arcs are ordered as follows: Arc e is ordered before e' iff the set of players on e is lexicographically less than the set of players on e' according to the ordering on the players just defined. Finally, ties are broken by choosing the first arc in this order, among the arcs in E'.

Example 1. As an example to clarify the tie-breaking rule, consider the following situation: Say player 5 has to choose 2 arcs among arc set $\{a, b, c, d\}$, which are chosen by the smallest number of players. Players 1 and 3 have deviated from S. Player 1 has chosen (among arcs $\{a, b, c, d\}$) arcs b and c, player 2 has chosen arcs c and d, player 3 has chosen arcs a and d, and player 4 has chosen arcs a and b. Thus, the players are ordered $3, 1, 4, 2$ and the arcs are ordered d, a, c, b, so player 5 chooses arcs a and d.   ◁

It is straightforward to see that S is a well-defined sequential strategy profile, i.e., whenever any of the actions *Greedy, Copy, Fill,* or *Punish(j)* is prescribed by S, it is possible for a player to choose this action. A detailed discussion of this is deferred to the full version of the paper.

Cost of the Outcome A of S. If each player i chooses the action prescribed by S_i, then the social cost is at least $(p\sqrt{x})(\sqrt{x}\sqrt{x}) + 3x^2 + (2x^2)2 = p\sqrt{x}x + 7x^2$. We see that $\lim_{p\to\infty} C(A)/C(A^*) = \lim_{p\to\infty}(px\sqrt{x} + 7x^2)/(p\sqrt{x} + 7x^2) = x$.

Checking that S is a Subgame Perfect Equilibrium. For a state $(A_{<i}, i)$, an action A_i is said to be *optimal with respect to a strategy profile S* iff choosing A_i minimizes i's cost when players $1, \ldots, i-1$ play $A_{<i}$, and players $i+1, \ldots n$ play according to S.

Lemma 1. *For each state $(A_{<i}, i)$ of Γ_x, action $S_i(A_{<i})$ is optimal with respect to S.*

Proof. For each of the possible actions *Greedy, Fill, Punish(j)* (where $j \in [i-1]$), and *Copy*, that S_i may prescribe to player i in state $(A_{<i}, i)$, we prove that deviating from this prescription will not decrease the cost of player i, on the assumption that all succeeding players $i+1, \ldots, n$ play according to S.

- Suppose player i is prescribed by S_i to play *Fill* or *Copy*. Then no player in $[i-1]$ has deviated from S. Therefore, (assuming that all succeeding players play according to S as well) the cost of player i when she does not deviate is x. If player i does deviate, then the subsequent players will play *Punish(i)*, which makes sure that in each segment one of the arcs chosen by i gets chosen by at least x players. Her utility will therefore be at least x. Thus, deviating is not beneficial for player i.
- Suppose player i is prescribed by S_i to play *Greedy*. Then (assuming that players $i+1, \ldots, n$ all play according to S) observe that by definition of S, players $i+1, \ldots, n$ play *Greedy*, even if player i deviates from playing *Greedy*. We denote by A^* the outcome that results if i does not deviate from S_i. We show that if i does deviate, then in each segment, i's costs at least as high as in A^*. Let $j \in [x]$ and consider segment R_j. Let e_i and e_n denote the arcs from R_i chosen by respectively player i and player n in A^*. Denote by R^* the set of arcs in R_j chosen by players $i, \ldots, n$ in A^*.

We denote by c the latency of e_n in A^*. Any arc $e \in R^*$ has latency either c/x or $(c-1)/x$. (If it were higher, then the last player who chose e would have chosen e_n, because she plays greedily.) Specifically the latency of e_i is at most c/x. Also, any arc $e \in R_i$ that is not in R^* is chosen by at least $c-1$ players of $[i-1]$. (If this were false, then in A^* player n would have chosen e instead of e_n.)

Now consider outcome A' which occurs when player i deviates from S_i. If player i chooses any arc e'_i that is not in R_*, then this arc has latency at least c/x. We now show that if e'_i is in R^*, then it has latency at least c/x as well. In that case, if any player $i' \in \{i+1, \ldots, n\}$ chooses an arc not in R^* then all arcs in R^* would yield cost at least c/x. (Because, if there would be an arc $e' \in R^*$ with cost $(c-1)/x$, then the tie breaking rule dictates that i' would have chosen e'_i instead of e'.) However, if all players $i, \ldots, n$ choose an arc in R^*, then player n has cost at least c/x. Combining this with the tie-breaking rule, we conclude that e'_i has a latency of at least c/x as well. Therefore, in all cases the costs of player n' do not decrease by deviating.

- Suppose player i is prescribed by S_i to play $Punish(j)$ for some $j \in [i-1]$. Let us compute first the cost of i if she would follow this prescription (assuming that players $i+1, \ldots, n$ all play according to S). Then observe that by definition of S, there is a number of other players succeeding i that play $Punish(j)$ as well. Let k be this number of players. So: $\{j+1, \ldots, i+k\}$ is the set of players that play $Punish(j)$. Let $\ell = |\{j+1, \ldots, i+k\}|$. Players $\{i+k+1, \ldots, n\}$ play $Greedy$, again by definition of S. Players in $[j-1]$ together occupy at most $j - 2 + \sqrt{x}$ arcs in each segment. Player j occupies at most x^2 arcs in each segment. Players $j+1, \ldots, i+k$ all choose $Punish(j)$, so they each occupy 1 arc per segment. The total number of arcs occupied per segment by players in $[i+k]$ is therefore $j - 2 + x^2 + \ell + \sqrt{x}$. Therefore, there are at least $F := (p-1)\sqrt{x} + 3x^2 - j - \ell + 2$ free arcs per segment after the first $i+k$ players have chosen their action. The set $i+k+1, \ldots, n$ is of size $G := p\sqrt{x} + 5x^2 - j - \ell$. We see that $G/F \leq 2$ so the $Greedy$ players will choose only those free arcs. (I.e., by the tie-breaking rule the $Greedy$ players will not choose arcs of player i). Therefore, player i's utility is exactly $2 - 1/x$ if she plays $Punish(j)$. (This holds because in $x - 1$ segments, i chooses 1 free arc that will not be chosen by any of her successors as we have shown. In the remaining segment, i chooses an arc that player j has chosen, which will be chosen by precisely x players.)

  Suppose next that i deviates from playing $Punish(j)$. In that case, all succeeding players will play $Greedy$. We prove that in each segment, i's costs are at least $2/x$, so that her total cost is at least 2. All players in $[j-1]$ together occupy at least $j-1$ arcs per segment. This implies that in state $(A_{<i}, i)$ in each segment there are at least $j-1$ occupied arcs and at most $p\sqrt{x} + 4x^2 - j + 1$ free arcs. The number of players succeeding i is $p\sqrt{x} + 5x^2 - i \geq p\sqrt{x} + 4x^2 - j + x$, where the inequality holds because $i \leq j + x^2 - x$ (because by the definition of S, there are at most $x^2 - x$ players choosing $Punish(j)$). Therefore, there exist players among the $Greedy$ players who choose in each segment an arc that is occupied by at least one player. The tie-breaking rule for the $Greedy$ action

then makes sure that the first such a *Greedy* player chooses in each segment an arc on which i is the sole player, in case such an arc exists. Therefore, when i deviates, her cost in each segment is at least $2/x$. □

It follows from Lemma 1 that S is a subgame perfect equilibrium of Γ_x. That concludes the proof of Theorem 2. Although the *SPoA* is not bounded by any constant, it is not hard to see that it is trivially upper bounded by the number of players n. In fact our construction shows a lower bound of $SPoA \geq \Omega(\sqrt{n})$. To see this, we choose $p = x\sqrt{x}$. Then $n = x^2 + 5x^2 = 6x^2$ which yields $x = \sqrt{n/6}$. Now, $SPoA \geq (x^3 + 7x^2)/(x^2 + 7x^2) \geq x^3/(8x^2) = x/8 = \sqrt{n}/(8\sqrt{6})$.

4.1 The Price of Anarchy

In this section we focus on the regular (i.e., non-sequential) price of anarchy of symmetric network routing games with linear latencies, and show that it equals $5/2$. This resolves an open question regarding the price of anarchy of congestion games [4]. Surprisingly, the lower bound that we provide is conceptually simpler than the one previously provided for the more general class of (non-network) affine congestion games [6].

Theorem 3. *The price of anarchy of symmetric linear and affine network routing games is* $5/2$.

Proof Sketch. It is known that the price of anarchy for affine (non-symmetric) congestion games is $5/2$ [3,6]. Thus, it suffices to prove that the price of anarchy of symmetric linear network routing games is at least $5/2$.

To this end we construct the following family of instances. For 3 players, the instance (along with the optimal and equilibrium strategies) is depicted in Fig. 4. In general, let n be the number of players and consider an instance in which there are n *principal* disjoint paths from the source s to the sink t. These paths are all composed of $2n - 1$ arcs (and thus $2n$ nodes, s being the first and t being the last), so we denote by $e_{i,j}$ the j-th arc of the i-th path, for $i = 1, \ldots, n$ and $j = 1, \ldots, 2n - 1$, and by $v_{i,j}$ the j-th node of the i-th path, for $i = 1, \ldots, n$ and $j = 1, \ldots, 2n$. There are $n \cdot (n - 1)$ additional *connecting* arcs that connect these paths: there is an arc from $v_{i,2k+1}$ to $v_{i-1,2k}$ for $k = 1, \ldots, n - 1$, where $i - 1$ is taken   mod (n). This defines the network. The latencies on the arcs are set as follows. Arcs $e_{i,1}$ (that start from s) have latency 2, arcs $e_{i,2n-1}$ (that end in t) have latency 2, while arcs $e_{i,j}$ with $1 < j < 2n - 1$ have latency 1. All connecting arcs have latency zero.

It is easy to check that the optimal solution in this instance is to route one player in each of the principal paths, as demonstrated in the top part of Fig. 4. On the other hand, a Nash equilibrium arises if each of the players make use of the connecting arcs and use only a segment of at most three consecutive arcs on each principal path, as demonstrated in the bottom part of Fig. 4. Straightforward calculations then show that this Nash equilibrium has a social cost that is $5/2$

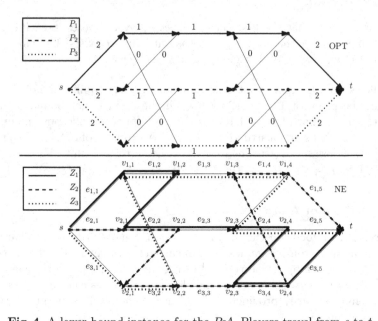

Fig. 4. A lower bound instance for the *PoA*. Players travel from s to t.

times worse than the optimal social cost, when we take n (i.e., both the number of principal paths and the number of players) to infinity.   □

5   Discussion and Open Problems

The central result of this paper states that the sequential price of anarchy is unbounded for symmetric affine network routing games. One property that stands out in our constructions is that they admit multiple subgame perfect equilibria. In fact, there even exists a subgame perfect equilibrium that induces an optimal strategy profile, and the existence of a poorly performing subgame perfect equilibrium relies crucially on tie breaking: Whenever a player is indifferent between two strategies, we essentially let the player choose the strategy that results in the worst social welfare. However, if we consider generic games, i.e., admitting a unique subgame perfect equilibrium, we do not know whether the sequential price of anarchy can be made arbitrarily high.

As for our bound on the (regular) price of anarchy: We emphasize that the existing upper bound of 5/2 for general affine congestion games holds even for coarse correlated equilibria, which contains the sets of pure, mixed, and correlated equilibria. Therefore, our last result on the price of anarchy implies that also for symmetric affine network routing games, the price of anarchy for mixed, correlated, and coarse correlated equilibria is 5/2. An open problem is to characterize the pure price of anarchy for symmetric network affine congestion games on *undirected* graphs.

Full length paper. For a version of this paper including all proofs, see [7].

Acknowledgments. We thank Mathieu Faure for stimulating discussions and particularly for pointing out a precursor of the instance depicted in Fig. 2. We thank Marco Scarsini and Victor Verdugo for discussions on the price of anarchy of the symmetric atomic network game. We thank the reviewers for some helpful comments. We also thank Éva Tardos for allowing us to (partially) recycle their paper title from [14].

References

1. Angelucci, A., Bilò, V., Flammini, M., Moscardelli, L.: On the sequential price of anarchy of isolation games. In: Du, D.-Z., Zhang, G. (eds.) COCOON 2013. LNCS, vol. 7936, pp. 17–28. Springer, Heidelberg (2013)
2. Anshelevich, E., Dasgupta, A., Kleinberg, J., Tardos, É., Wexler, T., Roughgarden, T.: The price of stability for network design with fair cost allocation. In: Proceedings of the 45th FOCS, pp. 295–304. IEEE (2004)
3. Awerbuch, B., Azar, Y., Epstein, A.: The price of routing unsplittable flow. In: Proceedings of the 37th STOC, pp. 57–66 (2005)
4. Bhawalkar, K., Gairing, M., Roughgarden, T.: Weighted congestion games: the price of anarchy, universal worst-case examples, and tightness. ACM Trans. Economics Comput. 2(4), Article 14 (2014)
5. Bilo, V., Flammini, M., Monaco, G., Moscardelli, L.: Some anomalies of farsighted strategic behavior. In: Proceedings of the 10th WAOA, pp. 229–241 (2013)
6. Christodoulou, G., Koutsoupias, E.: The price of anarchy of finite congestion games. In: Proceedings of the 37th STOC, pp. 67–73 (2005)
7. Corréa, J., de Keijzer, B., de Jong, J., Uetz, M.: The curse of sequentiality in routing games. Technical Report CTIT, University of Twente (2015). http://eprints.eemcs.utwente.nl/26301/
8. de Jong, J., Uetz, M., Wombacher, A.: Decentralized throughput scheduling. In: Spirakis, P.G., Serna, M. (eds.) CIAC 2013. LNCS, vol. 7878, pp. 134–145. Springer, Heidelberg (2013)
9. de Jong, J., Uetz, M.: The sequential price of anarchy for atomic congestion games. In: Liu, T.-Y., Qi, Q., Ye, Y. (eds.) WINE 2014. LNCS, vol. 8877, pp. 429–434. Springer, Heidelberg (2014)
10. Koutsoupias, E., Papadimitriou, C.: Worst-case equilibria. In: Meinel, C., Tison, S. (eds.) STACS 1999. LNCS, vol. 1563, pp. 404–413. Springer, Heidelberg (1999)
11. Kreps, D.M., Wilson, R.B.: Sequential equilibria. Econometrica 50, 863–894 (1982)
12. Kuhn, H.W.: Extensive games and the problem of information. Ann. Math. Stud. 28, 193–216 (1953)
13. Milchtaich, I.: Crowding games are sequentially solvable. Int. J. Game Theory 27, 501–509 (1998)
14. Leme, R.P., Syrgkanis, V., Tardos, É.: The curse of simultaneity. In: Proceedings of the 3rd ITCS, pp. 60–67 (2012)
15. Osborne, M.J.: An Introduction to Game Theory. Oxford University Press, Oxford (2003)
16. Selten, R.: Spieltheoretische Behandlung eines Oligopolmodells mit Nachfrageträgheit: Teil 1: Bestimmung des Dynamischen Preisgleichgewichts. Zeitschrift für die gesamte Staatswissenschaft 121(2), 301–324 (1965)

Adaptive Rumor Spreading

José Correa[1], Marcos Kiwi[2]([✉]), Neil Olver[3],
and Alberto Vera[1]

[1] Department of Industrial Engineering, Universidad de Chile, Santiago, Chile
[2] Department of Mathematical Engineering and Center for Mathematical Modelling,
Universidad de Chile, Santiago, Chile
mkiwi@dim.uchile.cl
[3] Department of Econometrics and Operations Research, Vrije Universiteit
Amsterdam; and CWI, Amsterdam, The Netherlands

Abstract. Motivated by the recent emergence of the so-called opportunistic communication networks, we consider the issue of adaptivity in the most basic continuous time (asynchronous) rumor spreading process. In our setting a rumor has to be spread to a population; the service provider can push it at any time to any node in the network and has unit cost for doing this. On the other hand, as usual in rumor spreading, nodes share the rumor upon meeting and this imposes no cost on the service provider. Rather than fixing a budget on the number of pushes, we consider the cost version of the problem with a fixed deadline and ask for a minimum cost strategy that spreads the rumor to every node. A non-adaptive strategy can only intervene at the beginning and at the end, while an adaptive strategy has full knowledge and intervention capabilities. Our main result is that in the homogeneous case (where every pair of nodes randomly meet at the same rate) the benefit of adaptivity is bounded by a constant. This requires a subtle analysis of the underlying random process that is of interest in its own right.

1 Introduction

A basic question in the study of social networks concerns the diffusion of a rumor, which may refer to adopting a new technology, updating content on a cell phone, or buying a new product or service. In this setting we are given a network in which vertices represent agents and edges represent social links. Initially, a single agent knows the rumor and we would like to estimate the time by which the full network is informed. The flow of information is governed by a certain stochastic process which may evolve in discrete or continuous time. The most widely studied discrete time models are the push model and the pull model; in the former at each time step a vertex knowing the rumor pushes it to a random neighbor, while in the latter a vertex not knowing pulls the rumor from

Supported by the Millennium Nucleus Information and Coordination in Networks ICM/FIC RC130003, CONICYT via Basal in Applied Mathematics, and an NWO Veni grant.

E. Markakis and G. Schäfer (Eds.): WINE 2015, LNCS 9470, pp. 272–285, 2015.
DOI: 10.1007/978-3-662-48995-6_20

a random neighbor. On the other hand, in the continuous time (asynchronous) model, every pair of connected vertices meet at random times following a Poisson process. The latter model was first formulated by Boyd et al. [8] as a way around the unrealistic time-synchronization issue implicit in discrete time models.

The diffusion of information through a social network has also posed a number of fundamental algorithmic questions, particularly in so-called *viral marketing campaigns*. The study of how the initial selection of vertices (who adopt a new product or gets it for free) influences further adoption through a cascading effect was pioneered by Domingos and Richardson [14], and rigorously addressed by Kempe et al. [20], who designed approximation algorithms for the influence maximization problem subject to a budget constraint on the number of initial nodes to which the rumor is pushed. Interestingly, these viral marketing ideas have permeated not only the technological industry, but also more traditional markets like automotive ones [4]. Unlike the rumor spreading process described above, Kempe et al. [20] consider "static" diffusion models, particularly the *independent cascade model*, in which the spread of the rumor is probabilistic, but time plays no role. An alternative approach, which we take in this paper (see also [17]), is to keep the standard rumor spreading process, but rather than fixing a budget on the initial set of selected vertices, fix a time horizon and only account for the vertices that receive the rumor within this time.

When working with a dynamic diffusion model a new problem pops up, namely that of *adaptivity*. Already Domingos and Richardson [14] identify this issue and state that: *A more sophisticated alternative would be to plan a marketing strategy by explicitly simulating the sequential adoption of a product by customers given different interventions at different times, and adapting the strategy as new data on customer response arrives.* Along these lines, Seeman and Singer [24] consider a two stage extension of the Kempe et al. model.

The central concern of this paper is that of adaptivity when speeding up rumor spreading on a social network. More precisely, we are given a network and a fixed deadline. As stated above, the diffusion model is the standard for asynchronous rumor spreading. Thus, every pair of connected vertices meet at random times following a Poisson process, and the rumor is spread whenever, upon meeting, one vertex knows the rumor and the other one does not. Along the way we are able to *push* the rumor to any vertex in order to speed up the diffusion process. We consider both the profit maximization and the cost minimization versions of the problem, which are equivalent from an optimization viewpoint. When maximizing profit we get zero profit for every vertex to which the rumor is pushed (say because we are giving the product for free) and get unit profit whenever a vertex gets the rumor through the diffusion process. The objective is thus to maximize the number of vertices that got the rumor through the diffusion process within the time horizon. The cost minimization version is exactly the opposite; every time we push the rumor we make a unit payment and if a vertex gets the rumor through the diffusion process this cost is not realized. The goal is to minimize the total payment made by the time horizon subject to the constraint that everyone should be informed by then.

A non-adaptive policy does not track the evolution of the process and therefore can only push the rumor at the starting time (and also at the deadline in the cost minimization version). In contrast, an adaptive policy may monitor the evolution of the diffusion process and intervene by pushing the rumor to additional vertices. The main contribution of this paper is to show that the advantage of adaptivity is small (in terms of cost or profit) in the setting of *homogeneous* networks, where interactions occur at the same rate between any pair of nodes. While the homogeneous case seems unrealistic, it is already highly nontrivial and we believe it will be a useful first step towards tackling more general situations.

The seemingly less natural cost minimization version of our problem actually constitutes our main motivation and finds its roots in opportunistic communication networks. The widespread adoption of networked mobile devices and the deployment of new technologies (3G, 4G), through which ever increasing data intensive services can be delivered, has generated an explosion of mobile data traffic. This trend is likely to continue, thus exacerbating current cellular network data overload [12]. Therefore, it is critical for operators and service providers to design networks and communication mechanisms that can not only handle the current traffic overload, but also allow rapid data dissemination that will be required by next-generation mobile-enabled devices and applications. A promising, less than a decade old, proposal to address the cellular network data overload consists in offloading traffic through so-called opportunistic communications. The key idea is for service providers to push mobile application content to a small subset of interested users through the cellular network and let them opportunistically spread the content to other interested users upon meeting them. Opportunistic communication can occur when mobile device users are (temporarily) in each others proximity, making it possible for their devices to establish local peer-to-peer connections (e.g., via Wifi or Bluetooth). Opportunistic communication based services have been proposed across several domains. For instance, studies have been done in the dissemination of dynamic content such as news using real world data sets, as well as that of traffic update information using dataset of the municipality of Bologna [19,25]. In several of the aforementioned application scenarios, the usefulness and/or relevance of the disseminated content crucially depends on it being opportunely delivered. Moreover, quality of service contractual obligations in subscription based contexts might entail strict deadlines for the delivery of data. Another key issue that arises in this context is the presumed feedback capabilities of network nodes and the service providers' knowledge of how data has propagated through the network up to a given moment of time.

Whitbeck et al. [25] were the first to study a fixed deadline scenario. They propose a *Push-and-Track* framework, where a subset of users receive the content from the infrastructure and start disseminating it epidemically. The main feature of *Push-and-Track* is the closed control loop, this supervises the injection of copies of the content via the infrastructure whenever it estimates that the ad hoc mode alone will fail to achieve full dissemination within the target horizon. Upon reaching the deadline, the system enters into a "panic zone" and pushes the content to all nodes that have not yet received it. Sciancalepore et al. [23]

initiate a more rigorous analysis of *Push-and-Track* type proposals. In particular, they derive formulas (although not explicit algebraic expressions) for the optimal number of nodes to initially push data in order to minimize the overall number of pushes. Furthermore, they propose a control theory based adaptive heuristic.

Our work is thus motivated by the natural question left open by Sciancalepore et al. [23]: whether or not an adaptive strategy, that harnesses the accrued information of how data has propagated through the network up to any given instant, can actually outperform an optimal non-adaptive strategy, and to what extent.

Model and Main Contributions. Consider a network of n nodes labeled by the elements of $[n] := \{1, \ldots, n\}$. Nodes are presumed mobile and such that the encounters of any two nodes i and j, $i \neq j$, are governed by a Poisson process of rate $\lambda_{i,j}$. Thus the time elapsed between two consecutive encounters of i and j is distributed as an exponential random variable of rate $\lambda_{i,j}$, henceforth denoted by $\mathrm{Exp}(\lambda_{i,j})$. All these random variables are independent, including those associated to distinct inter-encounter intervals for the same pair of nodes. As usual, if upon encountering each other one node is informed (i.e., is active) and the other is not, then the information is spread. We refer to the case where all the rates are identical as the *homogeneous* case; our main positive result will be for this setting.

We assume that there is a service provider who wishes to cost efficiently disseminate one unit of information to all nodes within a *deadline* of time $\tau > 0$. The set of nodes that posses the unit of information at time t will be denoted $S(t) \subseteq [n]$ and referred to as the set of *active notes at time* t. Initially (at time $t = 0$), the service provider selects a set of nodes $S(0)$ and *activates* them by pushing to them the unit of data of interest. Subsequently, nodes become active by either one of the following two mechanisms:

- Opportunistic communication: If nodes $i \neq j$ encounter each other at time t and either i or j belong to $S(t^-)$, say j, then i becomes informed at time t, i.e., i belongs to $S(t')$ for all $t' \geq t$. Here we used the convention $S(t^-) := \cup_{0 \leq t' < t} S(t')$. When a node becomes informed via opportunistic communication, it signals the service provider that his state has changed.
- Pushes: Because of the network's feedback capabilities, at any time $0 \leq t \leq \tau$, the service provider has full knowledge of the evolution of the set of active nodes, i.e., of $(S(t'))_{0 \leq t' < t}$, and based on this knowledge she decides whether or not, and which nodes to activate. Formally, at time t a set of nodes $C(t) \subseteq [n] \setminus S(t^-)$ is chosen and added to $S(t)$, in which case we say that at time t the nodes in $C(t)$ are activated and $|C(t)|$ pushes performed.

For most of the paper we deal with the cost minimization version of the problem, in which the service provider incurs a unit cost for activating a single node, independent of the nodes label and the time it happens. When the deadline is reached, all nodes not in $S(\tau)$ must be activated at a total cost of $n - |S(\tau)|$. Note that this is equivalent, from an optimization perspective, to the maximization problem where the service provider gets unit profit for each node informed via opportunistic communication and zero profit for the pushes she makes. We say that the service provider's strategy is *non-adaptive* if it can only activate nodes at time $t = 0$ and $t = \tau$. Otherwise, we say that its strategy is *adaptive*.

Of course, the cost of an optimal adaptive strategy is at most that of a non-adaptive strategy that initially activates an optimal number of nodes. A natural question is thus to determine the *adaptivity gap*, defined as the ratio between the cost of an optimal non-adaptive strategy and that of an optimal adaptive strategy. This question is certainly of practical significance – if the adaptivity gap turns out to be close to 1 for realistic ranges of the relevant parameters (n, τ, and $\lambda_{i,j}$'s), then at least from a purely cost effective point of view there is no justification for incurring the overhead of relying in the network's feedback capabilities, nor the extra cost required to implement a more computational demanding adaptive on-line strategy. Our main result is the following.

Theorem 1. *In the homogeneous case, i.e., $\lambda_{i,j} = \lambda$ for all $i, j \in [n]$, the adaptivity gap is bounded by a small constant, irrespective of n and the deadline.*

We also show that the adaptivity gap with respect to the profit objective, defined in the obvious way, is at most $(1 + o(1))$.

From a technical viewpoint, our analysis turns out to be significantly different for small, intermediate and large values of τ.

- For sufficiently large values of τ (say $\tau \geq \frac{1}{\lambda n}(2 + o(1)) \log n$), activating a single node initially will cause, with high probability, the entire network to be active by the deadline. This follows from classical work on stochastic epidemic models (for an overview, see [16] and [1]). So the optimal nonadaptive policy pays essentially 1, and the advantage of adaptivity is negligible.
- For the case of small τ we use a coupling argument to formalize the intuition that the process is "too deterministic" for adaptivity to win much. This already implies a $(1 + o(1))$ bound on the adaptivity gap of the profit maximization version of the problem.
- The case of intermediate values of τ is by far the most challenging. Unlike in the case of small τ, the number of nodes initially activated by an optimal nonadaptive policy is relatively small. The implication of this is that the behaviour of the process is initially *not* very concentrated, moreover, fluctuations in the rumor spreading behaviour in this initial phase can have a large impact on the cost. Since the optimal adaptive policy can be rather complicated, we consider a relaxation of an adaptive strategy which may push for free when certain conditions (which are always satisfied by an optimal adaptive strategy) are met; understanding the optimal behaviour of this relaxation turns out to be more tractable. The analysis then involves understanding an underlying martingale accounting for the expected final cost given the current situation.

The result holds also in the synchronous setting with the push protocol, in fact the argument is substantially easier than for our asynchronous setting. The reason for this is that the spreading process is much more predictable, even in the initial phase where few nodes are active. For example, the time required to activate all nodes starting from a single active node is very tightly concentrated [13], whereas, in the asynchronous case, even the time needed to go from one to two active nodes has substantial variance. As such, the result follows along the same lines as the small τ case in the asynchronous model. We defer further discussion on the synchronous model to the full version of this paper.

More General Networks. The adaptivity gap cannot be bounded by a constant in the general inhomogeneous setting. It can be shown that taking a 2-level, k-regular tree, with unit rates on the tree and all other rates 0, and a deadline of $k \log(k \log k)$, yields an adaptivity gap of $\Omega(\log k / \log \log k)$. A slight variant of this construction, with higher rates on the edges adjacent to the leaves, yields an adaptivity gap of $\Omega(\sqrt{n})$. Despite this, there remains a large scope for better understanding what network features affect the adaptivity gap. In particular, we leave bounds on the adaptivity gaps in the following settings as open questions.

- **Good expansion:** communicating pairs are described by a graph with good expansion, and all communicating pairs interact at the same rate. The lower bound constructions crucially exploit very poor connectivity.
- **Metric constraints:** the inverse rate $\lambda_{i,j}^{-1}$ describing the expected time between interactions between i and j satisfy the triangle inequality. This captures the natural idea that if i and j are frequently in the same vicinity, and likewise for j and k, then i and k are likely also frequently nearby.

.**Further Related Work.** The existing literature on rumor spreading is vast, particularly in the discrete time (synchronous) model. The natural problem here is to estimate the time at which every node in the network has the rumor. This question is quite well understood and extremely precise estimates are known when the network is a complete graph [13]. These estimates state that the time is highly concentrated around a logarithmic function of n, depending on the specific protocol. The arguably more realistic continuous time (asynchronous) model is not as well understood [8]. This is possibly due to inherent additional randomness of this process, particularly in the beginning, although logarithmic estimates for the expected time to activate the whole network, starting with one node, have been obtained for various classes of graphs [6].

Viral marketing is also an area of much interest, where models for the diffusion of information have received a lot of attention [22]. Closest to our work is the influence maximization problem, in which the goal is to find a subset, of at most k nodes, maximizing the total final number of informed nodes. The most studied underlying diffusion model is that of *independent cascades*: when node v becomes informed it has a single chance of informing each currently uninformed neighbor w and succeeds with probability p_{vw}. This problem was studied by Kempe et al. [20], who showed that the underlying optimization problem is a monotone submodular maximization problem, and therefore can be approximated efficiently within a factor of $1 - 1/e$. A long list of follow-up papers have studied the problem (see e.g. [7,9,10,21]) as well as several variations (see e.g. [2,11,15,18,24]).

Note. Due to lack of space, proofs are omitted from this extended abstract.

2 Preliminaries

In this section we further specify the model and the notation we will work with. While introducing the model we try to build some intuition and elicit how it behaves. We also establish some basic facts, which both capture some of the aforementioned intuition and will be needed in subsequent sections.

Recall that our study concerns the homogeneous case, i.e., when the rate $\lambda_{i,j}$ is a fixed value independent of the pair of nodes $i \neq j$. Moreover, everything is invariant if the rates and the deadline τ are both scaled by the same amount, so we assume $\lambda_{i,j} = 1/n$ for all $i \neq j$.

Because of symmetry considerations, the specific labels of active nodes is irrelevant and only their number at any given time matters. We henceforth denote by $K(t)$ the number of active nodes at time t for a non-adaptive scheme. Observe that $K(\cdot)$ is right continuous (i.e., $K(t^-) \leq K(t) = K(t^+)$). Also, define

$$u_k(t) := \mathbb{E}(n - K(\tau)|K(t) = k),$$

i.e., the expected number of pushes to be made at the end of the process given $K(t) = k$. We will need some information about the optimal non-adaptive choice $k_N := k_N(\tau)$ for the number of initially active nodes, i.e., the value of $k \in [n-1]$ that minimizes $k + u_k(0)$. For small values of n, one can compute k_N and $u_{k_N}(0)$ explicitly. To do so, it is convenient to consider the elapsed time between the i-th and $(i+1)$-th node activation, henceforth denoted X_i. Since X_i is the minimum of $i(n-i)$ random variables distributed according to $\text{Exp}(1/n)$, well known facts imply that X_i is distributed as $\text{Exp}(\lambda_i)$ for $\lambda_i := i(n-i)/n$.

Analogously, let $K^*(t)$ be the number of active nodes at time t for an *optimal* adaptive scheme, still assuming an explicit deadline τ. Since exponentially distributed random variables are memoryless, the optimal adaptive scheme is completely determined by a sequence $0 \leq t_0^* \leq \ldots \leq t_{n-1}^* \leq \tau$ (depending on τ), so if at time $t \in \{t_0^*, \ldots, t_{n-1}^*\}$ it holds that $K^*(t^-) \leq k$, where k is the largest index for which $t_k^* = t$, then the optimal scheme makes $k + 1 - K^*(t^-)$ pushes at time t. We interpret t_k^* as the first time when it is optimal to push more than k rumors. Let $P(t)$ denote the number of pushes performed by the optimal scheme up to (and including) time t, but excluding pushes made at the deadline τ (so $P(\tau) = P(\tau^-)$). Clearly, the cost of an optimal adaptive scheme must equal $\mathbb{E}(n - K^*(\tau^-) + P(\tau^-))$. Hence, on average $\mathbb{E}(K^*(\tau^-) - P(\tau^-))$ nodes are activated via opportunistic communication. ,

We can now start formally stating results that will be useful later on. Our first claim is that an optimal adaptive scheme will not perform pushes once roughly half the network's nodes become active. The intuition is that making a push when $i \geq n/2$ nodes are active *reduces*, to something less than λ_i, the rate at which nodes become activated (implying higher expected time between successive node activations). Thus, in expectation, there will be less than 1 more active node at time τ^-, a saving that is less than the cost of the push.

Proposition 1. *Optimal non-adaptive never starts with more than $\lceil (n-1)/2 \rceil$ active nodes. Furthermore, an optimal adaptive strategy never pushes at some time t if $K^*(t) \geq \lfloor (n-1)/2 \rfloor$, i.e., $t_k^* = \tau$ for all $k \geq \lfloor (n-1)/2 \rfloor$.*

We can think of the optimal adaptive scheme as having a target minimum number of active nodes that depends only on the current time t. Our next result essentially says that this target is not larger than $k_N(\tau - t)$, the number of initial pushes for the optimal non-adaptive strategy with deadline $\tau - t$.

Proposition 2. *Let $k \in [n-1]$ and $0 \leq t < \tau$. If $u_k(t) - u_{k+1}(t) < 1$, then $t_k^* > t$, i.e., adaptive will not push at time t if $K^*(t^-) \geq k$.*

To prove these two propositions, we need some information about the optimal adaptive strategy. For this purpose, it is useful to consider $u_k^*(t)$, defined as the expected cost incurred by an optimal adaptive scheme $K^*(\cdot)$ in the remaining time, conditioned on $K^*(t^-) = k$. By exploiting certain recurrences involving the u_k^*'s and their derivatives, we are able to show that $u_k^*(t) - u_{k+1}^*(t) \leq u_k(t) - u_{k+1}(t)$ for all $k \in [n-1], 0 \leq t \leq t_k^*$. Intuitively, this is explained by the enhanced control an adaptive scheme has over the underlying process, since it could choose to push immediately after time t, hence the benefit of being given this extra active node for free at time t is not more than one. From this, the above propositions follow fairly easily.

3 Estimates on the Evolution of the Non-adaptive Process

In this section we give a number of useful estimates on the evolution of the non-adaptive process, as well as characterize the optimal non-adaptive strategy and its cost.

Proposition 3. *If $t \in [0, \tau]$ and $k \in [n-1]$, then $u_k(t) = \dfrac{(1 + o(1))n}{1 + \frac{k}{n-k} \cdot e^{\tau - t}} + o(1)$.*

This result is essentially well-known (see e.g., [3,5]), so we only briefly sketch its proof. The evolution of the process starting from (say) $n/\log n$ active nodes, and all the way until all but $n/\log n$ nodes are active, is highly concentrated. With very high probability, it closely follows the solution of the deterministic differential equation $\frac{dx}{dt} = x(1 - x)$, where $x(t)$ denotes the proportion of active nodes at time t. This yields the logistic curve of the above proposition. When there are very few active nodes, $\lambda_i \approx i$, and the process is well approximated by a linear birth process, for which exact analytic results are available. A similar approximation holds when there are very few inactive nodes; stitching together these estimates yields Proposition 3.

We need some more refined estimates on how $u_k(t)$ varies with k and t. These do not follow from Proposition 3, but notice that they would follow immediately if $u_k(t)$ was *exactly* described by the logistic curve.

Lemma 1. *For all $k < n/2$, $u_k(t + h) \le u_k(t)e^h$ for all $h \le \tau - t$. Also, if $\tau - t = \omega(1)$, then $u_{k+1}(t) = u_k(t)\big(1 - \frac{1+o(1)}{\lambda_k}\big)$.*

The expected cost of a non-adaptive strategy starting with k pushes is $k + u_k(0)$. This cost is in fact a convex function of k. Again, this would follow immediately if $u_k(t)$ was precisely described by the logistic formula.

Lemma 2. *For every $t \in [0, \tau]$ the sequence $\{k + u_k(t)\}_{k \in [n]}$ is convex. As a consequence, k_N can be taken to be the smallest k such that $u_k(0) - u_{k+1}(0) < 1$.*

Now we obtain an estimate of the optimal non-adaptive strategy, i.e., the number k_N of nodes activated at the start. The rates λ_k are unimodal (increasing until $n/2$, and then decreasing). Intuitively then, the optimal non-adaptive strategy aims to have $n/2$ active nodes at time $\tau/2$, so that the rates are on average as large as possible during the evolution. The expected amount paid at the end should be roughly the same as the cost paid at the start; cf. [23] (the proof follows immediately by optimizing using the estimate of Proposition 3).

Proposition 4. *Given a deadline τ, the optimal non-adaptive pick is such that*

$$k_N = (1 + o(1))\frac{n}{1 + e^{\tau/2}} \quad and \quad u_{k_N}(0) = k_N(1 + o(1)).$$

Thus, the total expected cost of the optimal non-adaptive strategy is $2(1+o(1))k_N$.

4   Additive Gap for Small τ

In this section we consider the case in which τ is small, specifically, $\tau \le 2 \log \log n$. In this situation, thanks to Proposition 4, the optimal non-adaptive strategy activates $k_N = k_N(\tau) = (1 + o(1))\frac{n}{1+e^{\tau/2}} = \Omega(\frac{n}{\log n})$ nodes initially. This implies that the non-adaptive evolution is highly concentrated. Intuitively, this should be enough to conclude that adaptive cannot obtain a significant advantage; we use a coupling argument to make this precise.

Let $\mathcal{S}$ be any countable collection of points on $\mathbb{R}_+$, with an infinite number of points, with at least one point at 0, and denote by $\mathcal{S}_i$ the position of the i-th point. Associate to $\mathcal{S}$ a counting process $(K^{\mathcal{S}}(t))_{0 \le t \le \tau}$ as follows. Let $X_i^{\mathcal{S}} := (\mathcal{S}_{i+1} - \mathcal{S}_i)/\lambda_i$, and let $T_i^{\mathcal{S}} := \sum_{j=1}^{i-1} X_j^{\mathcal{S}}$, so $T_1^{\mathcal{S}} = 0$. Then, for $i \le n - 1$, set $K^{\mathcal{S}}(t) = i$ for all $t \in [T_i^{\mathcal{S}}, T_{i+1}^{\mathcal{S}}) \cap [0, \tau]$, and $K^{\mathcal{S}}(t) = n$ for all $t \in [T_n^{\mathcal{S}}, \tau]$.

Now, let $\mathcal{N}$ be a Poisson point process of unit intensity, and let $\mathcal{N}'$ be obtained by adding k_N additional points at the origin to $\mathcal{N}$. Since the inter-activation times $X_i^{\mathcal{N}'}$ are 0 for $i < k_N$, and distributed exponentially of rate λ_i for $k_N \le i < n$, we have that the law of $(K^{\mathcal{N}'}(t) : t \in [0, \tau])$ is precisely that of the evolution of the non-adaptive process with k_N pushes at time 0.

We can interpret an adaptive strategy directly in this perspective. For each $s \in \mathbb{R}_+$, it can decide whether to add a new point at position s, but based only on $\mathcal{N} \cap [0, s]$. In other words, it is a map φ that takes a set of points $\mathcal{S}$ and returns $\varphi(\mathcal{S}) \supseteq \mathcal{S}$, with $0 \in \varphi(\mathcal{S})$, and where $\varphi(\mathcal{S}) \cap [0, t]$ depends only on $\mathcal{S} \cap [0, t]$, for any t.

The resulting evolution is simply $K^{\varphi(\mathcal{S})}$, where the points in $\varphi(\mathcal{S}) \setminus \mathcal{S}$ correspond to pushes. One can see that this has the correct law of an evolution of an adaptive process, and that any adaptive strategy can be so described.

This provides a (somewhat non-obvious, but natural) coupling between the evolution of non-adaptive and adaptive. To exploit this, we relax the provision that adaptive may only look at the past when making its decisions. We define a *clairvoyant* strategy as *any* function ξ where $\xi(\mathcal{S}) \supseteq \mathcal{S}$ and $0 \in \xi(\mathcal{S})$. Clearly the optimal clairvoyant strategy has lower cost than the optimal adaptive one.

Lemma 3. *There is an optimal clairvoyant strategy which adds points only at the origin.*

So we are comparing the optimal non-adaptive strategy, which picks some number k_N of initial pushes without any knowledge of $\mathcal{N}$, to the optimal clairvoyant strategy, which picks some optimal number of initial pushes based on $\mathcal{N}$. A concentration argument shows that the extra information is very unlikely to be useful. More precisely, we argue that for any number k of initial pushes, the probability that the total cost paid is less than $2k_N - O(\sqrt{n} \text{ polylog}(n))$ is polynomially small, and then apply a union bound to conclude the following result.

Lemma 4. *Let $\frac{\log^2 n}{\sqrt{n}} < \tau \le 2 \log \log n$. Then the expected cost of the optimal clairvoyant strategy applied to $\mathcal{N}$ is $2k_N - O(\sqrt{n} \text{ polylog}(n))$.*

Consequences for the Profit Maximization Version. Note that the previous result already implies that for the profit maximization version of the problem the adaptivity gap is $1 + o(1)$. Indeed, if $\tau \ge 2 \log \log n$ then $k_N = o(n)$ and thus the profit of non-adaptive is $n - o(n)$ while adaptive certainly gets at most n. On the other hand, if $\tau \le 2 \log \log n$ is at least a constant, Lemma 4 implies that the activations that adaptive and non-adaptive make differ by a sub linear term, and since both get a profit which is linear in n the ratio is $1 + o(1)$. Finally, if $\tau = o(1)$ then $k_N = n/2 - o(n)$ so that throughout the rumor spreading process the λ_i's equal $4/(n + o(n))$ and this cannot be changed by an adaptive strategy.

5   Bounding the Adaptivity Gap

We now consider the case where $\tau \ge 2 \log \log n$. Here we need to do more than exploit the concentration of the evolution of the process. If the optimal non-adaptive scheme starts with relatively few pushes at the starting time, there will be a substantial amount of randomness at the beginning of the process, before the epidemic phase transition. Our goal is to show that an adaptive scheme cannot substantially exploit this.

The optimal adaptive strategy is difficult to handle, for example, the optimal choices of t_k^* are determined via an intricate recurrence. As in the last section, it will be very useful to rely on a more tractable lower bound, however, the lower bound we use here is quite different from the clairvoyant lower bound of Sect. 4.

As seen in Proposition 2, conditioned on $K^*(t^-) = k$, adaptive does not push at t if $u_k(t) - u_{k+1}(t) < 1$. We consider a modified set of rules for adaptive. Suppose it may push *for free*, however, if there are k active nodes at some time t, it may only push if $k < n/2$ and $u_k(t) - u_{k+1}(t) \geq 1$. Obviously the optimal adaptive strategy satisfies these restrictions, hence just pays less under these new rules. So the optimal strategy under these modified rules pays even less. The optimal "modified adaptive" strategy is very simple to describe: Since pushes are free, it will push whenever it is allowed to. We will show that the cost of the optimal modified adaptive strategy, which is simply the expected number of inactive nodes at time τ, is within a constant factor of the cost of non-adaptive.

Let $\tilde{K}(t)$ be the number of active nodes at time t using this optimal modified adaptive strategy. Let $\tilde{T}_i := \min\{t : \tilde{K}(t) \geq i\}$, and let $\tilde{P}(t)$ denote the number of pushes up to and including time t. Observe that $\tilde{K}(0) = k_N$, since k_N is the first k such that $u_k(t) - u_{k+1}(t) < 1$ by Lemma 2.

If one considers the non-adaptive evolution $K(t)$, $u_{K(t)}(t)$ is precisely the Doob martingale for the number of inactive nodes at time τ. It will be useful to look at a variant of this for the modified adaptive process. Specifically at $\tilde{U}(t) := u_{\tilde{K}(t)}(t)$, i.e., the expected number of inactive nodes at time τ, *given that no pushes are made between time t and τ*. It is a supermartingale, rather than a martingale, since any pushes made by the modified adaptive strategy will decrease the end payment. Since $\tilde{U}(0) = u_{k_N}(0) = (1 + o(1))k_N$, we will be interested in how much smaller (in expectation) $\tilde{U}(\tau)$ is compared to $\tilde{U}(0)$.

Define $\tilde{t}_k := \inf\{t \in [0, \tau] : u_k(t) - u_{k+1}(t) \geq 1\}$, or $\tilde{t}_k = \tau$ in the case the infimum is taken over the empty set. Then, if $\tilde{K}(t) = k$ and $t < \tilde{t}_k$, the modified adaptive strategy clearly cannot push, so $\tilde{t}_k$ is the first time when it is convenient to activate more than k nodes. Since $u_k(t)$ is a strictly increasing function of t, we can equivalently state this as: No push will occur at time t if $\tilde{U}(t^-) < \phi_{\tilde{K}(t^-)}$, where $\phi_k := u_k(\tilde{t}_k)$ for all k. Conversely, if $\tilde{U}(t^-) = \phi_{\tilde{K}(t^-)}$, then the optimal modified adaptive strategy will certainly push. This causes $\tilde{U}(t)$ to jump down by precisely 1 unit. So we will refer to the values ϕ_k as *thresholds*; the process $\tilde{U}(t)$ is always below the current threshold $\phi_{\tilde{K}(t)}$ and, if the threshold is reached, a push will be performed. Moreover, by convexity of the sequence $\{u_k(\cdot)\}_k$, the times $\tilde{t}_k$ are increasing, so only a single push occurs at any moment in time.

The following proposition connects the number of pushes made, i.e., the number of times $\tilde{U}$ reaches the current threshold, with the cost saved by the modified adaptive strategy.

Proposition 5. *The optimal modified adaptive strategy saves one unit of cost with respect to non-adaptive for each push after $t = 0$, i.e.,*

$$\mathbb{E}(n - \tilde{K}(\tau)) = u_{k_N}(0) - \mathbb{E}(\tilde{P}(\tau) - \tilde{P}(0)).$$

We use this as follows. Suppose $\mathbb{E}(\tilde{P}(\tau) - \tilde{P}(0)) \leq C$. Call the non-adaptive cost $c_N := k_N + u_{k_N}(0)$, from Proposition 4 we know that $c_N = 2(1 + o(1))u_{k_N}(0)$.

The adaptivity gap $\rho(n, \tau)$ is clearly bounded by $c_N / \mathbb{E}(n - \tilde{K}(\tau))$, so by Proposition 5

$$\rho(n,\tau) \leq \frac{c_N}{u_{k_N}(0) - \mathbb{E}(\tilde{P}(\tau) - \tilde{P}(0))} = 2(1 + o(1)) \left(1 + \frac{1}{u_{k_N}(0)/C - 1}\right).$$

We also have the trivial upper bound $\rho(n, \tau) \leq c_N = 2(1 + o(1))k_N$, just because an adaptive strategy will certainly need to push at least once. The required constant bound on $\rho(n, \tau)$ for the case $\tau \geq 2 \log \log n$ thus follows.

The aim for the rest of the section is to bound $\mathbb{E}(\tilde{P}(\tau) - \tilde{P}(0))$ by a constant. To exploit the characterization of modified adaptive we will use some of the estimates on $u_k(t)$ that we derived in Sect. 3. Recall Lemma 1, which states that $u_{k+1}(t) = u_k(t)(1 - \frac{1 + o(1)}{\lambda_k})$ and $u_k(t + h) \leq u_k(t)e^h$ for $t + h \leq \tau$. This has a very straightforward interpretation in terms of $\tilde{U}(t)$: Between activations $\tilde{U}(t)$ grows sub-exponentially, but if at time t there was an activation, then roughly $\tilde{U}(t)$ is multiplied by the factor $1 - 1/\lambda_{\tilde{K}(t-)}$.

We have now all the ingredients to bound how many times the process $\tilde{U}(t)$ hits the thresholds $\phi_{\tilde{K}(t)}$, which is exactly the number of pushes. We proceed by transforming the process in a number of ways. First, given the (sub)exponential growth, taking logarithms yields a process that between jumps grows no faster than a linear function with unit slope. Secondly, we locally shift the resulting process so that the threshold at any moment of time is moved to zero. This process will always be non-positive; we will be interested in the number of times that it hits zero. Finally, we locally rescale time, as in Sect. 4, so that the distribution of the times of random activations are described by a Poisson point process of unit intensity. We locally rescale the value at the same time, so that still the process increases linearly at unit rate in between jumps. Formally, we define the following transformed process $H(s)$:

$$H(L(t)) := \lambda_{\tilde{K}(t)}\left(\log \tilde{U}(t) - \log \phi_{\tilde{K}(t)}\right), \qquad \text{where } L(t) := \int_0^t \lambda_{\tilde{K}(x)} dx.$$

An illustration of the evolution of $\tilde{U}$ and the corresponding transformed process H is shown in Fig. 1. Notice how upon each random activation or push the threshold increases, while $\tilde{U}$ jumps down.

We are interested in the number of times that H reaches 0, as this corresponds to the number of pushes after time 0. It is convenient to consider instead

$$H'(L(t)) := H(L(t)) + \tilde{P}(t) - \tilde{P}(0).$$

The process H jumps down immediately whenever it reaches 0, and in fact the size of this jump is larger than 1. Very roughly speaking, H' cancels out these jumps (actually it may still jump down, but by a smaller amount), while the jumps corresponding to random activations are unaffected. It can easily be shown that the number of pushes is bounded by $\max\{0, 1 + \sup_{0 \leq r \leq t} H'(L(r))\}$.

So all that remains is to bound the expected supremum of $\tilde{H}'$. The reason that this is possible is very simple: Through most of its evolution, the process has

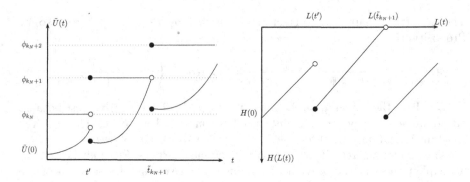

Fig. 1. A sample evolution of $\tilde{U}$, and the corresponding evolution of the transformed process H. A random activation occurred at time t', and a push occurred at time $\tilde{t}_{k_N+1}$.

a negative drift. More precisely, we have the following proposition. Here and for the remainder of this section, N will denote a Poisson process of unit intensity.

Proposition 6. *For any constant $0 < c \leq 1/2$, there is a $c' = (1 + o(1))c$ so that for any $\sigma \leq L(\tilde{T}_{cn})$, $H'(s + \sigma) - H'(\sigma)$ is stochastically dominated by $s - 2(1 - c')N(s)$ on the interval $[0, L(\tilde{T}_{cn}) - \sigma]$.*

Note in particular that $s - 2(1 - c)N(s)$ has negative drift for any $c < 1/2$.

Unfortunately, the bound we obtain on the drift gets worse as the number of active nodes increases. So some care is required; here we will sketch the argument. First, on the interval $[0, L(\tilde{T}_{n/4})]$, $H'(s)$ is dominated by $s - \frac{3}{2}N(s)$. This process starts from 0 and has negative drift (since $\mathbb{E}(s - \frac{3}{2}N(s)) = -s/2$). Proving that its expected maximum value is constant reduces to a straightforward concentration bound. Since $\tau \geq 2 \log \log n$, so that $k_N = O(n/\log n)$, this negative drift also implies that the process will be very negative (below say $-n/64$) at time $L(\tilde{T}_{n/4})$, with very high probability. On the interval $[L(\tilde{T}_{n/4}), L(\tilde{T}_{n/2})]$, $H'(s)$ can be dominated by $s - N(s)$ conditioned on $N(L(\tilde{T}_{n/4})) = H'(L(\tilde{T}_{n/4})) \leq -n/64$. Again a concentration argument shows that with very high probability the process remains negative in this interval, and so the expected number of further pushes is again at most a constant.

Acknowledgments. We thank Albert Banchs, Antonio Fernández, Domenico Giustiniano, Nicole Immorlica, Julia Komjáthy, Brendan Lucier, and Yaron Singer for stimulating discussions and helpful pointers to the literature.

References

1. Andersson, H., Britton, T.: Stochastic Epidemic Models and Their Statistical Analysis. Lecture Notes in Statistics. Springer, New York (2000)
2. Asadpour, A., Nazerzadeh, H., Saberi, A.: Stochastic submodular maximization. In: Papadimitriou, C., Zhang, S. (eds.) WINE 2008. LNCS, vol. 5385, pp. 477–489. Springer, Heidelberg (2008)

3. Bailey, N.: A simple stochastic epidemic. Biometrika **37**, 193–202 (1950)
4. Barry, K.: Ford bets the fiesta on social networking, September 2009. www.wired. com/2009/04/how-the-fiesta/
5. Bartlett, M.: An Introduction to Stochastic Processes, with Special Reference to Methods and Applications. Cambridge University Press, Cambridge (1978)
6. Bollobás, B., Kohayakawa, Y.: On Richardson's model on the hypercube. In: Bollobás, B., Thomason, A. (eds.) Combinatorics, Geometry and Probability. Cambridge University Press, Cambridge (1997)
7. Borgs, C., Brautbar, M., Chayes, J., Lucier, B.: Maximizing social influence in nearly optimal time. In: SODA (2014)
8. Boyd, S., Arpita, G., Prabhakar, B., Shah, D.: Randomized gossip algorithms. IEEE Trans. Inf. Theory **52**(6), 2508–2530 (2006)
9. Chen, N.: On the approximability of influence in social networks. In: SODA (2008)
10. Chen, W., Wang, C., Wang, Y.: Scalable influence maximization for prevalent viral marketing in large-scale social networks. In: KDD (2010)
11. Chen, Y., Krause, A.: Near-optimal batch mode active learning and adaptive submodular optimization. In: ICML (2013)
12. Cisco: VNI: Global mobile data traffic forecast update, 2013–2018 (2014). www.cisco.com/c/en/us/solutions/collateral/service-provider/visual-networking-index-vni/white_paper_c11-520862.html. Accessed 28 October 2014
13. Doerr, B., Künnemann, M.: Tight analysis of randomized rumor spreading in complete graphs. In: ANALCO (2014)
14. Domingos, P., Richardson, M.: Mining the network value customers. In: KDD (2001)
15. Golovin, D., Krause, A.: Adaptive submodularity: theory and applications in active learning and stochastic optimization. J. Artif. Intell. Res. **42**, 427–486 (2011)
16. Grimmett, G.: Probability on Graphs: Random Processes on Graphs and Lattices. Cambridge University Press, Cambridge (2010). Institute of Mathematical Statistics Textbooks
17. Han, B., Hui, P., Kumar, A., Marathe, M., Pei, G., Srinivasan, A.: Cellular traffic offloading through opportunistic communications: a case study. In: CHANTS (2010)
18. Horel, T., Singer, Y.: Scalable methods for adaptively seeding a social network. In: WWW (2015)
19. Ioannidis, S., Chaintreau, A., Massoulie, L.: Optimal and scalable distribution of content updates over a mobile social network. In: IEEE INFOCOM (2009)
20. Kempe, D., Kleinberg, J., Tardos, E.: Maximizing the spread of influence through a social network. In: SIGKDD (2003)
21. Kempe, D., Kleinberg, J., Tardos, E.: Influential nodes in a diffusion model. In: ICALP (2005)
22. Kleinberg, J.: Cascading behavior in social and economic networks. In: ACM EC (2013)
23. Sciancalepore, V., Giustiniano, D., Banchs, A., Picu, A.: Offloading cellular traffic through opportunistic communications: analysis and optimization. ArXiv preprint 1405.3548 (2014)
24. Seeman, L., Singer, Y.: Adaptive seeding in social networks. In: FOCS (2013)
25. Whitbeck, J., Amorim, M., Lopez, Y., Leguay, J., Conan, V.: Relieving the wireless infrastructure: When opportunistic networks meet guaranteed delays. In: WOW-MOM (2011)

Privacy and Truthful Equilibrium Selection for Aggregative Games

Rachel Cummings[1]([✉]), Michael Kearns[2], Aaron Roth[2], and Zhiwei Steven Wu[2]

[1] Department of Computing and Mathematical Sciences,
California Institute of Technology, Pasadena, USA
rachelc@caltech.edu
[2] Department of Computer and Information Science,
University of Pennsylvania, Philadelphia, USA
{mkearns,aaroth,wuzhiwei}@cis.upenn.edu

Abstract. We study a very general class of games — multi-dimensional aggregative games — which in particular generalize both anonymous games and weighted congestion games. For any such game that is also *large*, we solve the equilibrium selection problem in a strong sense. In particular, we give an efficient *weak mediator*: a mechanism which has only the power to listen to reported types and provide non-binding suggested actions, such that (a) it is an asymptotic Nash equilibrium for every player to truthfully report their type to the mediator, and then follow its suggested action; and (b) that when players do so, they end up coordinating on a particular asymptotic pure strategy Nash equilibrium of the induced complete information game. In fact, truthful reporting is an *ex-post* Nash equilibrium of the mediated game, so our solution applies even in settings of incomplete information, and even when player types are arbitrary or worst-case (i.e. not drawn from a common prior). We achieve this by giving an efficient differentially private algorithm for computing a Nash equilibrium in such games. The rates of convergence to equilibrium in all of our results are inverse polynomial in the number of players n. We also apply our main results to a multi-dimensional market game.

Our results can be viewed as giving, for a rich class of games, a more robust version of the Revelation Principle, in that we work with weaker informational assumptions (no common prior), yet provide a stronger solution concept (ex-post Nash versus Bayes Nash equilibrium). In comparison to previous work, our main conceptual contribution is showing that weak mediators are a game theoretic object that exist in a wide variety of games – previously, they were only known to exist in traffic routing games. We also give the first weak mediator that can implement an equilibrium optimizing a linear objective function, rather than implementing a possibly worst-case Nash equilibrium.

Keywords: Differential privacy · Equilibrium computation · Mechanism design

The full version of this extended abstract can be found on arXiv [9].

© Springer-Verlag Berlin Heidelberg 2015
E. Markakis and G. Schäfer (Eds.): WINE 2015, LNCS 9470, pp. 286–299, 2015.
DOI: 10.1007/978-3-662-48995-6_21

1   Introduction

Games with a large number of players are almost always played, but only sometimes modeled, in a setting of *incomplete information*. Consider, for example, the problem of selecting stocks for a 401 k portfolio among the companies listed in the S&P500. Because stock prices are the result of the aggregate decisions of millions of investors, this is a large multi-player strategic interaction, but it is so decentralized that it is implausible to analyze it in a complete information setting (in which every player knows the types or utilities of all of his opponents), or even in a Bayesian setting (in which every agent shares common knowledge of a prior distribution from which player types are drawn). How players will behave in such interactions is unclear; even under settings of complete information, there remains the potential problem of coordinating or selecting a particular equilibrium among many.

One solution to this problem, recently proposed by Kearns et al. [20] and Rogers and Roth [28], is to modify the game by introducing a *weak mediator*, which essentially only has the power to listen and to give advice. Players can ignore the mediator, and play in the original game as they otherwise would have. Alternately, they can use the mediator, in which case they can report their type to it (although they have the freedom to lie). The mediator provides them with a suggested action that they can play in the original game, but they have the freedom to disregard the suggestion, or to use it in some strategic way (not necessarily following it). The goal is to design a mediator such that *good behavior* – that is, deciding to use the mediator, truthfully reporting one's type, and then faithfully following the suggested action – forms an ex-post Nash equilibrium in the mediated game, and that the resulting play forms a Nash equilibrium of the original *complete information* game, induced by the actual (but unknown) player types. A way to approximately achieve this goal – which was shown in Kearns et al. [20], Rogers and Roth [28] – is to design a mediator which computes a Nash equilibrium of the game defined by the reported player types under a stability constraint known as *differential privacy* [12]. Prior to our work, this was only known to be possible in the special case of large, unweighted congestion games.

In this paper, we extend this approach to a much more general class of games known as *multi-dimensional aggregative games* (which among other things, generalize both anonymous games and weighted congestion games). In such a game, there is a vector of linear functions of players' joint actions called an *aggregator*. Each player's utility is then a possibly *non-linear* function of the aggregator vector and their own action. For example, in an investing game, the imbalance between buyers and sellers of a stock, which is a linear function of actions, may be used in the utility functions to compute prices, which are a non-linear function of the imbalances (see the full version). In an anonymous game, the aggregator function represents the number of players playing each action. In a weighted congestion game, the aggregator function represents the total weight of players on each of the facilities. Our results apply to any *large* aggregative game, meaning that any player's unilateral change in action can have at most a bounded influence on the utility of any other player, and the bound on this influence should be a diminishing function in the number of players in the game. Conceptually, our paper is the first to show that *weak mediators* are a game-theoretic object that exists in a large, general class of games: previously, although defined, weak mediators were only known to exist in traffic routing games [28].

This line of work can be viewed as giving robust versions of the Revelation Principle, which can implement Nash equilibria of the complete information game using a "direct revelation mediator," but without needing the existence of a prior type distribution. Compared to the Revelation Principle, which generically requires such a distribution and implements a Bayes Nash equilibrium, truth-telling forms an ex-post Nash equilibrium in our setting. We include a comparison to previous work in Table 1.

Finally, another important contribution of our work is that we are the first to demonstrate the existence of weak mediators (in *any* game) that have the power to optimize over an arbitrary linear function of the actions, and hence able to implement near optimal equilibria under such objective functions, rather than just implementing worst-case Nash equilibria.

Table 1. Summary of truthful mechanisms for various classes of games and solution concepts. Note that a "weak" mediator does not require the ability to verify player types. A "strong" mediator does. Weak mediators are preferred.

Mechanism	Class of games	Common prior?	Mediator strength	Equilibrium implemented
Revelation principle [24]	Any finite game	Yes	Weak	Bayes Nash
Kearns et al. [20]	Any large game	No	Strong	Correlated
Rogers and Roth [28]	Large congestion games	No	Weak	Nash
This work	Aggregative games	No	Weak	Nash

1.1 Our Results and Techniques

Our main result is the existence of a mediator which makes truthful reporting of one's type and faithful following of the suggested action (which we call the "good behavior" strategy) an ex-post Nash equilibrium in the mediated version of any aggregative game, thus implementing a Nash equilibrium of the underlying game of complete information. Unlike the previous work in this line [20,28], we do not have to implement an arbitrary (possibly worst-case) Nash equilibrium, but can implement a Nash equilibrium which optimizes any linear objective (in the player's actions) of our choosing. We here state our results under the assumption that any player's action has influence bounded by $(1/n)$ on other's utility, but our results hold more generally, parameterized by the "largeness" of the game.

Theorem 1 (Informal). *In a d-dimensional aggregative game of n players and m actions, there exists a mediator that makes good behavior an η-approximate ex-post Nash equilibrium, and implements a Nash equilibrium of the underlying complete information game that optimizes any linear objective function to within η, where*

$$\eta = O\left(\frac{\sqrt{d}}{n^{1/3}} \cdot \text{polylog}(n, m, d)\right).$$

It is tempting to think that the fact that players only have small influence on one another's utility function is sufficient to make any algorithm that computes a Nash equilibrium of the game a suitable weak mediator, but this is not so (see Kearns et al. [20] for an example). What we need out of a mediator is that any single agent's report should have little effect on the *algorithm* computing the Nash equilibrium, rather than on the payoffs of the other players.

The underlying tool that we use is differential privacy, which enforces the stability condition we need on the equilibrium computation algorithm. Our main technical contribution is designing a (jointly) differentially private algorithm for computing approximate Nash equilibria in aggregative games. The algorithm that we design runs in time polynomial in the number of players, but exponential in the dimension of the aggregator function. We note that since aggregative games generalize anonymous games, where the dimension of the aggregator function is the number of actions in the anonymous game, this essentially matches the best known running time for computing Nash equilibria in anonymous games, even non-privately [10]. Computing exact Nash equilibria in these games is known to be PPAD-complete [7]. Recent work of Barman and Ligett [5] showed that the equilibrium selection problem is also hard, even for more general solution concepts — it is NP-hard to compute a coarse correlated equilibrium that achieves a non-trivial approximation to the optimal welfare.

In the process of proving this result, we develop several techniques which may be of independent interest. First, we give the first algorithm for computing equilibria of multi-dimensional aggregative games (efficient for constant dimensional games) even in the absence of privacy constraints — past work in this area has focused on the single dimensional case [4,19]. Second, in order to implement this algorithm privately, we develop the first technique for solving a certain class of linear programs under the constraint of joint differential privacy.

We also give similar results for a class of one-dimensional aggregative games that permit a more general aggregation function and rely on different techniques, and we show how our main result can be applied to equilibrium selection in a multi-commodity market. The details are deferred to the full version [9].

1.2 Related Work

Conceptually, our work is related to the classic Revelation Principle of Myerson [24], in that we seek to implement equilibrium behavior in a game via a "mediated" direct revelation mechanism. Our work is part of a line, starting with Kearns et al. [20] and continuing with Rogers and Roth [28], that attempts to give a more robust reduction, without the need to assume a prior on types. Kearns et al. [20] showed how to privately compute correlated equilibria (and hence implement this agenda) in arbitrary large games. The private computation of correlated equilibrium turns out to give the desired reduction to a direct revelation mechanism only when the mediator has the power to verify types. Rogers and Roth [28] rectified this deficiency by privately computing Nash equilibria, but their result is limited to large unweighted congestion games. In this paper, we substantially generalize the class of games in which we can privately compute Nash equilibria (and hence solve the equilibrium selection problem with a direct-revelation mediator).

This line of work is also related to "strategyproofness in the large," introduced by Azevedo and Budish [3], which has similar goals. In comparison to this work, we do not require that player types be drawn from a distribution over the type-space, do not require any smoothness condition on the set of equilibria of the game, are algorithmically constructive, and do not require our game to be nearly as large. Generally, their results require the number of agents n to be larger than the size of the action set and the size of the type set. In contrast, we only require n to be as large as the *logarithm* of the number of actions, and require no bound at all on the size of the type space (which can even be infinite).

Our work is also related to the literature on mediators in games [22,23]. In contrast to our main goal (which is to implement solution concepts of the complete information game in settings of incomplete information), this line of work aims to modify the equilibrium structure of the complete information game. It does so by introducing a mediator, which can coordinate agent actions if they choose to opt in using the mediator. Mediators can be used to convert Nash equilibria into dominant strategy equilibria [22], or implement equilibrium that are robust to collusion [23]. Ashlagi et al. [2] considers mediators in games of incomplete information, in which agents can misrepresent their type to the mediators. Our notion of a mediator is related, but our mediators require substantially less power than the ones from this literature. For example, our mechanisms do not need the power to make payments [22], or the power to enforce suggested actions [23]. Like the mediators of Ashlagi et al. [2], ours are designed to work in settings of incomplete information and so do not need the power to verify agent types — but our mediators are weaker, in that they can only make suggestions (i.e. players do not need to cede control to our weak mediators).

The computation of equilibria in aggregative games (also known as summarization games) was studied in Kearns and Mansour [19], which gave efficient algorithms and learning dynamics converging to equilibria in the 1-dimensional case. Babichenko [4] also studies learning dynamics in this class of games and shows that in the 1-dimensional setting, sequential best response dynamics converge quickly to equilibrium. Our paper is the first to give algorithms for equilibrium computation in the multi-dimensional setting, which generalizes many well studied classes of games, including anonymous games. The running time of our algorithm is polynomial in the number of players n and exponential in the dimension of the aggregation function d, which essentially matches the best known running time for equilibrium computation in anonymous games [10].

We use a number of tools from differential privacy [12], as well as develop some new ones. In particular, we use the advanced composition theorem of Dwork et al. [14], the exponential mechanism from McSherry and Talwar [21], and the sparse vector technique introduced by Dwork et al. [13] (refined in Hardt and Rothblum [16] and abstracted into its current form in Dwork and Roth [11]). We introduce a new technique for solving linear programs under joint differential privacy, which extends a line of work (solving linear programs under differential privacy) initiated by Hsu et al. [17].

Finally, our work relates to a long line of work initiated by McSherry and Talwar [21] using differential privacy as a tool and desideratum in mechanism design. In addition to works already cited, this includes Blum et al. [6], Chen et al. [8], Ghosh and Ligett [15], Kannan et al. [18], Nissim et al. [25,26], Xiao [29] among others. For a survey of this area see Pai and Roth [27].

2 Model and Preliminaries

2.1 Aggregative Games

Consider an n-player game with action set $\mathcal{A}$ consisting of m actions and a (possibly infinite) type space $\mathcal{T}$ indexing utility functions. Let $\boldsymbol{x} = (x_i, \boldsymbol{x}_{-i})$ denote a strategy profile in which player i plays action x_i and the remaining players play strategy profile $\boldsymbol{x}_{-i}$. Each player i has a utility function, $u \colon \mathcal{T} \times \mathcal{A}^n \to [-1,1]$, where a player with type t_i experiences utility $u(t_i, \boldsymbol{x})$ when players play according to $\boldsymbol{x}$. When it is clear from context, we will use shorthand and write $u_i(\boldsymbol{x})$ to denote $u(t_i, \boldsymbol{x})$, the utility of player i at strategy profile $\boldsymbol{x}$.

The utility functions in *aggregative games* can be defined in terms of a multi-dimensional *aggregator* function $S \colon \mathcal{A}^n \to [-W, W]^d$, which represents a compact "sufficient statistic" to compute player utilities. In particular, each player's utility function can be represented as a function only of her own action x_i and the aggregator of the strategy profile $\boldsymbol{x}$: $u_i(\boldsymbol{x}) = u_i(x_i, S(\boldsymbol{x}))$. We also assume W to be polynomially bounded by n and m. In aggregative games, the function S_k for each coordinate $k \in [d]$, is an additively separable function: $S_k(\boldsymbol{x}) = \sum_{i=1}^n f_i^k(x_i)$.[1]

Similar to the setting of Kearns and Mansour [19] and Babichenko [4], we focus on γ-*aggregative games*, in which each player has a *bounded influence* on the aggregator:

$$\max_i \max_{x_i, x_i' \in \mathcal{A}} \|S(x_i, \boldsymbol{x}_{-i}) - S(x_i', \boldsymbol{x}_{-i})\|_\infty \le \gamma, \text{ for all } \boldsymbol{x}_{-i} \in \mathcal{A}^{n-1}.$$

---

[1] In the economics literature, aggregative games have more restricted aggregator function: $S_k(\boldsymbol{x}) = \sum_{i=1}^n x_i$. The games we study are more general, and sometimes referred to as *generalized aggregative games*.

That is, the greatest change a player can unilaterally cause to the aggregator is bounded by γ. With our motivation to study large games, we assume γ diminishes with the population size n. We also assume that all utility functions are 1-Lipschitz with respect to the aggregator: for all $x_i \in A$, $|u_i(x_i, s) - u_i(x_i, s')| \leq \|s - s'\|_\infty$.[2]

For γ-aggregative games, we can express the aggregator more explicitly as

$$S_k(\boldsymbol{x}) = \gamma \sum_{i=1}^{n} f_i^k(x_i),$$

where $f_i^k(x_i)$ is the influence of player i's action x_i on the k-th aggregator function, and also $|f_i^k(x_i)| \leq 1$ for all actions $i \in [n]$ and $x_i \in A$. Let $f_{ij}^k = f_i^k(a_j)$, where a_j denotes the j-th action in A.

We say that player i is playing an η-best response to $\boldsymbol{x}$ if $u_i(\boldsymbol{x}) \geq u_i(x_i', \boldsymbol{x}_{-i}) - \eta$, for all $x_i' \in A$. A strategy profile $\boldsymbol{x}$ is an η-pure strategy Nash equilibrium if all players are playing an η-best response in $\boldsymbol{x}$. We also consider mixed strategies, which are defined by probability distributions over the action set. For any profile of mixed strategies, given by a product distribution $\boldsymbol{p}$, we can define expected utility $u_i(\boldsymbol{p}) = \mathbb{E}_{\boldsymbol{x} \sim \boldsymbol{p}} u_i(\boldsymbol{x})$ and the expected aggregator

$$S_k(\boldsymbol{p}) = \mathop{\mathbb{E}}_{\boldsymbol{x} \sim \boldsymbol{p}} S_k(\boldsymbol{x}) = \gamma \sum_{i=1}^{n} \sum_{j=1}^{m} f_{ij}^k \, p_{ij} = \gamma \langle f^k, \boldsymbol{p} \rangle. \tag{1}$$

The support of a mixed strategy p, denoted $\mathrm{Supp}(\boldsymbol{p}_i)$, is the set of actions that are played with non-zero probabilities. A mixed strategy profile $\boldsymbol{p}$ is a mixed strategy Nash equilibrium if $u_i(\boldsymbol{p}) \geq \mathbb{E}_{\boldsymbol{x}_{-i} \sim \boldsymbol{p}_{-i}} u_i(x_i', \boldsymbol{x}_{-i})$ for all $i \in [n]$ and $x_i' \in A$.

For each aggregator s, we define the aggregative best response[3] for player i to s as $\mathrm{BA}_i(s) = \arg\max_{x_i \in A} \{u_i(x_i, s)\}$, breaking ties arbitrarily. We define the η-aggregative best response set for player i to s as

$$\eta\text{-}\mathrm{BA}_i(s) = \{x_i \in A \,|\, u_i(x_i, s) \geq \max_{x_i'} u_i(x_i', s) - \eta\}$$

to be the set of all actions that are at most η worse than player i's exact aggregative best response.

Remark 1. Note that best response is played against the other players' actions x_{-i}, but aggregative best response is played against the aggregator value s. Aggregative best response ignores the effect of the player's action on the aggregator, which is bounded by γ; the player reasons about the utility of playing

---

[2] Note that the influence that any single player's action has on the utility of others is also bounded by γ. If $\gamma = o(1/n)$, then any player's utility is essentially independent of other players' actions. Therefore, we further assume that $\gamma = \Omega(1/n)$ for the problem to be interesting. This will also simplify some statements.

[3] Sometimes called best react [4], and apparent best response [19].

different actions as if the aggregator value were *promised* to be s. Nevertheless, aggregative best response and best response can translate to each other with only an additive loss of γ in the approximation factor. Furthermore, aggregative best responses to different aggregators can translate to each other as long as the corresponding aggregators are close. If $\|s - s'\|_\infty \leq \alpha$, then the actions in η-BA(s) are also in $(\eta + 2\alpha)$-BA(s').

2.2 Mediated Games

We now define games modified by the introduction of a mediator. A mediator is an algorithm $M : (\mathcal{T} \cup \{\bot\})^n \to \mathcal{A}^n$ which takes as input reported types (or $\bot$ for any player who declines to use the mediator), and outputs a suggested action to each player. Given an aggregative game G, we construct a new game G_M induced by the mediator M. Informally, in G_M, players have several options: they can *opt-out* of the mediator (i.e. report $\bot$) and select an action independently of it. Alternately they can *opt-in* and report to it some type (not necessarily their true type), and receive a suggested action r_i. They are free to follow this suggestion or use it in some other way: they play an action $f_i(r_i)$ for some arbitrary function $f_i : \mathcal{A} \to \mathcal{A}$. Formally, the game G_M has an action set $\mathcal{A}_i$ for each player i defined as $\mathcal{A}_i = \mathcal{A}'_i \cup \mathcal{A}''_i$, where

$$\mathcal{A}'_i = \{(t_i, f_i) : t_i \in \mathcal{T}, f_i : \mathcal{A} \to \mathcal{A}\} \quad \text{and} \quad \mathcal{A}''_i = \{(\bot, f_i) : f_i \text{ is constant}\}.$$

Players' utilities in the mediated game are simply their expected utilities induced by the actions they play in the original game. Formally, they have utility functions u'_i: $u'_i(t, f) = \mathbb{E}_{x \sim M(t)}[u_i(f(x))]$. We are interested in finding mediators such that *good behavior* is an ex-post Nash equilibrium in the mediated game. We first define an ex-post Nash equilibrium.

Definition 1 (Ex-Post Nash Equilibrium). *A collection of strategies $\{\sigma_i : \mathcal{T} \to \mathcal{A}_i\}_{i=1}^n$ forms an η-approximate ex-post Nash equilibrium if for every type vector $t \in \mathcal{T}^n$, and for every player i and action $x_i \in \mathcal{A}_i$:*

$$u'_i(\sigma_i(t_i), \sigma_{-i}(t_{-i})) \geq u'_i(x_i, \sigma_{-i}(t_{-i})) - \eta$$

That is, it forms an η-approximate Nash equilibrium for every possible vector of types.

Note that ex-post Nash equilibrium is a very strong solution concept for incomplete information games because it does not require players to know a prior distribution over types.

In a mediated game, we would like players to truthfully report their type, and then faithfully follow the suggested action of the mediator. We call this *good behavior*. Formally, the good behavior strategy is defined as $g_i(t_i) = (t_i, \text{id})$ where id : $\mathcal{A} \to \mathcal{A}$ is the identity function – i.e. it truthfully reports a player's type to the mediator, and applies the identity function to its suggested action.

In order to achieve this, we use the notion of *joint differential privacy* defined in Kearns et al. [20] (adapted from *differential privacy*, defined in Dwork et al. [12]), as a privacy measure for mechanisms on agents' private data (types). Intuitively, it guarantees that the output to all other agents excluding player i is insensitive to i's private type, so the mechanism protects i's private information from arbitrary coalitions of adversaries.

Definition 2 (Joint Differential Privacy [20]). *Two type profiles t and t' are i-neighbors if they differ only in the i-th component. An algorithm $\mathcal{M} : T^n \to \mathcal{A}^n$ is (ε, δ)-joint differentially private if for every i, for every pair of i-neighbors $t, t' \in T^n$, and for every subset of outputs $\mathcal{S} \subseteq \mathcal{A}^{n-1}$,*

$$\Pr[\mathcal{M}(t)_{-i} \in \mathcal{S}] \le \exp(\varepsilon) \Pr[\mathcal{M}(t')_{-i} \in \mathcal{S}] + \delta.$$

If $\delta = 0$, we say that $\mathcal{M}$ is ε-jointly differentially private.

We here quote a theorem of Rogers and Roth [28], inspired by Kearns et al. [20] which motivates our study of private equilibrium computation.

Theorem 2 ([20,28]). *Let M be a mechanism satisfying (ε, δ)-joint differential privacy, that on any input type profile t with probability $1 - \beta$ computes an α-approximate pure strategy Nash equilibrium of the complete information game $G(t)$ defined by type profile t. Then the "good behavior" strategy $g = (g_1, \ldots, g_n)$ forms an η-approximate ex-post Nash equilibrium of the mediated game G_M for*

$$\eta = \alpha + 2(2\varepsilon + \beta + \delta).$$

Our private equilibrium computation relies on two private algorithmic tools, sparse vector mechanism (called Sparse) and exponential mechanism (called EXP), which allows us to access agents' types in a privacy-preserving manner.

3   Private Equilibrium Computation

Let G be a d-dimensional γ-aggregative game, and $L \colon \mathcal{A}^n \to \mathbb{R}$ be a γ-Lipschitz linear loss function:

$$L(\boldsymbol{x}) = \gamma \sum_i \ell_i(x_i) \quad \text{and} \quad L(\boldsymbol{p}) = \gamma \mathop{\mathbb{E}}_{\boldsymbol{x} \sim \boldsymbol{p}} L(\boldsymbol{x}) = \gamma \sum_i \langle p_{ij}, \ell_{ij} \rangle.$$

where $0 \le \ell_i(a_j) \le 1$ for all actions $a_j \in \mathcal{A}$, and $\ell_{ij} = \ell_i(a_j)$.

Given any $\zeta \ge \gamma \sqrt{8n \log(2mn)}$, let $\mathcal{E}(\zeta)$ be the set of ζ-approximate pure strategy Nash equilibria in the game G,[4] and let

$$\text{OPT}(\zeta) = \min\{L(\boldsymbol{x}) \mid \boldsymbol{x} \in \mathcal{E}(\zeta)\}.$$

We give the following main result:

---

[4] We show that $\mathcal{E}(\zeta)$ is non-empty for $\zeta \ge \gamma \sqrt{8n \log(2mn)}$ in the full version.

Theorem 3. *For any $\zeta \geq \gamma\sqrt{8n\log(2mn)}$, there exists a mediator M that makes good behavior an $(\zeta + \eta)$-approximate ex-post Nash equilibrium of the mediated game G_M, and implements an approximate pure strategy Nash equilibrium $\boldsymbol{x}$ of the underlying complete information game with $L(\boldsymbol{x}) \leq \mathrm{OPT}(\zeta) + \eta$, where*

$$\eta = O\left(n^{1/3}\gamma^{2/3}\sqrt{d} \cdot \mathrm{polylog}(n, m, d)\right).$$

Recall that the quantity γ is diminishing in n; whenever $\gamma = O(1/n^{1/2+\varepsilon})$ for $\varepsilon > 0$, the approximation factor η tends towards zero as n grows large. Plugging in $\gamma = 1/n$ and $\zeta = \gamma\sqrt{8n\log(2mn)}$ recovers the bound in Theorem 1.

This result follows from instantiating Theorem 2 with an algorithm that computes an approximate equilibrium under joint differential privacy as PRESL (Private Equilibrium Selection).[5] We give here an informal description of our algorithm, absent privacy concerns, and then describe how we implement it privately, deferring the formal treatment to the full version.

The main object of interest in our algorithm is the set-valued function

$$\mathcal{V}_\xi(\hat{s}) = \{S(\boldsymbol{p}) \mid \text{for each } i, \mathrm{Supp}(p_i) \subseteq \xi\text{-BA}_i(\hat{s})\},$$

which maps aggregator values $\hat{s}$ to the set of aggregator values that arise when players are randomizing between ξ-aggregative best responses to $\hat{s}$. An approximate equilibrium will yield an aggregator $\hat{s}$ such that $\hat{s} \in \mathcal{V}_\xi(\hat{s})$, so we wish to find such a fixed point for $\mathcal{V}_\xi$ (the value of ξ will be determined in the analysis, see the full version). Note that *pure strategy* Nash equilibria correspond to such fixed points, but a-priori, it is not clear that fixed points of this function (which may involve mixed strategies) are mixed strategy Nash equilibria. This is because player utility functions need not be linear in the aggregator, and so a best response to the expected value of the aggregator need not be a best response to the corresponding distribution over aggregators. However, as we will show, we can safely round such fixed points to approximate pure strategy Nash equilibria, because the aggregator will be well concentrated under rounding.

For every fixed value $\hat{s}$, the problem of determining whether $\hat{s} \in \mathcal{V}_\xi(\hat{s})$ is a linear program (because the aggregator is linear), and although $\mathrm{Supp}(p_i) \subseteq \xi\text{-BA}_i(\hat{s})$ is not a convex constraint in $\hat{s}$, the aggregative best responses are fixed for each fixed value of $\hat{s}$. The first step of our algorithm simply searches through a discretized grid of all possible aggregators $X = \{-W, -W + \alpha, \ldots, W - \alpha\}^d$, and solves this linear program to check if some point $\hat{s} \in \mathcal{V}_\xi(\hat{s})$. This results in a set of aggregators S that are induced by the approximate equilibria of the game. Let p_{ij} denote the probability that player i plays the j-th action.

---

[5] In the full version of this paper, we also present details of the non-private algorithm to compute equilibrium for aggregative games.

Then the linear program we need to solve is as follows:

$$\forall k \in [d], \qquad \hat{s}_k - \alpha \le \gamma \sum_{i=1}^{n} \sum_{j=1}^{m} f_{ij}^k p_{ij} \le \hat{s}_k + \alpha$$

$$\forall i \in [n], \qquad \forall j \in \xi\text{-BA}_i(\hat{s}), \qquad 0 \le p_{ij} \le 1 \tag{2}$$

$$\forall i \in [n], \qquad \forall j \notin \xi\text{-BA}_i(\hat{s}), \qquad p_{ij} = 0$$

Next, we need to find a particular equilibrium (an assignment of actions to players) that optimizes our cost-objective function L. This is again a linear program (since the objective function is linear) for each $\hat{s}$. Hence, for each fixed point $\hat{s} \in \mathcal{V}_\xi(\hat{s})$ we simply solve this linear program, and out of all of the candidate equilibria, output the one with the lowest cost. Finally, this results in mixed strategies for each of the players, and we round this to a pure strategy Nash equilibrium by sampling from each player's mixed strategy. This does not substantially harm the quality of the equilibrium; because of the low sensitivity of the aggregator, it is well concentrated around its expectation under this rounding. The running time of this algorithm is dominated by the grid search for the aggregator fixed point $\hat{s}$, which takes time exponential in d. Solving each linear program can be done in time polynomial in all of the game parameters.

Making this algorithm satisfy joint differential privacy is more difficult. There are two main steps. The first is to identify the fixed point $\hat{s} \in \mathcal{V}_\xi(\hat{s})$ that corresponds the lowest cost equilibrium. There are exponentially in d many candidate aggregators to check, and with naive noise addition we would have to pay for this exponential factor in our accuracy bound. However, we take advantage of the fact that we only need to *output* a single aggregator – the one corresponding to the lowest objective value equilibrium – and so the *sparse vector mechanism* Sparse (described in the full version) can be brought to bear, allowing us to pay only linearly in d in the accuracy bound.

The second step is more challenging, and requires a new technique: we must actually solve the linear program corresponding to $\hat{s}$, and output to each player the strategy they should play in equilibrium. The output strategy profile must satisfy joint differential privacy. To do this, we give a general method for solving a class of linear programs (containing in particular, LPs of the form (2)) under joint differential privacy, which may be of independent interest. This algorithm, which we call DistMW (described in the full version), is a distributed version of the classic multiplicative weights (MW) technique for solving LPs [1]. The algorithm can be analyzed by viewing each agent as controlling the variables corresponding to their own mixed strategies, and performing their multiplicative weights updates in isolation (and ensuring that their mixed strategies always fall within their best response set $\xi\text{-BA}_i(\hat{s})$). At every round, the algorithm aggregates the current solution maintained by each player, and then identifies a coordinate in which the constraints are far from being satisfied. The algorithm uses the *exponential mechanism* EXP (described in the full version) to pick such a coordinate while maintaining the privacy of the players' actions. The identification of such a coordinate is sufficient for each player to update their own variables. Privacy then follows by combining the privacy guarantee of the exponential mechanism with a bound on the convergence time of the multiplicative weights update rule. The fact that we can solve this LP in a distributed manner to get joint differential privacy (rather than standard differential privacy) crucially depends on the fact that the sensitivity γ of the aggregator is small. The algorithm DistMW will find a set of strategies that approximately satisfy the linear program – the violation on each coordinate is bounded by

$$E = O\left(\frac{n\gamma^2}{\varepsilon}\operatorname{polylog}\left(n, m, d, \frac{1}{\beta}, \frac{1}{\delta}\right)\right)^{1/2}.$$

The algorithm PRESL has the following guarantee:

Theorem 4. *Let $\zeta \geq \gamma\sqrt{8n\log(2mn)}, \varepsilon, \delta, \beta \in (0,1)$. PRESL $(t, \zeta, L, \varepsilon, \delta, \beta)$ satisfies $(2\varepsilon, \delta)$-joint differential privacy, and, with probability at least $1 - \beta$, computes a $(\zeta + 12\alpha)$-approximate pure strategy equilibrium x such that $L(x) <$ OPT$(\zeta) + 5\alpha$, where*

$$\alpha = O\left(\frac{(\sqrt{n\varepsilon} + d)\gamma}{\varepsilon}\operatorname{polylog}\left(n, m, d, 1/\beta, 1/\delta\right)\right).$$

We defer the full proof and technical details to the full version.

Remark 2. The running time of this algorithm is exponential in d, the dimension of the aggregative game. For games of fixed dimension (where d is constant), this yields a polynomial time algorithm. This exponential dependence on the dimension matches the best known running time for (non-privately) computing equilibrium in anonymous games by [10], which is a sub-class of aggregative games.

Theorem 3 follows by instantiating Theorem 2 with PRESL $\left(t, \zeta, L, n^{1/3}\gamma^{2/3}d^{1/2}, \frac{1}{n}, \frac{1}{n}\right)$ – i.e. by setting $\varepsilon = n^{1/3}\gamma^{2/3}d^{1/2}$ and $\delta = \beta = \frac{1}{n}$.

Acknowledgments. Research supported in part by NSF grants 1253345, 1101389, CNS-1254169, US-Israel Binational Science Foundation grant 2012348, Simons Foundation grant 361105, a Google Faculty Research Award, and the Alfred P. Sloan Foundation. Research performed while the first author was visiting the University of Pennsylvania.

References

1. Arora, S., Hazan, E., Kale, S.: The multiplicative weights update method: a meta-algorithm and applications. Theory Comput. **8**(1), 121–164 (2012)
2. Ashlagi, I., Monderer, D., Tennenholtz, M.: Mediators in position auctions. Games Econ. Behav. **67**(1), 2–21 (2009)
3. Azevedo, E.M., Budish, E.: Strategyproofness in the large as a desideratum for market design. In: Proceedings of the 13th ACM Conference on Electronic Commerce, EC 2012, p. 55 (2012)

4. Babichenko, Y.: Best-reply dynamic in large aggregative games. SSRN (2013). abstract 2210080

5. Barman, S., Ligett, K.: Finding any nontrivial coarse correlated equilibrium is hard. In: Proceedings of the Sixteenth ACM Conference on Economics and Computation, EC 2015, pp. 815–816 (2015)

6. Blum, A., Morgenstern, J., Sharma, A., Smith, A.: Privacy-preserving public information for sequential games. In: Proceedings of the 2015 Conference on Innovations in Theoretical Computer Science, ITCS 2015, pp. 173–180 (2015)

7. Chen, X., Durfee, D., Orfanou, A.: On the complexity of Nash equilibria in anonymous games. In: Proceedings of the Forty-Seventh Annual ACM on Symposium on Theory of Computing, STOC 2015, pp. 381–390 (2015)

8. Chen, Y., Chong, S., Kash, I.A., Moran, T., Vadhan, S.: Truthful mechanisms for agents that value privacy. In: Proceedings of the 14th ACM Conference on Electronic Commerce, EC 2013, pp. 215–232 (2013)

9. Cummings, R., Kearns, M., Roth, A., Wu, Z.S.: Privacy and truthful equilibrium selection for aggregative games, CoRR, abs/1407.7740 (2014)

10. Daskalakis, C., Papadimitriou, C.H.: Discretized multinomial distributions and Nash equilibria in anonymous games. In: Proceedings of the 49th Annual IEEE Symposium on Foundations of Computer Science, FOCS 2008, pp. 25–34 (2008)

11. Dwork, C., Roth, A.: The algorithmic foundations of differential privacy. Found. Trends Theoret. Comput. Sci. **9**(3–4), 211–407 (2014)

12. Dwork, C., McSherry, F., Nissim, K., Smith, A.: Calibrating noise to sensitivity in private data analysis. In: Halevi, S., Rabin, T. (eds.) TCC 2006. LNCS, vol. 3876, pp. 265–284. Springer, Heidelberg (2006)

13. Dwork, C., Naor, M., Reingold, O., Rothblum, G.N., Vadhan, S.: On the complexity of differentially private data release: efficient algorithms and hardness results. In: Proceedings of the 41st Annual ACM Symposium on Theory of Computing, STOC 2009, pp. 381–390 (2009)

14. Dwork, C., Rothblum, G.N., Vadhan, S.: Boosting and differential privacy. In: Proceedings of the 51st Annual IEEE Symposium on Foundations of Computer Science, FOCS 2010, pp. 51–60 (2010)

15. Ghosh, A., Ligett, K.: Privacy and coordination: computing on databases with endogenous participation. In: Proceedings of the 14th ACM Conference on Electronic Commerce, EC 2013, pp. 543–560 (2013)

16. Hardt, M., Rothblum, G.N.: A multiplicative weights mechanism for privacy-preserving data analysis. In: Proceedings of the 51st Annual IEEE Symposium on Foundations of Computer Science, FOCS 2010, pp. 61–70 (2010)

17. Hsu, J., Roth, A., Roughgarden, T., Ullman, J.: Privately solving linear programs. In: Esparza, J., Fraigniaud, P., Husfeldt, T., Koutsoupias, E. (eds.) ICALP 2014. LNCS, vol. 8572, pp. 612–624. Springer, Heidelberg (2014)

18. Kannan, S., Morgenstern, J., Roth, A., Wu, Z.S.: Approximately stable, school optimal, and student-truthful many-to-one matchings (via differential privacy). In: Proceedings of the Twenty-Sixth Annual ACM-SIAM Symposium on Discrete Algorithms, SODA 2015, pp. 1890–1903 (2015)

19. Kearns, M., Mansour, Y.: Efficient Nash computation in large population games with bounded influence. In: Proceedings of the 18th Conference on Uncertainty in Artificial Intelligence, UAI 2002, pp. 259–266 (2002)

20. Kearns, M., Pai, M., Roth, A., Ullman, J.: Mechanism design in large games: incentives and privacy. In: Proceedings of the 5th Conference on Innovations in Theoretical Computer Science, ITCS 2014, pp. 403–410 (2014)

21. McSherry, F., Talwar, K.: Mechanism design via differential privacy. In: Proceedings of the 48th Annual IEEE Symposium on Foundations of Computer Science, FOCS 2007, pp. 94–103 (2007)
22. Monderer, D., Tennenholtz, M.: k-implementation. In: Proceedings of the 4th ACM Conference on Electronic Commerce, EC 2003, pp. 19–28 (2003)
23. Monderer, D., Tennenholtz, M.: Strong mediated equilibrium. Artif. Intell. **173**(1), 180–195 (2009)
24. Roger, R.B.: Optimal auction design. Math. Oper. Res. **6**(1), 58–73 (1981)
25. Nissim, K., Orlandi, C., Smorodinsky, R.: Privacy-aware mechanism design. In: Proceedings of the 13th ACM Conference on Electronic Commerce, EC 2012, pp. 774–789 (2012)
26. Nissim, K., Smorodinsky, R., Tennenholtz, M.: Approximately optimal mechanism design via differential privacy. In: Proceedings of the 3rd Innovations in Theoretical Computer Science Conference, ITCS 2012, pp. 203–213 (2012)
27. Pai, M.M., Roth, A.: Privacy and mechanism design. SIGecom Exch. **12**(1), 8–29 (2013)
28. Rogers, R.M., Roth, A.: Asymptotically truthful equilibrium selection in large congestion games. In: Proceedings of the 15th ACM Conference on Economics and Computation, EC 2014, pp. 771–782 (2014)
29. Xiao, D.: Is privacy compatible with truthfulness? In: Proceedings of the 4th Conference on Innovations in Theoretical Computer Science, ITCS 2013, pp. 67–86 (2013)

Welfare and Revenue Guarantees for Competitive Bundling Equilibrium

Shahar Dobzinski[1], Michal Feldman[2], Inbal Talgam-Cohen[3], and Omri Weinstein[4]([⊠])

[1] Weizmann Institute of Science, 76100 Rehovot, Israel
[2] Tel-Aviv University, P.O.B. 39040, Tel Aviv, Israel
[3] Stanford University, Stanford, CA 94305, USA
[4] Princeton University, Princeton, NJ 08544, USA
omriw@cs.nyu.edu

Abstract. Competitive equilibrium, the central equilibrium notion in markets with indivisible goods, is based on pricing each good such that the demand for goods equals their supply and the market clears. This equilibrium notion is not guaranteed to exist beyond the narrow case of substitute goods, might result in zero revenue even when consumers value the goods highly, and overlooks the widespread practice of pricing bundles rather than individual goods. Alternative equilibrium notions proposed to address these shortcomings have either made a strong assumption on the ability to withhold supply in equilibrium, or have allowed an exponential number of prices.

In this paper we study the notion of *competitive bundling equilibrium* – a competitive equilibrium over the market induced by partitioning the goods into bundles. Such an equilibrium is guaranteed to exist, is succinct, and satisfies the fundamental economic condition of market clearance. We establish positive welfare and revenue guarantees for this solution concept: For welfare we show that in markets with homogeneous goods, there always exists a competitive bundling equilibrium that achieves a logarithmic fraction of the optimal welfare. We also extend this result to establish nontrivial welfare guarantees for markets with heterogeneous goods. For revenue we show that in a natural class of markets for which competitive equilibrium does not guarantee positive revenue, there always exists a competitive bundling equilibrium that extracts as revenue a logarithmic fraction of the optimal welfare. Both results are tight.

1 Introduction

Competitive equilibria play a fundamental role in market theory and design – they capture the market's steady states, in which each participant maximizes

S. Dobzinski—Incumbent of the Lilian and George Lyttle Career Development Chair. Supported in part by the I-CORE program of the planning and budgeting committee and the Israel Science Foundation 4/11 and by EU CIG grant 618128.

M. Feldman—partially supported by the European Research Council under the European Union's Seventh Framework Programme (FP7/2007–2013)/ERC grant agreement number 337122.

© Springer-Verlag Berlin Heidelberg 2015
E. Markakis and G. Schäfer (Eds.): WINE 2015, LNCS 9470, pp. 300–313, 2015.
DOI: 10.1007/978-3-662-48995-6_22

his profit at equilibrium prices, and supply equals demand such that the market clears [14, Parts III and IV].

This paper focuses on the well-known *combinatorial markets* model, which consists of a set M of m indivisible goods (or *items*), and a set N of n consumers. We consider both the extensively studied case of homogeneous goods, and its generalization to heterogeneous goods. Each consumer i has a *valuation* v_i : $2^M \to \mathbb{R}^+$ over bundles of goods. The standard assumptions are that each v_i is normalized ($v_i(\emptyset) = 0$) and monotone non-decreasing. A *competitive equilibrium* is an allocation of the goods to the consumers, denoted by $(S_1, \ldots, S_n)$, together with supporting item prices, denoted by p_j for good j, such that the following two conditions hold:

1. **Profit Maximization:** The profit of every consumer i is maximized by his allocation S_i; i.e., for every alternative set of goods T, $v_i(S_i) - \sum_{j \in S_i} p_j \geq v_i(T) - \sum_{j \in T} p_j$.
2. **Market Clearance:** All items are allocated; i.e., $\bigcup_i S_i = M$.

Unfortunately, despite their fundamental role, competitive equilibria are only guaranteed to exist in limited classes of combinatorial markets, most notably those in which all valuations are *gross substitutes* [10,16]. Intuitively this means that consumers do not view the goods as complementary, so that if the price of one good rises, the demand for other goods does not decline (see Sect. 2 for a formal definition of gross substitutes and other valuation classes).

The standard market model described above implicitly assumes that the goods on the market are exogenously determined, yet in many markets this is not true. For example, there is no inherent reason for beer to be sold in 6-packs rather than, say, 8-packs. This practice of *bundling* is ubiquitous in real-life markets: It is a well-known method for revenue extraction [13] – e.g., for this reason many airlines set the price of a one-way ticket to be equal to the price of a round-trip ticket. It is also a common mean for avoiding the "exposure" problem due to complementarities – e.g., in the online market for concert ticket resale StubHub.com, a seller holding two tickets may prohibit their separate resale so that if there is no demand for both, she may still enjoy the concert with a friend.

In this paper we study the role of bundling in steadying the market. We will see that bundling introduces new equilibria, and thus can recover stability in markets that lack a competitive equilibrium. *The main challenge is whether "good" bundlings exist, i.e., those which result in nearly optimal social efficiency and/or revenue extraction.*

1.1   Related Work and Definition of Our Equilibrium Concept

There have been several suggestions in the literature as to how to extend competitive equilibria to accommodate bundling. One direction initiated in [3] is the study of competitive equilibria over bundles that are supported by 2^m non-linear bundle prices, possibly personalized per consumer.[1] Auctions that reach

---

[1] In the full version available on arXiv.org we use linear programming to show that our solution concept actually applies more widely than this one.

such equilibria with personalized prices are studied in [1, 20], while anonymous supporting prices and auctions that reach them for classes of valuations are studied in [11, 22] (see also [18, 19, 21, 24]).

In contrast to this direction, consider perhaps the simplest possible extension of competitive equilibria, in which supporting prices are linear and anonymous. We define a *competitive bundling equilibrium* to consist of a partition of the goods into bundles, denoted by $\mathcal{B} = (B_1, \ldots, B_{m'})$ and referred to as a *bundling*, in addition to an allocation $(S_1, \ldots, S_n)$ of the bundles to the consumers, and a price p_{B_j} for each bundle B_j. Similar to competitive equilibria, two conditions must hold:

1. **Profit Maximization:** For every consumer i and alternative set of bundles T, $v_i(S_i) - \sum_{B_j \in S_i} p_{B_j} \geq v_i(T) - \sum_{B_j \in T} p_{B_j}$.
2. **Market Clearance:** $\bigcup_i S_i = M$.

An advantage of competitive bundling equilibria is that they always exist. This can be seen, e.g., by naively bundling all goods together and allocating the bundle to the highest-valuing consumer. The social efficiency, however, may reach only a $1/n$-fraction of that achieved by the optimal allocation. Hence we will mainly be occupied with seeking better competitive bundling equilibria than the naive one.

A previously studied notion that is closely related to, and inspires, our competitive bundling equilibrium notion is the solution concept known as *combinatorial Walrasian equilibrium*, introduced in [8]. In this solution concept, only the profit maximization condition must hold and not the market clearance one, and in this sense it is closer in spirit to an algorithmic pricing solution than to a classic market-stabilizing equilibrium.[2] The main result in [8] is that if one ignores the requirement to sell all the goods, then there is always a bundling together with anonymous and linear bundle prices that achieve at least half of the optimal welfare. The main open problem posed in [8] is whether their result can be extended under the market clearance requirement, and in this paper we address this open question, among others. The open question of [8] received partial treatment also in [9], but with respect to only two restricted classes of valuations. In this work we consider a much more general setting, and in passing generalize the results obtained in [9] for these two classes.[3]

---

[2] To emphasize this distinction we propose a different name – competitive bundling equilibrium – for the solution concept we focus on in this paper.

[3] The classes addressed in [9] are a strict subclass of *budget-additive* valuations, and *superadditive* valuations. We provide a general treatment of budget-additive valuations in the full version of this paper [6], and re-derive the result for super-additive valuations as a corollary of a more general argument in the full version: It was observed in [19] that the linear program introduced by [3] has an integrality gap of 1 for superadditive valuations, guaranteeing the existence of non-linear supporting prices; we show that an integrality gap of 1 also implies the existence of a competitive bundling equilibrium with optimal welfare.

Applicability. The notion of competitive bundling equilibrium is applicable when bundling is legally or effectively irrevocable. In the seminal paper of [3] this property is referred to as "crates cannot be opened". Retail markets where bundles are explicitly marked as "not for individual sale" are one example of markets with this property, as is the market for air tickets mentioned above. Sterilized products are another case in which the physical packaging cannot be opened. As another example consider attraction passes – companies like Disney sell bundles of several day passes which are activated upon first entrance, when identification is required from the visitor; the rest of the passes can then only be used by the same visitor. It is even common practice for different producers to bundle together their goods; for example, a travel website can offer bundles of air tickets, hotel rooms and car rental.[4]

A second condition for the applicability of competitive bundling equilibrium is that the market clears. This condition is sometimes violated in markets dominated by a single monopolist, who may attempt to enforce an outcome in which goods are withheld despite positive demand (note this will only succeed if, despite the fact that such markets are often regulated, the monopolist can credibly commit not to sell the goods in the future – see the classic paper of [5] for failure of a monopolist to do so). In competitive markets, however, market clearance is necessary for stability. The standard argument is that in uncleared markets, competing producers have incentive to undercut prices, thus leaving the market unstable (for a thorough discussion see Sect. 10.B of [14]). In addition, markets for resources like spectrum or public land will necessarily clear, since the governmental seller cannot withhold supply arbitrarily.

1.2 Our Results

We establish the existence of competitive bundling equilibria that are approximately optimal with respect to social efficiency and also revenue, where the approximation factors depend on the size of the combinatorial market's shorter side $\mu = \min\{n, m\}$. While our main focus is on existence of good equilibria, our results are constructive and often tractable.

(I) Welfare for Homogeneous Goods. We refer to combinatorial markets with homogeneous goods as *multi-unit settings*; in such settings the consumers' values depend on the number of units they receive. Multi-unit settings have been studied extensively in the literature (e.g., [2,4,15,23]), since this model captures important goods like Treasury bills, electricity and telecommunications spectrum, as well as online advertising. For a recent survey dedicated to multi-unit settings see [17].

The classic result of Vickrey [23] shows that if the consumers' valuations exhibit *decreasing marginal utilities* (i.e., the added value from adding a single unit decreases in the size of the existing bundle), then there always exists a

---

[4] An interesting future direction is trying to better understand the market processes leading to such bundling.

competitive equilibrium. This no longer holds in the more general case where valuations exhibit *complements* among units. We give a complete analysis for competitive bundling equilibria in multi-unit settings and establish the following theorem.

Theorem [Main] (Sect. 4): For every multi-unit market there exists a competitive bundling equilibrium that provides an $O(\log \mu)$-approximation to the optimal welfare. There exist such markets in which the approximation ratio of every competitive bundling equilibrium is $\Omega(\log \mu)$.

The lower bound in this theorem applies even if all valuations but one exhibit decreasing marginal utilities, and all are *subadditive* (see Sect. 2 for a definition).[5]

(II) Welfare for Heterogeneous Goods. Our techniques developed for homogeneous goods apply to combinatorial markets with heterogeneous goods as well, yielding the following results.

Theorem [General Markets] (Sect. 5): For every combinatorial market there exists a competitive bundling equilibrium that provides an $\tilde{O}(\sqrt{\mu})$-approximation to the optimal welfare. There exist such markets in which the approximation ratio of every competitive bundling equilibrium is $\Omega(\log \mu)$.

Theorem [Budget-Additive Consumers] (Sect. 6): For every combinatorial market with budget-additive valuations there exists a competitive bundling equilibrium that provides an $O(\log \mu)$-approximation to the optimal welfare. There exist such markets in which the approximation ratio of every competitive bundling equilibrium is no better than $\frac{5}{4}$.

The gap between the upper and lower bounds is one of two main open questions that arise from this paper.

Theorem [Two Consumers] (See the Full Version [6]): For every combinatorial market with $n = 2$ consumers there exists a competitive bundling equilibrium that provides a $\frac{3}{2}$-approximation to the optimal welfare. There exist such markets in which the approximation ratio of every competitive bundling equilibrium is no better than $\frac{3}{2}$.

(III) Revenue. We use our techniques to show a positive result for revenue, in order to highlight the role of bundling in revenue extraction and to demonstrate how the competitive bundling equilibrium notion can be useful even in gross

---

[5] This shows that even for slightly more complicated valuations than those considered by Vickrey, not only is it the case that a standard competitive equilibrium may not exist, but also no competitive bundling equilibrium necessarily provides a constant fraction of the optimal welfare. In the full version we also show that randomization – in the form of correlated lotteries – does not improve the logarithmic approximation factor. This tight and robust bound can be seen as a kind of price of stability result, which establishes a clear separation between the optimal welfare and the optimal *stable* welfare in markets with indivisible items.

substitutes markets (for which standard equilibria are guaranteed to exist but may possibly extract zero revenue).

Theorem [Revenue] (See the Full Version [6]): For every combinatorial market with valuations belonging to a subclass of gross substitutes there exists a competitive bundling equilibrium that extracts as revenue an $\Omega(1/\log \mu)$-fraction of the optimal welfare. There exist such markets in which the revenue of every competitive bundling equilibrium is an $O(1/\log \mu)$-fraction of the optimal welfare.

The second main open question that arises from this paper is whether the above result extends to the entire class of gross substitutes.

2 Preliminaries

We briefly address the standard issue of valuation representation and review certain classes of valuations. The expert reader can safely skip this section.

A naïve representation of a valuation is exponential; the standard assumption is thus that valuations are accessed by a succinct oracle. One standard kind of oracle is a *demand* oracle: Given a consumer valuation v and item prices $\boldsymbol{p}$, an item set T is in the consumer's *demand set* if it maximizes his profit, i.e., $v(T) - \sum_{j \in T} p_j = \max_{U \subseteq M} \{v(U) - \sum_{j \in U} p_j\}$. A demand query returns a member of the demand set under the given prices. The other standard kind of oracle is a *value* oracle: For a given valuation and bundle, a value query returns the value of the bundle. Even in multi-unit settings, demand queries are known to be strictly stronger than value queries.

An important class of valuations is *gross substitutes* valuations. A valuation v is a gross substitutes valuation if for every pair of price vectors $\boldsymbol{q} \geq \boldsymbol{p}$, for every item set T in the demand set of $\boldsymbol{p}$, there exists an item set U in the demand set of $\boldsymbol{q}$ such that U includes every item $j \in T$ whose price did not increase. A *unit-demand* valuation v is a kind of gross substitutes valuation for which the value of every bundle T is the maximum value of some item in T. A superclass of gross substitutes valuations is *subadditive* valuations. A valuation v is subadditive if for every two bundles T, U it holds that $v(T) + v(U) \geq v(T \cup U)$. A class of subadditive valuations that are not gross substitutes is *budget-additive* valuations. A valuation v is budget-additive if there exists b such that for every bundle T we have that $v(T) = \min\{\sum_{j \in T} v(\{j\}), b\}$.

3 Technical Tools

In this section we prepare our main workhorses: Lemmas 1 and 2. These lemmas identify structures – described in Definitions 1 and 2 – from which a competitive bundling equilibrium with certain welfare guarantees can be found in polynomial time utilizing a result of [8]. Subsequent sections of the paper devise and analyze algorithms to construct these structures. We first present the lemmas and then prove them in Sect. 3.1.

Definition 1. *A* high-demand priced bundling *consists of a bundling* $\mathcal{B}$ *and bundle prices* $\boldsymbol{p}$, *such that for every bundle* $B \in \mathcal{B}$, *there is a set* N_B *of at least* $|\mathcal{B}|$ *consumers for whom* $v_i(B) - p_B > 0$, *i.e.,* B *is strictly profitable.*

Lemma 1. *For every high-demand priced bundling* $(\mathcal{B}, \boldsymbol{p})$, *there exists a competitive bundling equilibrium with allocation* $S' = (S'_1, \ldots, S'_n)$, *whose welfare* $\sum_{i \in N} v_i(S'_i)$ *is at least the sum of prices* $\sum_{B \in \mathcal{B}} p_B$. *Moreover, it can be found in* $\text{poly}(m, n)$ *time using demand queries given* $(\mathcal{B}, \boldsymbol{p})$.

Definition 2. *A bundling* $\mathcal{B}$, *bundle prices* $\boldsymbol{p}$ *and allocation* S *over* $\mathcal{B}$ *form a* partial competitive bundling equilibrium *if they constitute a competitive bundling equilibrium for a consumer subset* $N' \subseteq N$.

Lemma 2. *For every partial competitive bundling equilibrium with bundling* $\mathcal{B}$ *and prices* $\boldsymbol{p}$, *there exists a competitive bundling equilibrium with allocation* $S' = (S'_1, \ldots, S'_n)$, *whose welfare* $\sum_{i \in N} v_i(S'_i)$ *is at least the revenue* $\sum_{B \in \mathcal{B}} p_B$. *Moreover, it can be found in* $\text{poly}(m, n)$ *time using demand queries given the partial competitive bundling equilibrium.*

3.1   Proofs

The proofs of Lemmas 1 and 2 use the following result that is a reinterpretation of [8].

Theorem 1 ([8]). *In a combinatorial market with general, possibly non-monotone valuations, let* $\mathcal{B}$ *be a bundling with bundle prices* $\boldsymbol{p}$. *There exist a further bundling* $\mathcal{B}'$ *over bundles in* $\mathcal{B}$[6], *and prices* $\boldsymbol{p}'$ *and an allocation* $S = (S_1, \ldots, S_n)$ *over bundling* $\mathcal{B}'$, *such that:*

1. *For every* i, *let* $T_i \subseteq \mathcal{B}$ *be the set of original bundles* S_i *is combined from, and let* $U_i \subseteq \mathcal{B}'$ *be the set of new bundles* S_i *is combined from. Then* $\sum_{B \in U_i} p'_B \geq \sum_{B \in T_i} p_B$.
2. S_i *is in consumer* i's *demand set given prices* p', *i.e.,* $v_i(S_i) - \sum_{B \in U_i} p'_B$ *maximizes* i's *profit among all subsets of* $\mathcal{B}'$.
3. *For every bundle* $B \in \mathcal{B}'$ *unallocated in* S, $B \in \mathcal{B}$ *and* $p'_B = p_B$.

The bundling, prices and allocation can be found in $\text{poly}(m, n)$ *time using demand queries given* $\mathcal{B}$ *and* $\boldsymbol{p}$.

Note that the bundling, prices and allocation guaranteed to exist by the above theorem do *not* form a competitive bundling equilibrium, as they do not satisfy the market clearance condition.

We will also need the following lemma, which formulates a standard argument and is proven for completeness in the full version of this paper [6].

---

[6] i.e., a coarser bundling of the original m items.

Lemma 3 (Bucketing – folklore). *For every allocation* $S = (S_1, \ldots, S_n)$ *of items* M, *there exists a value* v *and an allocation* $S' = (S'_1, \ldots, S'_n)$ *of* $M' \subseteq M$, *such that* $v_i(S'_i) \in [v, 2v)$ *for every* $S'_i \neq \emptyset$, *and a logarithmic fraction of the welfare is maintained, i.e.,* $\sum_{i \in N} v_i(S'_i) \geq \frac{1}{2(\log \mu + 2)} \sum_{i \in N} v_i(S_i)$ *where* $\mu = \min\{m, n\}$. *The value* v *and allocation* S' *can be found in* poly(m, n) *time using value queries given* S.

Proof (Lemma 1). We will show that after applying Theorem 1 to the high-demand priced bundling, all bundles in $\mathcal{B}$ are allocated. Assume towards contradiction there is a bundle $B \in \mathcal{B}$ that is not allocated. Then its price remains unchanged by property (3) of Theorem 1. But there are at least $|\mathcal{B}|$ consumers for which B is profitable at this price, hence for property (2) to hold, each of these consumers must be allocated an alternative bundle (otherwise their profit would be 0). However, there are only $|\mathcal{B}| - 1$ bundles except for B, a contradiction. □

Proof (Lemma 2). Fix some $\epsilon > 0$. Consider the partial competitive bundling equilibrium and denote its consumer subset by N'. For every consumer $i \in N'$ for which S_i is not empty, define a new valuation v_i^ϵ that is identical to v_i except for a shift of ϵ in the value of S_i, i.e., $v_i^\epsilon(S_i) = v_i(S_i) + \epsilon$ (the new valuation may no longer be monotone). For every other consumer i simply set $v_i^\epsilon = v_i$. Observe that the partial competitive bundling equilibrium is still a partial competitive bundling equilibrium with respect to the v_i^ϵ's. Now apply Theorem 1 to get a bundling $\mathcal{B}'$, allocation $(S'_1, \ldots, S'_n)$ and prices $\boldsymbol{p}'$. We show that since we started with a partial competitive bundling equilibrium, these bundling, allocation and prices form a competitive bundling equilibrium with respect to the v_i^ϵ's, that is, all bundles in $\mathcal{B}$ are allocated. By the latter fact and by properties (1) and (2) of Theorem 1, $\sum_{i \in N} v_i(S'_i) \geq \sum_{B \in \mathcal{B}} p_B$.

To show that we get a competitive bundling equilibrium, since property (2) of Theorem 1 is guaranteed, the only missing component is to show market clearance, i.e., that $\cup_i S_i = M$. Suppose towards a contradiction that there is a bundle $B \in \mathcal{B}$ that was not allocated. Let i be the consumer that was allocated that bundle in the partial competitive bundling equilibrium. Observe that under the prices of the partial competitive bundling equilibrium, B is the most profitable bundle of i. Now since B is unallocated, its price remaines the same by property (3) of Theorem 1, while the prices of the other bundles can only increase by properties (1) and (3). Thus B is *the* most preferred bundle for i (with valuation v_i it is only *a* most preferred bundle), and by property (2) consumer i must be allocated this bundle.

We would now like to show the existence of a competitive bundling equilibrium with respect to the v_i's and not just with respect to the v_i^ϵ's. When taking ϵ to 0, we get an infinite sequence of allocations and prices. Since the number of allocations is finite and since all prices are bounded between 0 and $\max\{\max_i v_i(M), \max_B(p_B)\}$, there exists a subsequence in which one allocation $\tilde{S}$ repeats and the prices converge to a price vector $\tilde{p}$. Note that it still holds that $\sum_{i \in N} v_i(\tilde{S}_i) \geq \sum_{B \in \mathcal{B}} p_B$.

To finish the proof we now claim that this allocation $\tilde{S}$ and prices $\tilde{p}$ are a competitive bundling equilibrium with respect to the v_i's. Observe that for every ϵ in the converging subsequence, if consumer i receives $\tilde{S}_i$, then $\tilde{S}_i$ is the *unique* bundle that maximizes his profit; otherwise, for smaller values of ϵ, $\tilde{S}_i$ is no longer the most profitable bundle for i in contradiction to the assumption that we have a competitive bundling equilibrium for the v_i^ϵ's. Since this is true for every $\epsilon > 0$, for $\epsilon = 0$ we get that $\tilde{S}_i$ is one of the most profitable bundles for i, which is enough to prove that $\tilde{S}$ and $\tilde{p}$ form a competitive bundling equilibrium with respect to the bundling $\mathcal{B}$.   □

4   Welfare for Homogeneous Goods

This section focuses on multi-unit markets, where units of a single good may be treated by different consumers as both substitutes and complements. We show existence of a competitive bundling equilibrium that logarithmically approximates the optimal social welfare; this result is tight even when all valuations are subadditive.

Theorem 2. *For every multi-unit market with n consumers and m items, there exists a competitive bundling equilibrium that provides an $O(\log\mu)$-approximation to the optimal welfare OPT, where $\mu = \min\{m, n\}$. Moreover, it can be found in $\mathrm{poly}(\log m, n)$ time using value queries.*[7]

Proof. Our goal is to show there exists a high-demand priced bundling whose aggregate price is an $O(\log\mu)$-approximation to OPT; the proof of existence then follows by applying Lemma 1.

Consider a welfare-optimal allocation $(O_1, \ldots, O_n)$. We begin by applying Lemma 3, by which there exist a value v and an allocation $(O_1', \ldots, O_n')$ of an item subset M', such that (1) for every consumer i with non-empty allocation, $v_i(O_i') \in [v, 2v]$; (2) a logarithmic fraction of the welfare is preserved, i.e., $\sum_{i \leq n} v_i(O_i') \geq \mathrm{OPT}/\Theta(\log\mu)$.

Without loss of generality assume $|O_1'| \geq \cdots \geq |O_n'|$, and let n' be the largest index such that $O_{n'}' \neq \emptyset$. If $n' = 1$, the proof is complete by allocating the grand bundle to consumer 1 for price $v_1(M)$. Assuming from now on $n' > 1$, we show how allocation O' can be used to construct a high-demand priced bundling.

Let $\mathcal{B}$ be a partition of all m items into $k := \lfloor n'/2 \rfloor \geq 1$ bundles of roughly equal size – if k does not divide m, place leftover items in one of the bundles arbitrarily. Observe that every bundle $B \in \mathcal{B}$ has size at least $|O_{k+1}'|$. Set the price of every such bundle to $p_B = v - \epsilon$ for some small $\epsilon = \epsilon(v)$. Then all k bundles are strictly profitable for consumers $k+1, \ldots, n'$, and since there are at least k such consumers we have a high-demand priced bundling.

It remains to show that the aggregate price, $k(v-\epsilon)$, is a logarithmic fraction of OPT: By choosing $\epsilon \leq v/5$, we have that $k(v-\epsilon) \geq v(2k+2)/5$. Plugging in

---

[7] A tractable algorithm for multi-unit market runs in time $\mathrm{poly}(\log m, n)$ and not $\mathrm{poly}(m, n)$ – see [7].

$2k + 2 \geq n'$ we get $k(v - \epsilon) \geq 2vn'/10$, which by the fact that $v_i(O_i') < 2v$ is at least $\frac{1}{10} \sum_{i \leq n} v_i(O_i') \geq \text{OPT}/\Theta(\log \mu)$, completing the existence proof.

We now sketch the proof of the computational result. Assuming $m \gg n$ (the case of interest), we add a preprocessing stage in which units are partitioned into equal-sized bundles of size m/n^2 (ignoring leftovers for simplicity). [7] show that the welfare-optimal allocation of such bundles achieves a 2-approximation to the welfare of an optimal unconstrained allocation. Further recall that in any multi-unit setting, there is a computationally efficient constant-approximation to OPT. Together with the computational guarantees of Lemmas 1 and 3, this implies that a competitive bundling equilibrium which provides an $O(\log \mu)$-approximation to the optimal welfare can be found in $\text{poly}(\log m, n)$ time using demand queries. It is left to show that a demand query in the new setting can be simulated by $\text{poly}(\log m, n)$ value queries. Since the total number of original units in any bundle is now a multiple of m/n^2, one can use dynamic programming to simulate a demand query as required. □

Theorem 3. *There exists a multi-unit market where $m = n$ and valuations are subadditive, such that every competitive bundling equilibrium has welfare that is a $1/\Omega(\log m)$-fraction of the optimal.*

Table 1. A lower bound for multi-unit markets with subadditive consumers

Consumer	Valuation
$i = 1$	$v_i(B) = \begin{cases} 2(1 + \epsilon) & \text{if } B = M \\ 1 + \epsilon & \text{if } B \subset M \end{cases}$
$i \in \{2, \ldots, n\}$	$v_i(B) = 1/i$ for every $B \subseteq M$

Proof. Consider the multi-unit market in which $m = n$ and consumer valuations are as described in Table 1. Note that all consumers are subadditive (in fact consumers 2 to n are unit-demand). The optimal allocation in this market allocates each of the consumers a single unit, achieving welfare $\Omega(\log m)$. We will show that in every competitive bundling equilibrium, consumer 1 is allocated all units, and thus the welfare is only $2 + 2\epsilon$.

Consider a competitive bundling equilibrium with allocation S over bundling $\mathcal{B}$, and bundle prices p. Let i' be the smallest index of a consumer whose allocation is non-empty. Observe that all bundles $B \in \mathcal{B}$ must have a common price $p \leq 1/i'$: Clearly consumer i' cannot be charged more than $1/i'$, and if the price of any bundle $> 1/i'$, then some consumer's profit is not being maximized. Now assume towards contradiction that $i' > 1$. By the market clearance property, $|\mathcal{B}| \leq i'$, and so the total price for all bundles in $\mathcal{B}$ is at most 1. Consumer 1 will thus strictly increase his profit by buying all bundles in $\mathcal{B}$, in contradiction to the profit maximization property of a competitive bundling equilibrium. This completes the proof. □

In the full version we strengthen Theorem 3 by showing it holds even when lotteries are allowed.

5   Welfare for Heterogeneous Goods

In this section we consider general combinatorial markets, and show that for every such market there exists a competitive bundling equilibrium whose allocation achieves a $\tilde{O}(\sqrt{\mu})$-approximation to the optimal welfare. In the full version of this paper, we also address computational aspects (see [6]).

Theorem 4. *For every combinatorial market with n consumers and m goods, there exists a competitive bundling equilibrium with welfare that is a $\tilde{O}(\sqrt{\mu})$-approximation to the optimal welfare OPT, where $\mu = \min\{m, n\}$.*

Proof. Apply Lemma 3 to the welfare-optimal allocation $(O_1, \ldots, O_n)$ to get a value v and an allocation $S = (S_1, \ldots, S_n)$ of items $M' \subseteq M$. Without loss of generality, assume that exactly the first r allocated parts in S are non-empty (r must be $\leq \mu$), so that (1) for every consumer $i \in [r]$, $v_i(S_i) \in [v, 2v)$; (2) a logarithmic fraction of the welfare is preserved, i.e., $\sum_{i \in [r]} v_i(S_i) \geq \frac{1}{O(\log \mu)} \sum_i v_i(O_i)$. By applying Lemma 4 below to the value v and allocation S, we get an $O(\sqrt{r}) = O(\sqrt{\mu})$-approximation to the welfare of S, which is an $\tilde{O}(\sqrt{\mu})$-approximation to OPT.   □

The proof of Theorem 4 relies on the following lemma.

Lemma 4. *Let v be a value and $S = (S_1, S_2, \ldots, S_r)$ an allocation of items $M' \subseteq M$ to the first r consumers, such that $\forall i \in [r]$, $S_i \neq \emptyset$ and $v_i(S_i) \in [v, 2v)$. Then there is a competitive bundling equilibrium that achieves an $O(\sqrt{r})$-approximation to the welfare of S.*

Proof. We show how to construct a high-demand priced bundling whose aggregate price is an $O(\sqrt{r})$-approximation to the welfare of S. The proof is then established by invoking Lemma 1.

Beginning with $S_1, \ldots, S_r$, we create new bundles by joining sets of $\lceil \sqrt{r} \rceil$ S_is, adding leftovers to the last bundle (which contains between $\lceil \sqrt{r} \rceil$ and $2\lceil \sqrt{r} \rceil$) S_is). Items in $M \setminus M'$ are also added to the last bundle. Let $\mathcal{B}$ denote the resulting bundling, and set a price $p_B = v$ for every $B \in \mathcal{B}$. The pair $(\mathcal{B}, \boldsymbol{p})$ is a high-demand priced bundling: By construction, $|\mathcal{B}| \leq \sqrt{r}$, and every bundle $B \in \mathcal{B}$ is profitable for at least $\sqrt{r}$ consumers (those who were originally allocated the S_is it contains).

Now, recall that $\sum_{i \in [r]} v_i(S_i) \leq 2vr$. Since no bundle $B \in \mathcal{B}$ contains more than $2\lceil \sqrt{r} \rceil \leq 2\sqrt{r} + 2$ S_is, then $|\mathcal{B}| \geq r/(2\sqrt{r} + 2)$, and so the proof is complete:

$$\sum_{B \in \mathcal{B}} p_B = |\mathcal{B}| v \geq \frac{rv}{2\sqrt{r} + 2} \geq \frac{1}{4\sqrt{r} + 4} \sum_{i \in [r]} v_i(S_i).$$

□

6   Welfare for Budget-Additive

In Sect. 5 we showed that every combinatorial market admits a competitive bundling equilibrium that provides an approximation ratio of $\tilde{O}(\sqrt{m})$ to the social welfare. A natural next step is to understand whether we can get better approximation ratios for specific subclasses. We make progress towards this goal by showing that if the valuations are all budget-additive then we can get a logarithmic approximation. The best lower bound we currently know shows that no market with budget-additive valuations can achieve an approximation ratio better than $\frac{5}{4}$ (for the proof, see the full version of this paper [6]).

Theorem 5. *In every combinatorial market with budget-additive valuations, there is a competitive bundling equilibrium with welfare that is an $O(\log m)$-approximation to the optimal welfare.*

We henceforth describe a high-level proof of Theorem 5 (see the full version of this paper [6] for details). Our proof is constructive and proceeds as follows: We first compute an allocation $A = (A_1, \ldots, A_n)$ by running (a slight variation of) the greedy algorithm of [12] for submodular valuations. This algorithm considers the items one by one in an arbitrary order, and allocates each item to a consumer that maximizes the marginal value for it given the items he received until now. This allocation is known to achieve a constant approximation to OPT for submodular valuations.

Next, our goal is to identify a subset of the bundles (A_i's) that can be converted into a *partial equilibrium* on the one hand, and give (at least) a logarithmic approximation to OPT on the other hand. Given such a subset, we can then apply Lemma 2 to complete the proof. The non-trivial challenges are: (1) how to allocate the items that do not belong to the identified subset of bundles, and (2) how to price the bundles to ensure profit maximization for the corresponding consumers.

To address these challenges, we distinguish between two cases: If most of the welfare of A comes from consumers who have exhausted their budgets (i.e., $v_i(A_i) = b_i$), we identify a subset of the buyers S_b with budgets $b_i \in (b, 2b]$ for some b, who contribute a logarithmic fraction of the welfare of A. Then, for every consumer $i \in S_b$, we bundle up the items in A_i into a bundle, and add the remaining items (i.e., those not belonging to any of the consumers in S_b) to an arbitrary consumer in S_b. We then price each bundle at a uniform price of b. It is not too difficult to verify that this is a partial competitive bundling equilibrium with respect to the consumers in S_b.

If most of the welfare of A comes from consumers who have not exhausted their budgets (i.e., $v_i(A_i) < b_i$), for every non-exhaustive consumer i, we bundle up the items in A_i into a bundle, and add the remaining items, T, to the non-exhaustive consumer i^* who maximizes the value $v_i(A_i \cup T)$. We charge every consumer exactly his valuation for the allocated bundle (so each one gains zero utility). The greediness of the initial algorithm ensures that every non-exhaustive consumer maximizes his profit over all allocations A_i given to other non-exhaustive consumers. It remains to show that this holds also with respect to the consumer i^*. But this is clear by the definition of i^* as the consumer who maximizes the valuation for this bundle. It follows that this is a partial competitive bundling equilibrium with respect to the non-exhaustive consumers.

References

1. Ausubel, L., Milgrom, P.: Ascending auctions with package bidding. B.E. J. Theoret. Econ. **1**(1), 1–44 (2002)
2. Ausubel, L.M.: An efficient ascending-bid auction for multiple objects. Am. Econ. Rev. **94**, 1452–1475 (2004)
3. Bikhchandani, S., Ostroy, J.M.: The package assignment model. J. Econ. Theory **107**, 377–406 (2002)
4. Borgs, C., Chayes, J., Immorlica, N., Mahdian, M., Saberi, A.: Multi-unit auctions with budget-constrained bidders. In: Proceedings of 7th ACM Conference on Electronic Commerce (EC) (2005)
5. Coase, R.H.: Durability and monopoly. J. Law Econ. **15**(1), 143–149 (1972)
6. Dobzinski, S., Feldman, M., Talgam-Cohen, I., Weinstein, O.: Welfare and revenue guarantees for competitive bundling equilibrium. CoRR abs/1406.0576 (2014). http://arxiv.org/abs/1406.0576
7. Dobzinski, S., Nisan, N.: Mechanisms for multi-unit auctions. J. Artif. Intell. Res. **37**, 85–98 (2010)
8. Feldman, M., Gravin, N., Lucier, B.: Combinatorial walrasian equilibrium. In: Proceedings of 44th ACM Symposium on Theory of Computing (STOC), pp. 61–70 (2013)
9. Feldman, M., Lucier, B.: Clearing markets via bundles. In: Lavi, R. (ed.) SAGT 2014. LNCS, vol. 8768, pp. 158–169. Springer, Heidelberg (2014)
10. Gul, F., Stacchetti, E.: Walrasian equilibrium with gross substitutes. J. Econ. Theory **87**(1), 95–124 (1999)
11. Lahaie, S., Parkes, D.C.: Fair package assignment. In: Das, S., Ostrovsky, M., Pennock, D., Szymanksi, B. (eds.) AMMA 2009. LNICST, vol. 14, pp. 92–92. Springer, Heidelberg (2009)
12. Lehmann, B., Lehmann, D.J., Nisan, N.: Combinatorial auctions with decreasing marginal utilities. Games Econ. Behav. **55**(2), 270–296 (2006)
13. Manelli, A.M., Vincent, D.R.: Bundling as an optimal selling mechanism for a multiple-good monopolist. J. Econ. Theory **127**, 1–35 (2006)
14. Mas-Collel, -A., Whinston, M.D., Green, J.R.: Microeconomic Theory. Oxford University Press, Oxford (1995)
15. Maskin, E., Riley, J.: Optimal multi-unit auctions. In: Hahn, F. (ed.) The Economics of Missing Markets, Information, and Games, pp. 312–335. Oxford University Press, Oxford (1989)
16. Milgrom, P.: Putting Auction Theory to Work. Cambridge University Press, Cambridge (2004)
17. Nisan, N.: Survey: algorithmic mechanism design (through the lens of multi-unit auctions). In: Young, P., Zamir, S. (eds.) Handbook of Game Theory, vol. 4, chap. 9. Elsevier, Amsterdam (2014)
18. Parkes, D.: Iterative combinatorial auctions. In: Cramton, P., Shoham, Y., Steinberg, R. (eds.) Combinatorial Auctions, chap. 2. MIT Press, Cambridge (2006)

19. Parkes, D.C.: iBundle: an efficient ascending price bundle auction. In: Proceedings of 1st ACM Conference on Electronic Commerce (EC), pp. 148–157 (1999)
20. Parkes, D.C., Ungar, L.H.: Iterative combinatorial auctions: theory and practice. In: Proceedings of the 17th National Conference on Artificial Intelligence, AAAI 2000, pp. 74–81 (2000)
21. Parkes, D.C., Ungar, L.H.: An ascending-price generalized Vickrey auction. In: Proceedings of Stanford Institute for Theoretical Economics Workshop on The Economics of the Internet, Stanford, CA (2002)
22. Sun, N., Yang, Z.: An efficient and incentive compatible dynamic auction for multiple complements. J. Polit. Econ. (2014, to appear)
23. Vickrey, W.: Counterspeculation, auctions, and competitive sealed tenders. J. Finance **16**, 8–37 (1961)
24. Vohra, R.V.: Mechanism Design: A Linear Programming Approach (Econometric Society Monographs). Cambridge University Press, Cambridge (2011)

Often Harder than in the Constructive Case: Destructive Bribery in CP-nets

Britta Dorn[1]([⊠]), Dominikus Krüger[2], and Patrick Scharpfenecker[2]

[1] Department of Computer Science, Faculty of Science,
University of Tübingen, Tübingen, Germany
britta.dorn@uni-tuebingen.de
[2] Institute of Theoretical Computer Science, Ulm University, Ulm, Germany
{dominikus.krueger,patrick.scharpfenecker}@uni-ulm.de

Abstract. We study the complexity of the destructive bribery problem (an external agent tries to prevent a disliked candidate from winning by bribery actions) in voting over combinatorial domains, where the set of candidates is the Cartesian product of several issues. This problem is related to the concept of the margin of victory of an election which constitutes a measure of robustness of the election outcome and plays an important role in the context of electronic voting. In our setting, voters have conditional preferences over assignments to these issues, modelled by CP-nets. We settle the complexity of all combinations of this problem based on distinctions of four voting rules, five cost schemes, three bribery actions, weighted and unweighted voters, as well as the negative and the non-negative scenario. We show that almost all of these cases are $\mathcal{NP}$-complete or $\mathcal{NP}$-hard for weighted votes while approximately half of the cases can be solved in polynomial time for unweighted votes.

Keywords: Computational social choice · Voting · Bribery · CP-nets · Destructive

1 Introduction

Voting in an election is a common procedure to aggregate the preferences of the parties involved, the voters, over a set of alternatives, the candidates, in order to find one or more winning alternatives. In many settings, the set of candidates is the Cartesian product of several issues. One might think of a referendum, where voters have to approve or disapprove of each issue, or the individual configuration of a product consisting of several components for each of which several options can be chosen, such as a car where the consumer can choose between different options for the model, equipment, color, and various other features, or a computer where different options are available for the operation system, hardware and software components. The number of possible candidates (available choices, outcomes) is hence exponential in the number of issues or components,

P. Scharpfenecker—Supported by DFG grant TO 200/3-1.

E. Markakis and G. Schäfer (Eds.): WINE 2015, LNCS 9470, pp. 314–327, 2015.
DOI: 10.1007/978-3-662-48995-6_23

and it may be an impossible task for voters to express their preferences over the whole set of available choices by ranking them all.

Additionally, voters might have conditional preferences over the candidates. The typical example is a meal consisting of several components, such as a main dish (fish or meat), a side dish (chips or rice), and a drink (beer or wine). A voter might (unconditionally) prefer meat to fish, and he might prefer wine to beer given that fish is served for the main dish. In the car example, a consumer might prefer a station wagon to a hatchback, and he might prefer a black car to a white one, but only if it is equipped with an air conditioning system.

In view of applications such as e-commerce and other settings on the web and internet where one has to deal with very large populations, one is interested in a compact description and efficient communication and aggregation of these conditional preferences in combinatorial domains. One approach is given by CP-nets, a graphical model introduced by Boutilier et al. [1] that incorporates *ceteris paribus* (cp) statements describing the conditional dependencies. Preference aggregation in CP-nets was studied by Rossi et al. [24] and various other authors (e.g., [5, 22, 27]).

Besides the problem of determining a winner, a central topic in the computational social choice literature is the study of the computational complexity of voting problems such as strategic voting (manipulation), election control and bribery [3, 10]. In the bribery problem, initially introduced by Faliszewski et al. [12] (see also [9, 11, 13]), voters can be bribed to change their preferences. In the *constructive* bribery problem, one asks whether a briber can make his favorite candidate win the election with these changes, subject to a budget constraint.

Mattei et al. [19] considered several procedures for determining a winner in voting with CP-nets and investigated the bribery problem in this context. They introduced and adapted several natural cost schemes for the bribery problem in the setting of CP-nets and determined the computational complexity of the problem under the various voting rules and cost schemes, also considering the level of dependency the briber can affect with his changes. In most of these cases, they obtained that the bribery problem is solvable in polynomial time. Dorn and Krüger [7] answered open cases and considered the weighted and negative versions of the problem. Further investigations of bribery in CP-nets deal with interaction and influence among voters [18] and with representation of the voters' preferences via soft constraints [21].

In this work, we study the complexity of the *destructive* bribery problem in CP-nets, which asks whether a disliked candidate can be prevented from winning the election by bribery actions. The study of destructive bribery is also related to the concept of the *margin of victory* [17, 23, 26] of an election. Given a voting rule and a set of votes, the margin of victory is the minimum number of votes that must be modified in order to change the winner(s) of the election. If the voting rule selects a unique winner, then the problem of deciding whether this number is larger than a given threshold corresponds to the destructive bribery problem introduced by Faliszewski et al. [12]. The margin of victory is a measure of robustness of the outcome of an election, specifying the number of errors that may occur in an election—be it due to inference or due to fraud—without having an effect on the outcome. It is of particular interest in the setting of electronic voting where post-election audits are executed to verify the correctness of the electronical record [20]. An audit samples ballots and measures the discrepancy of the sampled electronic votes with respect to their paper record. Risk-limiting post-election audits seek to minimize the size of the audit when the outcome is correct [25]. The margin of victory is an important parameter used to determine the size of an audit for this method.

We study all combinations of voting rules, cost schemes, and bribery actions considered by Mattei *et al.* [19], as well as weighted voters and the negative scenario. The destructive variant has been investigated in various voting problems [4,6,15], including bribery [13] without combinatorial domains. In all these settings, for the unique-winner case, the destructive version is at most as hard as the constructive one. We think that our work might be interesting for several reasons: First, in our setting, destructive bribery turns out to be harder than constructive bribery in many cases. Second, the problems we use for our reductions (two variants of the SATISFIABILITY problem and the KNAPSACK problem) are not the typical ones that are often used in the context of voting problems. An overview of our results is given in Table 1 on page 6.

2   Preliminaries

Almost all our notations and definitions can be found in greater detail and exemplified in the articles by Mattei *et al.* [19] and by Dorn and Krüger [7], who analyzed the constructive case of the same scenario.

This section is structured as follows. First, we present the $\mathcal{NP}$-complete problems we use for our reductions. Afterwards we define CP-nets and introduce related notation. This is followed by the introduction of the voting rules we will work with. We are then ready to define the bribery problem in the setting of CP-nets and introduce the different cost schemes and allowed bribery actions. Finally, we give an overview (Table 1) of the results obtained in this paper and close this section with an example.

For our reductions we use the following $\mathcal{NP}$-complete problems.

(NOT-ALL-EQUAL) 3-SATISFIABILITY, (NAE-)3SAT [14]
Given: A set U of $\mathfrak{n}$ variables ν_i, collection C of $\mathfrak{m}$ clauses over U such that each clause $\gamma \in C$ is a subset of U with $|\gamma| = 3$.
Question: Is there a truth assignment for U such that each clause in C has at least one true (and one false) literal?

KNAPSACK [14]
Given: A set U of n objects $(w_i, v_i) \in \mathbb{N}^2$ of weight $w_i \in \mathbb{N}$ and value $v_i \in \mathbb{N}$, positive integers $k, b \in \mathbb{N}$.
Question: Is there a subset $U' \subseteq U$ of objects with total weight at most b and total value at least k?

CP-nets. In our setting, we are given a set of m *issues* $M = \{X_1, \ldots, X_m\}$, and each $X_i \in M$ has a binary *domain* $D(X_i) = \{x_i, \overline{x_i}\}$. A complete assignment to all issues is called a *candidate*, so there are 2^m different candidates. Each of the n voters has (possibly) conditional preferences over the values assigned to the issues; if the preference of an issue X depends on one or more other issues (called the *parents* $Pa(X)$), we call this issue *dependent*, and *independent* otherwise. Formally, a CP-net is defined by a directed graph (with the issues as its vertices and directed edges going from $Pa(X)$ to X) modeling these dependencies, and a table for each issue, containing the preference over the assignment to this issue for each different complete assignment to its parents; each combination of an assignment to the parents and the corresponding preference over the issue is called a *cp-statement*. For example, for a CP-net with issues X and Y, the cp-statement $x > \overline{x}$ means that the assignment $X = x$ is unconditionally preferred to $X = \overline{x}$, while the statements $x : \overline{y} > y$ and $\overline{x} : y > \overline{y}$ express that the assignment $Y = \overline{y}$ is only preferred to $Y = y$ in the case that $X = x$ (hence, $Pa(Y) = \{X\}$ here). The collection of CP-nets of all voters is called a *profile*.

CP-nets only define a partial order over the candidates, i.e., some candidates are incomparable. One way to expand this to strict total orders over the candidates is to give a strict total order over all issues such that no issue depends on any issue following it in this order. If the same order $\mathcal{O}$ works out for all CP-nets of a profile, the profile is called $\mathcal{O}$-*legal* [16].

Throughout this work, we assume that the voters' preferences on the set of issues are given by *compact* (the number of parents of each issue is bounded by a constant) and *acyclic* (the corresponding graph is acyclic) CP-nets. For acyclic CP-nets, the most preferred candidate of a voter can be determined efficiently [1].

An example of a profile consisting of three CP-nets is given in Table 2 at the end of this section. The CP-nets encode conditional preferences for the alternative options of a menu consisting of a main dish, a side dish and a drink. Alice's choice for the drink is dependent of the choice for the main dish: She prefers beer to wine in case meat is served, and wine to beer if fish is served as a main dish.

Voting. A voting rule maps a profile to a set of candidates. With **One-Step-k-Approval** (OK), only the k most preferred candidates of each voter obtain 1 point each. The winner of the election is the candidate with the most points (or all candidates with those points). In particular, we consider the special cases **One-Step-Plurality** (OP), where $k = 1$, and **One-Step-Veto** (OV), where $k = 2^m - 1$. With **Sequential majority** (SM), given a total order $\mathcal{O}$ for which the profile is $\mathcal{O}$-legal, we follow this order issue by issue, and execute a majority vote for each issue. The voters fix the winning value of the corresponding issue in their CP-net and then go on to the next issue. The winning candidate is the combination of the winners of the individual steps taken. These rules are also used by Mattei *et al.* [19].

Interestingly, it is $\mathcal{NP}$-hard to determine the winner for the voting rule One-Step-k-Approval (OK) if k is part of the input [8]. Therefore, we restrict our analysis to efficient cases of OK where k has a value which is polynomial in n

and m or for $\mathcal{O}$-legal profiles where k is a power of 2. We denote these cases by OK^{eff}. Here, the winner can be determined in polynomial time using results by Brafman *et al.* [2, Theorem 9] and Mattei *et al.* [19, Lemma 1]. We refer to the longer version [8] of this work including explanations and concrete algorithms. In the rest of this work we will focus on OK^{eff} instead of OK.

Bribery. We consider the problem that an external agent, the *briber*, who knows the CP-nets of all voters, asks them to execute changes in their cp-statements. We distinguish the cases that the briber can ask for a change in cp-statements of independent issues only (IV), dependent issues only (DV), or in all $(IV + DV)$ issues [19]. We consider the following five cost schemes [19]:

- C_{EQUAL}: Any amount of change in a single CP-net has the same unit cost.
- C_{FLIP}: The cost of changing a CP-net is the total number of individual cp-statements that must be flipped to obtain the desired change.
- C_{LEVEL}: The cost of a bribery is computed[1] as $\sum_{X_j \in M'} (k + 1 - \text{level}(X_j))$, where $M' \subseteq M$ is the set of bribed issues for this voter, k is the number of levels in the CP-net, and $\text{level}(X_j)$ corresponds to the depth of issue X_j in the dependency graph. More precisely,
$$\text{level}(X_j) = \begin{cases} 1 & \text{if } X_j \text{ is an independent issue} \\ i + 1 & \text{else, with } i = max\{\text{level}(X_k) \mid X_k \text{ is a parent of } X_j\}. \end{cases}$$
- C_{ANY}: The cost is the sum of the flips, each weighted by a specific cost.
- C_{DIST}: This cost scheme requires a fixed order of the issues for each voter (not necessarily the same for each of them), inducing a strict total order over all candidates. The cost to bribe a voter to make c his top candidate is the number of candidates which are better ranked than c in this order.

Additionally, these cost schemes are extended by a cost vector $\mathbf{Q} \in (\mathbb{N})^n$ for an individual cost factor for each voter. The factor for voter v_i is denoted by $Q[i]$ and is multiplied with the costs calculated by the used cost scheme to obtain the amount that the briber has to pay to bribe v_i.

The destructive (D, A, C)-bribery problem is then defined in the following way:

(D, A, C)-DESTRUCTIVE-BRIBERY (DB)

Given: A profile of n CP-nets over m common binary issues, a winner determination voting rule $D \in \{SM, OP, OV, OK^{\mathit{eff}}\}$, a cost scheme $C \in \{C_{\text{EQUAL}}, C_{\text{FLIP}}, C_{\text{LEVEL}}, C_{\text{ANY}}, C_{\text{DIST}}\}$, a bribery action $A \in \{IV, DV, IV + DV\}$, a cost vector $\mathbf{Q} \in (\mathbb{N})^n$, a budget $\beta \in \mathbb{N}$, and a disliked ('hated') candidate h. With SM, we also require $\mathcal{O}$-legality for one common order and with C_{DIST}, and OK^{eff} up to n given total orders over the issues.

Question: Is it possible to change the cp-statements of the voters such that the candidate h is not in the set of winners of the bribed election, without exceeding β?

---

[1] The formula given here differs from the one of Mattei *et al.* [19]. See the argument of Dorn and Krüger [7, Remark 1] why both are equivalent.

We also consider WEIGHTED-(D, A, C)-DB, which is defined in the same way, but with weighted voters, which is a typical variant for bribery problems (see the overview of Faliszewski *et al.* [12]).

Moreover, we also consider weighted and unweighted (D, A, C)-NEGATIVE-DB. The notion of negative bribery was introduced by Faliszewski *et al.* [12] for the constructive case (to make a candidate win the election) in order to cover a more inconspicuous way of bribery: the briber wants to make his preferred candidate p win by not bribing any voter to vote directly for p, therefore just redistributing the votes for the other candidates through bribery. For the destructive case we consider in this work, the analogue restriction is to prohibit bribing voters to vote against the disliked candidate if they have not done so before (recall that with OK and OV, a voter votes for several candidates).

Sometimes we show that a result holds for both, the negative and the non-negative case. We indicate this by ⟨NEGATIVE⟩ in the problem name.

For all our hardness results we prove only $\mathcal{NP}$-hardness for the corresponding problems, but immediately obtain $\mathcal{NP}$-completeness due to obvious membership in $\mathcal{NP}$ for all of the problems. Due to space constraints we omit some proofs and the analysis of some restricted cases and refer to a longer version of this paper [8].

Table 1. Complexity results ($\mathcal{P}$ stands for solvability in polynomial time, $\mathcal{NP}$-c for $\mathcal{NP}$-completeness) for variants of the destructive bribery problem in CP-nets shown in this paper. These variants are specified by a cost scheme (C_{EQUAL}, C_{FLIP}, C_{LEVEL}, C_{ANY}, C_{DIST}), given at the top of the corresponding column, and a voting scheme ($SM, OP, OV, OK^{\text{eff}}$) at the beginning of the corresponding row. The unweighted variants are given in the top half of the table, the weighted ones are listed in the bottom half. The given results all hold for the bribery actions IV, DV, and $IV + DV$, if not stated differently. The cases that are solvable in polynomial time, if the entry in the cost vector is identical for every voter, are not included.

			C_{EQUAL}	C_{FLIP}	C_{LEVEL}	C_{ANY}	C_{DIST}	negative	non-negative
unweighted	SM		$\mathcal{P}$	$\mathcal{P}$	$\mathcal{P}$	$\mathcal{P}$	$\mathcal{P}$	Thm. 1	Thm. 1
	OP	IV	$\mathcal{NP}$-c	$\mathcal{NP}$-c	$\mathcal{NP}$-c	$\mathcal{NP}$-c	$\mathcal{NP}$-c	Thm. 2	Cor. 1
		DV	$\mathcal{NP}$-c	$\mathcal{NP}$-c	$\mathcal{NP}$-c	$\mathcal{NP}$-c	$\mathcal{NP}$-c	Thm. 2	Cor. 1
		$IV + DV$	$\mathcal{P}$	$\mathcal{NP}$-c	$\mathcal{NP}$-c	$\mathcal{NP}$-c	$\mathcal{NP}$-c	Thm. 3/2	Cor. 2/1
	OV		$\mathcal{P}$	$\mathcal{P}$	$\mathcal{P}$	$\mathcal{P}$	$\mathcal{P}$	Thm. 4	Thm. 4
	OK^{eff}	IV	$\mathcal{NP}$-c	$\mathcal{NP}$-c	$\mathcal{NP}$-c	$\mathcal{NP}$-c	$\mathcal{NP}$-c	Thm. 2	Cor. 1
		DV	$\mathcal{NP}$-c	$\mathcal{NP}$-c	$\mathcal{NP}$-c	$\mathcal{NP}$-c	$\mathcal{NP}$-c	Thm. 2	Cor. 1
		$IV + DV$	$\mathcal{P}$	$\mathcal{NP}$-c	$\mathcal{NP}$-c	$\mathcal{NP}$-c	$\mathcal{NP}$-c	Thm. 3/2	Cor. 2/1
weighted	SM		$\mathcal{NP}$-c	$\mathcal{NP}$-c	$\mathcal{NP}$-c	$\mathcal{NP}$-c	$\mathcal{NP}$-c	Thm. 7	Thm. 7
	OP		$\mathcal{NP}$-c	$\mathcal{NP}$-c	$\mathcal{NP}$-c	$\mathcal{NP}$-c	$\mathcal{NP}$-c	Thm. 5	Thm. 5
	OV	IV	$\mathcal{NP}$-c	$\mathcal{NP}$-c	$\mathcal{NP}$-c	$\mathcal{NP}$-c	$\mathcal{NP}$-c	Thm. 6	Thm. 6
		DV	$\mathcal{NP}$-c	$\mathcal{NP}$-c	$\mathcal{NP}$-c	$\mathcal{NP}$-c	$\mathcal{P}$	Thm. 6/8	Thm. 6/8
		$IV + DV$	$\mathcal{NP}$-c	$\mathcal{NP}$-c	$\mathcal{NP}$-c	$\mathcal{NP}$-c	$\mathcal{NP}$-c	Cor. 6	Thm. 6
	OK^{eff}		$\mathcal{NP}$-c	$\mathcal{NP}$-c	$\mathcal{NP}$-c	$\mathcal{NP}$-c	$\mathcal{NP}$-c	Thm. 5	Thm. 5

Table 2. Example for three CP-nets of the voters Alice, Bob, and Charlie over the three issues *main dish*, *side dish*, and *drink* with the domains $D(main) = \{fish, meat\}$, $D(side) = \{rice, chips\}$, and $D(drink) = \{wine, beer\}$.

	main	side	drink
Alice (A)	fish > meat	rice > chips	meat: beer > wine
(main > drink > side)			fish: wine > beer
Bob (B)	fish > meat	chips > rice	meat, chips: beer > wine
(main > side > drink)			meat, rice: beer > wine
			fish, chips: beer > wine
			fish, rice: wine > beer
Charlie (C)	chips: fish > meat	chips > rice	chips: beer > wine
(side > drink > main)	rice: meat > fish		rice: wine > beer

Example. Table 2 shows the CP-nets of the voters Alice (A), Bob (B), and Charlie (C) over the three issues *main dish*, *side dish*, and *drink* with the domains $D(main) = \{fish, meat\}$, $D(side) = \{rice, chips\}$, and $D(drink) = \{wine, beer\}$ to vote for a joined meal. Additionally, the individual orders of the issues are given for each voter, implying the following total orders as expansions of the partial orders defined by the CP-nets:

A: (*fish,rice,wine*) > (*fish,chips,wine*) > (*fish,rice,beer*) > (*fish,chips,beer*) ...
B: (*fish,chips,beer*) > (*fish,chips,wine*) > (*fish,rice,wine*) > (*fish,rice,beer*) ...
C: (*fish,chips,beer*) > (*meat,chips,beer*) > (*fish,chips,wine*) > (*meat,chips,wine*) ...

Using the voting rule OK with $k = 3$, the candidate (*fish, chips, wine*) wins the election, because it is the only one receiving a point from each of the three voters. With the voting rule OP, the candidate (*fish, chips, beer*) is the winner, thanks to the two points from Charlie and Bob. The same candidate wins with the voting rule SM with respect to the order $\mathcal{O}$: *side* > *main* > *drink*, for which the given profile is $\mathcal{O}$-legal. In the majority vote on the issue *side*, Bob and Charlie prefer *chips*, so *chips* is chosen as a *side dish*. Because of this Charlie votes—like the other two voters—for *fish* in the majority vote for the second issue of order $\mathcal{O}$. Finally Alice gets outvoted in the last issue, so the candidate (*fish, chips, beer*) is the winner with the voting rule SM, too. With the voting rule OV, Alice casts her veto against (*meat, chips, wine*), Bob casts his veto against (*meat, rice, wine*), and Charlie casts his veto against (*fish, rice, beer*). Therefore, the remaining five candidates are the winning candidates with this rule. If a unique winner was needed, a tie-breaking rule could be used.

If the briber wanted to prevent candidate $h = (fish, chips, beer)$ from winning with voting rule OP, it would be sufficient to bribe Bob to flip his preference in issue *side* to *rice* > *chips*. This would make (*fish, rice, wine*) the top candidate of Bob and, since Alice is voting for the very same candidate in the first place, the winner of the election. So the briber can reach his goal by this bribery.

Note that this flip is only possible with the bribery action IV or $IV + DV$, because *side* is an independent issue for Bob.

How much does the briber have to pay for this requested flip? For the cost scheme C_{ANY}, this directly depends on the input, since each flip can have its own costs. With C_{FLIP}, the cost factor is 1, since only one cp-statement has to be flipped. With C_{LEVEL}, it is 2 for this flip, because Bob's CP-net has two levels and the issue *side* is an independent one. The costs are the same for C_{DIST}, because Bob only prefers the candidates (*fish, rice, beer*) and (*fish, chips, wine*) to (*fish, rice, wine*). Finally, with C_{EQUAL}, bribing Bob has the same prize (assuming equal cost vector entries) as bribing any other arbitrary voter with a different top candidate to vote for (*fish, rice, wine*). To obtain the final costs the briber has to pay, each of these values is then multiplied by the corresponding entry of the cost vector $\mathbf{Q}$. Therefore, the briber might sometimes be cheaper off to bribe a voter with a small cost vector entry to flip a lot of cp-statements, than one with only a few flips required but having a huge cost vector entry.

3   The Unweighted Case

In this section, we investigate the case where voters are unweighted.

Theorem 1. (SM, A, C)-$\langle\text{NEGATIVE}\rangle$-DB *with bribery action* $A \in \{IV, DV, IV + DV\}$ *and cost scheme* $C \in \{C_{\text{EQUAL}}, C_{\text{ANY}}, C_{\text{FLIP}}, C_{\text{LEVEL}}, C_{\text{DIST}}\}$ *is solvable in polynomial time.*

Proof. We start with the negative case. For each issue find the minimum costs to spend for reaching a majority against h. This can be done by collecting all costs for one issue, sorting and summing up. Finally the issue which is cheapest to bribe is chosen. This works for each cost scheme for which it is easy to calculate the bribery costs for a single flipped issue, which is the case for all cost schemes used here[2].

Note that depending on the allowed bribery action, not every voter may be bribable in each issue. Similarly, we have to ignore voters who initially vote for h but who do not vote for h any more after a bribery of the considered issue. These voters have to be taken into account in the non-negative case, though. However, this is the only modification needed for this case. $\square$

Theorem 2. (D, A, C)-NEGATIVE-DB *with* $D \in \{OP, OK^{\text{eff}}\}$, $A \in \{IV, DV\}$, $C \in \{C_{\text{EQUAL}}, C_{\text{FLIP}}, C_{\text{LEVEL}}, C_{\text{ANY}}, C_{\text{DIST}}\}$ *is* $\mathcal{NP}$-*complete. In addition,* $(D, IV + DV, C')$-NEGATIVE-DB *is* $\mathcal{NP}$-*complete for* $C' \in \{C_{\text{FLIP}}, C_{\text{LEVEL}}, C_{\text{ANY}}, C_{\text{DIST}}\}$. *All these results hold even if all entries in the cost vector are identical.*

Proof. We give a base reduction from NAE-3SAT to prove Theorem 2. For this reduction we claim that some properties hold, which we will then show to hold

---

[2] This is most unintuitive for C_{DIST}, but identifying the top candidate after bribing one issue and determining the respective cost can be done in polynomial time as described by Mattei *et al.* [19, Theorem 3].

for the various combinations of allowed bribery action, cost scheme and voting rule.

Assume we are given an NAE-3SAT instance with m clauses and n variables. For each variable ν_i, we create one issue X_i. Since each issue has exactly two possible assignments, we can establish a one-to-one relation between the full assignments of the issues X_i and the one of the variables ν_i. For this relation we say the assignment of x_i to X_i corresponds to the assignment of 1 to variable ν_i. We will later on—in the extensions of the base case—create additional gadget-issues.

For each clause γ with the variables ν_q, ν_r, ν_s and each of the six different satisfying assignments to these variables for γ, we create one voter with the following preferences: He prefers x_i over $\overline{x_i}$ for each issue with $i \notin \{q, r, s\}$. For the remaining three issues he prefers x_l over $\overline{x_l}$, if 1 is assigned to variable ν_l in the satisfying assignment of γ; he prefers $\overline{x_l}$ over x_l otherwise. For the considered clause with variables ν_q, ν_r, ν_s, we obtain 6 voters which we refer to as qrs-voters. Doing this for all clauses, we obtain $6m$ voters. Finally, we create $m-1$ additional voters, all having h—an arbitrary candidate—as their top-candidate . We set the entry in the cost vector for each voter to 1.

We assume that the following property holds:

(i) No voter who is voting for h can be bribed to vote against h.

We further assume that for all other voters the following properties apply:

(ii) The preferences within issues associated with the clause the voter was created for cannot be changed.
(iii) Changing the preferences within a gadget-issue does not help the briber.
(iv) The preferences within all remaining issues can be bribed freely.

Assuming that properties (i) to (iv) are fulfilled, it is easy to see that the bribery instance can be solved if and only if the corresponding NAE-3SAT-formula is satisfiable. Given a satisfying assignment to the formula, we can translate it to the winning candidate by following the one-to-one relation. Since the assignment satisfies each clause, the briber can bribe one of the six voters for each clause to vote for the winning candidate (following (ii) and (iv)). Therefore this candidate will have m votes in the end, while there are still only $m - 1$ votes for h. The other direction can be shown analogously with the help of property (iii).

We will now show the extensions to this base reduction to prove Theorem 2. Property (i) is fulfilled because we are looking at the negative case.

(OP, IV, C): We need one gadget-issue X^*. Each voter is set to prefer x^* over $\overline{x^*}$. For each qrs-voter, the issues X_q, X_r, X_s are changed to depend on X^*, keeping their original preferences for x^* and inverted preferences for $\overline{x^*}$. Here we utilize that NAE-3SAT is closed under complement, therefore the assignment of X^* is not important at all. This modification ensures the properties (ii)–(iv) to hold for each cost scheme, since no costs are involved.

(OP, DV, C): Once again we need one gadget-issue X^*. Each voter is set to prefer x^* over $\overline{x^*}$. Complementary to the case before, for each qrs-voter, each issue X_i with $i \notin \{q, r, s\}$ is changed to depend on X^*, keeping their original preferences for x^* and inverted preferences for $\overline{x^*}$. This modification ensures the properties (ii)–(iv) to hold for each cost scheme, since no costs are involved.

$(OP, IV + DV, C_{\text{ANY}})$: With $IV + DV$ we need costs and an appropriate budget to ensure that the issues X_q, X_r, X_s cannot be bribed for a qrs-voter. With C_{ANY} we can simply set the costs to bribe these issues to 1, and for each remaining issue to 0. With the budget set to $\beta = 0$ this ensures the properties (ii)–(iv) hold.

$(OP, IV + DV, C_{\text{FLIP}})$: We need to add $\mathfrak{m}^2(\mathfrak{n} - 3)$ gadget-issues, which we denote as $X^*_{a,b}$ with $1 \leq a \leq \mathfrak{m}$ and $1 \leq b \leq \mathfrak{m}(\mathfrak{n} - 3)$. For each qrs-voter, each gadget-issue $X^*_{j,b}$ with $1 \leq b \leq \mathfrak{m}(\mathfrak{n} - 3)$ depends on the issues X_q, X_r, X_s corresponding to the variables of the clause γ_j. With the most preferred assignment in these three issues the preference in the gadget-issue is set to $x^*_{j,b} > \overline{x^*_{j,b}}$, and for each other assignment to $\overline{x^*_{j,b}} > x^*_{j,b}$. Finally we set the budget to $\beta = \mathfrak{m}(\mathfrak{n} - 3)$. This ensures the properties (ii)–(iv) to hold.

$(OP, IV + DV, C_{\text{LEVEL}})$: We need $\mathfrak{mn}$ gadget-issues X^*_b for $b \in \{1, \ldots, \mathfrak{mn}\}$. In contrast to C_{FLIP}, these issues do not depend on the issues X_q, X_r, X_s corresponding to the variables of the clause γ_j in *parallel*, but in a *queue*. So X^*_1 depends on the issues X_q, X_r, X_s for a qrs-voter. Only for the most preferred assignment within these three issues we set the preference of issue X^*_1 to $x^*_1 > \overline{x^*_1}$; in all other cases we set it to $\overline{x^*_1} > x^*_1$. For each subsequent issue X^*_b of this queue with $2 \leq b \leq \mathfrak{mn}$, we set the preferences $x^*_{b-1} : x^*_b > \overline{x^*_b}$ and $\overline{x^*_{b-1}} : \overline{x^*_b} > x^*_b$. In addition, for all issues X_i with $i \in \{1, \ldots, \mathfrak{n}\} \setminus \{q, r, s\}$, we set $x^*_{\mathfrak{mn}} : x_i > \overline{x_i}$ and else $\overline{x_i} > x_i$. Finally we set the budget to $\beta = \mathfrak{m}(\mathfrak{n} - 3)$. This ensures the properties (ii)–(iv).

$(OP, IV + DV, C_{\text{DIST}})$: We add $\lceil \log \mathfrak{m} \rceil + 1$ gadget-issues, denoted by X^*_a. Every voter prefers $x^*_a > \overline{x^*_a}$ for each such issue. For each voter we set the (not necessarily identical) order as follows. The most important issues are the three issues corresponding to the variables of the clause γ_j associated with it, followed by the gadget-issues, and then by the remaining $\mathfrak{n} - 3$ issues. The exact order within these three blocks is not important. We set the budget to $\beta = \mathfrak{m} \cdot 2^{\mathfrak{n}-3} - 1$. This ensures that the briber can bribe all of the least important $\mathfrak{n} - 3$ issues for $\mathfrak{m}$ voters, while the budget is still too low to bribe even only one of the three most important issues of just one voter. Note that bribing such an issue costs at least $2^{\mathfrak{n}-3+\lceil \log \mathfrak{m} \rceil} > \beta$. Therefore the properties (ii)–(iv) hold.

Since OP is the special case of OK^{eff} with $k = 1$, the $\mathcal{NP}$-completeness results shown so far carry over to OK^{eff}.   □

We can achieve the same results for the non-negative cases, but for this we have to drop the constraint that these problems are $\mathcal{NP}$-complete even if the cost vector contains only identical values.

Corollary 1. (D, A, C)-DB *is $\mathcal{NP}$-complete for each combination of a voting rule $D \in \{OP, OK^{\text{eff}}\}$, a cost scheme $C \in \{C_{\text{EQUAL}}, C_{\text{FLIP}}, C_{\text{LEVEL}}, C_{\text{ANY}}, C_{\text{DIST}}\}$ and a bribery action $A \in \{IV, DV\}$. Moreover, $(D, IV + DV, C')$-DB with cost scheme $C' \in \{C_{\text{FLIP}}, C_{\text{LEVEL}}, C_{\text{ANY}}, C_{\text{DIST}}\}$ is $\mathcal{NP}$-complete, too.*

Proof. This follows by the same techniques used in the proof of Theorem 2. The only difference is that property (i) is not automatically satisfied. We can achieve unbribability of the voters voting for h by setting the entries of the cost vector for these voters to $\beta + 1$. Therefore, $\mathcal{NP}$-completeness follows.   □

For the remaining cases, we omit the proofs and refer to the long version of this paper [8]. Interestingly, the voting rule OV is another special case of OK which can be evaluated and solved in polynomial time.

Theorem 3. $(D, IV + DV, C_{\text{EQUAL}})$-NEGATIVE-DB *with* $D \in \{OP, OK^{\textit{eff}}\}$ *is solvable in polynomial time.*

Corollary 2. $(D, IV + DV, C_{\text{EQUAL}})$-DB *is solvable in polynomial time with* $D \in \{OP, OK^{\textit{eff}}\}$.

Theorem 4. (OV, A, C)-⟨NEGATIVE⟩-DB *with bribery action* $A \in \{IV, DV, IV + DV\}$ *and cost scheme* $C \in \{C_{\text{EQUAL}}, C_{\text{FLIP}}, C_{\text{LEVEL}}, C_{\text{ANY}}, C_{\text{DIST}}\}$ *is in* $\mathcal{P}$.

We remark that the combinations of OP, $IV + DV$, and $C_{\text{FLIP}}, C_{\text{LEVEL}}, C_{\text{DIST}}$, and OP, IV, C_{FLIP}, each for the unweighted, non-negative case and if all entries of the cost vector are the same, can be solved in polynomial time, too (see [8]). This is in line with the observations of Mattei *et al.* [19, Theorem 7] and Dorn and Krüger [7, Theorem 7] that sometimes the bribery problem can be solved in polynomial time when the cost vector has only identical entries.

4 The Weighted Case

In this section, we consider the case of weighted voters, which turns out to be $\mathcal{NP}$-complete for almost all combinations—with two exceptions. The reductions we give here are all from the KNAPSACK problem.

Theorem 5. WEIGHTED-(D, A, C)-⟨NEGATIVE⟩-DB *is* $\mathcal{NP}$-complete *with voting rule* $D \in \{OP, OK^{\textit{eff}}\}$, *bribery action* $A \in \{IV, DV, IV + DV\}$, *and cost scheme* $C \in \{C_{\text{EQUAL}}, C_{\text{FLIP}}, C_{\text{LEVEL}}, C_{\text{ANY}}, C_{\text{DIST}}\}$.

Proof. For the voting rule OP, we use a reduction from KNAPSACK. Given a KNAPSACK instance $(\{(w_1, v_1), \ldots, (w_n, v_n)\}, k, b)$ we construct a voting scheme such that a successful bribery against the candidate h is possible if and only if the given Knapsack problem can be solved.

First, we use two issues X_1 and X_2, set $h = \overline{x_1}\,\overline{x_2}$, and the budget $\beta = b$. We create one voter preferring h with weight $l + k - 1$, where $l = \sum_{i=1}^{n} v_i$. We set the entry of the cost vector for this voter to $\beta + 1$ in order to make him unbribable (just to make this proof hold for the non-negative case, too).

For every object (w_i, v_i) we add a voter preferring $x_1\overline{x_2}$ to all other candidates. This voter is weighted by v_i and his entry in the cost vector is set to w_i. Last, we add a single voter of weight l preferring $x_1 x_2$ to all other candidates. Since this candidate should win the election, we make this voter unbribable by setting the entry in the cost vector to $\beta + 1$.

For the bribery actions IV and $IV + DV$ all issues are independent. For the bribery action DV we let the preferences of the issue X_2 depend on the assignment of issue X_1 for each voter created for an object (w_i, v_i). Such a voter will therefore prefer x_1 independently over $\overline{x_1}$, and the preference for X_2 will be $x_1 : \overline{x_2} > x_2$, $\overline{x_1} : x_2 > \overline{x_2}$.

By construction, only voters created for objects can be bribed, and only changing their favorite candidate to x_1x_2 is helpful here. Each bribery of one such voter results in a value of 1 for each cost scheme, which is then multiplied by the entry of the cost vector. Note that a cost of 1 is obvious in all cases except C_{DIST}. Here, we have to ensure that the target of our bribery is on the second position. As the prefered candidate is $x_1\overline{x_2}$ and our potential winner is x_1x_2, we have to set X_1 before X_2 in the order on the issues. The claimed cost of 1 for this bribery action then holds. It is easy to see that there must exist a solution to the underlying KNAPSACK instance in order to be able to prohibit the candidate h from winning.

Since the briber cannot bribe the single voter voting for h by construction, this reduction holds for the negative case as well. Because OP is a special case of OK^{eff}, this result automatically carries over. □

The main idea of this reduction can be adjusted to show $\mathcal{NP}$-completeness for the voting rules OV and SM, too. In contrast, the combination of OV with C_{DIST} can be solved in polynomial time by a greedy-algorithm because all possible bribery actions are for free. See the long version [8] for the proofs.

Theorem 6. WEIGHTED-(OV, A, C)-⟨NEGATIVE⟩-DB *with a bribery action* $A \in \{IV, DV, IV + DV\}$ *and cost scheme* $C \in \{C_{\text{EQUAL}}, C_{\text{FLIP}}, C_{\text{LEVEL}}, C_{\text{ANY}}, C_{\text{DIST}}\}$ *is* $\mathcal{NP}$-*complete, except for the combination* $A = DV$ *and* $C = C_{\text{DIST}}$.

Theorem 7. WEIGHTED-(SM, A, C)-⟨NEGATIVE⟩-DB *is* $\mathcal{NP}$-*complete for cost scheme* $C \in \{C_{\text{EQUAL}}, C_{\text{FLIP}}, C_{\text{LEVEL}}, C_{\text{ANY}}, C_{\text{DIST}}\}$ *and allowed bribery action* $A \in \{IV, DV, IV + DV\}$.

Theorem 8. WEIGHTED-$(OV, DV, C_{\text{DIST}})$-⟨NEGATIVE⟩-DB *is solvable in polynomial time.*

5 Conclusion

We extensively studied destructive bribery for the weighted, unweighted, negative and non-negative variations on all cost-, bribery- and evaluation-schemes introduced by Mattei *et al.* [19]. Table 1 summarizes our results. The main differences can be observed between the weighted and unweighted cases, while the negative and non-negative cases are very similar—we remark that these cases may behave differently, however, if the cost vector assigns the same value to each of the voters. The cost vector is also the tool to mimic the restriction of the negative setting in the non-negative case: with its help, one can make a voter unbribable (one cannot use it to affect the bribery actions though).

It is also interesting to observe that—for an arbitrary cost vector—all combinations in the weighted case for the *constructive* bribery problem turned out to be $\mathcal{NP}$-complete [7], whereas in the destructive setting, we have identified two tractable cases. They occur due to the strange side effect of the combination of DV and C_{DIST} that sets all reasonable bribery free of charge.

The most interesting observation might be that in the unweighted setting, only the combination of SM and C_{EQUAL} was shown to be $\mathcal{NP}$-complete for constructive bribery [7,19], while almost half of the combinations for destructive bribery turned out to be computationally hard - this behavior is rather unusual for voting problems and is due to the combinatorial structure of the set of candidates. If the number of candidates is part of the input (which is the case for many of the common settings for voting problems), the constructive case of a voting problem can be directly used to solve the destructive counterpart: If it is known how to make a designated candidate the only winner of the election, one can simply run this procedure for all candidates and find out which of them is the cheapest solution. In the setting of combinatorial domains where the number of candidates is exponential in the size of the input, one cannot simply check which of the exponentially many candidates should be chosen to be made the winner. It might turn out, nevertheless, that the destructive version is not computationally harder, but in our case, we have seen that precluding an alternative might be more difficult than pushing it through.

References

1. Boutilier, C., Brafman, R.I., Domshlak, C., Hoos, H.H., Poole, D.: Cp-nets: a tool for representing and reasoning with conditional ceteris paribus preference statements. J. Artif. Intell. Res. **21**, 135–191 (2004)
2. Brafman, R., Rossi, F., Salvagnin, D., Venable, K.B., Walsh, T.: Finding the next solution in constraint- and preference-based knowledge representation formalisms. In: 12th International Conference: Principles of Knowledge Representation and Reasoning, pp. 425–433. AAAI Press (2010)
3. Brandt, F., Conitzer, V., Endriss, U.: Computational social choice. In: Multiagent Systems, pp. 213–283. MIT Press (2013)
4. Conitzer, V., Lang, J., Sandholm, T.: How many candidates are needed to make elections hard to manipulate? In: 9th Conference on Theoretical Aspects of Rationality and Knowledge, pp. 201–214. ACM (2003)
5. Conitzer, V., Lang, J., Xia, L.: Hypercubewise preference aggregation in multi-issue domains. In: 22nd International Joint Conference on Artificial Intelligence, pp. 158–163. AAAI Press (2011)
6. Conitzer, V., Sandholm, T.: Complexity of manipulating elections with few candidates. In: 18th National Conference on Artificial Intelligence, pp. 314–319. AAAI Press (2002)
7. Dorn, B., Krüger, D.: On the hardness of bribery variants in voting with CP-nets. Ann. Math. Artif. Intell., 1–29 (2015). doi:10.1007/s10472-015-9469-3
8. Dorn, B., Krüger, D., Scharpfenecker, P.: Often harder than in the constructive case: destructive bribery in CP-nets, pp. 1–22 (2015). CoRR abs/1509.08628
9. Elkind, E., Faliszewski, P., Slinko, A.: Swap bribery. In: Mavronicolas, M., Papadopoulou, V.G. (eds.) Algorithmic Game Theory. LNCS, vol. 5814, pp. 299–310. Springer, Heidelberg (2009)

10. Faliszewski, P., Hemaspaandra, L., Hemaspaandra, E., Rothe, J.: A richer understanding of the complexity of election systems. In: Fundamental Problems in Computing: Essays in Honor of Professor Daniel J. Rosenkrantz, 1st edn., chap. 14, pp. 375–406. Springer (2009)
11. Faliszewski, P.: Nonuniform bribery. In: 7th International Joint Conference on Autonomous Agents and Multiagent Systems, vol. 3, pp. 1569–1572. IFAAMAS (2008)
12. Faliszewski, P., Hemaspaandra, E., Hemaspaandra, L.A.: How hard is bribery in elections? J. Artif. Intell. Res. **35**(2), 485–532 (2009)
13. Faliszewski, P., Hemaspaandra, E., Hemaspaandra, L.A., Rothe, J.: Llull and copeland voting computationally resist bribery and constructive control. J. Artif. Intell. Res. (JAIR) **35**, 275–341 (2009)
14. Garey, M.R., Johnson, D.S.: Computers and Intractability: A Guide to the Theory of NP-Completeness. W. H. Freeman, New York (1979)
15. Hemaspaandra, E., Hemaspaandra, L.A., Rothe, J.: Anyone but him: the complexity of precluding an alternative. Artif. Intell. **171**(5–6), 255–285 (2007)
16. Lang, J.: Vote and aggregation in combinatorial domains with structured preferences. In: 20th International Joint Conference on Artificial Intelligence, pp. 1366–1371 (2007)
17. Magrino, T.R., Rivest, R.L., Shen, E.: Computing the margin of victory in IRV elections. In: Electronic Voting Technology Workshop / Workshop on Trustworthy Elections. USENIX Association (2011)
18. Maran, A., Maudet, N., Pini, M.S., Rossi, F., Venable, K.B.: A framework for aggregating influenced CP-nets and its resistance to bribery. In: 27th AAAI Conference on Artificial Intelligence, pp. 668–674. AAAI Press (2013)
19. Mattei, N., Pini, M.S., Rossi, F., Venable, K.B.: Bribery in voting with CP-nets. Ann. Math. Artif. Intell. **68**(1–3), 135–160 (2013)
20. Norden, L., Burstein, A., Hall, J.L., Chen, M.: Post-election audits: restoring trust in elections. Brennan Center for Justice at New York University, Tech. report (2007)
21. Pini, M., Rossi, F., Venable, K.: Bribery in voting with soft constraints. In: 27th AAAI Conference on Artificial Intelligence, pp. 803–809. AAAI Press (2013)
22. Purrington, K., Durfee, E.H.: Making social choices from individuals' CP-nets. In: 6th International Joint Conference on Autonomous Agents and Multiagent Systems, pp. 1122–1124. IFAAMAS (2007)
23. Reisch, Y., Rothe, J., Schend, L.: The margin of victory in schulze, cup, and copeland elections: complexity of the regular and exact variants. In: 7th European Starting AI Researcher Symposium, pp. 250–259. IOS Press (2014)
24. Rossi, F., Venable, K.B., Walsh, T.: mCP Nets: representing and reasoning with preferences of multiple agents. In: 19th National Conference on Artificial Intelligence, pp. 729–734. AAAI Press (2004)
25. Stark, P.B.: Risk-limiting post-election audits: P-values from common probability inequalities. IEEE Trans. Inf. Forensics Secur. **4**, 1005–1014 (2009)
26. Xia, L.: Computing the margin of victory for various voting rules. In: ACM Conference on Electronic Commerce, EC 2012, Valencia, Spain, 4–8 June 2012, pp. 982–999. ACM (2012)
27. Xia, L., Conitzer, V., Lang, J.: Voting on multiattribute domains with cyclic preferential dependencies. In: 23rd AAAI Conference on Artificial Intelligence, pp. 202–207. AAAI Press (2008)

Improving Selfish Routing
for Risk-Averse Players

Dimitris Fotakis[1]([✉]), Dimitris Kalimeris[2], and Thanasis Lianeas[3]

[1] School of Electrical and Computer Engineering,
National Technical University of Athens, Athens, Greece
fotakis@cs.ntua.gr
[2] Department of Informatics and Telecommunications,
National and Kapodistrian University of Athens, Athens, Greece
sdi1000049@di.uoa.gr
[3] Department of Electrical and Computer Engineering,
University of Texas at Austin, Austin, TX 78701, USA
tlianeas@austin.utexas.edu

Abstract. We investigate how and to which extent one can exploit risk-aversion and modify the perceived cost of the players in selfish routing so that the Price of Anarchy (PoA) is improved. We introduce small random perturbations to the edge latencies so that the expected latency does not change, but the perceived cost of the players increases due to risk-aversion. We adopt the model of γ-modifiable routing games, a variant of routing games with restricted tolls. We prove that computing the best γ-enforceable flow is NP-hard for parallel-link networks with affine latencies and two classes of heterogeneous risk-averse players. On the positive side, we show that for parallel-link networks with heterogeneous players and for series-parallel networks with homogeneous players, there exists a nicely structured γ-enforceable flow whose PoA improves fast as γ increases. We show that the complexity of computing such a γ-enforceable flow is determined by the complexity of computing a Nash flow of the original game. Moreover, we prove that the PoA of this flow is best possible in the worst-case, in the sense that the re are instances where (i) the best γ-enforceable flow has the same PoA, and (ii) considering more flexible modifications does not lead to any further improvement.

1 Introduction

Routing games provide an elegant and practically useful model of selfish resource allocation in transportation and communication networks and have been extensively studied (see e.g., [17]). The majority of previous work assumes that the

This research was supported by the project Algorithmic Game Theory, co-financed by the European Union (European Social Fund) and Greek national funds, through the Operational Program "Education and Lifelong Learning" of the National Strategic Reference Framework - Research Funding Program: THALES, investing in knowledge society through the European Social Fund, and by grant NSF CCF 1216103.

E. Markakis and G. Schäfer (Eds.): WINE 2015, LNCS 9470, pp. 328–342, 2015.
DOI: 10.1007/978-3-662-48995-6_24

players select their routes based on precise knowledge of edge delays. In practical applications however, the players cannot accurately predict the actual delays due to their limited knowledge about the traffic conditions and due to unpredictable events that affect the edge delays and introduce uncertainty (see e.g., [1,12–14] for examples). Hence, the players select their routes based only on delay estimations and are aware of the uncertainty and the potential inaccuracy of them. Therefore, to secure themselves from increased delays, whenever this may have a considerable influence, the players select their routes taking uncertainty into account (e.g., people take a safe route or plan for a longer-than-usual delay when they head to an important meeting or to catch a long-distance flight).

Recent work (see e.g., [1,12,13,15] and the references therein) considers routing games with *stochastic delays* and *risk-averse players*, where instead of the route that minimizes her expected delay, each player selects a route that guarantees a reasonably low actual delay with a reasonably high confidence. There have been different models of stochastic routing games, each modeling the individual cost of risk-averse players in a slightly different way. In all cases, the actual delay is modeled as a random variable and the perceived cost of the players is either a combination of the expectation and the standard deviation (or the variance) of their delay [12,13] or a player-specific quantile of the delay distribution [1,14] (see also [4,18] about the perceived cost of risk-averse players).

No matter the precise modeling, we should expect that stochastic delays and risk-aversion cannot improve the network performance at equilibrium. Interestingly, [13,15] indicate that in certain settings, stochastic delays and risk-aversion can actually improve the network performance at equilibrium. Motivated by these results, we consider routing games on parallel-link and series-parallel networks and investigate how one can exploit risk-aversion in order to modify the perceived cost of the (possibly heterogeneous) players so that the PoA is significantly improved.

Routing Games. To discuss our approach more precisely, we introduce the basic notation and terminology about routing games. A (non-atomic) *selfish routing game* (or instance) is a tuple $\mathcal{G} = (G(V,E), (\ell_e)_{e \in E}, r)$, where $G(V,E)$ is a directed network with a source s and a sink t, $\ell_e : \mathbb{R}_{\geq 0} \to \mathbb{R}_{\geq 0}$ is a non-decreasing delay (or latency) function associated with edge e and $r > 0$ is the traffic rate. We let $\mathcal{P}$ denote the set of simple $s - t$ paths in G. We say that G is a parallel-link network if each $s - t$ path is a single edge (or link).

A (feasible) *flow* f is a non-negative vector on $\mathcal{P}$ such that $\sum_{p \in \mathcal{P}} f_p = r$. We let $f_e = \sum_{p:e \in p} f_p$ be flow routed by f on edge e. Given a flow f, the latency of each edge e is $\ell_e(f) = \ell_e(f_e)$, the latency of each path p is $\ell_p(f) = \sum_{e \in p} \ell_e(f)$ and the latency of f is $L(f) = \max_{p:f_p > 0} \ell_p(f)$.

The traffic r is divided among infinitely many players, each trying to minimize her latency. A flow f is a *Wardrop-Nash flow* (or a *Nash flow*, for brevity), if all traffic is routed on minimum latency paths, i.e., for any $p \in \mathcal{P}$ with $f_p > 0$ and for all $p' \in \mathcal{P}$, $\ell_p(f) \leq \ell_{p'}(f)$. Therefore, in a Wardrop-Nash flow f, all players incur a minimum common latency $\min_p \ell_p(f) = L(f)$. Under weak assumptions on delay functions, a Nash flow exists and is essentially unique (see e.g., [17]).

The efficiency of a flow f is measured by the *total latency* $C(f)$ of the players, i.e., by $C(f) = \sum_{e \in E} f_e \ell_e(f)$. The *optimal flow*, denoted o, minimizes the total latency among all feasible flows. The *Price of Anarchy* (PoA) quantifies the performance degradation due to selfishness. The $\text{PoA}(\mathcal{G})$ of a routing game $\mathcal{G}$ is the ratio $C(f)/C(o)$, where f is the Nash flow and o is the optimal flow o of $\mathcal{G}$. The PoA of a class of routing games is the maximum PoA over all games in the class. For routing games with latency functions in a class $\mathcal{D}$, the PoA is equal to $\text{PoA}(\mathcal{D}) = (1 - \beta(\mathcal{D}))^{-1}$, where $\beta(\mathcal{D}) = \sup_{\ell \in \mathcal{D}, x \geq y \geq 0} \frac{y(l(x) - l(y))}{xl(x)}$ only depends on the class of latency functions $\mathcal{D}$ [3,17].

Using Risk-Aversion to Modify Edge Latencies. The starting point of our work is that in some practical applications, we may intentionally introduce variance to edge delays so that the expected delay does not change, but the risk-averse cost of the players increases. E.g., in a transportation network, we can randomly increase or decrease the proportion of time allocated to the green traffic light for short periods or we can open or close an auxiliary traffic lane. In a communication network, we might randomly increase or decrease the link capacity allocated to a particular type of traffic or change its priority. At the intuitive level, we expect that the effect of such random changes to risk-averse players is similar to that of refundable tolls (see e.g., [6,11]), albeit restricted in magnitude due to the bounded variance in edge delays that we can afford.

E.g., let e be an edge with latency $\ell_e(x)$ where we can increase the latency temporarily to $(1 + \alpha_1)\ell_e(x)$ and decrease it temporarily to $(1 - \alpha_2)\ell_e(x)$. If we implement the former change with probability p_1 and the latter with probability $p_2 < 1 - p_1$, the latency function of e becomes a random variable with expectation $[p_1(1 + \alpha_1) + p_2(1 - \alpha_2) + (1 - p_1 - p_2)]\ell_e(x)$. Adjusting p_1 and p_2 (and possibly α_1 and α_2) so that $p_1\alpha_1 = p_2\alpha_2$, we achieve an expected latency of $\ell_e(x)$. However, if the players are (homogeneously) risk-averse and their perceived delay is given by an $(1 - p_1 + \varepsilon)$-quantile of the delay distribution (e.g., as in [1,14]), the perceived latency on e is $(1 + \alpha_1)\ell_e(x)$. Similarly, if the individual cost of the risk-averse players are given by the expectation plus the standard deviation of the delay distribution (e.g., as in [12]), the perceived latency is $(1 + \sqrt{p_1\alpha_1^2 + p_2\alpha_2^2})\ell_e(x)$. In both cases, we can achieve a significant increase in the delay perceived by risk-averse players, while the expected delay remains unchanged.

In most practical situations, the feasible changes in the latency functions are bounded (and relatively small). Combined with the particular form of risk-averse individual cost, this determines an upper bound γ_e on the multiplicative increase of the delay on each edge e. Moreover, the players may evaluate risk differently and be *heterogeneous* wrt. their risk-aversion factors. So, in general, the traffic rate r is partitioned into k risk-averse classes, where each class i consists of the players with risk-aversion factor a^i and includes a traffic rate r^i. If we implement a multiplicative increase γ_e on the perceived latency of each edge e, the players in class i have perceived cost $(1 + a^i\gamma_e)\ell_e(f)$ on each e and $\sum_{e \in p}(1 + a^i\gamma_e)\ell_e(f)$

on each path[1] p. If the players are *homogeneous* wrt. their risk-aversion, there is a single class of players with traffic rate r and risk-aversion factor $a = 1$.

Contribution. In this work, we assume a given upper bound γ on the maximum increase in the latency functions and refer to the corresponding routing game as a *γ-modifiable game*. We consider both homogeneous and heterogeneous risk-averse players. We adopt this model as a simple and general abstraction of how one can exploit risk-aversion to improve the PoA of routing games. Technically, our model is a variant of restricted refundable tolls considered in [2,9] for homogeneous players and in [10] for heterogeneous players. However, on the conceptual side and to the best of our knowledge, this is the first time that risk-aversion is proposed as a means of implementing restricted tolls, and through this, as a potential remedy to the inefficiency of selfish routing.

A flow f is *γ-enforceable* if there is γ_e-modification on each edge e, with $0 \leq \gamma_e \leq \gamma$, so that f is a Nash flow of the modified game, i.e., for each player class i, for every path p used by class i, and for all paths p', $\sum_{e \in p}(1+a^i\gamma_e)\ell_e(f) \leq \sum_{e \in p'}(1 + a^i\gamma_e)\ell_e(f)$. In this work, we are interested in computing either the best γ-enforceable flow, which minimizes total latency among all γ-enforceable flows, or a γ-enforceable flow with low PoA. We measure the PoA in terms of the total expected latency (instead of the total perceived delay of the players). In practical applications, the total expected latency is directly related to many crucial performance parameters (e.g., to the expected pollution in a transportation network or to the expected throughput in a communication network) and thus, it is the quantity that a central planner usually seeks to minimize.

In Sect. 3, we consider routing games on parallel links with homogeneous players and show that for every $\gamma > 0$, there is a nicely structured γ-enforceable flow whose PoA improves significantly as γ increases. More specifically, based on a careful rerouting procedure, we show that given an optimal flow o, we can find a γ-enforceable flow f (along with the corresponding γ-modification) that "mimics" o in the sense that if $f_e < o_e$, e gets a 0-modification, while if $f_e > o_e$, e gets a γ-modification (Lemma 1). The proof of Lemma 1 implies that given o, we can compute such a flow f and the corresponding γ-modification in time $O(|E|\, T_{\mathrm{NE}})$, where T_{NE} is the complexity of computing a Nash flow in the original instance. Generalizing the variational inequality approach of [3], similarly to [2, Sect. 4], we prove (Theorem 1) that the PoA of the γ-enforceable flow f constructed in Lemma 1 is at most $(1 - \beta_\gamma(\mathcal{D}))^{-1}$, where $\mathcal{D}$ is the class of latency functions and $\beta_\gamma(\mathcal{D}) = \sup_{\ell \in \mathcal{D}, x \geq y \geq 0} \frac{y(\ell(x)-\ell(y))-\gamma(x-y)\ell(x)}{x\ell(x)}$ is a natural generalization of the quantity $\beta(\mathcal{D})$ introduced in [3]. E.g., for affine latency functions, the PoA of the γ-enforceable f is at most $(1 - (1-\gamma)^2/4)^{-1}$ (Corollary 1), which is significantly less that $4/3$ even for small values of γ. We also show that the PoA of such γ-

---

[1] To simplify the model and make it easily applicable to general networks, we assume that the perceived cost of the players under latency modifications is separable. This is a reasonable simplifying assumption on the structure of risk-averse costs (see also [13,15]) and only affects the extension of our results to series-parallel networks.

enforceable flows is best possible in the worst-case for γ-modifiable games with latency functions in class $\mathcal{D}$ (Theorem 2).

In Sect. 4, we switch to parallel-link games with heterogenous players. We prove that computing the best γ-enforceable flow is NP-hard for parallel-link games with affine latencies and only two classes of heterogeneous risk-averse players (Theorem 3). The proof modifies the construction in [16, Sect. 6], which shows that the best Stackelberg modification of parallel-link instances is NP-hard. Our result significantly strengthens [10, Theorem 1], which establishes NP-hardness of best restricted tolls in general $s - t$ networks with affine latencies. On the positive side, we apply [10, Algorithm 1] and show (Theorem 5) that the γ-enforceable flow f of Lemma 1 can be turned into a γ-enforceable flow for parallel-link instances with heterogeneous players. Since only the γ-modifications are adjusted for heterogeneous players, but the flow itself does not change, the PoA of f is bounded as above and remains best possible in the worst case.

In Sect. 5, we extend our approach of finding a γ-enforceable flow that "mimics" the optimal flow to series-parallel networks. Series-parallel networks have received considerable attention in the literature of refundable tolls, see e.g., [5,7], but to the best of our knowledge, they have not been explicitly considered in the setting of restricted tolls. Extending the rerouting procedure of Lemma 1, we show that for routing games in series-parallel networks with homogeneous players, there is a γ-enforceable flow with PoA at most $(1 - \beta_\gamma(\mathcal{D}))^{-1}$ (Lemma 2 and Theorem 6). Such a γ-enforceable flow and the corresponding γ-modifications can be computed in time polynomially related to the time needed for computing Nash flows in series-parallel networks (Lemma 3).

In Sect. 6, we consider (p, γ)-modifiable games, where the p-norm of the edge modifications vector $(\gamma_e)_{e \in E}$ is at most γ. This generalization captures applications where the total variance introduced in the network should be bounded by γ and could potentially lead to an improved PoA. We prove that the worst-case PoA under (p, γ)-modifications is essentially identical to the worst-case PoA under $\gamma/\sqrt[p]{m}$-modifications (Theorem 8). Therefore, even for (p, γ)-modifiable games, the PoA of the $\gamma/\sqrt[p]{m}$-enforceable flow of Lemma 1 is essentially best possible. Due to space constraints, we only sketch the main ideas behind our results and defer the technical details to the full version of this work.

Previous Work. On the conceptual side, our work is closest to those considering the PoA of stochastic routing games with risk-averse players [1,12,15]. Nikolova and Stier-Moses [13] recently introduced the *price of risk-aversion* (PRA), which is the worst-case ratio of the total latency of the Nash flow for risk-averse players to the total latency of the Nash flow for risk-neutral players. Interestingly, PRA can be smaller than 1 and as low as $1 - \beta(\mathcal{D})$ for stochastic routing games on parallel-links (i.e., risk-aversion can improve the PoA to 1 for certain instances).

On the technical side, our work is closest to those investigating the properties of restricted refundable tolls for routing games [2,9,10]. Bonifaci et al. [2] proved that for parallel-link networks with homogeneous players, computing the best γ-enforceable flow reduces to the solution of a convex program. Moreover, they presented a tight bound of $(1 - \beta_\gamma(\mathcal{D}))^{-1}$ on the PoA of a γ-enforceable flow for routing games with latency functions in class $\mathcal{D}$. Jelinek et al. [10] considered restricted tolls for heterogeneous players and proved that computing the best γ-enforceable flow for $s - t$ networks with affine latencies is NP-hard. On the positive side, they proved that for parallel-link games with heterogeneous players, deciding whether a given flow is γ-enforceable (and finding the corresponding γ-modification) can be performed in polynomial time. Moreover, they showed how to compute the best γ-enforceable flow for parallel-link games with heterogeneous players if the maximum allowable modification on each edge is either 0 or infinite.

2   The Model and Preliminaries

The basic model of routing games is introduced in Sect. 1. Next, we introduce some more notation and the classes of γ-modifiable and (p, γ)-modifiable games.

γ-Modifiable Routing Games. A selfish routing game with heterogeneous players in k classes is a tuple $\mathcal{G} = (G(V, E), (\ell_e)_{e \in E}, (a^i)_{i \in [k]}, (r^i)_{i \in [k]})$, where G is a directed $s - t$ network with m edges, a^i is the aversion factor of the players in class i and r_i is the amount of traffic with aversion a^i. We assume that $a^1 = 1$ and $a^1 < a^2 < \ldots < a^k$. If the players are homogeneous, there is a single class with risk aversion $a^1 = 1$ and traffic rate r. Then, an instance is $\mathcal{G} = (G, \ell, r)$.

A flow f is a non-negative vector on $\mathcal{P} \times \{1, \ldots, k\}$. We let $f_p^{a^i}$ be the flow with aversion a^i on path p and $f_p = \sum_i f_p^{a^i}$ be the total flow on path p. Similarly, $f_e^{a^i} = \sum_{p:e \in p} f_p^{a^i}$ is the flow with aversion a^i on edge e and $f_e = \sum_i f_p^{a^i}$ is the total flow on edge e. We let $a_e^{\min}(f)$ (resp. $a_e^{\max}(f)$) be the smallest (resp. largest) aversion factor in e under f. If e is not used by f, we let $a_e^{\min}(f) = a_e^{\max}(f) = a^k$. We say that an edge e (resp. path p) is used by players of type a^i if $f_e^{a^i} > 0$ (resp. for all $e \in p$). To simplify notation, we may write ℓ_e, instead of $\ell_e(f)$.

We say that a routing game $\mathcal{G}$ is γ-*modifiable* if we can select a $\gamma_e \in [0, \gamma]$ for each edge e and change the edge latencies perceived by the players of type a^i from $\ell_e(x)$ to $(1 + a^i \gamma_e) \ell_e(x)$ using small random perturbations. Any vector $\Gamma = (\gamma_e)_{e \in E}$, where $\gamma_e \in [0, \gamma]$ for each edge e, is a γ-*modification* of $\mathcal{G}$. Given a γ-modification Γ, we let $\mathcal{G}^\Gamma$ denote the γ-modified routing game where the perceived cost of the players is changed according to the modification Γ.

A flow f is a *Nash flow* of $\mathcal{G}^\Gamma$, if for any path p and any type a^i with $f_p^{a^i} > 0$ and for all paths p', $\sum_{e \in p}(1 + a^i \gamma_e) \ell_e(f) \leq \sum_{e \in p'}(1 + a^i \gamma_{e'}) \ell_{e'}(f)$. Given a routing game $\mathcal{G}$, we say that a flow f is γ-*enforceable*, or simply *enforceable*, if there exists a γ-modification Γ of $\mathcal{G}$ such that f is a Nash flow of $\mathcal{G}^\Gamma$.

Our assumption is that γ-modifications do not change the expected latency. Therefore, the total latency of f in both $\mathcal{G}^\Gamma$ and $\mathcal{G}$ is $C(f) = \sum_{e \in E} f_e \ell_e(f)$. Hence, the optimal flow o of $\mathcal{G}$ is also an optimal flow of $\mathcal{G}^\Gamma$. A flow f is the *best γ-enforceable* flow of $\mathcal{G}$ if for any other γ-enforceable flow f' of $\mathcal{G}$, $C(f) \leq C(f')$. The Price of Anarchy PoA($\mathcal{G}^\Gamma$) of the modified game $\mathcal{G}^\Gamma$ is equal to $C(f)/C(o)$, where f is the Nash flow of $\mathcal{G}^\Gamma$. For a γ-modifiable game $\mathcal{G}$, the PoA of $\mathcal{G}$ under γ-modifications, denoted PoA$_\gamma(\mathcal{G})$, is $C(f)/C(o)$, where f is the best γ-enforceable flow of $\mathcal{G}$. For routing games with latency functions in class $\mathcal{D}$, $\mathrm{PoA}_\gamma(\mathcal{D})$ denotes the maximum $\mathrm{PoA}_\gamma(\mathcal{G})$ over all γ-modifiable games $\mathcal{G}$ with latencies in $\mathcal{D}$.

(p,γ)-Modifiable Routing Games. Generalizing γ-modifiable games, we select a modification $\gamma_e \geq 0$ for each edge e so that $\|(\gamma_e)_{e \in E}\|_p = \sqrt[p]{\sum_{e \in E} \gamma_e^p} \leq \gamma$, for some given integer $p \geq 1$, and change the perceived edge latencies as above. We refer to such games as (p, γ)-modifiable. All the notation above naturally generalizes to (p, γ)-modifiable games. The PoA of a game $\mathcal{G}$ under (p, γ)-modifications, denoted $\mathrm{PoA}_\gamma^p(\mathcal{G})$, is $C(f)/C(o)$, where f is the best (p, γ)-enforceable flow of $\mathcal{G}$. Similarly, $\mathrm{PoA}_\gamma^p(\mathcal{D})$ is the maximum PoA of all (p, γ)-modifiable games with latency functions in class $\mathcal{D}$.

Series-Parallel Networks. A directed $s - t$ network $G(V, E)$ is *series-parallel* if it either consists of a single edge (s, t) or can be obtained from two series-parallel networks with terminals (s_1, t_1) and (s_2, t_2) composed either in series or in parallel. In a *series composition*, t_1 is identified with s_2, s_1 becomes s, and t_2 becomes t. In a *parallel composition*, s_1 is identified with s_2 and becomes s, and t_1 is identified with t_2 and becomes t (see also [19] for computing the decomposition of a series-parallel network in linear time).

3   Modifying Routing Games in Parallel-Link Networks

In this section, we study γ-modifiable games on parallel-link networks with homogeneous risk-averse players. The following is a corollary of [2, Theorem 1] (see also the main result of [6,11]) and characterizes γ-enforceable optimal flows.

Proposition 1. *Let $\mathcal{G}$ be a γ-modifiable game on parallel links and let o be the optimal flow of $\mathcal{G}$. Then, o is γ-enforceable in $\mathcal{G}$ if and only if for any link e with $o_e > 0$ and all links $e' \in E$, $\ell_e(o) \leq (1 + \gamma)\ell_{e'}(o)$.*

Next, we show that for any instance $\mathcal{G}$, there exist a flow f mimicking o and a γ-modification enforcing f as the Nash flow of the modified instance.

Lemma 1. *Let $\mathcal{G} = (G, \ell, r)$ be a γ-modifiable instance on parallel-links with homogeneous risk-averse players and let o be the optimal flow of $\mathcal{G}$. There is a feasible flow f and a γ-modification Γ of $\mathcal{G}$ such that*
   (i) f is a Nash flow of the modified instance G^Γ.
   (ii) for any link e, if $f_e < o_e$, then $\gamma_e = 0$, and if $f_e > o_e$, then $\gamma_e = \gamma$.
Moreover, given o, we can compute f and Γ in time $O(mT_{NE})$, where T_{NE} is the complexity of computing the Nash flow of any given γ-modification of $\mathcal{G}$.

Proof sketch. The interesting case is where o is not γ-enforceable. Then, we use induction on the number of links. The base case is obvious. For the inductive step, let m be a used link with maximum latency in o. Removing m and decreasing the total traffic rate by $o_m > 0$, we obtain an instance $\mathcal{G}_{-m} = (G_{-m}, \ell, r - o_m)$ with one link less than $\mathcal{G}$. By induction hypothesis, there are a flow f' and a γ-modification $\Gamma' = (\gamma'_e)_{e \in E_{-m}}$ so that properties (i) and (ii) hold for $\mathcal{G}_{-m}$.

Now we restore link m and the traffic rate to r. The lemma follows directly from the hypothesis if there is a modification γ_m so that $(1 + \gamma_m)\ell_m(o) = L(f')$.

Otherwise, we have that $\ell_m(o) > L(f')$. Then, we carefully reroute flow from link m to the remaining links while maintaining properties (i) and (ii) in $\mathcal{G}_{-m}$. We do so until the latency of m becomes equal to the cost of the equilibrium flow that we maintain (under rerouting) in $\mathcal{G}_{-m}$. In order to maintain property (ii), we pay attention to links e where the flow f'_e reaches o_e for the first time and to links e' where $\gamma'_{e'}$ reaches γ for the first time. For the former, we stop increasing flow and start increasing γ'_e, so that the equilibrium property is maintained. For the latter, we stop increasing $\gamma'_{e'}$ and start increasing the flow again.

More formally, we partition the links in E_{-m} in three classes, according to property (ii) and to the current equilibrium flow f' and modification Γ'. We let $E_1 = \{e \in E_{-m} : f'_e < o_e \text{ and } \gamma'_e = 0\}$, $E_2 = \{e \in E_{-m} : f'_e = o_e \text{ and } \gamma'_e < \gamma\}$ and $E_3 = \{e \in E_{-m} : f'_e \geq o_e \text{ and } \gamma'_e = \gamma\}$. We let $L = (1 + \gamma'_e)\ell_e(f')$, where e is any link with $f'_e > 0$, be the cost of the current equilibrium flow f' in $\mathcal{G}_{-m}$. Moreover, we let $L_1 = \min_{e \in E_1} \ell_e(o)$ be the minimum cost of an equilibrium flow in $\mathcal{G}_{-m}$ that causes some links of E_1 to move to E_2, let $L_2 = \min_{e \in E_2}(1+\gamma)\ell_e(o)$ be the minimum cost of an equilibrium flow in $\mathcal{G}_{-m}$ that causes some links of E_2 to move to E_3, and let $L' = \min\{L_1, L_2\} \geq L$.

We reroute flow from link m to the links in $E_1 \cup E_3$ and increase γ'_e's for the links in E_2 so that we obtain an equilibrium flow in E_{-m} with cost L'. To this end, we let x_e be such that $L' = (1 + \gamma'_e)\ell_e(f'_e + x_e)$, for all $e \in E_1 \cup E_3$. Namely, x_e is the amount of flow we need to reroute to a link $e \in E_1 \cup E_3$ so that its cost becomes L'. For each link $e \in E_2$, we let $x_e = 0$ and increase its modification factor so that $L' = (1 + \gamma'_e)\ell_e(o)$. So the total amount of flow that we need to reroute from E_{-m} is $x = \sum_{e \in E_{-m}} x_e$. Next, we distinguish between different cases depending on the flow and the latency in link m after rerouting.

If $x < o_m$ and $\ell_m(o_m - x) \geq L'$, we update the flow on link m to $o_m - x$, the flow on each link $e \in E_{-m}$ to $f'_e + x_e$, and the modification factors of all links in E_2 and apply the rerouting procedure again (in fact, if $\ell_m(o_m - x) = L'$, the procedure terminates). By the definition of L', every time we apply the rerouting procedure, either some links e move from E_1 to E_2 (because after the update $f'_e = o_e$) or some links e' move from E_2 to E_3 (because after the update $\gamma'_e = \gamma$). Since links in E_3 cannot move to a different class, this rerouting procedure can be applied at most $2m$ times (in total, for all induction steps).

If $x < o_m$ and $\ell_m(o_m - x) < L'$, by continuity (see also [8, Sect. 3]), there is some $L'' \in (L, L')$ such that updating the flow and the modification factors with target equilibrium cost L'' (instead of L') reroutes flow $x' \leq x < o_m$ from link m to the links in E_{-m} so that $\ell_m(o_m - x') = L''$ and L'' is the cost of any used link in E_{-m}. Hence, we obtain the desired γ-enforceable flow f and the corresponding modification Γ. Such a value L'' can be found by computing the (unique) equilibrium flow f for the links in $E_1 \cup E_3 \cup \{m\}$ with total traffic rate $o_m + \sum_{e \in E_1 \cup E_3} f'_e$ and modifications $\gamma_e = 0$ for all links $e \in E_1 \cup \{m\}$ and $\gamma_e = \gamma$ for all links $e \in E_3$. Moreover, for all links $e \in E_2$, we let $f_e = o_e$ and set γ_e so that $L'' = (1 + \gamma_e)\ell_e(o_e)$, where $\gamma_e \leq \gamma$, because $L'' \leq L'$.

If $x = o_m$ and $\ell_m(0) < L'$, the target equilibrium cost L'' lies between L and L' and we apply the same procedure as above. If $x = o_m$ and $\ell_m(0) \geq L'$, we let $\gamma_m = 0$ and $f_m = 0$. Then, we apply rerouting as above and set $f_e = f'_e$ and $\gamma_e = \gamma'_e$ for the remaining links $e \in E_{-m}$.

If $x > o_m$ and $\ell_m(0) \geq L'$, the target equilibrium cost L'' lies between L and L' and link m is not used at equilibrium. So, we let $\gamma_m = 0$ and $f_m = 0$, compute the equilibrium flow f for the links in $E_1 \cup E_3$ with traffic rate $r - \sum_{e \in E_2} o_e$ and modifications $\gamma_e = 0$ for all $e \in E_1$ and $\gamma_e = \gamma$ for all $e \in E_3$. If $L'' \in (L, L')$ is the cost of this equilibrium flow, for all links $e \in E_2$, we let $f_e = o_e$ and set γ_e so that $L'' = (1 + \gamma_e)\ell_e(o_e)$. If $x > o_m$ and $\ell_m(0) < L'$, the target equilibrium cost L'' again lies between L and L', but now link m may be used at equilibrium. Hence, we apply the same procedure but with link m now included in E_1.   □

Price of Anarchy Analysis. We next prove an upper bound on the PoA of the γ-enforceable flow f of Lemma 1. This also serves as an upper bound on the PoA_γ of the best γ-enforceable flow. The approach is conceptually similar to that of [3] and exploits the properties (i) and (ii) of Lemma 1. The results are similar to the results in [2, Sect. 4], although our approach and the γ-modification that we consider here are different.

Theorem 1. *For γ-modifiable instances on parallel-links with latency functions in class $\mathcal{D}$, $\mathrm{PoA}_\gamma(\mathcal{D}) \leq (1 - \beta_\gamma(\mathcal{D}))^{-1}$, where*

$$\beta_\gamma(\mathcal{D}) = \sup_{\ell \in \mathcal{D}, x \geq y \geq 0} \frac{y(\ell(x) - \ell(y)) - \gamma(x - y)\ell(x)}{x\ell(x)}$$

Proof sketch. Let $\mathcal{G} = (G, \ell, r)$ be an instance on parallel-links with latency functions in class $\mathcal{D}$ and let o be the optimal solution of $\mathcal{G}$. We consider the γ-enforceable flow f and the corresponding modification $\Gamma = (\gamma_e)_{e \in E}$ of Lemma 1. By definition, $\mathrm{PoA}_\gamma(\mathcal{G}) \leq \mathrm{PoA}(\mathcal{G}^\Gamma)$. We next show an upper bound on $\mathrm{PoA}(\mathcal{G}^\Gamma)$.

Similarly to the proof of Lemma 1, we partition the links used by f into sets $E_1 = \{e \in E : 0 < f_e < o_e\}$, $E_2 = \{e \in E : f_e = o_e > 0\}$ and $E_3 = \{e \in E : f_e > o_e\}$. Using the fact that f is a Nash flow of $\mathcal{G}^\Gamma$, we obtain that

$$\sum_{e \in E} f_e \ell_e(f) \leq \sum_{e \in E} o_e \ell_e(o) + \sum_{e \in E_3} \left(o_e(\ell_e(f) - \ell_e(o)) - \gamma(f_e - o_e)\ell_e(f) \right) \quad (1)$$

Using the definition of $\beta_\gamma(\mathcal{D})$, we obtain that:

$$\sum_{e \in E} f_e \ell_e(f) \leq \sum_{e \in E} o_e \ell_e(o) + \beta_\gamma(\mathcal{D}) \sum_{e \in E_3} f_e \ell_e(f)$$

Therefore, $\mathrm{PoA}(\mathcal{G}^\Gamma) \leq (1 - \beta_\gamma(\mathcal{D}))^{-1}$.   □

Next we give upper bounds on the $\mathrm{PoA}_\gamma(\mathcal{D})$ for γ-modifiable instances with polynomial latency functions. These bounds apply to the γ-enforceable flow f of Lemma 1 and to the best γ-enforceable flow.

Corollary 1. *For γ-modifiable instances on parallel links with polynomial latency functions of degree d, we have that $\mathrm{PoA}_\gamma(d) = 1$, for all $\gamma \geq d$, and*

$$\mathrm{PoA}_\gamma(d) \leq \left(1 - d(\tfrac{\gamma+1}{d+1})^{\frac{d+1}{d}} + \gamma\right)^{-1}, \text{ for all } \gamma \in [0, d).$$

For affine latency functions, in particular, $\mathrm{PoA}_\gamma(1) = 1$, for all $\gamma \geq 1$, and

$$\mathrm{PoA}_\gamma(1) \leq \left(1 - (1-\gamma)^2/4\right)^{-1}, \text{ for all } \gamma \in [0, 1).$$

Furthermore, we can show that bounds on the PoA_γ of Theorem 1 and Corollary 1 are best possible in the worst-case.

Theorem 2. *For any class of latency functions $\mathcal{D}$ and for any $\epsilon > 0$, there is a γ-modifiable instance $\mathcal{G}$ on parallel links with homogeneous risk-averse players and latencies in class $\mathcal{D}$ so that $\mathrm{PoA}_\gamma(\mathcal{G}) \geq (1 - \beta_\gamma(\mathcal{D}))^{-1} - \epsilon$.*

4 Parallel-Link Games with Heterogeneous Players

In contrast to the case of homogeneous players, computing the best γ-enforceable flow for heterogeneous risk-averse players is NP-hard, even for affine latencies.

Theorem 3. *Given an instance $\mathcal{G}$ on parallel links with affine latencies and two classes of risk-averse players, a $\gamma > 0$ and a target cost $C > 0$, it is NP-hard to determine whether there is a γ-enforceable flow with total latency at most C.*

Proof sketch. The proof is a modification of the construction in [16, Sect. 6], which shows that the best Stackelberg modification for parallel links with affine latencies is NP-hard. Intuitively, the players with low aversion factor a^1 (resp. high aversion factor a^2) correspond to selfish (resp. coordinated) players in [16].

Specifically, we reduce $(1/3, 2/3)$-PARTITION to the best γ-enforceable flow. An instance of $(1/3, 2/3)$-PARTITION consists of n positive integers $s_1, s_2, \ldots, s_n$, so that $S = \sum_{i=1}^n s_i$ is a multiple of 3. The goal is to determine whether there exists a subset X so that $\sum_{i \in X} s_i = 2S/3$.

Given an instance $\mathcal{I}$ of $(1/3, 2/3)$-PARTITION, we create a routing game $\mathcal{G}$ with $n + 1$ parallel links and latencies $\ell_i(x) = (x/s_i) + 4$, $1 \leq i \leq n$, and $\ell_{n+1}(x) = 3x/S$. The traffic rate is $r = 2S$, partitioned into two classes with traffic $r^1 = 3S/2$ and $r^2 = S/2$. We set $\gamma = 2/17$. Working similarly to [16, Sect. 6], we show that if $a^1 = 0$ and $a^2 = 1$, $\mathcal{I}$ admits a $(1/3, 2/3)$-partition if and only if the routing game $\mathcal{G}$ admits a γ-enforceable flow f of total latency at most $35S/4$. We show that this holds if a^1 is small enough, e.g., if $a^1 = O(1/S^3)$. So, we can extend the NP-hardness proof to the case where $1 = a^1 < a^2$. $\square$

γ-Enforceable Flows with Good Price of Anarchy. Since the best enforceable flow is NP-hard, we next establish the existence of an enforceable flow that "mimics" the optimal flow o, as described by the properties (i) and (ii) in Lemma 1 and achieves a PoA as low as that in Theorem 1. In the following, we assume that the links are indexed in increasing order of $\ell_i(f)$, i.e. $i < j \Rightarrow \ell_i(f) \leq \ell_j(f)$, with ties broken in favor of links with $f_e > 0$. We start with a necessary and sufficient condition for a flow f to be γ-enforceable. [10, Algorithm 1] shows how to efficiently compute a γ-modification for any flow f that satisfies the following.

Theorem 4. *([10, Theorem 5]) Let $\mathcal{G}$ be a γ-modifiable instance on parallel links with heterogeneous players, let f be a feasible flow and let μ be the maximum index of a link used by f. Then, f is γ-enforceable if and only if (i) for any used link i, $\gamma \ell_i(f) \geq \sum_{l=i}^{\mu-1} \frac{\ell_{l+1}(f) - \ell_l(f)}{a_{l+1}^{\min}}$ and (ii) for all links i and j, if $\ell_i(f) < \ell_j(f)$, then $a_i^{\max}(f) \leq a_j^{\min}(f)$ (more risk-averse players use links of higher latency).*

To obtain a γ-enforceable flow f for an instance with heterogeneous players, we combine Lemma 1 with Theorem 4 and apply [10, Algorithm 1]. Specifically, we first ignore player heterogeneity and compute, using Lemma 1, a γ-enforceable flow f and the corresponding modification $\boldsymbol{\Gamma}$ so that f is a Nash flow of the modified game $\mathcal{G}^{\boldsymbol{\Gamma}}$ when all players have the minimum risk-aversion factor $a^1 = 1$. Assuming that the links are indexed in increasing order of their latencies in f, since f is γ-enforceable with risk-aversion factor $a^1 = 1$ for all players, Theorem 4 implies that for any used link i, $(1 + \gamma)\ell_i(f) \geq \ell_\mu(f)$.

Next, we greedily allocate the heterogeneous risk-averse players to f, taking their risk-averse factors into account, so that each link i receives flow f_i and property (ii) in Theorem 4 is satisfied. Finally, we use [10, Algorithm 1] and compute a γ-modification that turns f into an equilibrium flow for the modified instance with heterogeneous players. This is possible because, by construction, f satisfies condition (i) of Theorem 4. Moreover, since f satisfies the properties of (i) and (ii) in Lemma 1, the PoA of f can be bounded as in Theorem 1 and (in Corollary 1, for polynomial and affine latencies). Hence, we obtain the following.

Theorem 5. *Let $\mathcal{G}$ be a γ-modifiable instance on parallel-links with heterogeneous risk-averse players. Given the optimal flow of $\mathcal{G}$, we can compute a feasible flow f and a γ-modification $\boldsymbol{\Gamma}$ of $\mathcal{G}$ in time $O(m T_{NE})$, where T_{NE} is the complexity of computing the Nash flow of any given γ-modification of $\mathcal{G}$ with homogeneous risk-averse players. Moreover, the PoA_γ, under γ-modifications, achieved by f is upper bounded as in Theorem 1 and Corollary 1.*

5 Modifying Routing Games in Series-Parallel Networks

In this section, we consider γ-modifiable instances on series-parallel networks with homogeneous players and generalize the results of Sect. 3. We start with a sufficient and necessary condition for the optimal flow o to be γ-enforceable. The following generalizes Proposition 1 and is a corollary of [2, Theorem 1].

Proposition 2. *Let $\mathcal{G}$ be a γ-modifiable instance on a series-parallel network and let o be the optimal flow of $\mathcal{G}$. Then, o is γ-enforceable if and only if for any pair of internally vertex-disjoint paths p and q with common endpoints (possibly different from s and t) and with $o_e > 0$ for all edges $e \in p$, $\ell_p(o) \leq (1 + \gamma)\ell_q(o)$.*

We proceed to generalized Lemma 1 to series-parallel networks. The proof is based on an extension of the rerouting procedure in Lemma 1 combined with a continuity property of γ-enforceable flows in series-parallel networks.

Lemma 2. *Let $\mathcal{G} = (G, \ell, r)$ be a γ-modifiable instance with homogeneous risk-averse players on a series-parallel network G and let o be the optimal flow of $\mathcal{G}$. There is a feasible flow f and a γ-modification Γ of $\mathcal{G}$ such that*

(i) f is a Nash flow of the modified instance G^{Γ}.

(ii) for any edge e, if $f_e < o_e$, then $\gamma_e = 0$, and if $f_e > o_e$, then $\gamma_e = \gamma$.

Proof sketch. The proof is by induction on the series-parallel structure of G. For the base case of a single edge e, the lemma holds without any modifications.

The induction step follows directly from the induction hypothesis if G is obtained as a series composition of two series-parallel networks. The interesting case is where G is the result of a parallel composition of series-parallel networks G_1 and G_2. By induction hypothesis, for $i \in \{1, 2\}$, we let f_i be a γ-enforceable flow of rate r_i, with $r_1 + r_2 = r$, and Γ_i be a γ-modification of $\mathcal{G}_i$ such that f_i is the Nash flow of $G_i^{\Gamma_i}$. In the following, we let $L_i = L(f_i)$ be the equilibrium cost of flow f_i through network G_i with latency functions modified according to Γ_i.

If $L_1 = L_2$, the claim follows directly from the induction hypothesis. Otherwise, we assume wlog. that $L_1 > L_2$. In this case, we generalize the rerouting procedure of Lemma 1. Starting with f_1 and f_2, we reroute flow from used paths of $G_1^{\Gamma_1}$ to $G_2^{\Gamma_2}$, maintaining the equilibrium property on both $G_1^{\Gamma_1}$ and $G_2^{\Gamma_2}$ and trying to equalize their equilibrium cost. As in Lemma 1, we have also to maintain property (ii), by paying attention to edges e where f_e reaches o_e for the first time and to edges e' where $\gamma_{e'}$ reaches γ for the first time. For the former, we stop increasing the flow through any paths including e and start increasing γ_e, so that the equilibrium property is maintained. For the latter, we stop increasing $\gamma_{e'}$ and start increasing again the flow through paths that include e'.

The idea of the proof is similar to the induction step in Lemma 1. However, since G_1 and G_2 are general series-parallel networks connected in parallel, we need a continuity property, shown in [8, Sect. 3], about the changes in the equilibrium flow when the traffic rate slightly increases or decreases. $\square$

Using the properties (i) and (ii), we show that the upper bound on the PoA in Theorem 1 extends to the γ-enforceable flow f of Lemma 2 and to the PoA$_\gamma$ of the best γ-enforceable flow in series-parallel networks with homogeneous players.

Theorem 6. *For γ-modifiable instances on series-parallel networks with homogeneous players and latency functions in class $\mathcal{D}$, PoA$_\gamma(\mathcal{D}) \leq (1 - \beta_\gamma(\mathcal{D}))^{-1}$.*

Given the optimal flow of an instance $\mathcal{G}$ on a series-parallel network, we show how to compute a γ-enforceable flow f and the corresponding modification so that we achieve a PoA at most $(1 - \beta_\gamma(\mathcal{D}))^{-1}$. Given o, the running time is determined by the time required to compute a Nash flow of the original instance.

We first determine whether the optimal flow o is γ-enforceable. To this end, we remove from G all edges unused by o and check the feasibility of the following:

$$0 \le \gamma_e \le \gamma \qquad \forall \text{ used edges } e$$
$$\sum_{e \in p}(1 + \gamma_e)\ell_e(o) = \max_{p:o_p > 0} \ell_p(o) \quad \forall \text{ used paths } p \qquad (O_\gamma)$$

If the linear system (O_γ) is not feasible, then o is not γ-enforceable, by Proposition 2. Otherwise, using the solution of (O_γ) as γ_e's for the edges of G used by o and setting $\gamma_e = 0$ for the unused edges e, we enforce o as a Nash flow of the modified game G^Γ.

If (O_γ) is not feasible and o is not γ-enforceable, we exploit the constructive nature of the proof of Lemma 2 and find a γ-enforceable flow in time dominated by the time required to compute a Nash flow in series-parallel networks.

Lemma 3. *Let G be a γ-modifiable instance on a series-parallel network with homogeneous players. Given the optimal flow of G and any $\epsilon > 0$, we can compute a feasible flow f and a γ-modification Γ of G with the properties (i) and (ii) of Lemma 2 in time $O(m^2 T_{NE} \log(r/\epsilon))$, where T_{NE} is the complexity of computing the Nash flow of any given γ-modification of G and ϵ is an accuracy parameter.*

6 Parallel-Link Games with Relaxed Restrictions

In this section, we consider (p, γ)-modifiable games on parallel links with heterogeneous risk-averse players. Observing that any $\gamma/\sqrt[p]{m}$-modification is a (p, γ)-modification for a (p, γ)-modifiable game, we next show an upper bound on the PoA under such modifications.

Theorem 7. *For any (p, γ)-modifiable instance G on m parallel links with heterogeneous risk-averse players and latency functions in class $\mathcal{D}$, we have that $\text{PoA}_\gamma^p(G) \le \text{PoA}_{\gamma_0}(G) \le (1 - \beta_{\gamma_0}(\mathcal{D}))^{-1}$, where $\gamma_0 = \gamma/\sqrt[p]{m}$.*

The above bound is tight under weak assumptions on the class $\mathcal{D}$ of latency functions. More specifically, we say that a class of latency functions $\mathcal{D}$ is of the form $\mathcal{D}_0$ if (a) ℓ is continuous and twice differentiable in $(0, +\infty)$, (b) $\ell'(x) > 0, \forall x \in (0, +\infty)$ or ℓ is constant, (c) ℓ is semi-convex, i.e. $x\ell(x)$ is convex in $[0, +\infty)$ and (d) if $\ell \in \mathcal{D}$, then $(\ell + c) \in \mathcal{D}$, for all constants $c \in \mathbb{R}$ such that for all $x \in \mathbb{R}_{\ge 0}$, $\ell(x) + c \ge 0^2$. Then we obtain the following.

Theorem 8. *For any class $\mathcal{D}$ of the form $\mathcal{D}_0$ and any $\epsilon > 0$, there is an instance G on m parallel links with homogeneous players and latency functions in class $\mathcal{D}$, so that $\text{PoA}_\gamma^p(G) \ge (1 - \beta_{\gamma_0}(\mathcal{D}))^{-1} - \epsilon$, where $\gamma_0 = \gamma/\sqrt[p]{m}$.*

Proof sketch. We consider an instance $\mathcal{I}_m$, with m parallel links, where the first $m - 1$ links have the same latency function $\ell \in \mathcal{D}$ (to be fixed later) and link

---

[2] Property (d) requires that $\mathcal{D}$ should be closed under addition of constants, as long as the resulting function remains nonnegative.

m has constant latency $(1 + \gamma_1)\ell(\frac{r}{m-1})$, where $\gamma_1 = \gamma/\sqrt[p]{m-1}$. The instance has homogeneous risk-averse players with risk-aversion $a^1 = 1$. Also we let $\gamma_0 = \gamma/\sqrt[p]{m}$. The proof is an immediate consequence of the following three claims:

Claim 1. For every $m \geq 2$ and any latency function $\ell \in \mathcal{D}$ with $\ell(0) = 0$, $\mathrm{PoA}_\gamma^p(\mathcal{I}_m) = \mathrm{PoA}_{\gamma_1}(\mathcal{I}_m)$. i.e., Claim 1 states that the best (p, γ)-modification for the instance $\mathcal{I}_m$ is the modification that splits γ evenly among the first $m-1$ edges. The proof follows from an application of KKT optimality conditions.

Claim 2. For every $m \geq 2$ and any $\epsilon > 0$, there is a latency function $\ell_{\epsilon,m}$ with $\ell_{\epsilon,m}(0) = 0$ such that setting $\ell = \ell_{\epsilon,m}$ in the instance $\mathcal{I}_m$ results in $\mathrm{PoA}_{\gamma_1}(\mathcal{I}_m) \geq (1 - \beta_{\gamma_1}(\mathcal{D}))^{-1} - \epsilon/2$. The proof of Claim 2 is similar to the proof of Theorem 2.

Since $\ell_{\epsilon,m}(0) = 0$, we can combine claims 1 and 2 and obtain that for any $m \geq 2$ and any $\epsilon > 0$, $\mathrm{PoA}_\gamma^p(\mathcal{I}_m) \geq (1 - \beta_{\gamma_1}(\mathcal{D}))^{-1} - \epsilon/2$, if we use the latency function $\ell_{\epsilon,m}$.

Claim 3. For every class of latency functions $\mathcal{D}$, any $\epsilon > 0$ and any γ, there exists an $m_\epsilon \geq 2$ such that $(1 - \beta_{\gamma_1}(\mathcal{D}))^{-1} \geq (1 - \beta_{\gamma_0}(\mathcal{D}))^{-1} - \epsilon/2$.

The proof is based on the fact that γ_1 tends to γ_0 as the number of parallel links m grows. Therefore, for any $\epsilon > 0$, there are an m_ϵ and a latency function $\ell_{\epsilon,m_\epsilon}$ such that $\mathrm{PoA}_\gamma^p(\mathcal{I}_{m_\epsilon}) \geq (1 - \beta_{\gamma_0}(\mathcal{D}))^{-1} - \epsilon$. $\square$

References

1. Angelidakis, H., Fotakis, D., Lianeas, T.: Stochastic congestion games with risk-averse players. In: Vöcking, B. (ed.) SAGT 2013. LNCS, vol. 8146, pp. 86–97. Springer, Heidelberg (2013)
2. Bonifaci, V., Salek, M., Schäfer, G.: Efficiency of restricted tolls in non-atomic network routing games. In: Persiano, G. (ed.) SAGT 2011. LNCS, vol. 6982, pp. 302–313. Springer, Heidelberg (2011)
3. Correa, J.R., Schulz, A.S., Stier-Moses, N.E.: Selfish routing in capacitated networks. Math. Oper. Res. **29**(4), 961–976 (2004)
4. Fiat, A., Papadimitriou, C.: When the players are not expectation maximizers. In: Kontogiannis, S., Koutsoupias, E., Spirakis, P.G. (eds.) SAGT 2010. LNCS, vol. 6386, pp. 1–14. Springer, Heidelberg (2010)
5. Fleischer, L.: Linear tolls suffice: new bounds and algorithms for tolls in single source networks. Theoret. Comput. Sci. **348**, 217–225 (2005)
6. Fleischer, L., Jain, K., Mahdian, M.: Tolls for heterogeneous selfish users in multicommodity networks and generalized congestion games. In: Proceedings of the 45th IEEE Symposium on Foundations of Computer Science (FOCS 2004), pp. 277–285 (2004)
7. Fotakis, D.A., Spirakis, P.G.: Cost-balancing tolls for atomic network congestion games. In: Deng, X., Graham, F.C. (eds.) WINE 2007. LNCS, vol. 4858, pp. 179–190. Springer, Heidelberg (2007)
8. Hall, M.A.: Properties of the equilibrium state in transportation networks. Transp. Sci. **12**(3), 208–216 (1978)
9. Hoefer, M., Olbrich, L., Skopalik, A.: Taxing subnetworks. In: Papadimitriou, C., Zhang, S. (eds.) WINE 2008. LNCS, vol. 5385, pp. 286–294. Springer, Heidelberg (2008)

10. Jelinek, T., Klaas, M., Schäfer, G.: Computing optimal tolls with arc restrictions and heterogeneous players. In: Proceedings of the 31st Symposium on Theoretical Aspects of Computer Science (STACS 2014), LIPIcs 25, pp. 433–444 (2014)
11. Karakostas, G., Kolliopoulos, S.: Edge pricing of multicommodity networks for heterogeneous users. In: Proceedings of the 45th IEEE Symposium on Foundations of Computer Science (FOCS 2004), pp. 268–276 (2004)
12. Nikolova, E., Stier-Moses, N.E.: Stochastic selfish routing. In: Persiano, G. (ed.) SAGT 2011. LNCS, vol. 6982, pp. 314–325. Springer, Heidelberg (2011)
13. Nikolova, E., Stier-Moses, N.: The burden of risk aversion in mean-risk selfish routing. In: Proceedings of the 16th ACM Conference on Electronic Commerce (EC 2015), pp. 489–506 (2015)
14. Ordóñez, F., Stier-Moses, N.: Wardrop equilibria with risk-averse users. Transp. Sci. 44(1), 63–86 (2010)
15. Piliouras, G., Nikolova, E., Shamma, J.S.: Risk Sensitivity of price of anarchy under uncertainty. In: Proceedings of the 14th ACM Conference on Electronic Commerce (EC 2013), pp. 715–732 (2013)
16. Roughgarden, T.: Stackelberg scheduling strategies. SIAM J. Comput. 33(2), 332–350 (2004)
17. Roughgarden, T.: Selfish Routing and The Price of Anarchy. MIT press, Cambridge (2005)
18. Tversky, A., Kahneman, D.: Prospect theory: an analysis of decision under risk. Econometrica 47(2), 263–291 (1979)
19. Valdez, J., Tarjan, R.E., Lawler, E.L.: The recognition of series-parallel digraphs. SIAM J. Comput. 11(2), 298–313 (1982)

The VCG Mechanism for Bayesian Scheduling

Yiannis Giannakopoulos[1] and Maria Kyropoulou[2(✉)]

[1] University of Liverpool, Liverpool, UK
ygiannak@liverpool.ac.uk
[2] University of Oxford, Oxford, UK
kyropoul@cs.ox.ac.uk

Abstract. We study the problem of scheduling m tasks to n selfish, unrelated machines in order to minimize the makespan, where the execution times are independent random variables, identical across machines. We show that the VCG mechanism, which myopically allocates each task to its best machine, achieves an approximation ratio of $O\left(\frac{\ln n}{\ln \ln n}\right)$. This improves significantly on the previously best known bound of $O\left(\frac{m}{n}\right)$ for prior-independent mechanisms, given by Chawla et al. [STOC'13] under the additional assumption of Monotone Hazard Rate (MHR) distributions. Although we demonstrate that this is in general tight, if we do maintain the MHR assumption, then we get improved, (small) constant bounds for $m \geq n \ln n$ i.i.d. tasks, while we also identify a sufficient condition on the distribution that yields a constant approximation ratio regardless of the number of tasks.

1 Introduction

We consider the problem of scheduling tasks to machines, where the processing times of the tasks are *stochastic* and the machines are *strategic*. The goal is to minimize the expected completion time (a.k.a. *makespan*) of any machine, where the expectation is taken over the randomness of the processing times and the possible randomness of the mechanism. We are interested in the performance, i.e. the expected maximum makespan, of *truthful* mechanisms compared to the optimal mechanism that is not necessarily truthful. This problem, which we call the *Bayesian scheduling* problem, was previously considered by Chawla et al. [7].

Scheduling problems constitute a very rich and intriguing area of research [18]. In one of the most fundamental cases, the goal is to schedule m tasks to n parallel machines while minimizing the makespan, when the processing times of the tasks are selected by an adversary in an arbitrary way and can depend on the machine to which they are allocated. However, the assumption that the machines will blindly follow the instructions of a central authority (scheduler) was eventually challenged, especially due to the rapid growth of the Internet and

---

Supported by ERC Advanced Grant 321171 (ALGAME) and EPSRC grant EP/M008118/1. A full version of this paper can be found in [16].

Y. Giannakopoulos—A significant part of this work was carried out while the first author was a PhD student at the University of Oxford.

© Springer-Verlag Berlin Heidelberg 2015
E. Markakis and G. Schäfer (Eds.): WINE 2015, LNCS 9470, pp. 343–356, 2015.
DOI: 10.1007/978-3-662-48995-6_25

its use as a primary computing platform. This motivated a mechanism-design approach for the scheduling problem which Nisan and Ronen [26] introduced in their seminal paper: the processing times of the tasks are now private information of the machines, and each machine declares to the mechanism how much time it requires to execute each task. The mechanism then outputs the allocation of tasks to machines, as well as monetary compensations to the machines for their work, based solely on these declarations. In fact, the mechanism has to decide the output in advance, for any possible matrix of processing times the machines can report. Each machine is assumed to be rational and strategic, so, given the mechanism and the true processing times, its declarations are chosen in order to minimize the processing time/cost it has to spend for the execution of the allocated tasks minus the payment it will receive. In this scenario, the goal is to design a *truthful* mechanism that minimizes the makespan; truthful mechanisms define the allocation and payment functions so that the machines don't have an incentive to misreport their true processing-time capabilities. We will refer to this model as the *prior free* scheduling problem, as opposed to the stochastic model we discuss next.

In the Bayesian scheduling problem [7], the time a specific machine requires in order to process a task is drawn from a distribution. We consider one of the fundamental questions posed by the algorithmic mechanism design literature, which is about quantifying the potential performance loss of a mechanism due to the requirement for truthfulness. In the Bayesian scheduling setting, this question translates to: *What is the maximum ratio (for any distribution of processing times) of the expected maximum makespan of the best truthful mechanism over the expected optimal makespan (ignoring the requirement for truthfulness)?*

In this paper we tackle this question by considering a well known and natural truthful mechanism, the *Vickrey-Clarke-Groves mechanism (VCG)*. VCG greedily and myopically allocates each task to a machine that minimizes its processing time [10,17,31]. It is a well known fact that it is a truthful mechanism in a very strong sense; truth-telling is a dominant strategy for the machines. Because of the notorious lack of characterization results for truthfulness for restricted domains such as scheduling, VCG is the standard and obvious choice to consider for the Bayesian scheduling problem (or slightly more generally, affine maximizers). We stress here that for the scheduling domain (and for any additive domain) the VCG allocation and payments can be computed in polynomial time. Also, it is important to note that VCG is a *prior-independent* mechanism, i.e. it does not require any knowledge of the prior distribution from which the processing times are drawn.

Prior-independence is a very strong property, and is an important feature for mechanisms used in stochastic settings. Being robust with respect to prior distributions facilitates applicability in real systems, while at the same time bypassing the pessimistic inapproximability of worst case analysis. The idea is that we would like the mechanisms we use, without relying on any knowledge of the distribution of the processing times of the tasks, to still perform well compared to the optimal mechanism that is tailored for the particular distribution.

Chawla et al. [7] were the first to examine the Bayesian scheduling problem while considering the importance for prior-independence. They study the following two mechanisms:

Bounded overload with parameter c: Allocate tasks to machines such that the sum of the processing times of all tasks is minimized, subject to placing at most $c\frac{m}{n}$ tasks at any machine.

Sieve and bounded overload with parameters c, β, and δ: Fix a partition of the machines into two sets of sizes $(1 - \delta)n$ and δn. Ignoring all processing times which exceed[1] β (i.e. setting them equal to infinity), run VCG on the first set of machines. For the tasks that remain unallocated run the bounded overload mechanism with parameter c on the second set of machines.

The above mechanisms are inspired by maximal-in-range [27] (affine maximizers) and threshold mechanisms, as these are essentially the only non-trivial truthful mechanisms we know for the scheduling domain. One would expect that the simplest of those mechanisms, which is the VCG mechanism, would be the first to be considered. Indeed, VCG is the most natural, truthful, simple, polynomial time computable, and prior-independent mechanism. Still, the authors in [7] design the above mechanisms in an attempt to prevent certain bad behaviour that VCG exhibits on a specific input instance and don't examine VCG beyond that point. As we demonstrate in this paper, however, this is the worst case scenario for VCG and we can identify cases where VCG performs considerably better, either by placing a restriction on the number of tasks or by making some additional distributional assumptions.

Our Results. We prove an asymptotically tight bound of $\Theta\left(\frac{\ln n}{\ln \ln n}\right)$ for the approximation ratio of VCG for the Bayesian scheduling problem under the sole assumption that the machines are a priori identical. This bound is achieved by showing that the worst case input for VCG is actually one where the tasks are all of unit weight (point mass distributions). This resembles a balls-in-bins type scenario from which the bound is implied.

Whenever the processing times of the tasks are i.i.d. and drawn from an MHR continuous distribution, VCG is shown to be $2\left(1 + \frac{n \ln n}{m}\right)$-approximate for the Bayesian scheduling problem. This immediately implies a constant bound at most equal to 4 when $m \geq n \ln n$. We also get an improved bound of $1 + \sqrt{2}$ when $m \geq n^2$ using a different approach. For the complementary case of $m \leq n \ln n$, we identify a property of the distribution of processing times such that VCG again achieves a constant approximation. We observe that important representatives of the class of MHR distributions, that is the uniform distribution on $[0, 1]$ as well as exponential distributions, do satisfy this property, so for these distributions VCG is 4-approximate regardless of the number of tasks. We note however that this is not the case for all MHR distributions.

---

[1] Assume you run VCG on the first set of machines plus a dummy machine with processing time β on all tasks. The case where a task has processing time equal to β can be ignored without loss of generality for the case of continuous distributions.

The continuity assumption plays a fundamental role in the above results. In particular, we give a lower bound of $\Omega\left(\frac{\ln n}{\ln \ln n}\right)$ for the case of i.i.d. processing times that uses a discrete MHR distribution. Finally, we also consider the bounded overload and the sieve and bounded overload mechanisms that were studied by Chawla et al. [7], and present some instances that lower-bound their performance.

Related Work. One of the fundamental papers on the approximability of scheduling with unrelated machines is by Lenstra et al. [22] who provide a polynomial time algorithm that approximates the optimal makespan within a factor of 2. They also prove that it is NP-hard to approximate the optimal makespan within a factor of 3/2 in this setting. In the mechanism design setting, Nisan and Ronen [26] prove that the well known VCG mechanism achieves an n-approximation of the optimal makespan, while no truthful mechanism can achieve approximation ratio better than 2. Note that the upper bound immediately carries over to the Bayesian and the prior-independent scheduling case. The lower bound has been improved by Christodoulou et al. [9] and Koutsoupias and Vidali [20] to 2.61, while Ashlagi et al. [2] prove the tightness of the upper bound for anonymous mechanisms. In contrast to the negative result on the prior free setting presented in [2], truthful mechanisms can achieve sublinear approximation when the processing times are stochastic. In fact, we prove here that VCG can achieve a sublogarithmic approximation, and even a constant one for some cases, while similar bounds for other mechanisms have also been presented by Chawla et al. [7].

For the special case of related machines, where the private information of each machine is a single value, Archer and Tardos [1] were the first to give a 3-approximation truthful in expectation mechanism, while now truthful PTAS are known by the works of Christodoulou and Kovács [8] and Dhangwatnotai et al. [13]. Putting computational considerations aside, the best truthful mechanism in this single-dimensional setting is also optimal. Lavi and Swamy [21] managed to prove constant approximation ratio for a special, yet multi-dimensional scheduling problem; they consider the case where the processing time of each task can take one of two fixed values. Yu [32] then generalized this result to two-range-values, while together with Lu and Yu [24] and Lu [23], they gave constant (better than 1.6) bounds for the case of two machines.

Daskalakis and Weinberg [11] consider computationally tractable approximations of the best *truthful* mechanism when the processing times of the tasks follow distributions (with finite support) that are known to the mechanism designer. In fact the authors provide a reduction of this problem to an algorithmic problem. Chawla et al. [6] showed that there can be no approximation-preserving reductions from mechanism design to algorithm design for the makespan objective, however the authors in [11] bypass this inapproximability by considering the design of bi-criterion approximation algorithms.

Prior-independent mechanisms have been mostly considered in the problem of optimal auction design, where the goal is to design an auction mechanism that maximizes the seller's revenue. Inspired by the work of Dhangwatnotai et al. [14],

Devanur et al. [12] and Roughgarden et al. [29] independently provide approximation mechanisms for multi-dimensional settings. Moreover, Dughmi et al. [15] identify conditions under which VCG obtains a constant fraction of the optimal revenue, while Hartline and Roughgarden [19] prove Bulow-Klemperer type results for VCG. Prior robust optimization is also discussed by Sivan [30].

Chawla et al. [7] are the first to consider *prior-independent* mechanisms for the (Bayesian) scheduling problem. They introduce two variants of the VCG mechanism and bound their approximation ratios. In particular, the bounded overload mechanism is prior-independent and achieves a $O(\frac{m}{n})$ approximation of the expected optimal makespan when the processing times of the tasks are drawn from machine-identical MHR distributions. For the case where the processing times of the tasks are i.i.d. from an MHR distribution, the authors prove that sieve and bounded overload mechanisms can achieve an $O(\sqrt{\ln n})$ approximation of the expected optimal makespan, as well as an approximation ratio of $O((\ln \ln n)^2)$ under the additional assumption that there are at least $n \ln n$ tasks. We note that to achieve these improved approximation ratios, a sieve and bounded overload mechanism needs to have access to a small piece of information regarding the distribution of the processing times, in particular the expectation of the minimum of a certain number of draws; nevertheless, this still breaks the prior-independence requirement.

The VCG mechanism is strongly represented in the above works. Its simplicity and amenability to practise dictate a detailed analysis of its performance for the Bayesian scheduling problem.

Due to space limitations, some proofs have been omitted and can be found in the full version [16] of the paper.

2   Preliminaries and Notation

Assume that we have n unrelated parallel machines and $m \geq n$ tasks that need to be scheduled to these machines. Let t_{ij} denote the processing time of task j for machine i. In the Bayesian scheduling problem, each t_{ij} is independently drawn from some probability distribution $\mathcal{D}_{i,j}$. In this paper we mainly consider the machine-identical setting, that is the processing times of a specific task j are drawn from the same distribution D_j for all the machines. This is a standard assumption for the problem (see also [7]). We also consider the case where both machines and tasks are considered a priori identical, and the processing times t_{ij} are all i.i.d. drawn from the same distribution $\mathcal{D}$. The goal is to design a truthful mechanism that minimizes the expected makespan of the schedule.

We consider the VCG mechanism, the most natural and standard choice for a truthful mechanism. Thus, we henceforth assume that the machines always declare their true processing times. VCG minimizes the total workload by allocating each task to the machine that minimizes its processing time. So, if α denotes the allocation function of VCG (we omit the dependence on $\mathbf{t}$ for clarity of presentation) then, for any task j, $\alpha_{ij} = 1$ for some machine i such that $t_{ij} = \min_{i'} \{t_{i',j}\}$, otherwise $\alpha_{ij} = 0$. Without loss of generality we assume that in case of a tie, the machine is chosen uniformly at random[2]. The expected maximum makespan of VCG is then computed as

$$\mathbb{E}\left[\mathrm{VCG}(\mathbf{t})\right] = \mathbb{E}\left[\max_i \sum_{j=1}^{m} a_{ij} t_{ij}\right].$$

In what follows, we use variable $Y_{i,j}$ to denote the processing time of task j on machine i under VCG, that is $Y_{i,j} = \alpha_{ij} t_{ij}$. We also denote by $Y_i = \sum_{j=1}^{m} Y_{ij}$ the workload of machine i.

Note that in the machine-identical setting $\alpha_{ij} = 1$ with probability $\frac{1}{n}$ for any task j. So, VCG exhibits a balls-in-bins type behaviour in this setting, as the machine that minimizes the processing time of each task, and hence, the machine that will be allocated the task, is chosen uniformly at random for each task. We thus know from traditional balls-in-bins analysis, that the expected maximum number of tasks that will be allocated to any machine will be $\Theta\left(\frac{\ln n}{\ln \ln n}\right)$, whenever $m = \Theta(n)$. For more precise balls-in-bins type bounds see Raab and Steger [28]. We will use the following theorem to prove in Sect. 3 that this is actually the worst case scenario for VCG:

Theorem 1 (Berenbrink et al. [5]). *Assume two vectors $\mathbf{w} \in \mathbb{R}^m$, $\mathbf{w}' \in R^{m'}$ with $m \leq m'$ and their values in non-increasing order (that is $w_1 \geq w_2 \geq \ldots \geq w_m$ and $w_1' \geq w_2' \geq \ldots \geq w_{m'}'$). If the following two conditions hold:*

(i) $\sum_{j=1}^{m} w_j = \sum_{j=1}^{m'} w_j'$, and
(ii) $\sum_{j=1}^{k} w_j \geq \sum_{j=1}^{k} w_j'$ for all $k \in [m']$,

then the expected maximum load when allocating m balls with weights according to $\mathbf{w}$ is at least equal to the expected maximum load when allocating m' balls with weights according to $\mathbf{w}'$ to the same number of bins.

Following [5] we say that vector $\mathbf{w}$ *majorizes* $\mathbf{w}'$ whenever $\mathbf{w}$ and $\mathbf{w}'$ satisfy the conditions (i) and (ii) of Theorem 1.

Probability Preliminaries. We now give some additional notation regarding properties of distributions that will be used in the analysis.

Let T be a random variable following a probability distribution $\mathcal{D}$. Assuming we perform n independent draws from $\mathcal{D}$, we use $T[r : n]$ to denote the r-th order statistic (the r-th smallest) of the resulting values, following the notation from [7]. In particular, $T[1 : n]$ will denote the minimum of n draws from $\mathcal{D}$, while $T[1 : n][m : m]$ denotes the maximum value of m independent experiments where each one is the minimum of n draws from $\mathcal{D}$. Note that for $t_{ij} \sim \mathcal{D}_j$, the expected processing time of machine i for task j under VCG is

$$\mathbb{E}[Y_{i,j}] = \Pr\left[\alpha_{ij} = 1\right] \mathbb{E}\left[t_{ij} \mid \alpha_{ij} = 1\right] = \frac{1}{n}\mathbb{E}[T[1 : n]]. \tag{1}$$

---

[2] We note here that for continuous distributions, such events of ties occurs with zero probability.

In this work we also consider the class of probability distributions that have a *monotone hazard rate (MHR)*. A continuous distribution with pdf f and cdf F is MHR if its hazard rate $h(x) = \frac{f(x)}{1-F(x)}$ is a (weakly) increasing function. The definition of discrete MHR distributions is similar, only the hazard rate of a discrete distribution is defined as $h(x) = \frac{\Pr[X=x]}{\Pr[X\geq x]}$ (see e.g. Barlow et al. [4]). The following two technical lemmas demonstrate properties of MHR distributions.

Lemma 1. *If T is a continuous MHR random variable, then for every positive integer n, its first order statistic $T[1:n]$ is also MHR.*

Lemma 2. *For any continuous MHR random variable X and any positive integer r, $\mathbb{E}[X^r] \leq r!\mathbb{E}[X]^r$.*

We now introduce the notion of k-stretched distributions. The property that identifies these distributions plays an important role in the approximation ratio of VCG as we will see later in the analysis (Theorem 5).

Definition 1. *Given a function k over integers, we call a distribution k-stretched if its order statistics satisfy*

$$T[1:n][n:n] \geq k(n) \cdot T[1:n],$$

for all positive integers n.

We will use the following result by Aven to bound the expected maximum makespan of VCG.

Theorem 2 (Aven [3]). *If $X_1, X_2, \ldots, X_n$ are (not necessarily independent) random variables with mean μ and variance σ^2, then*

$$\mathbb{E}[\max_i X_i] \leq \mu + \sqrt{n-1}\sigma.$$

Finally, we use the notation introduced in the probability preliminaries to present some known bounds on the expected optimal makespan among all mechanisms, not necessarily truthful ones. So, if given a matrix of processing times **t** we denote its optimal makespan by OPT(**t**), we wish to bound $\mathbb{E}_{\mathbf{t}}[\text{OPT}(\mathbf{t})]$ (we omit dependence on **t** for clarity of presentation). Part of the notorious difficulty of the scheduling problem stems exactly from the lack of general, closed-form formulas for the optimal makespan. However, the following two easy lower bounds are widely used (see e.g. [7]):

Observation 1. *If the processing times are drawn from machine-identical distributions, then the expected maximum makespan of the optimal mechanism is bounded by*

$$\mathbb{E}[\text{OPT}] \geq \max\left\{ \mathbb{E}\left[\max_j T_j[1:n]\right], \frac{1}{n}\sum_{j=1}^{m} \mathbb{E}\left[T_j[1:n]\right] \right\},$$

where T_j follows the distribution corresponding to task j.

3   Upper Bounds

In this section we provide results on the performance of the VCG mechanism for the Bayesian scheduling problem for different assumptions on the number of tasks (compared to the machines), and different distributional assumptions on their processing times. Our first result shows that VCG is $O\left(\frac{\ln n}{\ln\ln n}\right)$–approximate in the general case, without assuming identical tasks or even MHR distributions. We then consider some additional assumptions under which VCG achieves a constant approximation of the expected optimal makespan. In what follows, an allocation where all machines have the same workload will be called *fully balanced*.

Theorem 3. *VCG is $O\left(\frac{\ln n}{\ln\ln n}\right)$-approximate for the Bayesian scheduling problem with n identical machines.*

As we will see later in Theorem 7, this result is in general tight. In order to prove Theorem 3 we will make use of the following lemma:

Lemma 3. *If VCG is ρ-approximate for the prior free scheduling problem with identical machines on inputs for which the optimal allocation is fully balanced, then VCG is ρ-approximate for the Bayesian scheduling problem where the machines are a priori identical.*

We are now ready to prove Theorem 3. Lemma 3 essentially reduces the analysis of VCG for the Bayesian scheduling problem for identical machines to that of a simple weighted balls-in-bins setting:

Proof (of Theorem 3). From Lemma 3, it is enough to analyze the performance of VCG on input matrices where the processing time of each task is the same across all machines and the optimal schedule is fully balanced. Without loss (by scaling) it can be further assumed that the optimal makespan is exactly 1. Then, since VCG is breaking ties uniformly at random, the problem is reduced to analyzing the expected maximum (weighted) load when throwing m balls with weights $(w_1, \dots, w_m) = \mathbf{w}$ (uniformly at random) into n bins, when $\sum_{i=1}^{m} w_i = n$. Then, by Theorem 1, that maximum load is upper bounded by the expected maximum load of throwing n (unit weight) balls into n bins, because the n-dimensional unit vector $\mathbf{1}_n$ majorizes $\mathbf{w}$: $\mathbf{1}_n$'s components sum up to n and also $w_j \leq 1$ for all $j \in [n]$ (due to the assumption that the optimal makespan is 1). By classic balls-in-bins results (see e.g. [25,28]), the expected maximum load of any machine is upper bounded by $\Theta\left(\frac{\ln n}{\ln\ln n}\right)$.   □

We now focus on the special but important case where both tasks and machines are a priori identical:

Theorem 4. *VCG is $2\left(1 + \frac{n\ln n}{m}\right)$-approximate for the Bayesian scheduling problem with i.i.d. processing times drawn from a continuous MHR distribution.*

Proof. Let T be a random variable following the distribution from which the execution times t_{ij} are drawn. Following the notation introduced in the Introduction, the workload of a machine i is given by the random variable $Y_i = \sum_{j=1}^m Y_{i,j}$. Then, for the expected maximum makespan $\mathbb{E}[\max_i Y_i]$ and any real $s > 0$ it holds that

$$e^{s \cdot \mathbb{E}[\max_i Y_i]} \leq \mathbb{E}[e^{s \max_i Y_i}] = \mathbb{E}[\max_i e^{sY_i}] \leq \sum_{i=1}^n \mathbb{E}[e^{sY_i}] = \sum_{i=1}^n \prod_{j=1}^m \mathbb{E}[e^{sY_{i,j}}]$$

$$= n\mathbb{E}[e^{sY_{1,1}}]^m, \tag{2}$$

where we have used Jensen's inequality through the convexity of the exponential function, and the fact that for a fixed machine i the random variables $Y_{i,j}$, $j = 1, \ldots, m$, are independent (the processing times are i.i.d. and VCG allocates each task independently of the others). We now bound the term $\mathbb{E}[e^{sY_{1,1}}]$:

$$\mathbb{E}[e^{sY_{1,1}}] = \mathbb{E}\left[\sum_{r=0}^\infty \frac{(sY_{1,1})^r}{r!}\right] = 1 + \sum_{r=1}^\infty s^r \frac{\mathbb{E}[Y_{1,1}^r]}{r!} = 1 + \frac{1}{n} \sum_{r=1}^\infty s^r \frac{\mathbb{E}[T[1:n]^r]}{r!}$$

$$\leq 1 + \frac{1}{n} \sum_{r=1}^\infty s^r \mathbb{E}[T[1:n]]^r,$$

where for the last inequality we have used the fact that the first order statistic of an MHR distribution is also MHR (Lemma 1) and Lemma 2. Then, by choosing $s = s^* \equiv \frac{1}{2 \cdot \mathbb{E}[T[1:n]]}$ we get that

$$\mathbb{E}[e^{s^* Y_{1,1}}] = 1 + \frac{1}{n} \sum_{r=1}^\infty \frac{1}{2^r} \leq 1 + \frac{1}{n},$$

and (2) yields

$$\mathbb{E}[\max_i Y_i] \leq \ln\left(n\mathbb{E}[e^{s^* Y_{1,1}}]^m\right) \frac{1}{s^*}$$

$$\leq 2\ln\left(n\left(1 + \frac{1}{n}\right)^m\right) \mathbb{E}[T[1:n]]$$

$$\leq 2\ln\left(ne^{m/n}\right) \mathbb{E}[T[1:n]]$$

$$= 2\left(\ln n + \frac{m}{n}\right) \mathbb{E}[T[1:n]]. \tag{3}$$

But from Observation 1 we know that $\mathbb{E}[OPT] \geq \frac{m}{n}\mathbb{E}[T[1:n]]$ for the case of i.i.d. execution times, and the theorem follows.   □

Notice that Theorem 4 in particular implies that VCG achieves a *small, constant* approximation ratio whenever the number of tasks is slightly more than that of machines:

Corollary 1. *VCG is 4-approximate for the Bayesian scheduling problem with $m \geq n \ln n$ i.i.d. tasks drawn from a continuous MHR distribution.*

The following theorem will help us analyze the performance of VCG for the complementary case to that of Corollary 1, that is when the number of tasks is $m \leq n \ln n$. Recall the notion of k-stretched distributions introduced in Definition 1.

Theorem 5. *VCG is* $4\frac{\ln n}{k(n)}$-*approximate for the Bayesian scheduling problem with* $m \leq n \ln n$ *i.i.d. tasks drawn from a* k-*stretched MHR distribution.*

In particular, we note that Theorem 5 yields a constant approximation ratio for VCG for the important special cases where the processing times are drawn independently from the uniform distribution on $[0,1]$ or any exponential distribution. Indeed, the uniform distribution on $[0,1]$ as well as any exponential distribution is ln-stretched. We get the following, complementing the results in Corollary 1:

Corollary 2. *VCG is* 4-*approximate for the Bayesian scheduling problem with i.i.d. processing times drawn from the uniform distribution on* $[0,1]$ *or an exponential distribution.*

We point out that the above corollary can not be generalized to hold for all MHR distributions, as the lower bound in Theorem 7 implies. For example, it is not very difficult to check that by taking $\varepsilon \to 0$ and considering the uniform distribution over $[1, 1+\varepsilon]$, no stretch factor $k(n) = \Omega(\ln n)$ can be guaranteed.

For our final positive result, we present an improved constant bound on the approximation ratio of VCG when we have many tasks:

Theorem 6. *VCG is* $1 + \sqrt{2}$-*approximate for the Bayesian scheduling problem with* $m \geq n^2$ *tasks with i.i.d. processing times drawn from a continuous MHR distribution.*

Proof. We use Theorem 2 to bound the performance of VCG in this setting. In order to do so, we first bound the expectation and the variance of the makespan of a single machine. From (1), for the workload Y_i of any machine i we have:

$$\mathbb{E}[Y_i] = \sum_{j=1}^{m} \mathbb{E}[Y_{i,j}] = \frac{1}{n} \sum_{j} \mathbb{E}[T[1:n]] = \frac{m}{n} \mathbb{E}[T[1:n]].$$

To compute the variance of the makespan of machine i, we note that the random variables $Y_{i,j}$ are independent with respect to j, for any fixed machine i and thus we can get

$$\mathrm{Var}[Y_i] = \sum_{j=1}^{m} \mathrm{Var}[Y_{i,j}] = \sum_{j=1}^{m} \left(\mathbb{E}[Y_{i,j}^2] - \mathbb{E}[Y_{i,j}]^2 \right)$$

$$\leq \sum_{j=1}^{m} \mathbb{E}[Y_{i,j}^2] = \sum_{j=1}^{m} \mathbb{E}[a_{ij}^2 t_{ij}^2] = \frac{1}{n} \sum_{j=1}^{m} \mathbb{E}[T[1:n]^2]$$

$$= \frac{m}{n} \mathbb{E}[T[1:n]^2].$$

We are now ready to use Theorem 2 and bound the expected maximum makespan:

$$\mathbb{E}[\max_i Y_i] \leq \mathbb{E}[Y_1] + \sqrt{n-1}\sqrt{\text{Var}[Y_1]}$$

$$\leq \frac{m}{n}\mathbb{E}[T[1:n]] + \sqrt{m}\sqrt{\mathbb{E}[T[1:n]^2]}$$

$$\leq \frac{m}{n}\mathbb{E}[T[1:n]] + \sqrt{2}\sqrt{m}\mathbb{E}[T[1:n]]$$

$$\leq (1+\sqrt{2})\frac{m}{n}\mathbb{E}[T[1:n]]$$

$$\leq (1+\sqrt{2})\mathbb{E}[\text{OPT}],$$

where the third inequality follows from Lemma 2 (and Lemma 1), for the fourth inequality we use the assumption that $m \geq n^2$ and to complete the proof, the last inequality uses a lower bound on $\mathbb{E}[\text{OPT}]$ from Observation 1. $\square$

4   Lower Bounds

In this section we prove some lower bounds on the performance of VCG under different distributional assumptions on the processing times. In an attempt for a clear comparison of VCG with the mechanisms that were previously considered for the Bayesian scheduling problem (in [7]), we provide instances that lower bound their performance as well.

Theorem 7. *For any number of tasks, there exists an instance of the Bayesian scheduling problem where VCG is not better than $\Omega\left(\frac{\ln n}{\ln \ln n}\right)$-approximate and the processing times are drawn from machine-identical continuous MHR distributions.*

Notice that when the number of tasks equals that of the machines, i.e. $m = n$, then the lower bound of Theorem 7 holds for an instance of i.i.d. distributions and not just for identical machines. However, if we restrict our focus only on discrete distributions, then we can strengthen that lower bound to hold for i.i.d. distributions for essentially any number of tasks and not only for $m = n$:

Theorem 8. *For any number of $m = O(ne^n)$ tasks, there exists an instance of the Bayesian scheduling problem where VCG is not better than $\Omega\left(\frac{\ln n}{\ln \ln n}\right)$-approximate and the tasks have i.i.d. processing times drawn from a discrete MHR distribution.*

Proof. Consider an instance with n identical machines and m tasks where the processing times t_{ij} are drawn from $\{0,1\}$ such that $t_{ij} = 1$ with probability $\left(\frac{n}{2m}\right)^{\frac{1}{n}} \equiv p$ and $t_{ij} = 0$ with probability $1 - p$. Notice that this is a well-defined distribution, since for all $m \geq n$ we have $p < 1$. Furthermore, it is easy to check that this distribution is MHR; its hazard rate at 0 is $\frac{\Pr[t_{ij}=0]}{\Pr[t_{ij}\geq 0]} = \frac{1-p}{1} = 1 - p$ and at 1 is $\frac{\Pr[t_{ij}=1]}{\Pr[t_{ij}\geq 1]} = \frac{p}{p} = 1$.

Next, let M be the random variable denoting the number of tasks whose best processing time in any machine is non-zero, that is

$$M = |\{j \mid \min_i t_{ij} = 1\}|.$$

Then M follows a binomial distribution with probability of success p^n and m trials, since the probability of a task having processing time 1 at *all* machines (success) is p^n, while there are m tasks in total. Given the definition for p, the average number of tasks that will end up requiring a processing time of 1 at any machine is $\mathbb{E}[M] = mp^n = \frac{n}{2}$. Also, we can derive that $\Pr[M > 3n] \le e^{-n}$ by using a Chernoff bound[3]. As we have argued before, we can use classical results from balls-in-bins analysis to bound the performance of VCG. So, when $M \le 3n$, we know that the expected maximum makespan (since each task has processing time at least 1) will be $\Omega\left(\frac{\ln n}{\ln \ln n}\right)$. That event happens almost surely, with probability $1 - e^{-n} = 1 - o(1)$.

On the other hand, we next show that the mechanism that simply balances the M "expensive" tasks across the machines (by allocating $\left\lceil\frac{M}{n}\right\rceil$ of them to every machine) achieves a constant maximum makespan, hence providing a constant upper-bound on the optimal maximum makespan:

$$\mathbb{E}[\text{OPT}] \le \Pr[M \le 3n] \cdot \frac{3n}{n} \cdot 1 + \Pr[M > 3n] \cdot \left\lceil\frac{m}{n}\right\rceil \cdot 1 \le 4 + \frac{m}{ne^n} = O(1).$$

$\square$

Notice however that Theorem 8 still leaves open the possibility for *continuous* MHR distributions to perform better (see also Theorem 6 and Corollary 1).

We finally conclude with a couple of simple observations, for the sake of completeness. First, our initial requirement (see Sect. 2) for identical machines (which is a standard one, see [7]) is crucial for guaranteeing any non-trivial approximation ratios on the performance of VCG:

Observation 2. *There exists an instance of the Bayesian scheduling problem where VCG is not better than n-approximate even when the tasks are identically distributed according to continuous MHR distributions.*

We now present some lower bounds on the performance of the mechanisms analyzed by Chawla et al. [7]. A definition of these mechanisms can be found in the Introduction. The following demonstrates that the analysis of the approximation ratio for the class of bounded overload mechanisms presented in [7] is asymptotically tight:

Observation 3. *For any number of $m \ge n$ tasks, there exists an instance of the Bayesian scheduling problem where a bounded overload mechanism with parameter c is not better than $\min\{c\frac{m}{n}, n-1\}$-approximate and the processing times are drawn from machine-identical continuous MHR distributions.*

---

[3] Here we use the following form, with $\beta = 1 + \sqrt{5}$: for any $\beta > 0$, $\Pr[X \ge (1 + \beta)\mu] \le e^{-\frac{\beta^2}{2+\beta}\mu}$ for any binomial random variable with mean μ.

The same instance can be used to bound the performance of the bounded overload mechanism with parameter c that breaks ties uniformly at random as well. Having sufficiently many tasks ($m = \Omega\left(\frac{n \ln n}{\ln \ln n}\right)$) implies that the mechanism behaves almost like the VCG mechanism while allocating the unit-cost tasks, assuming they are the first to be allocated. This gives a lower bound of $\Omega\left(\frac{\ln n}{\ln \ln n}\right)$ on the approximation ratio of this mechanism as well.

Similar instances can provide lower bounds on the performance of the class of sieve and bounded overload mechanisms with parameters c, β, and δ, even for the case of i.i.d. processing times. To see this notice that if all tasks have $t_{ij} = 1$ with probability 1 on any machine ($T[1 : k] = 1$ for any k), and we choose threshold $\beta < 1$ as is done in [7] for the case $m \leq n \ln n$, then a sieve and bounded overload mechanism with parameters $c, \beta \leq 1$, and δ immediately reduces to a bounded overload mechanism with parameter c on δn machines.

References

1. Archer, A., Tardos, É.: Truthful mechanisms for one-parameter agents. In: FOCS, pp. 482–491 (2001)
2. Ashlagi, I., Dobzinski, S., Lavi, R.: Optimal lower bounds for anonymous scheduling mechanisms. Math. Oper. Res. 37(2), 244–258 (2012)
3. Aven, T.: Upper (lower) bounds on the mean of the maximum (minimum) of a number of random variables. J. Appl. Probab. 22(3), 723–728 (1985)
4. Barlow, R.E., Marshall, A.W., Proschan, F.: Properties of probability distributions with monotone hazard rate. Ann. Math. Stat. 34(2), 375–389 (1963)
5. Berenbrink, P., Friedetzky, T., Hu, Z., Martin, R.: On weighted balls-into-bins games. Theor. Comput. Sci. 409(3), 511–520 (2008)
6. Chawla, S., Immorlica, N., Lucier, B.: On the limits of black-box reductions in mechanism design. In: STOC, pp. 435–448 (2012)
7. Chawla, S., Hartline, J.D., Malec, D., Sivan, B.: Prior-independent mechanisms for scheduling. In: STOC, pp. 51–60 (2013)
8. Christodoulou, G., Kovács, A.: A deterministic truthful PTAS for scheduling related machines. SIAM J. Comput. 42(4), 1572–1595 (2013)
9. Christodoulou, G., Koutsoupias, E., Vidali, A.: A lower bound for scheduling mechanisms. Algorithmica 55(4), 729–740 (2009)
10. Clarke, E.H.: Multipart pricing of public goods. Public Choice 11(1), 17–33 (1971)
11. Daskalakis, C., Weinberg, S.M.: Bayesian truthful mechanisms for job scheduling from bi-criterion approximation algorithms. In: SODA, pp. 1934–1952 (2015)
12. Devanur, N., Hartline, J., Karlin, A., Nguyen, T.: Prior-independent multi-parameter mechanism design. In: Chen, N., Elkind, E., Koutsoupias, E. (eds.) WINE 2011. LNCS, vol. 7090, pp. 122–133. Springer, Heidelberg (2011)
13. Dhangwatnotai, P., Dobzinski, S., Dughmi, S., Roughgarden, T.: Truthful approximation schemes for single-parameter agents. SIAM J. Comput. 40(3), 915–933 (2011)
14. Dhangwatnotai, P., Roughgarden, T., Yan, Q.: Revenue maximization with a single sample. Games Econ. Behav. 91, 318–333 (2015)
15. Dughmi, S., Roughgarden, T., Sundararajan, M.: Revenue submodularity. Theory Comput. 8(1), 95–119 (2012)

16. Giannakopoulos, Y., Kyropoulou, M.: The VCG mechanism for bayesian scheduling. CoRR, abs/1509.07455 (2015). http://arxiv.org/abs/1509.07455
17. Groves, T.: Incentives in teams. Econometrica **41**(4), 617–631 (1973)
18. Hall, L.A.: Approximation algorithms for scheduling. In: Hochbaum, D.S. (ed.) Approximation Algorithms for NP-Hard Problems, pp. 1–45. PWS, Boston (1997)
19. Hartline, J.D., Roughgarden, T.: Simple versus optimal mechanisms. In: EC, pp. 225–234 (2009)
20. Koutsoupias, E., Vidali, A.: A lower bound of $1 + \varphi$ for truthful scheduling mechanisms. Algorithmica **66**(1), 211–223 (2013)
21. Lavi, R., Swamy, C.: Truthful mechanism design for multidimensional scheduling via cycle monotonicity. Games Econ. Behav. **67**(1), 99–124 (2009)
22. Lenstra, J.K., Shmoys, D.B., Tardos, É.: Approximation algorithms for scheduling unrelated parallel machines. Math. Program. **46**, 259–271 (1990)
23. Lu, P.: On 2-player randomized mechanisms for scheduling. In: Leonardi, S. (ed.) WINE 2009. LNCS, vol. 5929, pp. 30–41. Springer, Heidelberg (2009)
24. Lu, P., Yu, C.: An improved randomized truthful mechanism for scheduling unrelated machines. In: STACS, pp. 527–538 (2008)
25. Motwani, R., Raghavan, P.: Randomized Algorithms. Cambridge University Press, Cambridge (1995)
26. Nisan, N., Ronen, A.: Algorithmic mechanism design. Games Econ. Behav. **35**(1/2), 166–196 (2001)
27. Nisan, N., Ronen, A.: Computationally feasible VCG mechanisms. J. Artif. Int. Res. **29**(1), 19–47 (2007)
28. Raab, M., Steger, A.: "Balls into Bins" — a simple and tight analysis. In: Luby, M., Rolim, J.D.P., Serna, M. (eds.) RANDOM 1998. LNCS, vol. 1518, pp. 159–170. Springer, Heidelberg (1998)
29. Roughgarden, T., Talgam-Cohen, I., Yan, Q.: Supply-limiting mechanisms. In: EC, pp. 844–861 (2012)
30. Sivan, B.: Prior Robust Optimization. Ph.D. thesis, University of Wisconsin-Madison (2013)
31. Vickrey, W.: Counterspeculation, auctions, and competitive sealed tenders. J. Finance **16**(1), 8–37 (1961)
32. Yu, C.: Truthful mechanisms for two-range-values variant of unrelated scheduling. Theor. Comput. Sci. **410**(21–23), 2196–2206 (2009)

Query Complexity of Approximate Equilibria in Anonymous Games

Paul W. Goldberg and Stefano Turchetta$^{(\boxtimes)}$

Department of Computer Science, University of Oxford, Oxford, UK
{paul.goldberg,stefano.turchetta}@cs.ox.ac.uk

Abstract. We study the computation of equilibria of two-strategy anonymous games, via algorithms that may proceed via a sequence of adaptive queries to the game's payoff function, assumed to be unknown initially. The general topic we consider is *query complexity*, that is, how many queries are necessary or sufficient to compute an exact or approximate Nash equilibrium.

We show that exact equilibria cannot be found via query-efficient algorithms. We also give an example of a 2-strategy, 3-player anonymous game that does not have any exact Nash equilibrium in rational numbers. Our main result is a new randomized query-efficient algorithm that finds a $O(n^{-1/4})$-approximate Nash equilibrium querying $\tilde{O}(n^{3/2})$ payoffs and runs in time $\tilde{O}(n^{3/2})$. This improves on the running time of pre-existing algorithms for approximate equilibria of anonymous games, and is the first one to obtain an inverse polynomial approximation in poly-time. We also show how this can be used to get an efficient PTAS. Furthermore, we prove that $\Omega(n \log n)$ payoffs must be queried in order to find any ϵ-well-supported Nash equilibrium, even by randomized algorithms.

1 Preliminaries

This paper studies two-strategy *anonymous* games, in which a large number of players n share two pure strategies, and the payoff to a player depends on the number of players who use each strategy, but not their identities. Due to the this property, these games have a polynomial-size representation. Daskalakis and Papadimitriou [13] consider anonymous games and graphical games to be the two most important classes of concisely-represented multi-player games. Anonymous games appear frequently in practice, for example in voting systems, traffic routing, or auction settings. Although they have polynomial-sized representations, the representation may still be inconveniently large, making it desirable to work with algorithms that do not require all the data on a particular game of interest.

Query complexity is motivated in part by the observation that a game's entire payoff function may be syntactically cumbersome. It also leads to new results

P.W.Goldberg – Supported by EPSRC project EP/K01000X/1.

S. Turchetta – Work based on MSc. thesis at the Technische Universität München.

E. Markakis and G. Schäfer (Eds.): WINE 2015, LNCS 9470, pp. 357–369, 2015.

DOI: 10.1007/978-3-662-48995-6_26

that distinguish the difficulty of alternative solution concepts. We assume that an algorithm has black-box access to the payoff function, via queries that specify an anonymized profile and return one or more of the players' payoffs.

1.1 Definitions and Notation

Anonymous Games. A k-strategy anonymous game is a tuple $(n, k, \{u_j^i\}_{i \in [n], j \in [k]})$ that consists of n players, k pure strategies per player, and a utility function $u_j^i : \{0, \ldots, n-1\} \longrightarrow [0,1]$ for each player $i \in [n]$ (where we use $[n]$ to denote the set $\{1, \ldots, n\}$) and every strategy $j \in [k]$, whose input is the number of other players who play strategy one if $k = 2$. The number of payoffs stored by a 2-strategy game is $2n^2$ (generally, $O(n^k)$). As indicated by u_j^i's codomain, we make a standard assumption that all payoffs are normalized into the interval $[0,1]$.

For all $i \in [n]$, let X_i be a random indicator variable being equal to one if and only if player i plays strategy one. For 2-strategy games, a mixed strategy for i is represented by the probability $p_i := \mathbb{E}[X_i]$ that player i plays strategy one. Let $X_{-i} := \sum_{\ell \in [n] \setminus \{i\}} X_\ell$ be the sum of all the random variables but X_i. The expected utility obtained by player $i \in [n]$ for playing pure strategy $j \in \{1, 2\}$ against X_{-i} is

$$\mathbb{E}[u_j^i(X_{-i})] := \sum_{x=0}^{n-1} u_j^i(x) \cdot \Pr[X_{-i} = x].$$

If i is playing a mixed strategy (i.e., $p_i \in (0,1)$) her expected payoff simply consists of a weighted average, i.e., $\mathbb{E}[u^i(X)] := p_i \cdot \mathbb{E}[u_1^i(X_{-i})] + (1 - p_i) \cdot \mathbb{E}[u_2^i(X_{-i})]$, where $X := (X_i, X_{-i})$. It is known that $\mathbb{E}[u_j^i(X_{-i})]$, which involves computing the p.m.f. of X_{-i} – a Poisson Binomial Distribution – can be computed in polynomial time (see e.g., [13]).

Exact and Approximate Nash Equilibria. With the above notation, we say that X_i is a best-response if and only if $\mathbb{E}[u^i(X)] \geq \mathbb{E}[u_j^i(X_{-i})]$ for all $j \in \{1, 2\}$. A *Nash equilibrium* (NE) requires the players to be best-responding to each other; therefore, the above best-response condition must hold for every $i \in [n]$. This can be also viewed as no player having an incentive to deviate from her strategy. We consider a relaxation of NE, the notion of an ϵ-*approximate Nash equilibrium* (ϵ-NE), where every player's incentive to deviate is at most $\epsilon > 0$. We say that $(X_i)_{i \in [n]}$, which represents a mixed-strategy profile, constitutes an ϵ-NE if for all $i \in [n]$ and all $j \in \{1, 2\}$,

$$\mathbb{E}[u^i(X)] + \epsilon \geq \mathbb{E}[u_j^i(X_{-i})].$$

This definition, however, does not prohibit allocating a small amount of probability to arbitrarily bad strategies. An ϵ-*approximate well-supported Nash equilibrium* (ϵ-WSNE) addresses this issue by forcing every player to place a positive amount of probability solely on ϵ-approximate best-responses, i.e., $(X_i)_{i \in [n]}$ constitutes an ϵ-WSNE if for all $i \in [n]$,

$$\mathbb{E}[u_1^i(X_{-i})] + \epsilon < \mathbb{E}[u_2^i(X_{-i})] \implies p_i = 0, \text{ and}$$
$$\mathbb{E}[u_2^i(X_{-i})] + \epsilon < \mathbb{E}[u_1^i(X_{-i})] \implies p_i = 1.$$

Although an ϵ-WSNE is also an ϵ-NE, the converse need not be true.

Query-Efficiency and Payoff Query Models. Our general interest is in polynomial-time algorithms that find solutions of anonymous games, while checking just a small fraction of the $2n^2$ payoffs of an n-player, 2-strategy game. The basic kind of query is a *single-payoff query* which receives as input a player $i \in [n]$, a strategy $j \in \{1, 2\}$, and the number $x \in \{0, \ldots, n-1\}$ of players playing strategy one, and it returns the corresponding payoff $u_j^i(x)$. The *query complexity* of an algorithm is the expected number of single-payoff queries that it needs in the worst case. Hence, an algorithm is query-efficient if its query complexity is $o(n^2)$.

A *profile query* (used in [15]) consists of an action profile $(a_1, \ldots, a_n) \in \{1, 2\}^n$ as input and outputs the payoffs that *every* player i obtains according to that profile. Clearly, a profile query can be simulated using n single-payoff queries. Finally, an *all-players query* consists of a pair (x, j) for $x \in \{0, \ldots, n-1\}$, $j \in \{1, 2\}$, and the response to (x, j) consists of the values $u_j^i(x)$ for all $i \in [n]$. We will consider the cost of a query to be equal to the number of payoffs it returns; hence, a profile or an all-players query costs n single-payoff queries. We find that an algorithm being constrained to utilize profile queries may incur a linear loss in query-efficiency[1]. Therefore, we focus on single-payoff and all-players queries, which better exploit the symmetries of anonymous games.

1.2   Related Work

In the last decade, there has been interest in the complexity of computing approximate Nash equilibria. A main reason is the **PPAD**-completeness results for computing an exact NE, for normal-form games [5,8] (the latter paper extends the hardness also to an FPTAS), and recently also for anonymous games with 7 strategies [6]. The **FIXP**-completeness results of [14] for multiplayer games show an algebraic obstacle to the task of writing down a useful description of an exact equilibrium. On the other hand, there exists a subexponential-time algorithm to find an ϵ-NE in normal-form games [20], and one important open question regards the existence of a PTAS for bimatrix games.

Daskalakis and Papadimitriou proved that anonymous games admit a PTAS and provided several improvements of its running time over the past few years. Their first algorithm [9] concerns two-strategy games and is based upon the quantization of the strategy space into nearby multiples of ϵ. This result was also

---

[1] Due to space constraints, we defer this discussion to the full version of the paper (http://arxiv.org/abs/1412.6455).

extended to the multi-strategy case [10]. Daskalakis [7] subsequently gave an efficient PTAS whose running time is $\text{poly}(n) \cdot (1/\epsilon)^{O(1/\epsilon^2)}$, which relies on a better understanding of the structure of ϵ-equilibria in two-strategy anonymous games: There exists an ϵ-WSNE where either a small number of the players – at most $O(1/\epsilon^3)$ – randomize and the others play pure strategies, or whoever randomizes plays the same mixed strategy. Furthermore, Daskalakis and Papadimitriou [11] proved a lower bound on the running time needed by any *oblivious* algorithm, which lets the latter algorithm be essentially optimal. In the same article, they show that the lower bound can be broken by utilizing a non-oblivious algorithm, which has the currently best-known running time for finding an ϵ-equilibrium in two-strategy anonymous games of $O(\text{poly}(n) \cdot (1/\epsilon)^{O(\log^2(1/\epsilon))})$. A complete proof is in [12].

In Sect. 3 we present a bound for λ-Lipschitz games, in which λ is a parameter limiting the rate at which $u_j^i(x)$ changes as x changes. Any λ-Lipschitz k-strategy anonymous game is guaranteed to have an ϵ-approximate *pure* Nash equilibrium, with $\epsilon = O(\lambda k)$ [1,13]. The convergence rate to a Nash equilibrium of best-reply dynamics in the context of two-strategy Lipschitz anonymous games is studied by [2,19]. Moreover, Brandt et al. [4] showed that finding a pure equilibrium in anonymous games is easy if the number of strategies is constant w.r.t. the number of players n, and hard as soon as there is a linear dependence.

In the last two years, several researchers obtained bounds for the query complexity for approximate equilibria in different game settings, which we briefly survey. Fearnley et al. [15] presented the first series of results: they studied bimatrix games, graphical games, and congestion games on graphs. Similar to our negative result for exact equilibria of anonymous games, it was shown that a Nash equilibrium in a bimatrix game with k strategies per player requires k^2 queries, even in zero-sum games. However, more positive results arise if we move to ϵ-approximate Nash equilibria. Approximate equilibria of bimatrix games were studied in more detail in [16].

The query complexity of equilibria of n-player games – a setting where payoff functions are exponentially-large – was analyzed in [3,17,18]. Hart and Nisan [18] showed that exponentially many deterministic queries are required to find a $\frac{1}{2}$-approximate correlated equilibrium (CE) and that any randomized algorithm that finds an exact CE needs $2^{\Omega(n)}$ expected cost. Notice that lower bounds on correlated equilibria automatically apply to Nash equilibria. Goldberg and Roth [17] investigated in more detail the randomized query complexity of ϵ-CE and of the more demanding ϵ-well-supported CE. Babichenko [3] proved an exponential-in-n randomized lower bound for finding an ϵ-WSNE in n-player, k-strategy games, for constant $k = 10^4$ and $\epsilon = 10^{-8}$. These exponential lower bounds do not hold in anonymous games, which can be fully revealed with a polynomial number of queries.

1.3   Our Results and Their Significance

Query-efficiency seems to serve as a criterion for distinguishing exact from approximate equilibrium computation. It applies to games having exponentially-large representations [18], also for games having poly-sized representations (e.g. bimatrix games [15]). Here we extend this finding to the important class of anonymous games. We prove that even in two-strategy anonymous games, an exact Nash equilibrium demands querying the payoff function exhaustively, even with the most powerful query model (Theorem 1). Alongside this, we provide an example of a three-player, two-strategy anonymous game whose unique Nash equilibrium needs all players to randomize with an irrational amount of probability (Theorem 2), answering a question posed in [13]. These results motivate our subsequent focus on approximate equilibria.

We exhibit a simple query-efficient algorithm that finds an approximate pure Nash equilibrium in Lipschitz games (Algorithm 1; Theorem 3), which will be used by our main algorithm for anonymous games.

Our main result (Theorem 4) is a new randomized approximation scheme[2] for anonymous games that differs conceptually from previous ones and offers new performance guarantees. It is query-efficient (using $o(n^2)$ queries) and has improved computational efficiency. It is the first PTAS for anonymous games that is polynomial in a setting where n and $1/\epsilon$ are polynomially related. In particular, its runtime is polynomial in n in a setting where $1/\epsilon$ may grow in proportion to $n^{1/4}$ and also has an improved polynomial dependence on n for all $\epsilon \geq n^{-1/4}$. In more detail, for any $\epsilon \geq n^{-1/4}$, the algorithm adaptively finds a $O(\epsilon)$-NE with $\tilde{O}(\sqrt{n})$ (where we use $\tilde{O}(\cdot)$ to hide polylogarithmic factors of the argument) all-players queries (i.e., $\tilde{O}(n^{3/2})$ single payoffs) and runs in time $\tilde{O}(n^{3/2})$. The best-known algorithm of [13] runs in time $O(\text{poly}(n) \cdot (1/\epsilon)^{O(\log^2(1/\epsilon))})$, where $\text{poly}(n) \geq O(n^7)$.

In addition to this, we derive a randomized logarithmic lower bound on the number of all-players queries needed to find any non-trivial ϵ-WSNE in two-strategy anonymous games (Theorem 5).

2 Exact Nash Equilibria

We lower-bound the number of single-payoff queries (the least constrained query model) needed to find an exact NE in an anonymous game. We exhibit games in which any algorithm must query most of the payoffs in order to determine what strategies form a NE. Difficult games are ones that only possess NE where $\Omega(n)$ players must randomize.

Example 1. Let G be the following two-strategy, n-player anonymous game. Let n be even, and let $\delta = 1/n^2$. Half of the players have a utility function as shown by the top side (a) of Fig. 1, and the remaining half as at (b).

Theorem 1. *A deterministic single-payoff query-algorithm may need to query $\Omega(n^2)$ payoffs in order to find an exact Nash equilibrium of an n-player, two-strategy anonymous game.*

---

[2] To make Theorem 4 easier to read, we state it only for the best attainable approximation (i.e., $n^{-1/4}$); however, it is possible to set parameters to get any approximation $\epsilon \geq n^{-1/4}$. For details, see the proof of Theorem 4.

x	0	1	2	$\ldots$	$n-2$	$n-1$
$u_1^i(x)$	$\frac{1}{2} - \left(\frac{n}{2} - \frac{1}{2}\right)\delta$	$u_1^i(0) - \left(\frac{n}{2} - \frac{3}{2}\right)\delta$	$u_1^i(1) - \left(\frac{n}{2} - \frac{5}{2}\right)\delta$	$\ldots$	$u_1^i(n-3) + \left(\frac{n}{2} - \frac{3}{2}\right)\delta$	$u_1^i(n-2) + \left(\frac{n}{2} - \frac{1}{2}\right)\delta$
$u_2^i(x)$	$\frac{1}{2}$	$u_1^i(0)$	$u_1^i(1)$	$\ldots$	$u_1^i(n-3)$	$u_1^i(n-2)$

(a) Payoff table for "majority-seeking" player i

x	0	1	2	$\ldots$	$n-2$	$n-1$
$u_1^i(x)$	$\frac{1}{2} + \left(\frac{n}{2} - \frac{1}{2}\right)\delta$	$u_1^i(0) + \left(\frac{n}{2} - \frac{3}{2}\right)\delta$	$u_1^i(1) + \left(\frac{n}{2} - \frac{5}{2}\right)\delta$	$\ldots$	$u_1^i(n-3) - \left(\frac{n}{2} - \frac{3}{2}\right)\delta$	$u_1^i(n-2) - \left(\frac{n}{2} - \frac{1}{2}\right)\delta$
$u_2^i(x)$	$\frac{1}{2}$	$u_1^i(0)$	$u_1^i(1)$	$\ldots$	$u_1^i(n-3)$	$u_1^i(n-2)$

(b) Payoff table for "minority-seeking" player i

Fig. 1. Majority-minority game G's payoffs. There are $\frac{n}{2}$ majority-seeking players and $\frac{n}{2}$ minority-seeking players. x denotes the number of players other than i who play 1.

The proof of Theorem 1 (in the full version of the paper) shows that in *any* NE of G, at least $n/2$ players must use mixed strategies. Consequently the distribution of the number of players using either strategy has support $\geq n/2$, so for a typical player it is necessary to check $n/2$ of his payoffs.

2.1   A Game Whose Solution Must Have Irrational Numbers

Daskalakis and Papadimitriou [13] note as an open problem, the question of whether there is a 2-strategy anonymous game whose Nash equilibria require players to mix with irrational probabilities. The following example shows that such a game does indeed exist, even with just 3 players. In the context of this paper, it is a further motivation for our focus on approximate rather than exact Nash equilibria.

Example 2. Consider the following anonymous game represented in normal-form in Fig. 2. It can be checked that the game satisfies the anonymity condition. In the unique equilibrium, the row, the column, and the matrix players must randomize respectively with probabilities

$$p_r = \frac{1}{12}(\sqrt{241} - 7), \quad p_c = \frac{1}{16}(\sqrt{241} - 7), \quad p_m = \frac{1}{36}(23 - \sqrt{241}).$$

	1	2
1	$(1,0,1)$	$(1,\frac{1}{2},0)$
2	$(0,0,0)$	$(\frac{1}{2},\frac{1}{4},0)$

1

	1	2
1	$(1,0,0)$	$(0,\frac{1}{4},\frac{1}{2})$
2	$(\frac{1}{2},1,\frac{1}{2})$	$(1,0,1)$

2

Fig. 2. The three-player two-strategy anonymous game in normal form. A payoff tuple (a,b,c) represents the row, the column, and the matrix players' payoff, respectively.

Theorem 2. *There exists a three-player, two-strategy anonymous game that has a unique Nash equilibrium where all the players must randomize with irrational probabilities.*

We show in the full version of the paper that Example 2 is a game that does indeed satisfy the conditions of Theorem 2.

3 Lipschitz Games

Lipschitz games are anonymous games where every player's utility function is Lipschitz-continuous, in the sense that for all $i \in [n]$, all $j \in \{1, 2\}$, and all $x, y \in \{0, \ldots, n-1\}$, $\left|u_j^i(x) - u_j^i(y)\right| \le \lambda |x - y|$, where $\lambda \ge 0$ is the Lipschitz constant. For games satisfying a Lipschitz condition with a small value of λ, we obtain a positive result (that we apply in the next section) for approximation and query complexity.

Definition 1. *Let* $(x \in \{0, \ldots, n-1\}, j \in \{1, 2\})$ *be the input for an all-players query. For* $\delta \ge 0$, *a* δ*-accurate all-players query returns a tuple of values* $(f_j^1(x), \ldots, f_j^n(x))$ *such that for all* $i \in [n]$, $\left|u_j^i(x) - f_j^i(x)\right| \le \delta$, *i.e., they are within an additive* δ *of the correct payoffs* $(u_j^1(x), \ldots, u_j^n(x))$.

Theorem 3. *Let G be an n-player, two-strategy λ-Lipschitz anonymous game. Algorithm 1 finds a pure-strategy $3(\lambda + \delta)$-WSNE with $4 \log n$ δ-accurate all-players payoff queries.*

The proof (in the full version of the paper) shows how a solution can be found via a binary search on $\{0, \ldots, n-1\}$. Existence of pure approximate equilibria is known already by [13] in the context of k-strategy games. Their proof reduces the problem to finding a Brouwer fixed point. Theorem 3 is used in the next section as part of an algorithm for general anonymous games.

4 General Two-Strategy Anonymous Games

First, we present our main result (Theorem 4). Next, we prove a lower bound on the number of queries that any randomized algorithm needs to make to find any ϵ-WSNE.

4.1 Upper Bound

Before going into technical lemmas, we provide an informal overview of the algorithmic approach. Suppose we are to solve an n-player game G. The first idea is to *smooth* every player's utility function, so that it becomes λ-Lipschitz continuous for some λ. We smooth a utility function by requiring every player to use some amount of randomness. Specifically, for some small ζ we make every player place probability either ζ or $1 - \zeta$ onto strategy one. Consequently, the expected payoff for player i is obtained by averaging her payoff values w.r.t. a sum of two binomial distributions, consisting of a discrete bell-shaped distribution whose standard deviation is at least $\zeta\sqrt{n}$.

We construct the smooth game $\bar{G}$ in the following manner. The payoff received in $\bar{G}$ by player i when x other players are playing strategy one is given by the expected payoff received in G by player i when x other players play one with probability $1 - \zeta$ and $n - 1 - x$ other players play one with probability ζ. This creates a λ-Lipschitz game $\bar{G}$ with $\lambda = O\left(1/\zeta\sqrt{n}\right)$.

Due to dealing with a two-strategy Lipschitz game, we can use the bisection method of Algorithm 1. If we were allowed to query $\bar{G}$ directly, a logarithmic number of all-players queries would suffice. Unfortunately, this is not the case; thus, we need to simulate a query to $\bar{G}$ with a small number of queries to the original game G. Those queries are randomly sampled from the mixed anonymous profile above, and we take enough samples to ensure we get good estimates of the payoffs in $\bar{G}$ with sufficiently high probability.

Thus, we are able to find an approximate pure Nash equilibrium of $\bar{G}$ with $\tilde{O}(\sqrt{n})$ all-players queries. This equilibrium is mapped back to G by letting the players who play strategy one in $\bar{G}$, play it with probability $1 - \zeta$ in G, and the ones who play strategy two in $\bar{G}$ place probability ζ on strategy one in G. The quality of the approximation is proportional to $\left(\zeta + (\zeta\sqrt{n})^{-1}\right)$.

Before presenting our main algorithm (Algorithm 2) and proving its efficiency, we state the following lemmas (proven in the full version and used in the proof of Theorem 4).

Lemma 1 [13]. *Let X, Y be two random variables over $\{0, \dots, n\}$ such that $\|X - Y\|_{\mathrm{TV}} \leq \delta$ (where $\|X - Y\|_{\mathrm{TV}}$ denotes the total variation distance between X and Y, i.e., $1/2 \cdot \sum_{x=0}^{n} |\Pr[X = x] - \Pr[Y = x]|$). Let $f : \{0, \dots, n\} \longrightarrow [0,1]$. Then,*

$$\sum_{x=0}^{n} f(x) \cdot (\Pr[X = x] - \Pr[Y = x]) \leq 2\delta.$$

Lemma 2 (Simulation of a query to $\bar{G}$ (Algorithm 2)). *Let $\delta, \tau > 0$. Let X be the sum of $n - 1$ Bernoulli random variables representing a mixed anonymous profile of an n-player game G. Suppose we want to estimate, with additive error δ, the expected payoffs $\mathbb{E}[u_j^i(X)]$ for all $i \in [n], j \in \{1, 2\}$. This can be done with probability $\geq 1 - \tau$ using $(1/2\delta^2) \cdot \log(4n/\tau)$ all-players queries.*

Lemma 3. *Let $X^{(j,n)} := \sum_{i \in [n]} X_i$ denote the sum of n independent 0-1 random variables such that $\mathbb{E}[X_i] = 1 - \zeta$ for all $i \in [j]$, and $\mathbb{E}[X_i] = \zeta$ for all $i \in [n] \setminus [j]$. Then, for all $j \in [n]$, we have that*

$$\left\| X^{(j-1,n)} - X^{(j,n)} \right\|_{\mathrm{TV}} \leq O\left(\frac{1}{\zeta\sqrt{n}}\right).$$

Definition 2. Let $G = (n, 2, \{u_j^i\}_{i \in [n], j \in \{1,2\}})$ be an anonymous game. For $\zeta > 0$, the ζ-smoothed version of G is a game $\bar{G} = (n, 2, \{\bar{u}_j^i\}_{i \in [n], j \in \{1,2\}})$ defined as follows. Let $X_{-i}^{(x)} := \sum_{j \neq i} X_i$ denote the sum of $n - 1$ Bernoulli random variables where x of them have expectation equal to $1 - \zeta$, and the remaining ones have expectation equal to ζ. The payoff $\bar{u}_j^i(x)$ obtained by every player $i \in [n]$ for playing strategy $j \in \{1, 2\}$ against $x \in \{0, \ldots, n-1\}$ is

$$\bar{u}_j^i(x) := \sum_{y=0}^{n-1} u_j^i(y) \cdot \Pr\left[X_{-i}^{(x)} = y\right] = \mathbb{E}\left[u_j^i\left(X_{-i}^{(x)}\right)\right].$$

Theorem 4. Let $G = (n, 2, \{u_j^i\}_{i \in [n], j \in \{1,2\}})$ be an anonymous game. For ϵ satisfying $1/\epsilon = O(n^{1/4})$, Algorithm 2 can be used to find (with probability $\geq \frac{3}{4}$) an ϵ-NE of G, using $O(\sqrt{n} \cdot \log^2 n)$ all-players queries (hence, $O(n^{3/2} \cdot \log^2 n)$ single-payoff queries) in time $O(n^{3/2} \cdot \log^2 n)$.

Algorithm 1. Approximate NE Lipschitz

Data: δ-accurate query access to utility function $\bar{u}$ of n-player λ-Lipschitz game $\bar{G}$.

Result: pure-strategy $3(\delta + \lambda)$-NE of $\bar{G}$.

begin

    Let $BR_1(i)$ be the number of players whose best response (as derived from the δ-accurate queries) is 1 when i of the other players play 1 and $n - 1 - i$ of the other players play 2.

    Define $\phi(i) = BR_1(i) - i$.   // by construction, $\phi(0) \geq 0$

                                                   // and $\phi(n-1) \leq 0$

    If $BR_1(0) = 0$, **return** all-1's profile.

    If $BR_1(n - 1) = n$, **return** all-2's profile.

    Otherwise,   // In this case, $\phi(0) > 0$ and $\phi(n-1) \leq 0$

    Find, via binary search, x such that $\phi(x) > 0$ and $\phi(x + 1) \leq 0$.

    Construct pure profile $\bar{p}$ as follows:

    For each player i, if $\bar{u}_1^i(x) - \bar{u}_2^i(x) > 2\delta$, let i play 1, and if $\bar{u}_2^i(x) - \bar{u}_1^i(x) > 2\delta$, let i play 2. (The $\bar{u}_j^i$'s are δ-accurate.) Remaining players are allocated either 1 or 2, subject to the constraint that x or $x + 1$ players in total play 1.

    return $\bar{p}$.

end

Proof. Set ζ equal to ϵ and let $\bar{G}$ be the ζ-smoothed version of G. We claim that $\bar{G}$ is a λ-Lipschitz game for $\lambda = O\left((\zeta\sqrt{n})^{-1}\right)$. Let $X_{-i}^{(x)}$ be as in Definition 2. By Lemma 3, $\left\|X_{-i}^{(x-1)} - X_{-i}^{(x)}\right\|_{\mathrm{TV}} \leq O\left(\frac{1}{\zeta\sqrt{n}}\right)$ for all $x \in [n-1]$. Then by Lemma 1, we have

$$\left|\bar{u}_j^i(x - 1) - \bar{u}_j^i(x)\right| \leq O\left(\frac{1}{\zeta\sqrt{n}}\right).$$

Theorem 3 shows that Algorithm 1 finds a pure-strategy $3(\lambda + \delta)$-WSNE of $\bar{G}$, using $O(\log n)$ δ-accurate all-players queries. Thus, Algorithm 1 finds a $O(\frac{1}{\zeta\sqrt{n}} + \delta)$-WSNE of $\bar{G}$, where δ is the additive accuracy of queries.

Algorithm 2. Approximate NE general

Data: ϵ; query access to utility function u of n-player anonymous game G;
  parameters τ (failure probability), δ (accuracy of queries).
Result: $O(\epsilon)$-NE of G.
begin
  Set $\zeta = \epsilon$. Let $\bar{G}$ be the ζ-smoothed version of G, as in Definition 2.
                         // By Lemma 1 and Lemma 3 it follows that
                         // $\bar{G}$ is λ-Lipschitz for $\lambda = O(1/\zeta\sqrt{n})$.
  Apply Algorithm 1 to $\bar{G}$, simulating each all-players δ-accurate query to $\bar{G}$
  using multiple queries according to Lemma 2.
  Let $\bar{p}$ be the obtained pure profile solution to $\bar{G}$.
  Construct p by replacing probabilities of 0 in $\bar{p}$ with ζ and probabilities of 1
  with $1 - \zeta$.
  return p.
end

Despite not being allowed to query $\bar{G}$ directly, we can simulate any δ-accurate query to $\bar{G}$ with a set of randomized all-players queries to G. This is done in the body of Algorithm 2. By Lemma 2, for $\tau > 0$, $(1/2\delta^2)\log(4n/\tau)$ randomized queries to G correctly simulate a δ-accurate query to $\bar{G}$ with probability $\geq 1 - \tau$.

In total, the algorithm makes $O\left(\log n \cdot (1/\delta^2) \cdot \log(n/\tau)\right)$ all-players payoff queries to G. With a union bound over the $4\log n$ simulated queries to $\bar{G}$, this works with probability $1 - 4\tau\log n$.

Once we find this pure-strategy $O\left(\frac{1}{\zeta\sqrt{n}} + \delta\right)$-WSNE of $\bar{G}$, the last part of Algorithm 2 maps the pure output profile to a mixed one where whoever plays 1 in $\bar{G}$ places probability $(1 - \zeta)$ on 1, and whoever plays 2 in $\bar{G}$ places probability ζ on 1. It is easy to verify that the regret experienced by player i (that is, the difference in payoff between i's payoff and i's best-response payoff) in G is at most ζ more than the one she experiences in $\bar{G}$.

The extra additive ζ to the regret of players means that we have an ϵ-NE of G with $\epsilon = O(\zeta + \delta + \frac{1}{\zeta\sqrt{n}})$. The query complexity thus is $O(\log n \cdot (1/\delta^2) \cdot \log(n/\tau))$.

Setting $\delta = 1/\sqrt[4]{n}$, $\zeta = 1/\sqrt[4]{n}$, $\tau = 1/16\log n$, we find an $O(1/\sqrt[4]{n})$-Nash equilibrium using $O(\sqrt{n} \cdot \log^2 n)$ all-players queries with probability at least $3/4$. We remark that the above parameters can be chosen to satisfy any given approximation guarantee $\epsilon \geq n^{-1/4}$, i.e., simply find solutions to the equation $\epsilon = \zeta + \delta + (\zeta\sqrt{n})^{-1}$. This allows for a family of algorithms parameterized by ϵ, for $\epsilon \in [n^{-1/4}, 1)$, thus an approximation scheme.

The runtime is equal to the number of single-payoff queries and can be calculated as follows. Calculating the value of $\phi(i)$ in Algorithm 1 takes $O(n\sqrt{n}\log n)$. We make $O(\sqrt{n}\log n)$ queries to G to simulate one in $\bar{G}$, and once we gather all the information, we need an additional linear time to count the number of players whose best response is 1. The fact that the above part is performed at every step of the binary search implies a total running time of $O(n^{3/2} \cdot \log^2 n)$

for Algorithm 1. Algorithm 2 simply invokes Algorithm 1 and only needs linear time to construct the profile p; thus, it runs in the same time.   □

4.2   Lower Bound

We use the minimax principle and thus define a distribution over instances that will lead to the lower bound on query complexity, for any deterministic algorithm. We specify a distribution over certain games that possess a unique pure Nash equilibrium. The n players that participate in any of these games are partitioned into $\log n$ groups, which are numbered from 1 to $\log n$. Group i's equilibrium strategy depends on what all the previous groups $\{1, \ldots, i-1\}$ play at equilibrium. Hence, finding out what the last group should play leads to a lower bound of $\Omega(\log n)$ all-players queries.

Lemma 4. *Let $\mathcal{G}_n$ be the class of n-player two-strategy anonymous games such that $u_1^i(x) = 1 - u_2^i(x)$ and $u_1^i(x) \in \{0, 1\}$, for all $i \in [n], x \in \{0, \ldots, n-1\}$. Then, there exists a distribution $\mathcal{D}_n$ over $\mathcal{G}_n$ such that every G drawn from $\mathcal{D}_n$ has a unique (pure-strategy) ϵ-WSNE.*

Theorem 5. *Let $\mathcal{G}_n$ be defined as in Lemma 4. Then, for any $\epsilon \in [0, 1)$, any randomized all-players query algorithm must make $\Omega(\log n)$ queries to find an ϵ-WSNE of $\mathcal{G}_n$ in the worst case.*

5   Conclusions and Further Work

Our interest in the query complexity of anonymous games has resulted in an algorithm that has an improved runtime-efficiency guarantee, although limited to when the number of strategies k is equal to 2. Algorithm 2 (Theorem 4) finds an ϵ-NE faster than the PTAS of [13], for any $\epsilon \geq 1/\sqrt[4]{n}$. In particular, for $\epsilon = 1/\sqrt[4]{n}$, their algorithm runs in subexponential time, while ours is just $\tilde{O}(n^{3/2})$; however, our ϵ-NE is not well-supported.

An immediate question is whether we can obtain sharper bounds on the query complexity of two-strategy games. There are ways to potentially strengthen the results. First, our lower bound holds for well-supported equilibria; it would be interesting to know whether a logarithmic number of queries is also needed to find an ϵ-NE for $\epsilon < \frac{1}{2}$. We believe this is the case at least for small values of ϵ. Second, the ϵ-NE found by our algorithm are not well-supported since all players are forced to randomize. Is there a query-efficient algorithm that finds an ϵ-WSNE? Third, we may think of generalizing the algorithm to the (constant) k-strategy case by letting every player be obliged to place probability either $\frac{\zeta}{k}$ or $1 - \frac{k-1}{k}\zeta$ and obtain a similar smooth utility function. However, in this case we cannot use a bisection algorithm to find a fixed point of the smooth game. As a consequence, the query complexity might be strictly larger.

References

1. Azrieli, Y., Shmaya, E.: Lipschitz games. Math. Oper. Res. **38**(2), 350–357 (2013)
2. Babichenko, Y.: Best-reply dynamics in large binary-choice anonymous games. Games Econ. Behav. **81**(1), 130–144 (2013)
3. Babichenko, Y.: Query complexity of approximate Nash equilibria. In: Proceedings of the 46th Annual ACM Symposium on Theory of Computing. pp. 535–544. STOC 2014. ACM, USA (2014)
4. Brandt, F., Fischer, F., Holzer, M.: Symmetries and the complexity of pure Nash equilibrium. J. Comput. Syst. Sci. **75**, 163–177 (2009)
5. Chen, X., Deng, X., Teng, S.: Settling the complexity of computing two-player Nash equilibria. J. ACM **56**(3), 1–57 (2009)
6. Chen, X., Durfee, D., Orfanou, A.: On the complexity of Nash equilibria in anonymous games. In: Proceedings of the Forty-Seventh Annual ACM on Symposium on Theory of Computing, STOC 2015, ACM, pp. 381–390 (2015)
7. Daskalakis, C.: An efficient PTAS for two-strategy anonymous games. In: Papadimitriou, C., Zhang, S. (eds.) WINE 2008. LNCS, vol. 5385, pp. 186–197. Springer, Heidelberg (2008)
8. Daskalakis, C., Goldberg, P.W., Papadimitriou, C.H.: The complexity of computing a Nash equilibrium. SIAM J. Comput. **39**(1), 195–259 (2009)
9. Daskalakis, C., Papadimitriou, C.H.: Computing equilibria in anonymous games. In: Proceedings of the 48th Symposium on Foundations of Computer Science (FOCS), pp. 83–93 (2007)
10. Daskalakis, C., Papadimitriou, C.H.: Discretized multinomial distributions and Nash equilibria in anonymous games. In: Proceedings of the 49th Symposium on Foundations of Computer Science (FOCS), pp. 25–34 (2008)
11. Daskalakis, C., Papadimitriou, C.H.: On oblivious PTAS's for Nash equilibrium. In: Proceedings of the 41st Annual ACM Symposium on Theory of Computing, STOC 2009, pp. 75–84. ACM, USA (2009)
12. Daskalakis, C., Papadimitriou, C.H.: Sparse covers for sums of indicators (2013). CoRR abs/1306.1265
13. Daskalakis, C., Papadimitriou, C.H.: Approximate Nash equilibria in anonymous games. J. Econ. Theory **156**, 207–245 (2015)
14. Etessami, K., Yannakakis, M.: On the complexity of Nash equilibria and other fixed points. SIAM J. Comput. **39**(6), 2531–2597 (2010)
15. Fearnley, J., Gairing, M., Goldberg, P.W., Savani, R.: Learning equilibria of games via payoff queries. In: Proceedings of the 14th ACM Conference on Electronic Commerce, EC 2013, pp. 397–414. ACM, USA (2013)
16. Fearnley, J., Savani, R.: Finding approximate Nash equilibria of bimatrix games via payoff queries. In: Proceedings of the Fifteenth ACM Conference on Economics and Computation, EC 2014, pp. 657–674. ACM, USA (2014)
17. Goldberg, P.W., Roth, A.: Bounds for the query complexity of approximate equilibria. In: Proceedings of the Fifteenth ACM Conference on Economics and Computation, EC 2014, pp. 639–656. ACM, USA (2014)
18. Hart, S., Nisan, N.: The query complexity of correlated equilibria. In: Vöckling, B. (ed.) SAGT 2013. LNCS, vol. 8146, p. 268. Springer, Heidelberg (2013)

19. Kash, I.A., Friedman, E.J., Halpern, J.Y.: Multiagent learning in large anonymous games. In: Eighth International Conference on Autonomous Agents and Multiagent Systems (AAMAS), pp. 765–772 (2009)
20. Lipton, R.J., Markakis, E., Mehta, A.: Playing large games using simple strategies. In: Proceedings of the 4th ACM Conference on Electronic Commerce, EC 2003, pp. 36–41. ACM, USA (2003)

Incentivizing Exploration with Heterogeneous Value of Money

Li Han, David Kempe, and Ruixin Qiang[✉]

University of Southern California, Los Angeles, USA
{li.han,dkempe,rqiang}@usc.edu

Abstract. Recently, Frazier et al. proposed a natural model for crowd-sourced exploration of different a priori unknown options: a principal is interested in the long-term welfare of a population of agents who arrive one by one in a multi-armed bandit setting. However, each agent is myopic, so in order to incentivize him to explore options with better long-term prospects, the principal must offer the agent money. Frazier et al. showed that a simple class of policies called time-expanded are optimal in the worst case, and characterized their budget-reward tradeoff. The previous work assumed that all agents are equally and uniformly susceptible to financial incentives. In reality, agents may have different utility for money. We therefore extend the model of Frazier et al. to allow agents that have heterogeneous and non-linear utilities for money. The principal is informed of the agent's tradeoff via a signal that could be more or less informative.

Our main result is to show that a convex program can be used to derive a signal-dependent time-expanded policy which achieves the best possible Lagrangian reward in the worst case. The worst-case guarantee is matched by so-called "Diamonds in the Rough" instances; the proof that the guarantees match is based on showing that two different convex programs have the same optimal solution for these specific instances.

Keywords: Multi-armed bandit problems · Mechanism design · Incentives

1 Introduction

The goal of mechanism design is to align incentives when different parties have conflicting interests. In the VCG mechanism, the mechanism designer wants to maximize social welfare whereas each bidder selfishly maximizes his own pay-off. In revenue maximization, the objectives are even more directly opposed, as any increase in the bidders' surplus hurts the revenue for the auctioneer. In all of these cases, it is the mechanism's task to trade off between the differing interests.

The phrase "trade off" is also frequently applied in the context of online learning and the multi-armed bandit (MAB) problem, where the "exploration

© Springer-Verlag Berlin Heidelberg 2015
E. Markakis and G. Schäfer (Eds.): WINE 2015, LNCS 9470, pp. 370–383, 2015.
DOI: 10.1007/978-3-662-48995-6_27

vs. exploitation tradeoff" is routinely referenced. However, in the traditional view of a single principal making a sequence of decisions to maximize long-term rewards, it is not clear what exactly is being traded off against what. Recent work by Frazier et al. [4] makes this tradeoff more explicit, by juxtaposing a principal (with a far-sighted goal of maximizing long-term rewards) with selfish and myopic agents. Thus, the principal wants to "explore," while the agents want to "exploit." In order to partially align the incentives, the principal can offer the agents monetary payments for pulling particular arms.

The framework of Frazier et al. [4] is motivated by many real-world applications, all sharing the property that the principal is interested in the long-term outcome of an exploration of different options, but cannot carry out the exploration herself[1]. Perhaps the most obvious fit is that of an online retailer with a large selection of similar products (e.g., cameras on amazon.com); in order to learn which of these products are best (and ensure that *future* buyers purchase the best product), the retailer needs to rely on customers to buy and review the products. Each customer prefers to purchase the best product for himself based on the current reviews, whereas the principal may want to obtain additional reviews for products that currently have few reviews, but may have the potential of being high quality. Customers can be incentivized to purchase such products by offering suitable discounts.

Frazier et al. [4] explore this tradeoff under the standard time-discounted Bayesian[2] multi-armed bandit model (described formally in Sect. 2). In each round, each arm i has a known posterior reward distribution v_i conditioned on its history so far, and one arm is pulled based on the current state of the arms. The principal's goal is to maximize the total expected time-discounted reward $R = \sum_{t=0}^{\infty} \gamma^t \mathbb{E}[v_{i_t}]$, where γ is the time discount factor. However, without incentives, each selfish agent would pull the *myopic* arm i maximizing the immediate expected reward $\mathbb{E}[v_i]$. When the principal offers payments c_i for pulling arms i, in [4], the agent's utility for pulling arm i is $\mathbb{E}[v_i] + c_i$, and a myopic agent will choose the arm maximizing this sum.

Implicit in this model is the assumption that all agents have the same (one-to-one) tradeoff between arm rewards and payments. In reality, different agents might have different and non-linear tradeoffs between these two, due to a number of causes. The most obvious is that an agent with a large money endowment may not value additional payments as highly as an agent with less endowment; this is generally the motivation for positing concave utility functions of money. In the case of an online retailer, another obvious reason is that different customers may intend to use the product for different amounts of time or with different intensity, making the optimization of quality more or less important. Concretely, a professional photographer may be much less willing to compromise on quality in return for a discount than an amateur.

---

[1] To avoid ambiguity, we consistently refer to the principal as female and the agents as male.

[2] Both Frazier et al. [4] and our work in fact consider a generalization in which each arm constitutes an independent Markov chain with Martingale rewards.

The main contribution of the present article is an extension of the model and analysis of Frazier et al. [4] to incorporate non-uniform and non-linear tradeoffs between rewards and money.

Related Work

The MAB problem was first proposed by Robbins [13] as a model for sequential experiments design. Under the Bayesian model with time-discounted rewards, the problem is solved optimally by the Gittins Index policy [7]; a further discussion is given in [5,6,9,17].

An alternative objective of MAB problem, often pursued in the CS literature, is regret-minimization, as initiated by Lai and Robbins [11] within a Bayesian arm reward setting. Auer et al. [1,2] gave an algorithm with regret bound for adversarial settings.

There is a rich literature that considers MAB problems when incentive issues arise. A common model is that a principal has to hire workers to pull arms, and both sides want to maximize their own utility. Singla and Krause [14] gave a truthful posted price mechanism. In [10,12], the reward history is only known by the principal, and she can incentivize workers by disclosing limited information about the reward history to the worker. Ho et al. [8] used the MAB framework as a tool to design optimal contracts for agents with moral hazard. Using the technique of discretization, they achieved sublinear regret for the net utility (reward minus payment) over the time horizon. For a review of more work in the area, see the position paper [15].

2   Preliminaries

2.1   Multi-armed Bandits

In a multi-armed bandits (MAB) instance, we are given N arms, each of which evolves independently as a known Markov chain whenever pulled. In each round[3] $t = 0, 1, 2, \cdots$, an algorithm can only pull one of the arms; the pulled arm will generate a random reward and then transition to a new state randomly according to the known Markov chain.

Formally, let $v_{t,i}$ be the random reward generated by arm i if it is pulled at time t. Let $S_{0,i}$ be the initial state of the Markov chain of the i-th arm and $S_{t,i}$ the state of arm i in round t. The distribution of $v_{t,i}$ is determined by $S_{t,i}$. Then, an MAB instance consists of N independent Markov chains and their initial states $\mathbf{S}_0 = (S_{0,i})_{i=1}^N$.

In this article, we are only interested in cases where the reward sequence for any single arm forms a *Martingale*, i.e.,

$$\mathbb{E}\left[\mathbb{E}\left[v_{t+1,i} \mid S_{t+1,i}\right] \mid S_{t,i}\right] = \mathbb{E}\left[v_{t,i} \mid S_{t,i}\right].$$

---

[3] We use the terms "round" and "time" interchangeably.

A *policy* $\mathcal{A}$ is an algorithm that decides which arm to pull in round t based on the history of observations and the current state of all arms. Formally, a policy is a (randomized) mapping $\mathcal{A} : (t, \mathcal{H}_t, \mathbf{S}_t) \mapsto i_t$, where $\mathbf{S}_t = (S_{t,i})_{i=1}^N$ is the vector of arms' states, $\mathcal{H}_t$ is the history up to time t, and i_t is the selected arm.

To evaluate the performance of a policy $\mathcal{A}$, we use standard time-discounting [7]. Let $\gamma \in (0, 1)$ be the time discount factor that measures the relative importance between future rewards and present rewards. The total expected time-discounted reward can be defined as $R^{(\gamma)}(\mathcal{A}) = \mathbb{E}_{\mathcal{A}} [\sum_{t=0}^{\infty} \gamma^t v_{t,i_t}]$, where $\mathbb{E}_{\mathcal{A}} [\cdot]$ denotes the expectation conditioned on the policy $\mathcal{A}$ being followed and the information it obtained, as in [4].

Given a time discount factor γ, we denote the optimal policy for that time discount (and also — in a slight overload of notation — its total expected time-discounted reward) by OPT_γ. This can be accomplished by the well-known Gittins Index policy [7], which computes an index for each arm i based on the state of the Markov chain, and then chooses the arm with largest index

We call the arm with the maximum *immediate* expected reward $\mathbb{E}[v_{t,i} \mid \mathbf{S}_t]$ the *myopic arm*. A policy is called myopic if it pulls the myopic arm in each round. The myopic policy only *exploits* with no *exploration*, so it is inferior to the optimum policy in general, especially when the time-discount factor γ is close to 1.

2.2 Selfish Agents

We label each agent by the time t when he arrives, and assume he has a monotone and concave money utility function $\mu_t : \mathbb{R}^+ \to \mathbb{R}^+$ mapping the payment he gets to his corresponding utility. The special case where $\mu_t(x) = x$ was studied in [4].

In round t, the Markov chain state $\mathbf{S}_t$ and $\mathbb{E}[v_{t,i} \mid \mathbf{S}_t]$ are publicly known by both the agent and the principal. When the principal offers a payment $c_{t,i}$ for pulling arm i, incentivized by these extra payments $c_{t,i}$, agent t now pulls the arm maximizing his utility $\mathbb{E}[v_{t,i} \mid \mathbf{S}_{t,i}] + \mu_t(c_{t,i})$. If arm i_t is pulled by agent t, the principal's reward from this pull is $\mathbb{E}[v_{t,i_t} \mid \mathbf{S}_{t,i_t}]$.

We assume a publicly known prior (whose cumulative distribution function is denoted by F) over the money utility function μ. When a new agent arrives, his money utility function is drawn from F independently of prior draws.

2.3 Signaling Scheme

In the presence of uncertainty about each individual's money utility function, an important question is how much the principal knows about μ_t at the time she chooses the payment vector $\mathbf{c}_t = (c_{t,i})_i$ to announce for the arm pulls. In the worst case, the principal may know nothing about agent t as he arrives. In that case, the payment vector $\mathbf{c}_t$ can only depend on F. At the other extreme, the principal may learn μ_t exactly. Reality will typically lie between these two extreme cases. Both financial endowments and intended use can be partially inferred from past searches and purchases in the case of an online retailer. This partial information will give the principal a more accurate estimate of the agent's money utility function than what could be learned from the prior distribution F alone, allowing her to better engineer the incentives.

We model the notion of partial information using the standard economic notion of an exogenous signaling scheme [16], which is given as input. When an agent with money utility function μ arrives, a signal $s \in \Sigma$ correlated to μ is revealed to the principal according to the signaling scheme φ; Σ is called the *signal space*, and we assume that it is countable.[4]

Formally, let $\varphi(\mu, s)$ be the probability that signal s is revealed when the agent's money utility function is μ. In this way, the signals are statistically correlated with the money utility function μ, and thus each signal reveals partial information about μ. After receiving the signal s, the principal updates her posterior belief of the agent's money utility function according to Bayes Law.

2.4   Linear Money Utility Functions

Justified by the following lemma (proved in the full version), for the rest of the article, we focus on linear money utility functions.

Lemma 1. *Given a distribution F over money utility functions, define a new distribution F' as follows: if μ is the result of a draw, output $\mu'(x) = r \cdot x$, where $r = \lim_{x \to \infty} \frac{\mu(x)}{x}$. Then the same approximation ratio (defined later) can be achieved for F and F', and this ratio is tight.*

Lemma 1 shows that in a certain sense, *linear* functions μ constitute the worst case for the principal.

Therefore, we will exclusively focus on the case of linear money utility functions $\mu_t(x) = r_t \cdot x$. We then identify the distribution F with a distribution over the values r_t, which we call the *conversion ratio* of agent t. For the remainder of this article, all distributions and signals are assumed to be over conversion ratios instead of money utility functions.

Now if the principal receives a signal s from a signaling scheme φ about the agent's conversion ratio r drawn from distribution with PDF[5] f, she will update her posterior belief of r:

$$f_s(r) = \frac{\varphi(r, s)f(r)}{p_s}, \tag{1}$$

where $f_s(r)$ is the PDF of the posterior belief and $p_s = \int_0^\infty \varphi(r, s)f(r)\mathrm{d}r$ is the probability that signal s is observed. For each signal $s \in \Sigma$, let F_s be the CDF

---

[4] When the signal space is uncountable, defining the posterior probability density requires the use of Radon-Nikodym derivatives, and raises computational and representational issues. In Sect. 6, we consider what is perhaps the most interesting special case: that the signal reveals the precise value of r to the principal.

[5] In Eq. (1), if the support of r is finite, $f(r)$ can be replaced by the probability mass function.

of the corresponding posterior belief. As a special case, if the signaling scheme reveals no information, then $F_s = F$.

Throughout, we focus on the case when the posterior distributions F_s satisfies a condition called *semi-regularity*, which is defined as follows:

Definition 1 (Semi-Regularity). *A distribution with CDF G is called* semi-regular *if $\frac{1-x}{G^{-1}(x)}$ is convex. (When G is not invertible, we define $G^{-1}(x) := \sup\{t \geq 0 : G(t) \leq x\}$.)*

Semi-regularity is a generalization of a well-known condition called regularity, defined as follows.

Definition 2 (Regularity). *A distribution with CDF G is regular if $G^{-1}(x) \cdot (1 - x)$ is concave.*

Lemma 2. *Let G be a CDF. If $G^{-1}(x) \cdot (1-x)$ is concave, then $\frac{1-x}{G^{-1}(x)}$ is convex. In particular, regularity implies semi-regularity.*

2.5 Policies with Selfish Agents and Partial Information

The previous definition we gave of a policy did not take information on the agent's type into account. In light of this additional information, we give a refined definition. In addition to deciding on which arm to pull, a policy may decide on the payment to offer the agents based on the partial information obtained from signals. Formally, a policy is now a randomized mapping $\mathcal{A} : (t, \mathcal{H}_t, \mathbf{S}_t, s_t) \mapsto \mathbf{c}_t$, where s_t is the signal revealed in round t, and $c_{t,i}$ is the extra payment offered for pulling arm i in round t. After $\mathbf{c}_t$ is announced, the myopic agent with conversion ratio r will pull the arm i_t that maximizes his own utility, causing that arm to transition according to the underlying Markov chain.

The expected payment of the principal is also time-discounted by the same[6] factor γ. When $\mathcal{A}$ is implemented, the total expected payment will be $C^{(\gamma)}(\mathcal{A}) = \mathbb{E}_\mathcal{A} \left[\sum_{t=0}^{\infty} \gamma^t c_{t,i_t} \right]$.

The principal faces two conflicting objectives: (a) maximizing the total expected time-discounted reward $R^{(\gamma)}(\mathcal{A})$; (b) minimizing the total expected time-discounted payment $C^{(\gamma)}(\mathcal{A})$; There are two natural ways of combining the two objectives: via a Lagrangian multiplier, or by optimizing one subject to a constraint on the other.

In the Lagrangian objective, the principal wishes to maximize $R^{(\gamma)}(\mathcal{A}) - \lambda C^{(\gamma)}(\mathcal{A})$ for some constant $\lambda \in (0, 1)$. Here, λ can also be regarded as the conversion ratio for the principal herself. Alternatively, the principal may be constrained by a budget b, and want to maximize $R^{(\gamma)}(\mathcal{A})$ subject to the constraint that $C^{(\gamma)}(\mathcal{A}) \leq b$.

In this paper, we perform a worst-case analysis with respect to the MAB instances in a similar way as [4], while keeping an exogenous signaling scheme φ and the prior F fixed.

---

[6] A natural justification for having the same discount factor is that after each round, with probability $1 - \gamma$, the game ends.

Definition 3. *For the Lagrangian objective of the problem, a policy $\mathcal{A}$ has approximation ratio α under the signaling scheme φ and prior F if for all MAB instances[7],*

$$R^{(\gamma)}(\mathcal{A}) - \lambda C^{(\gamma)}(\mathcal{A}) \geq \alpha \cdot \text{OPT}_\gamma. \tag{2}$$

Likewise, for the budgeted version, a policy has approximation ratio α respecting budget b if

$$R^{(\gamma)}(\mathcal{A}) \geq \alpha \cdot \text{OPT}_\gamma \qquad\qquad C^{(\gamma)}(\mathcal{A}) \leq b \cdot \text{OPT}_\gamma. \tag{3}$$

3   Our Results

Our first main theorem addresses the Lagrangian objective.

Theorem 1. *Let γ be the time discount factor. Given a semi-regular prior distribution F and signaling scheme φ, one can efficiently compute a policy **TES** and $p^*(\varphi)$ such that the Lagrangian reward of **TES** is a $(1 - p^*(\varphi)\gamma)$-approximation to OPT_γ. This bound is tight.*

Our second main theorem is for the budgeted version.

Theorem 2. *Given a semi-regular prior distribution F, signaling scheme φ and budget constraint b, there exists a policy **TES** whose total expected time-discounted reward is a $\min_\lambda\{1 - p^*(\varphi)\lambda + \lambda b\} - \epsilon$ approximation to OPT_γ, while spending at most $b\text{OPT}_\gamma$ in expectation. This bound is tight.*

In a sense, these theorems quantify the power of the signaling scheme φ in a single number $1 - p^*(\varphi)\gamma$, via the approximation guarantee that can be achieved using φ. If this number is meaningful, more informative signaling schemes should allow for better approximation ratios, which we address by the following theorem:

Theorem 3. *Let φ and φ' be two signaling schemes such that φ' is a garbling of φ. Then, $1 - p^*(\varphi)\gamma \geq 1 - p^*(\varphi')\gamma$.*

Here, *garbling* is defined as follows:

Definition 4. *Let φ, φ' be signaling schemes with respective signal spaces Σ, Σ'. φ' is a garbling of φ if for all conversion ratios r and signals $s \in \Sigma, s' \in \Sigma'$: $f_{s,s'}(r) = f_s(r)$, where $f_s(r)$ is the PDF of the conversion ratio r conditioned on signal s.*

Due to limited space, we will only give a proof sketch for Theorem 1 in Sects. 4 and 5. The complete proofs of all theorems and lemmas in the article can be found in the full version.

---

[7] Note that all $R^{(\gamma)}(\mathcal{A})$, $C^{(\gamma)}(\mathcal{A})$ and OPT_γ depend on the MAB instance.

4 Lower Bound: Time-Expanded Algorithm

In this section, we analyze *time-expanded algorithms* with a Lagrangian objective, in a generalization of the originally proposed notion of [4]. In a time-expanded algorithm, the principal randomizes between offering the agents no reward (having them play myopically), and offering the reward necessary to incentivize the agent to play the arm i_t^* according to a particular algorithm $\mathcal{A}$. In the presence of signals, the randomization probabilities for the different signals need to be chosen and optimized carefully, which is the main algorithmic contribution in this section. On the other hand, notice that if the posterior distribution of the conversion ratio conditioned on the signal is continuous, then the randomness in the user's type can instead be used as a randomization device, and the principal may be able to offer incentives deterministically.

More formally, Frazier et al. [4] define a time-expanded version $\mathbf{TE}_{p,\mathcal{A}}$ of a policy $\mathcal{A}$, parameterized by a probability p, as

$$\mathbf{TE}_{p,\mathcal{A}}(t) := \begin{cases} \mathcal{A}(\hat{\mathbf{S}}_t) & \text{if } Z_t = 1 \\ \text{argmax}_i \, \mathbb{E}\left[v_{t,i} \mid \mathbf{S}_t\right], & \text{otherwise} \end{cases}$$

where Z_t is a Bernoulli$(1-p)$ variable. $\hat{\mathbf{S}}_t$ is the arm status that couples the execution of the time-expanded policy and the policy $\mathcal{A}$, which we will formally define later. When $Z_t = 1$, with the uniform agents defined in [4], in order to incentivize an agent to pull the non-myopic arm, the principal has to offer a payment of $\max_i \mathbb{E}\left[v_{t,i} \mid \mathbf{S}_t\right] - \mathbb{E}\left[v_{t,i_t^*} \mid \mathbf{S}_t\right]$, where $i_t^* = \mathcal{A}(\hat{\mathbf{S}}_t)$.

A time-expanded version of policy $\mathcal{A}$ with signaling scheme φ works as follows: at time t, conditioned on the received signal s, the principal probabilistically offers a payment of c_{t,i_t^*} if the agent t pulls the arm i_t^*. Notice that only two options might maximize the agent's utility: pulling the myopic arm, or pulling the arm i_t^* and getting the payment. There is a direct correspondence between the payment c_{t,i_t^*} and the probability q_s that the agent chooses to pull the myopic arm. We will describe this correspondence below.

First, though, we discuss *which* arm i_t^* the principal is trying to incentivize the agent to pull. As in [4], it is necessary for the analysis that the execution of $\mathcal{A}$ and of its time-expanded version can be coupled. To achieve this, in order to evaluate which arm should be pulled next by $\mathcal{A}$, the principal must only take the information obtained from the *non-myopic pulls* into consideration. Formally, we define $\hat{\mathbf{S}}_t$ as follows: Define the random variable

$$Z_t := \begin{cases} 0 & \text{agent } t \text{ pulls the myopic arm} \\ 1 & \text{otherwise} \end{cases}$$

and $X_{t,i} = 1$ if arm i is pulled at time t and 0 otherwise. Notice that Z_t is a Bernoulli variable, and $\text{Prob}[Z_t = 0]$ depends on the received signal s and the payment offered by the principal. Let $N_{t,i} = \sum_0^{t-1} Z_t X_{t,i}$ be the number of non-myopic pulls of arm i before time t. Using this notation, we define $\hat{S}_{t,i}$ to be the state of the Markov chain of arm i after the first[8] $N_{t,i}$ pulls in the execution history of the time-expanded policy $\mathbf{TE}_{p,\mathcal{A}}$, and $\hat{\mathbf{S}}_t = (\hat{S}_{t,i})_i$.

Let F_s be the posterior CDF of the agent's conversion ratio. For simplicity of notation, define $x = \max_i \mathbb{E}\left[v_{t,i} \mid \mathbf{S}_t\right]$ to be the expected reward of the myopic arm and $y = \mathbb{E}\left[v_{t,i_t^*} \mid \mathbf{S}_t\right]$ to be the expected reward of arm i_t^*. If the principal offers a payment of c_{t,i_t^*}, then agents with conversion ratio $r < \frac{x-y}{c_{t,i_t^*}}$ will still choose the myopic arm. Assuming that agents break ties in favor of the principal, when $r \geq \frac{x-y}{c_{t,i_t^*}}$, they will prefer to pull arm i_t^*.

Conversely, in order to achieve a probability of q_s for pulling the myopic arm, the principal can choose a payment of $c_{t,i_t^*} = \inf\{c \mid F_s(\frac{x-y}{c}) \leq q_s\}$. If F_s is continuous at $\frac{x-y}{c_{t,i_t^*}}$, the probability of myopic play (conditioned on the signal) is exactly q_s, and c_{t,i_t^*} is the smallest payment achieving this probability. If there is a discontinuity at $\frac{x-y}{c_{t,i_t^*}}$, then for every $\epsilon > 0$, the probability of myopic play with payment $c_{t,i_t^*} + \epsilon$ is less than q_s. In that case, the principal offers a payment of c_{t,i_t^*} with probability $\frac{1-q_s}{1-F_s((x-y)/c_{t,i_t^*})}$ for pulling arm i_t^*, and no payment otherwise. Now, the probability of a myopic pull will again be exactly q_s.

To express the payment more concisely, we write $F_s^{-1}(q_s) = \sup\{r \mid F_s(r) \leq q_s\}$. Then, the payment can be expressed as $c_{t,i_t^*} = \frac{x-y}{F_s^{-1}(q_s)}$. In particular, when F_s is continuous, c_{t,i_t^*} will be offered deterministically; otherwise, the principal randomizes.

In summary, we have shown a one-to-one mapping between desired probabilities q_s for myopic play, and payments (and possibly probabilities, in the case of discontinuities) for achieving the q_s. We write $\mathbf{q} = (q_s)_{s \in \Sigma}$ for the vector of all probabilities. The unconditional (prior) probability of playing myopically is $\sum_{s \in \Sigma} p_s q_s$, and the expected payment $(x - y) \cdot \sum_{s \in \Sigma} p_s \frac{1-q_s}{F_s^{-1}(q_s)}$.

We can now summarize the argument above and give a formal definition of the time-expanded version of policy $\mathcal{A}$ with signal scheme φ,

Definition 5. *A policy $\mathbf{TES}_{\mathbf{q},\mathcal{A},\varphi}$ is a time-expanded version of policy $\mathcal{A}$ with signaling scheme φ, if at time t, after receiving the signal s about agent t's conversion ratio, the principal chooses (randomized) payments such that the myopic arm is pulled with probability q_s, and the arm $i_t^* = \mathcal{A}(\hat{\mathbf{S}}_t)$ is pulled with probability $1 - q_s$.*

This is achieved by offering, with probability $\frac{1-q_s}{1-\sup\{F_s(r)\mid F_s(r)\leq q_s\}}$, a payment of $\frac{\max_i \mathbb{E}[v_{t,i} \mid \mathbf{S}_t] - \mathbb{E}\left[v_{t,i_t^*} \mid \mathbf{S}_t\right]}{F_s^{-1}(q_s)}$ for pulling arm i_t^*.

Here, $F_s(r)$ is the CDF of the posterior distribution of the agent's conversion ratio conditioned on signal s.

---

[8] As in [4], in order to facilitate the analysis, this may include myopic and non-myopic pulls of arm i. For instance, if arm 1 was pulled as non-myopic arm at times 1 and 6, and a myopic pull of arm 1 occurred at time 3, then we would use the state of arm 1 after the pulls at times 1 and 3.

In order to analyze the expected Lagrangian reward, we adopt a heuristic that cancels the payment with the expected reward the principal gets from the myopic pull, leaving the reward from non-myopic pull alone in the expectation. Then we have the following lemma, which gives a sufficient condition on $\mathbf{q}$ that allows us to obtain a good approximation ratio of the Lagrangian to the optimum solely in terms of $\sum_s p_s q_s$.

Lemma 3. *Fix a signaling scheme* φ. *If* $\sum_{s\in\Sigma} p_s q_s \geq \lambda \sum_{s\in\Sigma} p_s \frac{1-q_s}{F_s^{-1}(q_s)}$ *is satisfied by* $\mathbf{q}$, *for* $\eta = \frac{(1-p)\gamma}{1-p\gamma}$, *where* $p = \sum_{s\in\Sigma} p_s q_s$, *the time-expanded policy* $\mathbf{TES}_{\mathbf{q},\mathrm{OPT}_\eta,\varphi}$ *satisfies*

$$R_\lambda^{(\gamma)}(\mathbf{TES}_{\mathbf{q},\mathrm{OPT}_\eta,\varphi}) \geq (1 - p\gamma) \cdot \mathrm{OPT}_\gamma. \tag{4}$$

Lemma 3 suggests a natural heuristic for choosing the myopic probabilities $\mathbf{q}$: maximize the approximation ratio $1 - p\gamma$ by minimizing p subject to the conditions of the lemma. The optimization of p can be carried out using the following non-linear program. Surprisingly, this naïve heuristic, motivated predominantly by the need to cancel out terms in the expected reward of a single round, actually gives us the optimal approximation ratio.

$$\begin{aligned}
\text{minimize} \quad & \sum_{s\in\Sigma} p_s q_s \\
\text{subject to} \quad & \sum_{s\in\Sigma} p_s q_s \geq \lambda \sum_{s\in\Sigma} p_s \frac{1-q_s}{F_s^{-1}(q_s)} \\
& 0 \leq q_s \leq 1, \quad \text{for all } s \in \Sigma.
\end{aligned} \tag{5}$$

First, notice that the optimization problem is feasible, because $\mathbf{q} = \mathbf{1}$ is a trivial solution. Whenever F_s is semi-regular, $\frac{1-x}{F_s^{-1}(x)}$ is convex. Therefore, the feasibility region of the optimization problem (5) is convex, and the problem can be solved efficiently [3].

Theorem 4. *Given a signaling scheme* φ, *let* $\mathbf{q}^*$ *be the optimal solution of the convex program (5), and* p^* *be the optimal value. Then, with* $\eta = \frac{(1-p^*)\gamma}{1-p^*\gamma}$, $\mathbf{TES}_{\mathbf{q}^*,\mathrm{OPT}_\eta,\varphi}$ *is a* $1 - p^*\gamma$ *approximation policy to* OPT_γ.

Notice in Theorem 4 that $\mathbf{q}$ can be determined without knowledge of the specific MAB instance; only the signaling scheme and the prior distribution of conversion ratios need to be known.

5   Upper Bound: Diamonds in the Rough

In this section, we show that the approximation ratio $1 - p^*\gamma$ is actually tight when the distribution F_s is semi-regular, where p^* is the value of convex program (5). For simplicity, when $\mathbf{q}^*$ is clear from the context, we let $\mathbf{TES}^*$ denote the policy $\mathbf{TES}_{\mathbf{q}^*,\mathrm{OPT}_\eta,\varphi}$ where $\eta = \frac{(1-p^*)\gamma}{1-p^*\gamma}$ (as in Theorem 4). We will show that on a class of MAB instances called Diamonds-in-the-rough [4], the *optimal policy with payments* (defined below) can achieve only a $(1 - p^*\gamma)$-fraction of OPT_γ. Therefore, not only is the analysis of $\mathbf{TES}^*$'s approximation ratio tight, but $\mathbf{TES}^*$ also has the *optimal* approximation ratio $1 - p^*\gamma$.

Definition 6. *The Diamonds-in-the-rough MAB instance $\Delta(B, \gamma)$ is defined as follows. Arm 1 has constant value $1 - \gamma$. All other (essentially infinitely many) arms have the following reward distribution:*

1. *With probability $1/M$, the arm's reward is a degenerate distribution of the constant $(1 - \gamma)B \cdot M$ (good state);*
2. *With probability $1 - 1/M$, the arm's reward is a degenerate distribution of the constant 0 (bad state).*

Note that if $B < 1$, then arm 1 is the myopic arm.

Since $\Delta(B, \gamma)$ is uniquely determined by B and is just one single instance, the optimal policy that maximizes the Lagrangian objective, i.e., $R^{(\gamma)}(\mathcal{A}) - \lambda C^{(\gamma)}(\mathcal{A})$, is well-defined[9]. We call the policy that maximizes the Lagrangian objective the *optimal policy with payments*, and denote it by $OPT_\lambda^{(\gamma)}(\Delta(B, \gamma))$.

We can solve for the optimal policy with payments using another convex program, which we next derive. Suppose that the optimal policy with payments has time-discounted Lagrangian objective V. In the first round, it only has two options: (a) let the agent play myopically (i.e., pull the constant arm); (b) incentivize him to play a non-constant arm.

If option (a) is chosen and the agent pulled the constant arm, then the principal learns nothing and faces the same situation in the second round. So conditioned on the constant arm being pulled, the principal will get $1 - \gamma + \gamma V$. If option (b) is chosen and a non-constant arm was pulled, then with probability $1/M$, the non-constant arm will be revealed to be in the good state, and the principal does not need to pay any agent again, obtaining value $(1 - \gamma)B \cdot M \sum_{i=0}^\infty \gamma^i = B \cdot M$; with probability $1 - 1/M$, the non-constant arm will be revealed to be in the bad state, and the principal faces the same situation in the second round, obtaining value γV. Recall that c_s is the payment needed to ensure that the myopic arm is played with probability at most q_s when signal s is revealed. To summarize, if we set the probabilities for myopic play to $(q_s)_{s \in \Sigma}$, then V satisfies the following equation:

$$V = (1 - \gamma + \gamma V) \sum_{s \in \Sigma} p_s q_s + \sum_{s \in \Sigma} p_s(1 - q_s)(\frac{1}{M} \cdot B \cdot M + (1 - \frac{1}{M})\gamma V - \lambda c_s). \quad (6)$$

Solving for V while taking $M \to \infty$, we get $(1 - \gamma)V = (1 - \gamma) \sum_{s \in \Sigma} p_s q_s + \sum_{s \in \Sigma} p_s(1 - q_s)(B - \lambda c_s)$. As the difference between the expected rewards of the myopic arm and the non-myopic arm is $(1 - \gamma) - (1 - \gamma)B$, we have $c_s = \frac{(1-\gamma)(1-B)}{F_s^{-1}(q_s)}$. The optimal policy with payments needs to choose the best myopic probabilities, which is equivalent to:

$$\text{maximize} \quad (1 - \gamma) \sum_{s \in \Sigma} p_s q_s + \sum_{s \in \Sigma} p_s(1 - q_s)(B - \lambda \frac{(1-\gamma)(1-B)}{F_s^{-1}(q_s)})$$
$$\text{subject to } 0 \le q_s \le 1, \quad \text{for all } s \in \Sigma. \qquad (7)$$

---

[9] This is in contrast to the case where the performance of a policy is evaluated on a *class* of instances rather than single instance.

Notice that the objective function of program (7) is concave, so the program is convex. Let $\hat{\mathbf{q}}$ be the optimal solution to the program (7). Denote by $\mathcal{A}(\hat{\mathbf{q}})$ the policy determined by $\hat{\mathbf{q}}$. Recall that $\mathbf{q}^*$ is the solution to the following convex program:

$$
\begin{aligned}
\text{minimize} \quad & \textstyle\sum_{s \in \Sigma} p_s q_s \\
\text{subject to} \quad & \textstyle\sum_{s \in \Sigma} p_s q_s \geq \lambda \sum_{s \in \Sigma} p_s \frac{1 - q_s}{F_s^{-1}(q_s)} \\
& 0 \leq q_s \leq 1, \forall s \in \Sigma.
\end{aligned}
\tag{5}
$$

Note that the $\hat{\mathbf{q}}$ are probabilities for choosing the myopic arm given by the above program and depend on a specific MAB instance, i.e., $\Delta(B, \gamma)$. On the other hand, $\mathbf{q}^*$ is *independent* of any MAB instance and only depends on the signaling scheme φ and F. Lemma 4 shows that for the right choice of B, $\hat{\mathbf{q}}$ and $\mathbf{q}^*$ actually coincide on the corresponding Diamonds-in-the-rough instance.

Lemma 4. *There exists a B such that the myopic probabilities given by the convex program (5) are equal to the myopic probabilities given by program (7).*

Based on this lemma, we now have the main theorem.

Theorem 5. *The policy TES^*, parameterized by $\mathbf{q}^*$, has optimal approximation ratio $1 - p^* \gamma$. In particular, there exists a* worst-case *MAB instance in which the optimal policy with payments achieves exactly a Lagrangian reward of a $(1 - p^* \gamma)$ fraction of the optimum.*

6 Full Information Revelation

Our main positive results hold for the case of countable or finite signal spaces, whereas uncountable signal spaces lead to technical challenges. However, one important special case of uncountable signal spaces is more easily handled, namely when the principal learns the exact conversion ratio r, i.e., $s = r$. We show that in that case, r itself can be used as the sole randomization device, leading to a *threshold policy*. In this section, we assume that the distribution F is continuous (an assumption that was not needed in Sect. 4).

6.1 Optimal Time-Expanded Policy

Our first goal will be to show that the optimal time-expanded policy fixes a threshold θ and only incentivizes agents whose conversion ratio lies above the threshold. Then, an optimization over threshold policies is easy to carry out.

Definition 7. *The threshold policy $TP_{\theta, \mathcal{A}}$ with threshold θ is defined as follows: When an agent with conversion ratio r arrives, he is incentivized with suitable payment to pull $i_t^* = \mathcal{A}(\hat{\mathbf{S}}_t)$ if and only if $r \geq \theta$.*

Lemma 5. *Consider a single arm pull, and a value $q \in [0, 1]$. Among all policies that have this arm pull be myopic with probability q, the one minimizing expected cost is a threshold policy.*

Lemma 6. *The Lagrangian objective of any time-expanded policy $\mathcal{P}$ of $\mathcal{A}$ is (weakly) dominated by that of a threshold policy.*

Based on Lemma 6, it suffices to study the optimal threshold policy, and determine the correct threshold. We choose the threshold θ carefully to cancel out the expected payment with the expected reward from myopic pulls, similar to the argument in Sect. 4. This gives us a $(1 - p^*\gamma)$ approximation ratio for the Lagrangian objective as for the case of discrete signals, where $p^* = F(\theta)$.

6.2 Upper Bound

As for discrete signals, we next give a "Diamond-in-the-Rough" instance $\mathbf{\Delta}(B, \gamma)$ on which the upper bound for any policy matches the approximation ratio of the threshold policy. Consider the choice of the policy in the first round; it allows the agent to play myopically with some probability q. By Lemma 5, the optimal way to implement this probability q is to choose a threshold[10] θ and offer incentives to the agent if and only if $r \geq \theta$. Now, similar to Eq. (6), the corresponding Lagrangian objective is

$$
V = (1 - \gamma + \gamma V)F(\theta)
$$
$$
+ \int_{\theta}^{\infty} \left(\frac{1}{M} \cdot B \cdot M + (1 - \frac{1}{M})\gamma V - \lambda \frac{(1 - \gamma)(1 - B)}{r} \right) dF(r). \tag{8}
$$

Letting $M \to \infty$ and solving Eq. (8), we obtain

$$
(1 - \gamma)V = (1 - \gamma - B)F(\theta) + B - \lambda(1 - \gamma)(1 - B) \int_{\theta}^{\infty} \frac{dF(r)}{r}. \tag{9}
$$

Taking a derivative of Eq. (9) with respect to θ suggests (note that the function may not be differentiable, so this is merely used as a tool to suggest a useful choice) that if we set B to solve $-\lambda(1 - \gamma)(1 - B) = (1 - \gamma - B)\theta^*$, the threshold policy with threshold θ^* will be optimal for the instance $\mathbf{\Delta}(B, \gamma)$ (proved in the full version).

Thus, on this particular instance, the ratio achieved by our threshold policy matches that of the best possible policy.

References

1. Auer, P., Cesa-Bianchi, N., Freund, Y., Schapire, R.E.: Gambling in a rigged casino: the adversarial multi-armed banditproblem. In: Proceedings of the 36th IEEE Symposium on Foundations of Computer Science, pp. 322–331 (1995)
2. Auer, P., Cesa-Bianchi, N., Freund, Y., Schapire, R.E.: The nonstochastic multi-armed bandit problem. SIAM J. Comput. **32**(1), 48–77 (2003)

---

[10] Note that a priori, it is not clear that this threshold will not change in subsequent rounds; hence, we cannot yet state that a threshold policy is optimal.

3. Boyd, S., Vandenberghe, L.: Convex Optimization. Cambridge University Press, Cambridge (2004)
4. Frazier, P., Kempe, D., Kleinberg, J., Kleinberg, R.: Incentivizing exploration. In: Proceedings of the 16th ACM Conference on Economics and Computation, pp. 5–22 (2014)
5. Gittins, J.C.: Multi-Armed Bandit Allocation Indices. Wiley, New York (1989)
6. Gittins, J.C., Glazebrook, K.D., Weber, R.: Multi-Armed Bandit Allocation Indices, 2nd edn. Wiley, New York (2011)
7. Gittins, J.C., Jones, D.M.: A dynamic allocation index for the sequential design of experiments. In: Gani, J. (ed.) Progress in Statistics, pp. 241–266 (1974)
8. Ho, C.J., Slivkins, A., Vaughan, J.W.: Adaptive contract design for crowdsourcing markets: bandit algorithms for repeated principal-agent problems. In: Proceedings of the 16th ACM Conf. on Economics and Computation, pp. 359–376 (2014)
9. Katehakis, M.N., Veinott Jr., A.F.: The multi-armed bandit problem: decomposition and computation. Math. Oper. Res. 12(2), 262–268 (1987)
10. Kremer, I., Mansour, Y., Perry, M.: Implementing the "wisdom of the crowd". In: Proceedings of the 15th ACM Conf. on Electronic Commerce, pp. 605–606 (2013)
11. Lai, T.L., Robbins, H.E.: Asymptotically efficient adaptive allocation rules. Adv. Appl. Math. 6(1), 4–22 (1985)
12. Mansour, Y., Slivkins, A., Syrgkanis, V.: Bayesian incentive-compatible bandit exploration. In: Proceedings of the 17th ACM Conference on Economics and Computation, pp. 565–582 (2015)
13. Robbins, H.E.: Some aspects of the sequential design of experiments. Bull. Am. Math. Soc. 58, 527–535 (1952)
14. Singla, A., Krause, A.: Truthful incentives in crowdsourcing tasks using regret minimization mechanisms. In: 22nd International World Wide Web Conference, pp. 1167–1178 (2013)
15. Slivkins, A., Wortman Vaughan, J.: Online decision making in crowdsourcing markets: theoretical challenges (position paper). ACM SIGecam Exch. 12(2), 4–23 (2013)
16. Spence, M.: Job market signaling. Q. J. Econ. 87, 355–374 (1973)
17. Whittle, P.: Multi-armed bandits and the Gittins index. J. Roy. Stat. Soc. Ser. B (Methodol.) 42(2), 143–149 (1980)

Bottleneck Routing with Elastic Demands

Tobias Harks[1], Max Klimm[2]([⊠]), and Manuel Schneider[2]

[1] Institute of Mathematics,
University of Augsburg, 86135 Augsburg, Germany
tobias.harks@math.uni-augsburg.de

[2] Department of Mathematics, Technische Universität Berlin, Berlin, Germany
klimm@math.tu-berlin.de

Abstract. Bottleneck routing games are a well-studied model to investigate the impact of selfish behavior in communication networks. In this model, each user selects a path in a network for routing their *fixed* demand. The disutility of a used only depends on the most congested link visited. We extend this model by allowing users to continuously vary the demand rate at which data is sent along the chosen path. As our main result we establish tight conditions for the existence of pure strategy Nash equilibria.

1 Introduction

Banner and Orda [1] and independently Caragiannis et al. [4] introduced selfish bottleneck routing games as a theoretical model of resource allocation in distributed communication networks. In these games, every user of the network is associated with a non-negative demand that they want to send from their source to the respective destination and their goal is to find a path that minimizes the congestion of the most congested link. It has been argued by several researchers (cf. [5,19,27]) that in the context of packet-switched communication networks, the performance of a path is closer related to the most congested link as compared to the classical sum-aggregation of costs (as in [18,28,33]). In particular, in the area of designing congestion control protocols (as alternatives to the current TCP), there are several proposals (cf. [25,34,36,37]) that postulate to replace the sum-aggregation of congestion costs with the max-aggregation, primary, because the max-aggregation leads to favorable properties of protocols in terms of their stability in presence of communication delays [36,38].

While the bottleneck model of [1,4] was an important step in terms of integrating routing decisions with bottleneck objectives, it lacks one fundamental tradeoff inherent in packet-switched communication networks: once a path is selected, a user increases the sending rate in case of low congestion and decreases it in case of high congestion.

M. Klimm—This research was carried out in the framework of MATHEON supported by Einstein Foundation Berlin.

In this paper, we address this tradeoff by introducing bottleneck congestion games with *elastic* demands, where users can continuously vary their demands. Formally, there is a finite set of resources and a strategy of a player corresponds to a tuple consisting of a subset of resources and a demand out of a prescribed interval of feasible demands. For the case that the allowable subsets of a player correspond to the set of routes connecting the player's source node to their terminal node, we obtain unsplittable bottleneck routings as in [1,4]. Resources have player-specific cost functions that are non-decreasing and strictly convex. Every user is associated with a non-decreasing strictly concave utility function measuring the received utility from sending at a certain demand rate. The goal of a user is to select both a subset of resources and a demand rate that maximizes the utility (from the demand rate) minus the congestion cost on the most expensive resource contained in the chosen resource set. Our model thus integrates as a special case (*i*) single-path routing (which is up to date standard as splitting packets over several routes leads to different packet inter-arrival times and synchronization problems) and (*ii*) congestion control via data rate adaption based on the maximum congestion experienced.

1.1   Our Results

As our main result we derive conditions for the cost functions so that the resulting bottleneck congestion game with elastic demands admits a pure Nash equilibrium (PNE). Our condition requires that for every player the player-specific resource cost functions are non-decreasing, strictly convex and equal up to resource specific shifts in their argument. While monotonicity and convexity are natural conditions, the last assumption seems limiting. We can show, however, that without this assumption there are examples without any PNE, even for the special case that all resource cost functions are not player-specific.[1] Moreover, we demonstrate that the our conditions on the resource cost functions are still general enough to model $M/M/1$ functions that are frequently used to model delays in communication networks.

We prove the existence result by devising an algorithm that computes a PNE. The main idea of the algorithm is as follows. Assume we are given a strategy profile with a fixed resource set (e.g., a path in the network setting) for every player. We consider a series of decoupled games on each resource contained in the set separately. Then, we compute an equilibrium for the decoupled game of each resource and call the resulting vector of demands a *distributed equilibrium*. Note that a distributed equilibrium is not a feasible strategy profile of the original game as a player may now have different demands on different resources along the chosen path. We then devise an algorithm that turns a distributed equilibrium into a feasible strategy profile with the property that no player can improve by changing the demand only. We call such a profile a *demand equilibrium*. Our main algorithm then iteratively (*i*) computes demand equilibria, and (*ii*) if an equilibrium is not reached yet, lets single players play a better and best response.

---

[1] We defer this counterexample to the full version of this paper.

We provide a lexicographical potential for this special dynamic and thus prove that the algorithm terminates.

In the interest of space, we defer most of the proofs to the full version of this paper.

1.2 Related Work

Bottleneck Routing. In bottleneck routing games with fixed demands strong equilibria have been shown to exist [14], even for more general classes of cost functions where the cost of a resource may depend on the sets of its users rather than the aggregated demand. The complexity of computing PNE and strong equilibria in these games was further investigated in [10]. For further works in this area considering the price of anarchy of PNE as well as the worst-case quality of strong equilibria we refer to [2,3,5,7,16,17].

In an independent line of research, Kukushkin [24] studied generalizations of the congestion game model of Rosentahl [28] in which the sum-aggregation is replaced by an arbitrary aggregation function. He proved that among the aggregation functions for which a cost increase on one resource always leads to an increased private cost for all of its users, sum-aggregation is the only aggregation function that guarantees the existence of a pure Nash equilibrium. Games with maximum-aggregation were also studied by Kukushkin [23], where he proved that any bottleneck congestion game with non-decreasing cost admits a strong equilibrium. Further related is our previous work [13], where we establish the existence of an equilibrium for a class of aggregative location games. This existence result implies the existence of a pure Nash equilibria for the present model assuming that the allowable sets of resource of players contain singletons only.

Combined Routing and Congestion Control. Integrated routing and congestion control has been studied in [9,20,21,31,32], where the existence of an equilibrium is proved by showing that it corresponds to an optimal solution of an associated convex utility maximization problem. These models, however, require that every user possibly splits the flow among up to an exponential number of paths. This issue has been addressed in [6,26], where controllable route splitting at routers is assumed which can effectively limit the resulting number of used routes. For all the above models, however, the end-to-end applications may suffer in service quality due to packet jitter caused by different path delays. Partly because of this issue, the standard TCP/IP protocol suite still uses single path routing. Also in contrast to our model, all these models assume that congestion feedback is aggregated via the sum instead of the max operator.

Yang et al. [35] introduced the so-called *MAXBAR*-games where users select a single path and adapt their sending rate. In their model, edges have fixed capacities and users (synchronously) increase their rate until the capacity of an edge is reached. After such an event all rates of users using this tight edge are fixed. Yang et al. showed that these games possess a PNE, and that the price of anarchy of pure Nash equilibria is n, where n is the number of players. Harks et al. [11]

generalized the model of Yang et al. by allowing for more general ways of increasing the rates of users. They derived the existence of strong equilibria for this more general model.

Harks and Klimm [12] introduced congestion games with variable demands that coincide with the present model except that the traditional sum-aggregation of costs is used. They showed that only affine and certain exponential cost functions lead to the existence of PNE. These results are in contrast to the results obtained in this paper because here we prove that general player-specific convex cost functions lead to PNE as long as they are equal up to resource specific shifts in their argument.

2   The Model

Congestion Model. The games considered in this paper are based on a *congestion model* defined as follows. Let $R = \{1, \ldots, m\}$ be a nonempty and finite set of $m \in \mathbb{N}$ resources, and let $N = \{1, \ldots, n\}$ be a nonempty and finite set of $n \in \mathbb{N}$ players. For every $i \in N$ let $X_i \subset 2^R \setminus \{\emptyset\}$ be a nonempty set of nonempty subsets of resources available to player i. Whenever a player i uses the resources in $x_i \in X_i$, we say that the resources in x_i are allocated to player i; we also call x_i an allocation of player i and we denote by $x = (x_i)_{i \in N}$ the overall allocation vector. For every player i and every resource $r \in R$ we are given a player-specific cost function $c_{i,r} : \mathbb{R}_{\geq 0} \to \mathbb{R}_{\geq 0}$ that maps the aggregated demand on r to a cost value for player i. We call the tuple $\mathcal{M} = \big(N, R, (X_i)_{i \in N}, (c_{i,r})_{r \in R, i \in N}\big)$ a congestion model.

Bottleneck Congestion Games with Elastic Demands. In a *bottleneck congestion game with elastic demands*, we are given a congestion model $\mathcal{M}$ and, for every player $i \in N$, a utility function $U_i : [\sigma_i, \tau_i] \to \mathbb{R}_{\geq 0}$ where $[\sigma_i, \tau_i] \subseteq \mathbb{R}_{\geq 0}$ with $\sigma_i \in \mathbb{R}_{\geq 0}$, $\tau_i \in \mathbb{R}_{\geq 0} \cup \{\infty\}$, $\sigma_i \leq \tau_i$ is the interval of feasible demands of player i. We denote by $d = (d_i)_{i \in N}$ the overall demand vector. A bottleneck congestion game with elastic demands is the maximization game $G = (N, \mathcal{S}, \pi)$ with $\mathcal{S}_i = X_i \times [\sigma_i, \tau_i]$ for all $i \in N$, where for $s = (x, d) \in \mathcal{S}$ we define

$$\pi_i(s) := U_i(d_i) - \max_{r \in x_i}\Big\{c_{i,r}\big(\ell_r(x, d)\big)\Big\} \quad \text{for all } i \in N.$$

Here, $\ell_r(s) = \ell_r(x, d) = \sum_{j \in N : r \in x_j} d_j$ is the *load* (or *aggregated demand*) of resource r under strategy profile $s = (x, d)$.

We impose the following assumptions on the utility and cost functions, respectively.

Assumption 1 *For every player $i \in N$, the utility function $U_i : [\sigma_i, \tau_i] \to \mathbb{R}_{\geq 0}$ is non-decreasing, differentiable and strictly concave.*

Strict concavity of the utility function in the demand is justified by application-specific characteristics such as the rate-control algorithm used in common congestion control protocols, see [18, 30].

Assumption 2 *For every resource $r \in R$ and player $i \in N$, the cost function $c_{i,r} : \mathbb{R}_{\geq 0} \to \mathbb{R}_{\geq 0}$ is non-decreasing, differentiable and strictly convex.*

Note that in many applications the considered cost functions are strictly convex, e.g., the polynomial delay functions considered in transportation networks (cf. [29]) and $M/M/1$ functions modeling queuing delays in telecommunication networks (cf. [31]).

Assumption 3 *There are functions $c_i : \mathbb{R}_{\geq 0} \to \mathbb{R}_{\geq 0}$ for all $i \in N$ and offsets $v_r \in \mathbb{R}_{\geq 0}$ for all $r \in R$ such that for every $i \in N$ and $r \in R$*

$$c_{i,r}(t) = c_i(t + v_r) \quad \text{for all } t \geq 0.$$

The above assumption implies that for every player, the maximum load (including offsets) experienced on the chosen subset of resources determines the bottleneck. While this assumption is certainly restrictive, we show in the full version of this paper that without it, there are games without a PNE. We remark that the important class of $M/M/1$ delay functions that are frequently used to model queueing delays (cf. [8,22] and references therein) still satisfy Assumption 3.

Observation 4 *Player-specific $M/M/1$ functions of the form $c_{i,r} : [0, c_r) \to \mathbb{R}_{\geq 0}$ with $c_{i,r}(t) = \frac{t_i}{z_r - t}, z_r > 0, t_i \geq 0$ satisfy Assumption 3.*[2]

Proof. Let $z_{\max} := \max_{r \in R}\{z_r\}$. For resource-specific offsets defined as $v_r := z_{\max} - z_r \geq 0$ for all $r \in R$ and player-specific functions $c_i(t) := t_i/(z_{\max} - t)$, we then obtain

$$c_{i,r}(t) = \frac{t_i}{z_r - t} = \frac{t_i}{z_{\max} - (t + v_r)} = c_i(t + v_r)$$

for all $r \in R$ and $i \in N$. □

For a strategy profile $s = (x, d)$ let

$$b_i(s) = b_i(x, d) = \max_{r \in x_i}\left\{\ell_r(x, d) + v_r\right\}$$

denote the maximal load or *bottleneck* that player i experiences and denote by $b_i^{-1}(x, d)$ the set of resources where player i experiences their bottleneck, i.e.,

$$b_i^{-1}(s) = b_i^{-1}(x, d) = \arg\max_{r \in x_i}\left\{\ell_r(x, d) + v_r\right\}.$$

---

[2] Technically, they do not satisfy Assumption 3 since their domain is only a subinterval of the non-negative reals. This, however, is not an issue as the functions diverge to ∞ as they approach the right boundary of their domain, so that no player has an incentive to raise its demand in a way that the load on a resource exceeds the domain of its cost function.

3   A Characterization of Pure Nash Equilibria

We now present a complete characterization of pure Nash equilibria in bottle-neck congestion games. Our characterization relies on the notion of a *demand equilibrium* which we define as a strategy profile with the property that no player can increase their payoff by unilaterally changing her demand only.

Definition 5 (Demand Equilibrium). *Let G be a bottleneck congestion game with elastic demands. We call (x, d) a demand equilibrium if for all $i \in N$*

$$\pi_i(x, d) \geq \pi_i(x, \tilde{d})$$

for all $\tilde{d} = (d_{-i}, \tilde{d}_i)$ with $\tilde{d}_i \in [\sigma_i, \tau_i]$.

We obtain the following immediate necessary condition for a PNE.

Lemma 6. *Every PNE of a bottleneck congestion game with elastic demands is a demand equilibrium.*

Before we proceed deriving optimality conditions for demand equilibria we demonstrate that in general (i.e., without Assumption 3), the payoff functions $\pi_i(x, d), i \in N$ are not necessarily differentiable with respect to d_i.

Example 7. Consider a game with one player i with utility function U_i and two resources r_1, r_2 equipped with two affine cost functions $c_{r_1}(t) = 1 + t$ and $c_{r_2}(t) = 2t$ for all $t \in \mathbb{R}_{\geq 0}$. The payoff to player i with resource allocation $x = x_i = \{r_1, r_2\}$ is not differentiable with respect to d_i at $d_i = 1$. To see this, first note that $c_{r_1}(t) \geq c_{r_2}(t)$ for all $t \leq 1$ and $c_{r_1}(t) \leq c_{r_2}(t)$ for all $t \geq 1$. We then obtain

$$\frac{\partial^- \pi_i(x, 1)}{\partial x} = \lim_{\epsilon \uparrow 0} \frac{\pi_i(x, 1 + \epsilon) - \pi_i(x, 1)}{\epsilon}$$

$$= \lim_{\epsilon \uparrow 0} \frac{U_i(1 + \epsilon) - U_i(1) - c_{r_1}(1 + \epsilon) + c_{r_1}(1)}{\epsilon} = U_i'(1) - 1,$$

$$\frac{\partial^+ \pi_i(x, 1)}{\partial x} = \lim_{\epsilon \downarrow 0} \frac{\pi_i(x, 1 + \epsilon) - \pi_i(x, 1)}{\epsilon}$$

$$= \lim_{\epsilon \downarrow 0} \frac{U_i(1 + \epsilon) - U_i(1) - c_{r_2}(1 + \epsilon) + c_{r_2}(1)}{\epsilon} = U_i'(1) - 2.$$

For cost functions that satisfy Assumption 3, however, we obtain the following necessary condition for demand equilibria.

Lemma 8 *Let (x, d) be a demand equilibrium. Then, for all $i \in N$ with $\sigma_i < \tau_i$ the following two conditions are satisfied:*

1. *$d_i < \tau_i \Rightarrow U_i'(d_i) - c_i'\big(\ell_{r^*}(x, d) + v_{r^*}\big) \leq 0$ for all $r^* \in b_i^{-1}(x, d)$.*
2. *$d_i > \sigma_i \Rightarrow U_i'(d_i) - c_i'\big(\ell_{r^*}(x, d) + v_{r^*}\big) \geq 0$ for all $r^* \in b_i^{-1}(x, d)$.*

3.1 Sensitivity Analysis of Demand Equilibria

Let $s = (x, d)$ be a demand equilibrium and let $\ell_r^{-i}(s) := \sum_{j \in N \setminus \{i\}} d_j$ denote the *residual load* of player i on some $r \in R$. Then, using that s is a demand equilibrium, we obtain

$$d_i \in \arg \max_{d_i' \in [\sigma_i, \tau_i]} U_i(d_i') - c_i(\ell_{r^*}^{-i}(s) + v_{r^*} + d_i') \quad \text{for all } r^* \in b_i^{-1}(s).$$

In the following, we investigate how a demand equilibrium is adapted if the residual load on a resource changes. This will be important later on in order to understand the effect of switching sets of resource when changing the strategy from some $x_i \in X_i$ to some $y_i \in X_i$. Given some fixed residual load α on a resource r, we analyze how the *best-response demand function*

$$d_i(\alpha) := \arg \max_{d_i \in [\sigma_i, \tau_i]} U(d_i) - c_i(\alpha + d_i)$$

depends on α. As for all $\alpha \in \mathbb{R}_{\geq 0}$ the function $f(y) := U_i(y) - c_i(\alpha + y)$ is strictly concave in y and $\lim_{y \to \infty} f(y) = -\infty$, the above optimization problem has a unique solution, hence, the best response demand function is well defined.

Lemma 9 (Individual Best Response Demands). *Let $\alpha, \beta \in \mathbb{R}_{\geq 0}$. Then, the following two conditions are equivalent:*

(i) $\alpha < \beta$.
(ii) $d_i(\alpha) + \alpha < d_i(\beta) + \beta$.

We are now in position to derive a complete characterization of PNE.

Theorem 10. *Let G be a bottleneck congestion game. A strategy profile $s = (x, d)$ is a PNE for G if and only if s is a demand equilibrium and*

$$r \in b_i^{-1}(s) \qquad \Rightarrow \qquad \ell_r^{-i}(s) + v_r \leq \max_{t \in y_i} \ell_t^{-i}(s) + v_t$$

for all $i \in N$ and $y_i \in X_i$.

Proof. "$\Rightarrow$": By Lemma 6 we get that s must be a demand equilibrium. For the second statement, assume there is a player i and $y_i \in X_i$ with $\ell_r^{-i}(s) + v_r > \max_{t \in y_i} \ell_t^{-i}(s) + v_t$. This implies that player i can increase their payoff by simply deviating from strategy $s_i = (x_i, d_i)$ to $\tilde{s}_i = (y_i, d_i)$.

"$\Leftarrow$": Define $s' = (s_{-i}, s_i')$ with $s_i' = (y_i, d_i')$ for some arbitrary $y_i \in X_i$ and $d_i' \in [\sigma_i, \tau_i]$. We obtain

$$\begin{aligned}
\pi_i(s') &= U_i(d_i') - c_i\left(\max_{t \in y_i} \{ \ell_t^{-i}(s') + v_t \} + d_i' \right) \\
&\leq U_i(d_i') - c_i\left(\ell_r^{-i}(s) + v_r + d_i' \right) \\
&\leq U_i(d_i) - c_i\left(\ell_r^{-i}(s) + v_r + d_i \right) \\
&= \pi_i(s),
\end{aligned}$$

where the first inequality holds due to $\ell_r^{-i}(s) + v_r \leq \max_{t \in y_i} \ell_t(s) + v_t$ and c_i is non-decreasing. The second inequality holds because s is a demand equilibrium and therefore d_i is player i's best response demand. $\qquad\qquad \square$

4   Computing Demand Equilibria

Theorem 10 shows that for computing a PNE we must be able to compute a demand equilibrium. In this section, we describe an algorithm that does exactly this. For the algorithm we first need the notion of *distributed equilibria* defined below.

4.1   Distributed Equilibria

Let $G = (N, \mathcal{S}, \pi)$ be a bottleneck congestion game, $M \subseteq N$, and let r be a resource. We define the restriction of G on M and r, written $G|_{(M,r)}$ as the bottleneck congestion game $G|_{(M,r)} = (M, \mathcal{S}', \pi')$ with $\mathcal{S}'_i = \{\{r\}\} \times [\sigma_i, \tau_i]$ for all $i \in M$ and $\pi'_i(x, d) = U_i(d_i) - c_i(\ell_r(x, d) + v_r)$.

Definition 11 (Distributed Equilibrium). *Let G be a bottleneck congestion game with elastic demands. For $x \in X$ define $N_r(x) := \{i \in N : r \in x_i\}$. A non-negative vector $\tilde{d} = (\tilde{d}_{i,r})_{r \in R, i \in N_r(x)}$ is called a* distributed equilibrium, *if for all $r \in R$ the strategy profile $(\tilde{d}_{i,r})_{i \in N_r(x)}$ is a PNE of $G|_{(N_r(x),r)}$.*

Note that every restricted game $G|_{(N_r(x),r)}$ is a concave game on a compact action space, thus, by Kakutani's fixed point theorem [15] the existence of a distributed equilibrium is guaranteed. Moreover, as we will show below, these equilibria are in fact unique. For a distributed equilibrium $\tilde{d}$ with respect to $x \in X$, we define $\tilde{\ell}_r(x, \tilde{d}) := \sum_{i \in N_r(x)} \tilde{d}_{i,r}$. We first need the following lemma.

Lemma 12. *Let $x, x' \in X$ and let $\tilde{d}$ and $\tilde{d}'$ be two respective distributed equilibrium demands. Then, $\tilde{d}_{i,r} \leq \tilde{d}'_{i,r'}$ for all $r, r' \in R$ with $\tilde{\ell}_r(x, \tilde{d}) + v_r \geq \tilde{\ell}_{r'}(x', \tilde{d}') + v_{r'}$ and all $i \in N_r(x) \cap N_{r'}(x')$.*

Using this lemma, we obtain the following immediate corollary.

Corollary 13 (Uniqueness). *Let $x, x' \in X$ and let $\tilde{d}$ and $\tilde{d}'$ be two respective distributed equilibrium demands. Then, the following two statements hold:*

1. *$\tilde{\ell}_r(x, \tilde{d}) \leq \tilde{\ell}_r(x', \tilde{d}')$ for all $r \in R$ with $N_r(x) \subseteq N_r(x')$.*
2. *$\tilde{\ell}_r(x, \tilde{d}) = \tilde{\ell}_r(x', \tilde{d}')$ for all $r \in R$ with $N_r(x) = N_r(x')$.*

We now derive a corollary that will be useful later on. It states the intuitive fact that when the feasible demand set of a player is fixed to a value not larger than the demand for resource r is a particular distributed equilibrium, then the total demand for r in after recomputing a distributed equilibrium may not increase.

Corollary 14 (Demand Restriction). *Let $x \in X$ and let $\tilde{d}$ be a corresponding distributed equilibrium demand. For $i \in N$ and $r \in x_i$, define a new game G' that differs only in the fact that $\tau'_i = \sigma'_i \leq \tilde{d}_{i,r}$. Then $\tilde{\ell}_r(x, \tilde{d}') \leq \tilde{\ell}_r(x, \tilde{d})$ for each distributed equilibrium demand $\tilde{d}'$ of G'.*

Algorithm 1. Computation of a demand equilibrium

Input: Bottleneck congestion game with elastic demands G and $x \in X$
Output: Demand equilibrium (x, d) of G

1 initialize $N' \leftarrow N, R' \leftarrow R$;
2 **while** $R' \neq \emptyset$ **do**
3   compute distributed equilibrium $(\tilde{d}_{i,r})_{i \in N', r \in R'}$;
4   choose an index-minimal $r \in \arg\max_{r \in R'} \sum_{i \in N_r(x)} \tilde{d}_{i,r} + \upsilon_r$;
5      $d_j \leftarrow \tilde{d}_{j,r}; \sigma_j \leftarrow d_j; \tau_j \leftarrow d_j$ for all $j \in N' \cap N_r(x)$;
6      $N' \leftarrow N' \setminus N_r(x); R' \leftarrow R' \setminus \{r\}$;
7 **end**

4.2 An Algorithm for Computing Demand Equilibria

We are now ready to propose an algorithm that takes as input an allocation $x \in X$ and computes a corresponding demand equilibrium $(x, d) \in \mathcal{S}$. The algorithm first computes a distributed equilibrium $(x, \tilde{d})$. Then, a resource r with maximum load is chosen and the demand of each player $i \in N_r(x)$ is fixed to the demand $\tilde{d}_{i,r}$. For the remaining players we recompute a distributed equilibrium and reiterate. The formal description is given in Algorithm 1.

We shall show that Algorithm 1 indeed outputs a demand equilibrium.

Theorem 15. *Algorithm 1 computes a demand equilibrium.*

Proof. The demand vector d computed by Algorithm 1 satisfies $d_i \in [\sigma_i, \tau_i]$ for all $i \in N$ and is, thus, feasible. We proceed to show that (x, d) is a demand equilibrium.

To this end, let us assume that $R = \{1, \dots, m\}$. Further, it is without loss of generality to assume that the resources R are ordered such that for each $k \in \{1, \dots, m\}$ in the k-th iteration of the algorithm resource k is chosen in line 4.

We proceed to show that $\ell_1(x, d) + \upsilon_1 \geq \cdots \geq \ell_m(x, d) + \upsilon_m$. To see this, let $k \in \{1, \dots, m-1\}$ be arbitrary. For an iteration counter $j \in \{1, \dots, m\}$, we denote by $\tilde{d}^j$ the distributed equilibrium demand computed in line 3 of the algorithm. As we choose in each iteration j a resource that maximizes $\tilde{\ell}_r(x, \tilde{d}^j) + \upsilon_r$, we obtain in particular $\tilde{\ell}_k(x, \tilde{d}^k) + \upsilon_k \geq \tilde{\ell}_{k+1}(x, \tilde{d}^k) + \upsilon_{k+1}$. Lemma 12 implies $\tilde{d}^k_{i,k} \leq \tilde{d}^k_{i,k+1}$ for all $i \in N$ with $\{k, k+1\} \subseteq x_i$. Corollary 14 gives the claimed result.

We derive that for each player i the bottleneck is attained at the resource $r \in x_i$ with minimal index. As the algorithm fixes the demand of each player i the first time one of the resources used by player i is considered (line 5), the demand vector d computed by Algorithm 1 is a demand equilibrium. $\qquad\square$

Remark 16. Note that for a given input G and x, Algorithm 1 computes the same demand equilibrium (x, d). This follows since the distributed demand equilibria are unique (see Corollary 13) and there is a fixed tie-breaking rule employed that determines the order in which resources are fixed (as specified in line 3).

Algorithm 2. Computation of a PNE

    Input: Bottleneck congestion game with elastic demands G
    Output: PNE (x, d) of G
1 $(x, d') \leftarrow$ arbitrary strategy profile;
2 **while** *true* **do**
3 | Compute a demand equilibrium (x, d) by Algorithm 1;   **Phase I**
4 | **if** *there is a player i with a better reply to (x, d)* **then**   **Phase II**
5 | | $(y_i, d'_i) \leftarrow$ best reply of player i to (x, d);
6 | | $(x, d') \leftarrow (y_i, x_{-i}, d'_i, d_{-i})$;
7 | **else**
8 | | return (x, d);
9 | **end**
10 **end**

---

5   An Algorithm for Computing PNE

We present an algorithm that computes a PNE. The algorithm starts with an arbitrary strategy profile and computes a demand equilibrium. Then, whenever there is a player that can improve, we let this player deviate (by Theorem 10 this implies that the player's resource set changes) and recompute a demand equilibrium. The technically involved part is to show that the algorithm terminates. In this section, we show the following theorem.

Theorem 17. *Algorithm 2 computes terminates and computes a PNE.*

Before we prove the above theorem, we derive several properties of intermediate strategy profiles during the execution of the algorithm.

5.1   Analysis of the Algorithm

For a strategy profile $s = (x, d)$, we consider $b_i(s) = \max_{r \in x_i} \{\ell_r(s) + v_r\}$ and $b(s) = (b_i(s))_{i \in N}$. We shall prove that $b(s)$ strictly decreases with respect to the sorted lexicographical order $\prec_{\text{lex}}$ that is defined as follows: For two vectors $u, v \in \mathbb{R}^n_{\geq 0}$ we say that u is *sorted lexicographically* smaller than v, written $u \prec_{\text{lex}} v$, if there is an index $k \in \{1, \ldots, n\}$ such that $u_{\pi(i)} = v_{\psi(i)}$ for all $i < k$ and $u_{\pi(k)} < v_{\psi(k)}$ where π and ψ are permutations that sort u and v non-increasingly, i.e., $u_{\pi(1)} \geq u_{\pi(2)} \geq \cdots \geq u_{\pi(n)}$ and $v_{\psi(1)} \geq v_{\psi(2)} \geq \cdots \geq v_{\psi(n)}$.

    We now derive a crucial lemma that relates two consecutive demand equilibria s and $\bar{s}$ (as computed in Phase I).

Lemma 18. *Let (x, d) be a demand equilibrium computed (by Algorithm 1) in Phase I of Algorithm 2. Let $s'_i = (x'_i, d'_i)$ be a better and best reply of player i and let $s' = (x', d')$, where $x' = (x'_i, x_{-i})$ and $d' = (d'_i, d_{-i})$. Denote by $\bar{s} = (x'_i, x_{-i}, \bar{d})$ the demand equilibrium that is computed in Algorithm 1 in Phase I in the following iteration. Then, $b(\bar{s}) \prec b(s)$.*

Proof. We first derive the following statements relating $b(s)$ with $b(\bar{s})$:

(i) $b_i(\bar{s}) < b_i(s)$.
(ii) For all $j \in N$ with $b_j(s) \geq b_i(s)$, it holds $b_j(\bar{s}) \leq b_j(s)$.
(iii) For all $j \in N$ with $b_j(s) < b_i(s)$, it holds $b_j(\bar{s}) < b_i(s)$.

We start proving (i). Let $r_1 \in b_i^{-1}(s)$ and $r_2 \in b_i^{-1}(s')$. We will first show that $\ell_{r_1}(s) + \upsilon_{r_1} > \ell_{r_2}(s') + \upsilon_{r_2}$ holds. Note that by using s_i' player i strictly improves, thus, Theorem 10 implies $\ell_{r_1}^{-i}(s) + \upsilon_{r_1} > \ell_{r_2}^{-i}(s') + \upsilon_{r_2}$, hence, we can use Lemma 9 to obtain (we define $\beta := \ell_{r_1}^{-i}(s) + \upsilon_{r_1}$ and $\alpha := \ell_{r_2}^{-i}(s') + \upsilon_{r_2}$):

$$\ell_{r_1}(s) + \upsilon_{r_1} = \ell_{r_1}^{-i}(s) + d_i(\beta) + \upsilon_{r_1} > \ell_{r_2}^{-i}(s') + d_i(\alpha) + \upsilon_{r_2} = \ell_{r_2}(s') + \upsilon_{r_2}.$$

We now claim that $\ell_{r_2}(s') + \upsilon_{r_2} \geq \ell_{r_2}(\bar{s}) + \upsilon_{r_2}$. Assume $\ell_{r_2}(s') + \upsilon_{r_2} < \ell_{r_2}(\bar{s}) + \upsilon_{r_2}$. Then there is $j \in N_{r_2}(s')$ with $\bar{d}_j > d_j'$. If $j = i$, we obtain

$$\begin{aligned} 0 &\geq U_i'(d_i') - c_i'(\ell_{r_2}(s') + \upsilon_{r_2}) \\ &> U_i'(\bar{d}_i) - c_i'(\ell_{r_2}(\bar{s}) + \upsilon_{r_2}) \\ &\geq 0, \end{aligned}$$

a contradiction. If, $j \neq i$, we have $\bar{d}_j > d_j' = d_j$ and also obtain

$$\begin{aligned} 0 &\geq U_j'(d_j) - c_j'(\ell_{r_2}(s') + \upsilon_{r_2}) \\ &> U_j'(\bar{d}_j) - c_j'(\ell_{r_2}(\bar{s}) + \upsilon_{r_2}) \\ &\geq 0, \end{aligned}$$

using that d_j was a demand equilibrium for player j under s.

Now we prove (ii). Assume by contradiction that there is $j \in N$ with $b_j(s) \geq b_i(s)$ and $b_j(\bar{s}) > b_j(s)$. Note that (using (i)) this implies

$$b_i(\bar{s}) < b_i(s) \leq b_j(s) < b_j(\bar{s}).$$

From $b_i(\bar{s}) < b_j(\bar{s})$ we obtain

$$b_j^{-1}(\bar{s}) \cap x_i' = \emptyset.$$

Let $r^1, \ldots r^m$ denote the order in which resources are fixed in Algorithm 1 (with input s'). Let $r^k \in b_j^{-1}(\bar{s})$ denote the index-minimal resource in $b_j^{-1}(\bar{s})$ that is fixed in the k-th iteration. Note that by Theorem 15 we have

$$\ell_{r^1}(\bar{s}) + \upsilon_{r^1} \geq \cdots \geq \ell_{r^k}(\bar{s}) + \upsilon_{r^k}, \tag{1}$$

thus, for all $l \in \{1, \ldots, k\}$ we have $r^l \notin x_i'$ and, hence, $N_{r^l}(x') \subseteq N_{r^l}(x)$. If $r^l \notin x_i$ for all $l \in \{1, \ldots, k\}$, then, we have

$$\ell_{r^l}(s) + \upsilon_{r^l} = \ell_{r^l}(\bar{s}) + \upsilon_{r^l} \text{ for all } l \in \{1, \ldots, k\},$$

which contradicts $b_j(\bar{s}) > b_j(s)$. Thus, let g be the smallest index such that $r^g \in x_i$. This implies $b_i(s) \geq \ell_{r^g}(s) + v_{r^g}$. Note that by (1) we get $b_j(\bar{s}) \leq \ell_{r^g}(\bar{s}) + v_{r^g}$. Using $N_{r^g}(x') \subset N_{r^g}(x)$ we further obtain by using Corollary 13

$$\ell_{r^g}(s) + v_{r^g} \geq \ell_{r^g}(\bar{s}) + v_{r^g}.$$

This is, however, a contradiction to

$$\ell_{r^g}(\bar{s}) + v_{r^g} \geq b_j(\bar{s}) > b_j(s) \geq b_i(s) \geq \ell_{r^g}(s) + v_{r^g}.$$

For (iii), assume there is $j \in N$ with $b_j(s) < b_i(s)$ and $b_j(\bar{s}) > b_i(s)$. The last strict inequality implies again

$$b_j^{-1}(\bar{s}) \cap x_i' = \emptyset,$$

and we get the same contradiction as in case (ii).

Now we have shown that (i) the i-th entry of $b(s)$ strictly decreases for $\bar{s}$, (ii) all entries that are above $b_i(s)$ only decrease for $\bar{s}$, and finally, (iii) all entries that increase for $\bar{s}$ (compared to the b-values under s) stay strictly below $b_i(s)$. This implies $b(\bar{s}) \prec b(s)$. □

We are now in position to prove Theorem 17.

Proof (Proof of Theorem 17). Lemma 18 shows that the vector $b(s)$ strictly lexicographically decreases during the execution of Algorithm 2. Thus, since the demand equilibrium computed by Algorithm 1 in Phase I of Algorithm 2 is always the same, no vector $x \in X$ is visited twice during the execution of Algorithm 2, and, hence, the algorithm terminates (as X contains only finitely many elements). □

Acknowledgements. Some of the proof techniques used in this paper appeared in the diploma thesis of the third author. We wish to thank three anonymous referees who found several (nontrivial) typos and suggested several improvements and new related references.

References

1. Banner, R., Orda, A.: Bottleneck routing games in communication networks. IEEE J. Sel. Area Commun. **25**(6), 1173–1179 (2007)
2. Busch, C., Kannan, R., Samman, A.: Bottleneck routing games on grids. In: Proceedings 2nd International ICST Conference on Game Theory for Networks, pp. 294–307 (2011)
3. Busch, C., Magdon-Ismail, M.: Atomic routing games on maximum congestion. Theor. Comput. Sci. **410**(36), 3337–3347 (2009)
4. Caragiannis, I., Galdi, C., Kaklamanis, C.: Network load games. In: Deng, X., Du, D.-Z. (eds.) ISAAC 2005. LNCS, vol. 3827, pp. 809–818. Springer, Heidelberg (2005)

5. Cole, R., Dodis, Y., Roughgarden, T.: Bottleneck links, variable demand, and the tragedy of the commons. In: Proceedings of the 17th Annual ACM-SIAM Symposium on Discrete Algorithms, pp. 668–677 (2006)
6. Cominetti, R., Guzman, C.: Network congestion control with markovian multipath routing. Math. Program. Ser. A **147**(1–2), 231–251 (2014)
7. de Keijzer, B., Schäfer, G., Telelis, O.A.: On the inefficiency of equilibria in linear bottleneck congestion games. In: Kontogiannis, S., Koutsoupias, E., Spirakis, P.G. (eds.) SAGT 2010. LNCS, vol. 6386, pp. 335–346. Springer, Heidelberg (2010)
8. Gai, Y., Liu, H., Krishnamachari, B.: A packet dropping mechanism for efficient operation of $M/M/1$ queues with selfish users. In: Proceedings of the 30th IEEE International Conference on Computer Communications, pp. 2687–2695 (2011)
9. Han, H., Shakkottai, S., Hollot, C.V., Srikant, R., Towsley, D.F.: Multi-path TCP: a joint congestion control and routing scheme to exploit path diversity in the internet. IEEE/ACM Trans. Netw. **16**(6), 1260–1271 (2006)
10. Harks, T., Hoefer, M., Klimm, M., Skopalik, A.: Computing pure nash and strong equilibria in bottleneck congestion games. Math. Program. Ser. A **141**, 193–215 (2013)
11. Harks, T., Hoefer, M., Schewior, K., Skopalik, A.: Routing games with progressive filling. In: Proceedings of the 33rd IEEE International Conference on Computer Communications, pp. 352–360 (2014)
12. Harks, T., Klimm, M.: Congestion games with variable demands. In: Apt, K. (ed.) Proceedings of the 13th Conference Theoretical Aspects of Rationality and Knowledge, pp. 111–120 (2011)
13. Harks, T., Klimm, M.: Equilibria in a class of aggregative location games. J. Math. Econom. (2015) forthcoming
14. Harks, T., Klimm, M., Möhring, R.: Strong equilibria in games with the lexicographical improvement property. Internat. J. Game Theory **42**(2), 461–482 (2012)
15. Kakutani, S.: A generalization of Brouwer's fixed point theorem. Duke Math. J. **8**(3), 457–458 (1941)
16. Kannan, R., Busch, C.: Bottleneck congestion games with logarithmic price of anarchy. In: Kontogiannis, S., Koutsoupias, E., Spirakis, P.G. (eds.) SAGT 2010. LNCS, vol. 6386, pp. 222–233. Springer, Heidelberg (2010)
17. Kannan, R., Busch, C., Vasilakos, A.V.: Optimal price of anarchy of polynomial and super-polynomial bottleneck congestion games. In: Jain, R., Kannan, R. (eds.) GAMENETS 2011. LNICST, vol. 75, pp. 308–320. Springer, Heidelberg (2012)
18. Kelly, F., Maulloo, A., Tan, D.: Rate control in communication networks: Shadow prices, proportional fairness, and stability. J. Oper. Res. Soc. **49**, 237–252 (1998)
19. Keshav, S.: An Engineering Approach to Computer Networking: ATM Networks, the Internet, and the Telephone Network. Addison-Wesley, Boston (1997)
20. Key, P.B., Massoulié, L., Towsley, D.F.: Path selection and multipath congestion control. Commun. ACM **54**(1), 109–116 (2011)
21. Key, P.B., Proutiere, A.: Routing games with elastic traffic. SIGMETRICS Perform. Eval. Rev. **37**(2), 63–64 (2009)
22. Korilis, Y., Lazar, A.: On the existence of equilibria in noncooperative optimal flow control. J. ACM **42**(3), 584–613 (1995)
23. Kukushkin, N.: Acyclicity of improvements in games with common intermediate objectives. Russian Academy of Sciences, Dorodnicyn Computing Center, Moscow (2004)
24. Kukushkin, N.: Congestion games revisited. Internat. J. Game Theory **36**, 57–83 (2007)

25. Miller, K., Harks, T.: Utility max-min fair congestion control with time-varying delays. In: Proceedings of the 27th IEEE International Conference on Computer Communicatins, pp. 331–335 (2008)
26. Paganini, F., Mallada, E.: A unified approach to congestion control and node-based multipath routing. IEEE/ACM Trans. Netw. **17**(5), 1413–1426 (2009)
27. Qiu, L., Yang, Y., Zhang, Y., Shenker, S.: On selfish routing in Internet-like environments. IEEE/ACM Trans. Netw. **14**(4), 725–738 (2006)
28. Rosenthal, R.: A class of games possessing pure-strategy Nash equilibria. Internat. J. Game Theory **2**(1), 65–67 (1973)
29. Sheffi, Y.: Urban Transp. Netw. Prentice-Hall, Upper Saddle River (1985)
30. Shenker, S.: Fundamental design issues for the future Internet. IEEE J. Sel. Area Commun. **13**, 1176–1188 (1995)
31. Srikant, R.: The Mathematics of Internet Congestion Control. Birkhäuser, Basel (2003)
32. Voice, T.: Stability of multi-path dual congestion control algorithms. IEEE/ACM Trans. Netw. **15**(6), 1231–1239 (2007)
33. Wardrop, J.: Some theoretical aspects of road traffic research. In: Proceedings of the Institute of Civil Engineers (Part II), vol. 1, pp. 325–378 (1952)
34. Wydrowski, B., Andrew, L.L.H., Zukerman, M.: Maxnet: a congestion control architecture for scalable networks. IEEE Commun. Lett. **7**, 511–513 (2003)
35. Yang, D., Xue, G., Fang, X., Misra, S., Zhang, J.: A game-theoretic approach to stable routing in max-min fair networks. IEEE/ACM Trans. Netw. **21**(6), 1947–1959 (2013)
36. Zhang, Y., Kang, S., Loguinov, D.: Delay-independent stability and performance of distributed congestion control. IEEE/ACM Trans. Netw. **15**(4), 838–851 (2007)
37. Zhang, Y., Leonard, D., Loguinov, D.: Jetmax: scalable max-min congestion control for high-speed heterogeneous networks. Comput. Netw. **52**(6), 1193–1219 (2008)
38. Zhang, Y., Loguinov, D.: On delay-independent diagonal stability of max-min congestion control. IEEE Trans. Autom. Control **54**(5), 1111–1116 (2009)

Mechanisms with Monitoring for Truthful RAM Allocation

Annamária Kovács[1], Ulrich Meyer[1 (✉)], and Carmine Ventre[2]

[1] Goethe University, Frankfurt am Main, Germany
{panni,umeyer}@cs.uni-frankfurt.de
[2] Teesside University, Middlesbrough, UK
c.ventre@tees.ac.uk

Abstract. Novel algorithmic ideas for big data have not been accompanied by advances in the way central memory is allocated to concurrently running programs. Commonly, RAM is poorly managed since the programs' trade offs between speed of execution and RAM consumption are ignored. This trade off is, however, well known to the programmers. We adopt mechanism design tools to truthfully elicit this (multidimensional) information with the aim of designing more clever RAM allocation algorithms. We introduce a novel paradigm wherein programs are bound to overbidding declarations of their running times. We show the limitations of this paradigm in the absence of transfers and prove how to leverage waiting times, as a currency, to obtain optimal money burning mechanisms for the makespan.

1 Introduction

With data volumes growing much faster than typical users' computing infrastructure, the role of efficient algorithms for big data becomes crucial. While it might be tempting to move all data to some huge commercial cloud service, legal and logistic issues will often force users or companies to keep their valuable data locally and either stick to their existing hardware or seek for a moderate cost-effective upgrade. In this situation, users are often rather willing to slowly run their programs concurrently on shared hardware and compete for resources instead of submitting their programs to an offline queuing system where they might experience unpredictable waiting times before their program is quickly executed. As a consequence the accumulated input data will typically not completely fit in the main memory (RAM) of the computer system at hand but has to reside on external storage like hard disks. External-memory (EM) algorithms [12,17] are especially tuned for this setting. They optimize the data access patterns and typically perform the better the more RAM they are assigned. However, there are huge differences: For example during a linear data scan, EM

Partially supported by the DFG grant ME 2088/3-1, EPSRC grant EP/M018113/1, and by MADALGO – Center for Massive Data Algorithmics, a Center of the Danish National Research Foundation.

E. Markakis and G. Schäfer (Eds.): WINE 2015, LNCS 9470, pp. 398–412, 2015.
DOI: 10.1007/978-3-662-48995-6_29

algorithms often can do with only a constant number of pages held in RAM, whereas for EM merge-sort of n items, the number of phases is bounded by $O(\log_x n)$ where x denotes the number of pages that can be kept in RAM simultaneously: obviously, the larger x the faster the sorting, but one essentially has to square x in order to halve the number of sorting phases.

In isolation, a typical EM algorithm prefers to take as much RAM as is available. For example, prior to running a program from the EM algorithms library STXXL [6] the default setting is to reserve a significant fraction of the available RAM for its execution. With several programs competing for a shared RAM the task becomes significantly harder, especially if these programs are not cooperating. Already from the simple example discussed above it is clear that assigning equally sized fixed shares of RAM to all programs will not necessarily optimize overall performance. Therefore, a common operating systems' solution is to apply online paging algorithms like LRU that dynamically decide which pages to keep in RAM. Unfortunately, even for a single program, online paging algorithms do not know about the future request sequences and are therefore prone to wrong decisions in the worst case [7].

We aim to use methods from Algorithmic Mechanism Design (AMD) in order to reasonably solve the RAM assignment problem for concurrently running programs: In principle, the best knowledge about the trade offs between usable RAM size and performance is with the designers/programmers of the individual algorithms. If they extend their programs with an interface in order to bid for individual RAM sizes, an operating system could use these bids within an appropriate mechanism in order to solve the RAM assignment problem for concurrently running programs. The obvious advantage of this setting would be that neither the *users* of programs (who are typically *not the programmers*) nor the operating system need knowledge about the RAM-performance footprints of the individual programs and yet obtain a reasonable RAM sharing. Of course the RAM assignment mechanism must be designed to motivate truthful requests, especially if the users do not have to pay money for the RAM their executed programs occupy. In the absence of money, in fact, selfish programmers could claim to need unreasonably large chunks of central memory for a "fast" execution of their programs.

Our contribution. In this work, we focus on the design of truthful mechanisms in the context of RAM allocation to programs, with the objective to minimize the makespan. Specifically, we concentrate on a feasibility study for the case of a *single* execution interval and ask the question of whether truthfulness can be enforced, and at what cost, in the single-shot case.

Monitoring. In our model, each programmer, also termed *agent* or *bidder* in this setting, controls one program/task, and has as a *type* a decreasing[1] *cost function* mapping an amount of allocated RAM to execution times. She declares

---

[1] For simplicity, throughout the paper we use 'decreasing' with the meaning 'non-increasing', and similarly we use 'increasing' instead of 'non-decreasing'.

to the RAM allocation mechanism a potentially different function as her cost.[2] Ideally, truthfulness of bidders' declarations should be guaranteed without the use of monetary transfers since there is no easy way to charge the programmers. However, very little can be done in mechanism design when money is out of the picture, e.g., as noted above, nothing prevents programmers from exaggerating their RAM needs by overbidding their execution times. Therefore, we look at a mechanism design framework wherein a bidder overbidding her execution time (her *cost* in mechanism design terminology), ends up with this augmented execution time (rather than her true execution time).[3] So a bidder's reported cost will be interpreted by the mechanism as a lower bound to her execution time: a bidder will be allowed to be slower than declared but not faster. This assumption was part of the model defined by Nisan and Ronen in [13][4], and has been later dubbed *monitoring* in [14]. We believe these mechanisms to make a feasible assumption that can be implemented in some real-life scenarios (such as the application that motivates us).

Monitoring and verification. Motivated by the recent advances in trading verification of bidders' behavior with money in mechanism design for CAs [8,9], in Sect. 4 we ask whether similar conclusions can be drawn in our scenario. *Verification*, in this context, means that a bidder cannot underbid her execution time, for otherwise the RAM would be preempted and the task would be aborted (see preliminaries for a formal definition). We call a mechanism using monitoring and verification, a *mechanism with strong monitoring*. We prove a quite interesting dichotomy.[5] To the algorithmic characterization of truthfulness, we pair the positive result that the optimum makespan can be computed truthfully when tasks have *known k-step* cost functions (i.e., the values of the cost functions are known to the mechanism but the discontinuity points are not) and the negative result that no approximation of the optimal makespan can be returned by a truthful mechanism when the k-step functions are *unknown* (i.e., value and discontinuity points are both unknown) even for $k = 1$.

Monitoring and transfers. Given the limitations of mechanisms with strong monitoring and no monetary transfers, in Sect. 5 we turn our attention to mechanisms using (some form of) transfers. Since, as observed above, currency is not available in the setting at hand, we interpret transfers as *waiting times* and focus on *money burning mechanisms* [10]. In details, the output of a mechanism will be a RAM chunk size and a waiting time for each bidder so that the bidder will

---

[2] When these functions have a "large" representation, oracle queries are used just like in the Combinatorial Auctions (CA) literature [3], see preliminaries for a discussion.

[3] This might be implemented by letting the mechanism hold back the results of the computation whenever the program terminates before the reported time.

[4] Specifically, Nisan and Ronen embedded the monitoring assumption in their 'mechanisms with verification', but here we use the term verification in a different sense.

[5] Even though we depart from much of the recent literature (see, e.g., [8,14] and references therein) on mechanisms with verification, which uses no monitoring, we remark that using similar arguments, one can prove the same dichotomy also in that weaker model.

be able to run her task using the amount of RAM allocated by the mechanism only *after* the waiting time. Since waiting times degrade the performance of the system, the objective function of interest must take transfers in consideration. In our case, the objective is the minimization of the maximum (over all bidders) of the *total cost*, where the total cost of a bidder is defined as the *sum* of her execution time and waiting time (transfer). This is a "money burning" version of the makespan objective function. Here we drop the verification assumption but keep the monitoring hypothesis, and call the mechanisms in this section *mechanisms with monitoring*. As a warm-up, we consider the case that RAM chunking is fixed and give a truthful mechanism with monitoring that returns solutions minimizing the makespan (for the fixed chunking), and the total costs of the tasks do not exceed the makespan. This mechanism is thus optimal not only for the classical makespan minimization objective, but also for the money burning objective function. We also show that its transfers (waiting times) are minimal, for the given allocation. We complement this result by showing how to maintain optimality, minimal transfers and truthfulness while computing the RAM chunking that gives the minimum possible makespan.

Following the preliminaries in Sects. 2 and 3 provides a graph-theoretic characterization of the algorithms that are truthful(ly implementable) in a mechanism with monitoring. In Sects. 4 and 5 we present our results for mechanisms with strong monitoring and no transfers, and for mechanisms with monitoring and transfers, respectively. Some proofs are omitted due to space limitations.

Related work. This study connects to a number of research agendas in (A)MD.

Mechanisms with verification (i.e., strong monitoring in our terminology) have been introduced in [13] for the problem of scheduling unrelated selfish machines. A stream of work has looked instead at verification without monitoring (i.e., assuming only no underbidding) in the presence of money [11,14,16] and without [8,9]. Money burning mechanisms are studied in [10] for single-parameter agents and *utilitarian* money burning objective functions. For multi-unit auctions, [10] shows that the largest relative loss due to money burning is logarithmic in the number of bidders. In contrast, we show that transfers do not add any cost to the *makespan*.

An interesting hybrid between verification and money burning mechanisms is [2], which considers exact but costly verification and seeks to maximize the social welfare minus the verification cost for truthful auctions of indivisible goods.

Mechanisms for selfish jobs are also relevant. [1,4] consider truthful mechanisms for selfish one-parameter tasks and makespan minimization on identical machines, with different definitions for the tasks' completion times, use of money, and definition of verification. Coordination mechanisms [5] deal with selfish tasks but focus on equilibrium approximation guarantee rather than truthfulness.

2   Preliminaries

We have one resource available in m copies and n selfish agents. Each selfish agent has a decreasing *cost function*, also called *type* $t_i : [m]_{>0} \to \mathbb{R}_{>0}$, where

$[m]_{>0}$ denotes the set of positive integers not larger than m. For $m' \in [m]_{>0}$, $t_i(m')$ is the cost paid by agent i if she is allowed to use m' copies of the resource. The type t_i is *private knowledge* of agent i. The set of all legal cost functions t_i is called the *domain* of agent i, and is denoted by D_i.

Assuming that each agent has reported or *bid* a (true or false) cost function $b_i \in D_i$, a *mechanism* determines an allocation $(o_1, \ldots, o_n)$ of the m copies of the resource to the n agents. The set $\mathcal{O}$ of all possible allocations contains tuples $(o_1, \ldots, o_n)$ such that $o_i \geq 1$ and $\sum_i o_i \leq m$. Furthermore, depending on (the allocation and) the b_i,'s, it may determine transfers to be paid by the agents. In our model without money, transfers are realized as waiting times, and will be denoted by $w = (w_1, \ldots, w_n)$. (When currency is involved, transfers are usually called payments in literature.) In summary, a *mechanism* is a pair (f, w), where $f : D_1 \times \ldots \times D_n \to \mathcal{O}$ is an algorithm (also termed *social choice function*) that maps agents' costs to a feasible solution in $\mathcal{O}$; and $w : D_1 \times \ldots \times D_n \to \mathbb{R}_{\geq 0}^n$ is a function mapping cost vectors to transfers from each agent i to the mechanism.

A *mechanism without transfers* is simply a social choice function f as above; sometimes, it is convenient to see a mechanism without transfers as a pair (f, w) where w is a constant function of value 0.

Given a vector $\mathbf{b} = (b_i, \mathbf{b}_{-i}) = (b_1, \ldots, b_n)$ of reported cost functions, and $f(\mathbf{b}) = (o_1, \ldots, o_n)$, we let $f_i(\mathbf{b}) = o_i$ denote the number of copies of the resource that the function f assigns to agent i on input $\mathbf{b}$. We assume no externalities, that is, the cost of an agent depends only on her own received number of copies. Therefore, $b_i(f(\mathbf{b})) = b_i(f_i(\mathbf{b}))$. For mechanism (f, w) (with or without transfers) let $cost_i^{(f,w)}(b_i, \mathbf{b}_{-i})$ denote the *total cost* (including transfer w_i) of agent i for the output computed by (f, w) on input $(b_i, \mathbf{b}_{-i})$. Since the types t_i are private knowledge of the agents, they might find it profitable to bid $b_i \neq t_i$. We are interested in mechanisms for which truthtelling is a dominant strategy for each agent.

Definition 1 (Truthful mechanisms). *A mechanism (f, w) (with or without transfers) is truthful if for any i, any bids $\mathbf{b}_{-i}$ of the agents other than i, and any $b_i \in D_i$, $cost_i^{(f,w)}(t_i, \mathbf{b}_{-i}) \leq cost_i^{(f,w)}(b_i, \mathbf{b}_{-i})$.*

Observe that the mechanisms we deal with, are not *individually rational*, in that the agents have a positive cost, and therefore negative valuation for any outcome. Also, not giving an agent any portion of the resource, is not a possible output for the mechanism. Formally, we could elaborate on this (i.e., make the mechanism individually rational), by assuming that an agent not performing her task incurs an infinitely high cost.

Commonly, $cost_i$ is defined as a linear combination of the transfer and the agent's *true* cost for the resource allocated by the algorithm. Here, we define a novel mechanism design paradigm, called *mechanisms with monitoring*, wherein this quasi-linear definition is maintained but costs paid by the agents for the allocated resource are more strictly tied to their declarations. Intuitively, monitoring means that those agents who are allocated a portion of the resource for that their reported cost was exaggerated ($b_i > t_i$), have to process their task up to time b_i instead of the true processing time t_i.

Definition 2 (Mechanism with monitoring). *In a mechanism with monitoring* (f, w)*, the bid* b_i *is a lower bound on agent* i*'s cost of using* $f_i(b_i, \mathbf{b}_{-i})$*, so an agent is allowed to have a real cost higher than* $b_i(f(\mathbf{b}))$ *but not lower. Formally, we have* $cost_i^{(f,w)}(b_i, \mathbf{b}_{-i}) := w_i(\mathbf{b}) + \max\{t_i(f(\mathbf{b})), b_i(f(\mathbf{b}))\}$.

This notion of monitoring is very much related to a concept introduced and termed 'mechanisms with verification' by Nisan and Ronen in [13]. The idea is that if costs are verifiable (e.g., they represent time) and if agents are monitored and claim that the cost using resource $f_i(\mathbf{b})$ is $b_i(f_i(\mathbf{b}))$, then either this is going to be their actual cost (whenever $b_i(f(\mathbf{b})) \geq t_i(f(\mathbf{b}))$), or it can be verified that this declaration is insincere since $b_i(f(\mathbf{b}))$ (which is smaller than $t_i(f(\mathbf{b}))$) is not going to be enough for them to complete their work with the resource (e.g., execute a job). The latter case assumes implicitly that the resource is preempted after $b_i(f(\mathbf{b}))$ time steps at which point the cost of the agent is simply ∞. In other words, agents will never underbid in the model of [13]. Our *mechanism with monitoring* model is much less restrictive and punitive for the agents as we allow them to complete the job, i.e., we do not preempt resources. Moreover, unlike [13], we do not tie transfers with observed costs but only with declarations.

However, for our partially negative results without transfers (Sect. 4), we allow the mechanisms to even use both monitoring (for overbidding agents) *and* verification (for underbidding agents), i.e., the resource is never provided longer than $b_i(f_i(\mathbf{b}))$ time for processing the task of i. Practically, in this model under-bidding is excluded. We call a mechanism with monitoring that also uses this verification for underbidding agents a *mechanism with strong monitoring*.

We say that a social choice function f is *implementable with (strong) monitoring* if there exists a suitable transfer function w such that (f, w) is a truthful mechanism with (strong) monitoring. In this case, we say that w (strongly) implements f. Given $\mathbf{b}$, we say that w_i^* *minimally implements* f at $\mathbf{b}$ *for agent* i if $w_i^*(\mathbf{b}) = \min_{w \text{ implements } f} w_i(\mathbf{b})$.

We consider mechanisms that run in time polynomial in n and $\log m$; however, the representation of the types might need time exponential in those parameters. We therefore assume, as in the related literature, that types are accessed through value or demand queries [3] depending on the algorithm at hand.

3 Graph-Theoretic Characterization of Truthful Mechanisms with Monitoring

In this section we show how to adapt the cycle-monotonicity technique to design truthful mechanisms with (strong) monitoring. The proofs are standard, and are omitted in this short version.

The central tools we use are defined next. An edge (a, b) in the defined graphs represents the option of bidding b instead of the true cost function a. The weight

$\delta_{a,b}$ of the edge represents the difference of actual costs when bidding b instead of a; thus when negative, its absolute value is a lower bound on the difference of truthful payments.

Definition 3. *Let f be a social choice function. For every i and $\mathbf{b}_{-i}$, the declaration graph associated to f has a node for each type in D_i and an additional source node called ω. The set of directed weighted edges is defined as follows. For every $a, b \in D_i$, $a \neq b$, add an edge (a, b) of weight $\delta_{a,b} :=$ $\max\{a(f(\mathbf{b})), b(f(\mathbf{b}))\} - a(f(a, \mathbf{b}_{-i}))$; for any $a \in D_i$, add an edge (ω, a) of weight 0.*

The verification graph associated to f is defined similarly but an edge (a, b) belongs to the graph only if $a(f(b, \mathbf{b}_{-i})) \leq b(f(b, \mathbf{b}_{-i}))$.

Note that the *declaration graph* will be useful for proving truthfulness with monitoring, and the *verification graph* can be used in case of strong monitoring. Since in the latter case underbidding is not an option, no edge (a, b) has to be considered if $b(f(b, \mathbf{b}_{-i})) < a(f(b, \mathbf{b}_{-i}))$.

The next theorem states that, in order to check that a social choice function is implementable with (strong) monitoring and with transfers, it suffices to check that all cycles of the associated graph(s) have a nonnegative weight. For implementation without transfers, instead, it suffices to look at the sign of every single edge. The argument is similar to that used for classical mechanisms [15,18] and mechanisms with verification and no monitoring [8,16].

Theorem 1. *A social choice function f is implementable with monitoring (resp., strong monitoring) when agents bid from finite domains, if and only if, for all i and declarations $\mathbf{b}_{-i}$, the declaration (resp., verification) graph associated to f does not have negative weight cycles.*

Moreover, f is implementable with monitoring (resp., strong monitoring) without transfers if and only if, for all i and $\mathbf{b}_{-i}$, the declaration (resp., verification) graph associated to f does not have negative weight edges (the size of the domains does not matter).

Given our interest in money burning mechanisms, we also prove here what form minimal transfers have.

Theorem 2. *Let f be a social choice function f implementable with monitoring (resp., strong monitoring). For any $\mathbf{b} = (b_i, \mathbf{b}_{-i})$ and any i, the transfer function that minimally implements f at $\mathbf{b}$ for agent i is $w_i^*(\mathbf{b}) = -\mathcal{SP}(\omega, b_i)$, where $\mathcal{SP}(\omega, b_i)$ is the length of the shortest path from ω to b_i in the declaration (resp., verification) graph associated to f.*

4   Mechanisms with Strong Monitoring and No Transfers

Here we give results on mechanisms with strong monitoring and no transfers.

4.1 Algorithmic Characterization

We begin by characterizing the class of algorithms that are truthful with strong monitoring in the case in which transfers are not allowed.

Theorem 3. *An algorithm f is truthful with strong monitoring and no transfers if and only if for all i, $\mathbf{b}_{-i}$, and $a, b \in D_i$, $a(f(b, \mathbf{b}_{-i})) \leq b(f(b, \mathbf{b}_{-i}))$ implies $b(f(b, \mathbf{b}_{-i})) \geq a(f(a, \mathbf{b}_{-i}))$.*

4.2 Known k-step Tasks

We now provide the characterization for a specific family of domains for selfish tasks.

Definition 4. *The task (agent) i has a* **known k-step function domain** *if, for some known $c_i^1 \geq \ldots \geq c_i^k$ and unknown $r_i^1 \leq \ldots \leq r_i^{k-1} (\leq r_i^k = m)$, her type satisfies*

$$t_i(m') = \begin{cases} c_i^1 \ \textit{if} \ 0 < m' < r_i^1 \\ c_i^j \ \textit{if} \ r_i^{j-1} \leq m' < r_i^j, \ 1 < j \leq k \end{cases}.$$

The cost function of such a task is then completely determined by the threshold values r_i^j; i.e., D_i can be assumed to consist of vectors in $[m]_{>0}^k$.

A *known k-step task* is a task with a known k-step function domain. Below, with a slight abuse of notation, $a \in D_i$ will both denote the $(k$-$)$tuple in D_i and the corresponding cost function. We define the property that characterizes truthful algorithms f in this context. Subsequently, we show that this property is a quite natural one, in the sense that for a large class of objective functions, the optimal allocation fulfils it.

Definition 5. *An algorithm f is* **k-step monotone** *if for any i, $\mathbf{b}_{-i}$, and $a = (r_i^j)_{j=1}^k, b = (\tilde{r}_i^j)_{j=1}^k \in D_i$, with $r_i^j \leq \tilde{r}_i^j$ for all $1 \leq j < k$, $f_i(a, \mathbf{b}_{-i}) < r_i^j$ implies $f_i(b, \mathbf{b}_{-i}) < \tilde{r}_i^j$.*

Lemma 1. *An algorithm f is truthful with strong monitoring and no transfers for known k-step tasks if and only if it is k-step monotone.*

For a bid vector $\mathbf{b}$ and feasible solution $o = (o_1, \ldots, o_n) \in \mathcal{O}$, let $\mu(\mathbf{b}, o)$ be a function increasing in every single cost $b_i(o_i)$, e.g., for the makespan $\mu(\mathbf{b}, o) = \max_i b_i(o_i)$. Define OPT_μ as the social choice function that on input $\mathbf{b}$ returns a solution minimizing μ using a tie-breaking rule independent of $\mathbf{b}$.

Theorem 4. *For any increasing cost function μ, OPT_μ is k-step monotone.*

4.3 Unknown Single-Step Tasks and Limitations of Mechanisms Without Transfers

Definition 6. *The task (agent) i has an* unknown single-step function domain *if her type satisfies*

$$t_i(m') = \begin{cases} h_i & \text{if } m' < r_i \\ l_i & \text{if } m' \geq r_i \end{cases},$$

for some unknown $h_i > l_i$ and unknown r_i. The cost function of such a task is then completely determined by the triple (r_i, h_i, l_i); i.e., D_i can be assumed to consist of vectors in $[m]_{>0} \times \mathbb{R}^2_{>0}$.

An *unknown single-step task* is a selfish task with an unknown single-step function domain. Given a cost function $a = (a_r, a_h, a_l) \in D_i$, a_r will denote the threshold in $[m]$, and a_h and a_l denote the high and low cost respectively.

Definition 7. *An algorithm f is* unknown single-step monotone *if for any i, $\mathbf{b}_{-i}$, and $a, b \in D_i$, such that $a_r \leq b_r$, $a_h \leq b_h$ and $a_l \leq b_l$, $f_i(a, \mathbf{b}_{-i}) < a_r$ implies $f_i(b, \mathbf{b}_{-i}) < b_r$.*

The property above characterizes truthfulness when costs are not known:

Lemma 2. *An algorithm f is truthful with strong monitoring and no transfers for unknown single-step tasks if and only if it is unknown single-step monotone.*

We now prove that these algorithms cannot return any reasonable approximation of the makespan.

Theorem 5. *For any $\alpha > 0$, there is no algorithm without transfers that is truthful with strong monitoring for unknown single-step tasks, and returns a better than α-approximation of the optimal makespan.*

Proof. Consider an instance with two unknown single-step tasks such that $r_1 = r_2 = r$, $l_1 = l_2 = 1$, $h_1 = \alpha(1+\delta)$ and $h_2 = 1+\delta$ for some $\delta > 0$. Set $r < m < 2r$ so that only one task can get the RAM she needs to be fast. Any better than α-approximate algorithm for the makespan will assign r to task 1 and some $\varepsilon > 0$ to task 2. Consider now a new instance wherein task 2 modifies h_2 as $h'_2 = \alpha^2(1+\delta)$. Since the algorithm is truthful then it must be unknown single-step monotone thus implying that the outcome of the algorithm cannot assign at least r to task 2, thus returning an α-approximation of the makespan (the optimum would indeed allocate r to task 2 and some $\varepsilon > 0$ to task 1).

We next show that by introducing transfers – in terms of waiting time for using the allocated RAM – we can indeed design better mechanisms for tasks with general cost functions.

5 Optimal Mechanisms with Monitoring Using Transfers

We begin with a general result. Quite interestingly, the next theorem shows that given monitoring, there is a truthful PTAS for scheduling unrelated machines (at least for finite domains), alternative to the compensation-and-bonus mechanism of [13], that does not need verification to be truthful.

Theorem 6. *For any social choice function f, there exists a transfer function w such that (f, w) is truthful with monitoring when agents bid from finite domains.*

For the applicability of the theorem, we need to bound and discretize the range of the admitted cost functions t_i, so we assume for the rest of the section that the $t_i(m')$ (and the bids $b_i(m')$) are integers from a given Interval $[0, T]$.

5.1 Optimal Mechanism for Makespan with Fixed Memory Chunks

Assume that n memory chunks of fixed size have to be allocated one-to-one among n agents, each of whom has a task to process. We identify the memory chunks with their sizes $m_1 \leq m_2 \leq \ldots \leq m_n$ in increasing order. Let $t_i(m_j)$ denote the (true) processing time of task i using a memory chunk of size m_j.

We consider a greedy allocation rule called *Best-Fit Procedure* that allocates the chunks in increasing order of size, as follows: m_1 is allocated to the task i with the minimum processing time given this amount of memory $t_i(m_1) = \min_k t_k(m_1)$; then iteratively, for every $j = 2, \ldots, n$, m_j is given to the remaining agent with the smallest reported processing time with memory of size m_j.

Best-Fit Allocation Procedure

> **Input:** matrix of processing times $\mathbf{t} = (t_1, t_2, \ldots, t_n)$

1. $N \leftarrow [n]$
2. **for** $j = 1 \ldots n$ **do**
   - (a) Let $i = \arg\min_{i \in N} t_i(m_j)$
   - (b) Set $f_i^W(\mathbf{t}) = m_j$
   - (c) $N \leftarrow N \setminus \{i\}$
3. **Output** $f^W = (f_1^W, \ldots, f_n^W)$.

We claim first, that without waiting times as transfers, this allocation rule is optimal for the makespan objective (maximum processing time over all tasks). Then we introduce waiting times w_i as payments, so that the resulting mechanism is truthful, and the waiting times do not increase the makespan, so the mechanism is both truthful and achieves optimal makespan.

Lemma 3. *The Best-Fit procedure achieves optimal makespan among all bijective allocations of the n memory chunks to the n agents.*

The proof goes by induction on n, and is based on a standard exchange-argument in a fixed optimal allocation turning it into the Best-Fit allocation while preserving optimality. In particular, if the smallest chunk is allocated to task i in Best-Fit, but to task k Opt, then exchanging the chunks between these two tasks in Opt does not increase the makespan.

Next we show that this allocation rule can be implemented by a truthful mechanism by using waiting times as payments by the agents. Given the allocation f^W, these waiting times are defined to be smallest possible (for each agent) such that in increasing order of chunk size the total costs (processing plus waiting time) of the respective agents become *increasing*.

In the code below we complement the Best-Fit Procedure to a mechanism by setting the waiting times w_i. The mechanism takes as input the matrix $\mathbf{b}$ of reported running times of the agents. Observe that c_j stands for the maximum processing time over chunks 1 to j after allocation step j. For bidder i, who gets chunk m_j, the payment in form of waiting time is $w_i = c_j - b_i(m_j)$.

Best-Fit Mechanism

   Input: matrix of reported processing times $\mathbf{b} = (b_1, b_2, \ldots, b_n)$

1. $N \leftarrow [n]$
2. $(c_1, \ldots, c_n) \leftarrow (0, \ldots, 0)$
3. **for** $j = 1 \ldots n$ **do**
   (a) Let $i = \arg \min_{i \in N} b_i(m_j)$
   (b) $f_i^W(\mathbf{b}) \leftarrow m_j$
   (c) $N \leftarrow N \setminus \{i\}$
   (d) $c_j \leftarrow \max\{c_{j-1}, b_i(m_j)\}$
   (e) $w_i \leftarrow c_j - b_i(m_j)$
4. **Output** $f^W = (f_1^W, \ldots, f_n^W)$, and $w = (w_1, w_2, \ldots, w_n)$.

Note that the total cost $cost_i(\mathbf{b})$ of the agent who gets m_j, is $\max\{c_j, t_i(m_j)\} \geq \max\{c_{j-1}, t_i(m_j)\}$ (here we use that the cost is always at least the true running time), and it is exactly $c_j = \max\{c_{j-1}, t_i(m_j)\}$ if $b_i(m_j) = t_i(m_j)$.

Theorem 7. *The Best-Fit Mechanism is truthful.*

Proof. Here we provide only a sketch of the proof. For some bidder i, let $b_i \neq t_i$ be an advantageous false bid with the minimum number of indices j such that $b_i(m_j) \neq t_i(m_j)$, and let ℓ be the smallest such index.

There are two nontrivial cases to consider. First, when i receives m_ℓ by bidding $t_i(m_\ell)$ and does not receive m_ℓ when bidding $b_i(m_\ell) > t_i(m_\ell)$. This occurs when there is a bidder k with bid $b_i(m_\ell) \geq b_k(m_\ell) \geq t_i(m_\ell)$, who gets m_ℓ. The total cost of i when bidding t_i would be $\max\{c_{\ell-1}, t_i(m_\ell)\}$. With bid b_i she gets a chunk with higher index, and her cost will be at least $c_\ell = \max\{c_{\ell-1}, b_k(m_\ell)\} \geq \max\{c_{\ell-1}, t_i(m_\ell)\}$ (where c_ℓ is meant with input b_i).

Second, consider the case when i receives m_ℓ by bidding $b_i(m_\ell)$ and does not receive m_ℓ when bidding $t_i(m_\ell) > b_i(m_\ell)$. Again, there must be a bid of some agent k so that $t_i(m_\ell) \geq b_k(m_\ell) \geq b_i(m_\ell)$. Now, if agent i bids b_i, then her cost is $\max\{c_{\ell-1}, t_i(m_\ell)\}$. If she bids $t_i(m_\ell)$ instead of $b_i(m_\ell)$ then she receives a (larger) chunk m_s, for total cost of $\max\{c_{s-1}, t_i(m_s)\}$. However, it can be shown that $\max\{c_{s-1}, t_i(m_s)\} \le \max\{c_{\ell-1}, t_i(m_\ell)\}$, implying that agent i could change her bid for chunk ℓ from $b_i(m_\ell)$ to $t_i(m_\ell)$.

Finally, we show that the waiting times used as transfers in the Best-Fit mechanism have further appealing features apart from truthfulness. First, these waiting times do not ruin the makespan-minimizing property of the mechanism; second, these waiting times correspond to the transfers that minimally implement the makespan minimizing allocation rule Best-Fit.[6]

Lemma 4. *The Best-Fit mechanism achieves minimum makespan (for any given input* $\mathbf{b}$*).*

Lemma 5. *For fixed memory chunks* $m_1, m_2, \ldots, m_n$*, the payments* $w_i = c_j - b_i(m_j)$ *used in the Best-Fit mechanism correspond to the transfer functions that minimally implement the Worst-Fit allocation rule* f^W*.*

5.2 Mechanism with Memory Chunking

In this section we treat the problem of optimally chunking a given total size $m \in \mathbb{N}$ of memory into n chunks $(m_1, m_2, \ldots, m_n)$ (s.t. $\sum_j m_j = m$, and $m_j \in \mathbb{N}$), and then determining a one-to-one allocation of the chunks, with the goal of minimizing the makespan over all chunkings and all bijections $f : [n] \to \{m_1, m_2, \ldots, m_n\}$. We call such a more complex algorithm a *chunking algorithm*, which then can be implemented by a *chunking mechanism*. Unfortunately it turns out that finding the optimal chunking, and applying the Best-Fit mechanism with this given chunking does not yield a truthful chunking mechanism.

Theorem 8. *For any algorithm that takes as input the (reported) cost functions* $b_i : [m] \to \mathbb{N}$*, then determines an optimal (makespan minimizing) chunking* $(m_1, m_2, \ldots, m_n)$*, and finally outputs the optimal allocation* f^W *and transfers* w *according to the Best-Fit mechanism with input* $(m_1, m_2, \ldots, m_n)$ *and* $\mathbf{b}$*, the resulting chunking mechanism is not truthful.*

Proof. Consider the following instance with $n = 3$ tasks, and total memory size $m = 6$. Let the true cost-functions be $t_1(m') = 1$ for all $m' \ge 1$; $t_2(1) = 5$, and $t_2(m') = 3$ for $m' \ge 2$; finally, $t_3(m') = 7$ for $m' < 4$, and $t_3 = 3$ for $m' \ge 4$. The optimal makespan is 5, achieved with the memory chunking $(1, 1, 4)$. In this optimal allocation task 2 has running time $t_2(1) = 5$, and no waiting time. However, if agent 2 bids $b_2(1) = 8$, and $b_2(m') = 3$ for $m' \ge 2$, then $(2, 2, 2)$ becomes the optimal chunking with makespan 7, and task 2 has running time $\max\{t_2(2), b_2(2)\} = 3$ and no waiting time. Thus agent 2 has an incentive to report false running times, so the mechanism is not truthful.

---

[6] For the definition of 'minimally implements', see the Preliminaries.

Nevertheless, we know from Theorem 6, that for any fixed optimal allocation algorithm with memory chunking, there do exist transfers that yield a truthful mechanism. Indeed, one such chunking mechanism is the following. Let the chunking algorithm determine an optimal chunking and allocation with makespan M_{opt}. A trivial truthful mechanism charges $w_i = M_{opt} - b_i(m^i)$ to agent i who gets chunk m^i, so that the total cost of each agent i becomes $cost_i = \max\{M_{opt}, t_i(m^i)\}$. (In fact, such mechanism is optimal and truthful also in case of any fixed chunking.) There is a slightly better truthful pricing rule, charging the above prices, *except* for agents who get a memory chunk of minimum size; these agents do not have waiting times. This slight modification of the transfer function may seem to be of little use. Observe though, that charging the makespan as total cost to *every* agent is a highly unrealistic solution, because with this rule the total cost of an agent can become by an arbitrary factor higher than her running time using *any* memory size. In contrast, in the Best-Fit mechanism, the total cost c_j of a truthful agent getting chunk m_j is either her own running time, or the running time of some task getting a smaller chunk, so that agent i would have had a higher running time than c_j with that chunk. That is, for each task her total cost is within the range of running times of this task.

We define a particular chunking mechanism that finds the optimal makespan by binary search, and charges waiting times according to the above rule. Subsequently, we show that the mechanism is truthful, and that the waiting times correspond to the minimum transfers that implement this particular allocation rule truthfully. Note however, that there might exist different *optimal* allocation rules with smaller truthful payments.

Binary-Chunking Mechanism

Input: reported functions of processing times $\mathbf{b} = (b_1, b_2, \ldots, b_n)$, where $b_i : [m] \to [T]$

1. $M \leftarrow \lfloor T/2 \rfloor$
2. do binary search for the optimum makespan M_{opt}
   (a) **for** $i = 1$ **to** n **do**
       find (with binary search) the minimum demand m^i of agent i in order to finish within M
   (b) **if** $\sum_i m^i > m$ **then** set M higher
   (c) **else** set M lower if possible, otherwise set $M_{opt} = M$
3. **for** $i = 1$ **to** n **do**
   (a) $f_i^C(\mathbf{b}) \leftarrow m^i$
   (b) **if** $m^i = 1$ **then** $w_i \leftarrow 0$
   (c) **else** $w_i \leftarrow M_{opt} - b_i(m^i)$
4. **Output** $f^C = (f_1^C, \ldots, f_n^C)$, and $w = (w_1, w_2, \ldots, w_n)$.

We note that the binary search for M_{opt} can also be carried out using demand queries. In this case, subsequently the $b_i(m^i)$ have to be queried as well (since $cost_i = M_{opt} - b_i(m^i) + b_i(m^i)$, there is no reason to report these non-truthfully).

Theorem 9. *The Binary-Chunking mechanism is truthful. The same holds for any chunking mechanism with an optimal chunking algorithm, and with the payments of Binary-Chunking.*

Theorem 10. *The waiting times used in the Binary-Chunking mechanism are the minimum transfers that make the allocation rule of Binary-Chunking truthful.*

6 Conclusions

We have started our research from a rather practical problem in the context of concurrent execution of memory-bound programs. Our first solutions presented here, deal with the static case where an appropriate RAM distribution has to be determined once, under the makespan objective.

From a more theoretical point of view, our work introduces an interesting new model of mechanism design wherein studying money burning objective functions is the right research challenge. In fact, we prove that *all* algorithms admit transfers that make them truthful with monitoring and therefore, also in light of the negative results in [10], this paradigm seems to be the right arena to study the optimal trade off between quality of allocation *and* transfers introduced.

We believe that our results pave the way to a number of interesting open questions, the main being the extent to which our positive results can be exported to more general models allowing repeated allocation mechanisms and/or stronger solution concepts (e.g., collusion-resistance for known coalitions). In our setting, the minimization of the sum of the *total* costs of the agents (i.e., the original utilitarian objective for money burning) needs to be explored.

References

1. Auletta, V., De Prisco, R., Penna, P., Persiano, G.: How to route and tax selfish unsplittable traffic. In: SPAA, pp. 196–205 (2004)
2. Ben-Porath, E., Dekel, E., Lipman, B.L.: Optimal allocation with costly verification. Am. Econ. Rev. **104**(12), 3779–3813 (2014)
3. Blumrosen, L., Nisan, N.: On the computational power of demand queries. SIAM J. Comput. **39**(4), 1372–1391 (2009)
4. Christodoulou, G., Gourvès, L., Pascual, F.: Scheduling selfish tasks: about the performance of truthful algorithms. In: Lin, G. (ed.) COCOON 2007. LNCS, vol. 4598, pp. 187–197. Springer, Heidelberg (2007)
5. Christodoulou, G., Koutsoupias, E., Nanavati, A.: Coordination mechanisms. In: Díaz, J., Karhumäki, J., Lepistö, A., Sannella, D. (eds.) ICALP 2004. LNCS, vol. 3142, pp. 345–357. Springer, Heidelberg (2004)
6. Dementiev, R., Kettner, L., Sanders, P.: STXXL: standard template library for XXL data sets. Softw. Pract. Exper. **38**(6), 589–637 (2008)
7. Fiat, A., Karp, R., Luby, M., McGeoch, L., Sleator, D., Young, N.: Competitive paging algorithms. J. Algorithms **12**(4), 685–699 (1991)
8. Fotakis, D., Krysta, P., Ventre, C.: Combinatorial auctions without money. In: AAMAS, pp. 1029–1036 (2014)

9. Fotakis, D., Krysta, P., Ventre, C.: The power of verification for greedy mechanism design. In: AAMAS, pp. 307–315 (2015)
10. Hartline, J.D., Roughgarden,T.: Optimal mechanism design and money burning. In: STOC, pp. 75–84 (2008)
11. Krysta, P., Ventre, C.: Combinatorial auctions with verification are tractable. Theoret. Comput. Sci. **571**, 21–35 (2015)
12. Meyer, U., Sanders, P., Sibeyn, J.F. (eds.): Algorithms for Memory Hierarchies. LNCS, vol. 2625. Springer, Heidelberg (2003)
13. Nisan, N., Ronen, A.: Algorithmic mechanism design. Games Econ. Behav. **35**, 166–196 (2001)
14. Penna, P., Ventre, C.: Optimal collusion-resistant mechanisms with verification. Games Econ. Behav. **86**, 491–509 (2014)
15. Rochet, J.-C.: A condition for rationalizability in a quasi-linear context. J. Math. Econ. **16**, 191–200 (1987)
16. Ventre, C.: Truthful optimization using mechanisms with verification. Theoret. Comput. Sci. **518**, 64–79 (2014)
17. Vitter, J.S.: Algorithms and data structures for external memory. Found. Trends Theoret. Comput. Sci. **2**(4), 305–474 (2006)
18. Vohra, R.V.: Mechanism Design: A Linear Programming Approach. Cambridge University Press, New York (2011)

Inverse Game Theory: Learning Utilities in Succinct Games

Volodymyr Kuleshov and Okke Schrijvers[✉]

Stanford University, Stanford, CA 94305, USA
kuleshov@stanford.edu, okkes@cs.stanford.edu

Abstract. One of the central questions in game theory deals with predicting the behavior of an agent. Here, we study the inverse of this problem: given the agents' equilibrium behavior, what are possible utilities that motivate this behavior? We consider this problem in arbitrary normal-form games in which the utilities can be represented by a small number of parameters, such as in graphical, congestion, and network design games. In all such settings, we show how to efficiently, i.e. in polynomial time, determine utilities consistent with a given correlated equilibrium. However, inferring both utilities and structural elements (e.g., the graph within a graphical game) is in general NP-hard. From a theoretical perspective our results show that rationalizing an equilibrium is computationally easier than computing it; from a practical perspective a practitioner can use our algorithms to validate behavioral models.

1   Introduction

One of the central and earliest questions in game theory deals with predicting the behavior of an agent. This question has led to the development of a wide range of theories and solution concepts — such as the Nash equilibrium — which determine the players' actions from their utilities. These predictions in turn may be used to inform economic analysis, improve artificial intelligence software, and construct theories of human behavior.

Perhaps equally intriguing is the *inverse* of the above question: given the observed behavior of players in a game, how can we infer the utilities that led to this behavior? Surprisingly, this question has received much less attention, even though it arises just as naturally as its more famous converse.

For instance, inferring or *rationalizing* player utilities ought to be an important part of experimental protocols in the social sciences. An experimentalist should test the validity of their model by verifying whether it admits any utilities that are consistent with observed data. More ambitiously, the experimentalist may wish to develop predictive techniques, in which one tries to forecast the agents' behavior from earlier observations, with utilities serving as an intermediary in this process.

Inferring utilities also has numerous engineering applications. In economics, one could design mechanisms that adapt their rules after learning the utilities of their users, in order for instance to maximize profits. In machine learning,

© Springer-Verlag Berlin Heidelberg 2015
E. Markakis and G. Schäfer (Eds.): WINE 2015, LNCS 9470, pp. 413–427, 2015.
DOI: 10.1007/978-3-662-48995-6_30

algorithms that infer utilities in a single-agent reinforcement learning setting are key tools for developing helicopter autopilots, and there exists ongoing research on related algorithms in the multi-agent setting.

1.1 Our Contributions

Previous work on computational considerations for rationalizing equilibria has mainly focused on specific types of games, such as matching [1] and network formation games [2]. Here, we instead take a top-down approach and consider the problem in an *arbitrary* normal-form game. Although our results hold generally, the problem becomes especially interesting when the normal-form game is succinct, meaning that player utilities (which are normally exponentially-sized objects) can be represented by a small number of parameters. A vast number of games studied in the literature — including congestion, graphical, scheduling, and network design games — have this property. Within large classes of succinct games, we establish the following two main results:

- When the structure of a game (e.g. the graph in a graphical game) is known, we can find utilities that rationalize the equilibrium using a small convex program. This program is polynomial rather than exponential in the number of players and their actions, and hence can be solved efficiently. We discuss these results in Sect. 4.
- If the structure of a succinct game is unknown, inferring both utilities and the correct game structure is NP-hard. We discuss these results in Sect. 5.

1.2 Related Work

Theoretical Computer Science. Kalyanaraman et al. studied the computational complexity of rationalizing stable matchings [1], and network formation [2]. In the latter case, they showed that game attributes that are local to a player can be rationalized, while other, more global, attributes cannot; this mirrors our observations on the hardness of inferring utilities versus inferring game structure. The forward direction of our problem — computing an equilibrium from utilities — is a central question within algorithmic game theory. Computing Nash equilibria is intractable [3] even for 2 player games [4] (and therefore may be a bad description of human behavior); correlated equilibria, however, are easy to compute in succinct games [5] and can be found using simple iterative dynamics [6,7]. Our results show that while a Nash equilibrium is hard to compute, it is easy to rationalize. For correlated equilibria, both computing and rationalizing it are feasible.

Economics. Literature on rationalizing agent behavior [8–10] far predates computational concerns. The field of revealed preference [11] studies an agent who buys different bundles of a good over time, thus revealing more information about its utilities. These are characterized by sets of linear inequalities, which become progressively more restrictive; we adopt this way of characterizing agent utilities in our work as well, but in addition we prove that solving the problem can be done in polynomial time.

Econometrics. Recently, Nekipelov et al. [12] discussed inferring utilities of bidders in online ad auctions, assuming bidders are using a no-regret algorithm for bidding. While no-regret learning agents do converge to a correlated equilibrium, the authors discuss a private-information game, rather than the full information games we consider.

The identification literature in econometrics [13–15] is closely related to our work. However, this literature does not entirely address many computational concerns, e.g. it studies the rationalization problem in arbitrary normal-form games in which utilities are exponentially-sized [15]. Our work instead highlights the importance of succinct games and offers efficient algorithms for a large class of such games. Moreover, rather than returning a single utility function, we explicitly characterize the set of valid utilities, which can be interpreted as a measure of the confidence of our prediction and thus may be useful for the practitioner.

Inverse Reinforcement Learning. Algorithms that infer the payoff function of an agent within a Markov decision process [16] are a key tool in building helicopter autopilots [17]. Our work establishes an analogous theory for multi-agent settings. In both cases, valid utilities are characterized via sets of linear inequalities. Inverse reinforcement learning has also been used to successfully predict driver behavior in a city [18,19]; unlike our work, these earlier methods do not directly learn the utilities of the game playing agents.

Inverse Optimization. Game theory can be interpreted as multi-player optimization, with different agents maximizing their individual objective functions. One of the central results in inverse optimization shows that one can recover the objective function of an optimization program from its solution by solving a related linear program [20]. Our work considers the analogous inverse problem for multiple players and solves it using a linear program as well.

2   Preliminaries

In a normal-form game $G \triangleq [(A_i)_{i=1}^n, (\mathbf{u}_i)_{i=1}^n]$, a player $i \in \{1, 2, ..., n\}$ has m_i actions $A_i \triangleq \{a_1^i, a_2^i, ..., a_{m_i}^i\}$ and utilities $\mathbf{u}_i \in \mathbb{R}^m$, where $m = \prod_{i=1}^n m_i$ is the cardinality of the joint-action space $A \triangleq \times_{i=1}^n A_i$. An $\mathbf{a} \in A$ is called a joint action of all the players and let $\mathbf{a}_{-i}$ be $\mathbf{a}$ with the action of player i removed. A mixed strategy of player i is a probability distribution $\mathbf{p}_i \in \mathbb{R}^{m_i}$ over the set of actions A_i. A correlated equilibrium (CE) of G is a probability distribution $\mathbf{p} \in \mathbb{R}^m$ over A that satisfies

$$\sum_{\mathbf{a}_{-i}} p(a_j^i, \mathbf{a}_{-i}) u(a_j^i, \mathbf{a}_{-i}) \geq \sum_{\mathbf{a}_{-i}} p(a_j^i, \mathbf{a}_{-i}) u(a_k^i, \mathbf{a}_{-i}) \tag{1}$$

for each player i and each pair of actions a_j^i, a_k^i. This equation captures the idea that no player wants to unilaterally deviate from their equilibrium strategy.

Correlated equilibria exist in every game, are easy to compute using a linear program, and arise naturally from the repeated play of learning players [6,7].

A (mixed) Nash equilibrium is a correlated equilibrium $\mathbf{p}$ that is a product distribution $p(\mathbf{a}) = p_1(a_1) \times \ldots \times p_n(a_n)$, where the $\mathbf{p}_i \in \mathbb{R}^{m_i}$ are mixed player strategies. In a Nash equilibrium, each player chooses their own strategy (hence the product form), while in a correlated equilibrium the players' actions can be viewed as coming from an outside mediator. A Nash equilibrium exists in every game, but is hard to compute even in the 2-player setting [4].

3   Succinct Games

In general, the dimension m of player i's utility $\mathbf{u}_i$ is exponential in the number of players: if each player has t actions, $\mathbf{u}_i$ specifies a value for each of their t^n possible combinations. Therefore, we restrict our attention to games G that have a special structure which allows the $\mathbf{u}_i$ to be parametrized by a small number of parameters $\mathbf{v}$; such games are called *succinct* [5].

A classical example of a succinct game is a *graphical game*, in which there is a graph H with a node for every player, and the utility of a player depends only on itself and the players on incident nodes in H. If k be the number of neighbors of i in H, then we only need to specify the utility of i for each combination of actions of $k+1$ players (rather than n). For bounded-degree graphs, this greatly reduces the number of parameters in the game. If the maximum degree in the graph is k and each player has at most t actions, then the total number of utility values per player is at most t^{k+1}, which is independent of n.

Definition 1. *A succinct game*

$$G \triangleq [(A_i)_{i=1}^n, (\mathbf{v}_i)_{i=1}^n, (F_i)_{i=1}^n]$$

is a tuple of sets of player actions A_i, parameters $\mathbf{v}_i \in \mathbb{R}^d$, and functions $F_i : \mathbb{R}^d \times A \to \mathbb{R}$ that efficiently compute the utility $u_i(\mathbf{a}) = F_i(\mathbf{v}_i, \mathbf{a})$ of a joint action $\mathbf{a}$.

We will further restrict our attention to succinct games in which the F_i have a particular linear form. As we will soon show, almost every succinct game in the literature is also linear. This definition will in turn enable a simple and unified mathematical analysis across all succinct games.

Definition 2. *A linear succinct game*

$$G \triangleq [(A_i)_{i=1}^n, (\mathbf{v}_i)_{i=1}^n, (O_i)_{i=1}^n]$$

is a succinct game in which the utilities $\mathbf{u}_i$ are specified by $\mathbf{u}_i = O_i \mathbf{v}_i$, where $O_i \in \{0,1\}^{m \times d}$ is an outcome matrix mapping parameters into utilities. We assume that the O_i have a compact representation and that each component of $O_i \mathbf{v}_i$ can be computed efficiently.

Note that a linear succinct game is a special case of Definition 1 with $F_i(\mathbf{v}_i, \mathbf{a}) = (O_i\mathbf{v}_i)_\mathbf{a}$, which is the component of $O_i\mathbf{v}$ corresponding to $\mathbf{a}$.

The outcome matrix O_i has an intuitive interpretation. We can think of a set of d distinct outcomes $\mathcal{O}_i$ that can affect the utility of player i. The parameters $\mathbf{v}_i$ specify a utility $\mathbf{v}_i(o)$ for each outcome $o \in \mathcal{O}_i$. When a joint action $\mathbf{a}$ occurs, it results in the realization of a subset $\mathcal{O}_i(\mathbf{a}) \triangleq \{o : (O_i)_{\mathbf{a},o} = 1\}$ of the outcomes, specified by the positions of the non-zero entries of matrix O_i. The utility $u_i(\mathbf{a}) = (O_i\mathbf{v}_i)_\mathbf{a}$ equals the sum of valuations of the realized outcomes:

$$u_i(\mathbf{a}) = \sum_{o \in \mathcal{O}_i(\mathbf{a})} v_i(o).$$

Graphical games, which we discussed above, are an example of a succinct game that is linear. In a graphical game with an associated graph H, outcomes correspond to joint actions $\mathbf{a}_{N(i)} = (a^{(k)})_{k \in N(i)}$ by i and its neighbors in H. A joint-action $\mathbf{a}$ activates the single outcome o that is associated to a $\mathbf{a}_{N(i)}$ in which the actions are specified by $\mathbf{a}$. The matrix O_i is defined as

$$(O_i)_{\mathbf{a},\mathbf{a}_{N(i)}} = \begin{cases} 1 & \text{if } \mathbf{a}, \mathbf{a}_{N(i)} \text{ agree on the actions of } N(i) \\ 0 & \text{otherwise.} \end{cases}$$

3.1 Succinct Representations of Equilibria

Since there is an exponential number of joint actions, a correlated equilibrium $\mathbf{p}$ (which is a distribution over joint actions) may require exponential space to write down. To make sure that the input is polynomial in n, we require that $\mathbf{p}$ be represented as a polynomial mixture of product distributions (PMP) $\mathbf{p} = \sum_{k=1}^{K} \mathbf{q}_k$, where K is polynomial in n, $q_k(\mathbf{a}) = \prod_{i=1}^{n} q_{ik}(a_i)$ and q_{ik} is a distribution over A_i. Correlated equilibria in the form of a PMP exist in every game and can be computed efficiently [5]. A Nash equilibrium is already a product distribution, so it is a PMP with $K = 1$.

The issue of representing equilibria also raises several practical questions, the most important of which concerns how the $\mathbf{p}$ are estimated. In principle, we allow the user to use any estimation strategy, such as recently proposed methods based on the maximum entropy principle [18,19]. Note, however, that $\mathbf{p}$ can also be a pure strategy equilibrium; we introduce below methods for rationalizing several $\mathbf{p}$ at once, which implies that our method directly accepts sequences of player actions as input. Another potential concern is that a correlated equilibrium may be only privately known by a mediator; this may indeed complicate the estimation of such equilibria, but does not limit the applicability of our methods to many other solution concepts, such as the Nash equilibrium. We present our results in the context of correlated equilibria simply because it is the most general solution concept that we can handle; in practice, however, our techniques are applicable to sequences of directly observed pure Nash equilibria within the same game.

3.2   What It Means to Rationalize an Equilibrium

Finding utilities consistent with an equilibrium $\mathbf{p}$ amounts to finding $\mathbf{u}_i$ that satisfy Eq. 1 for each player i and for each pair of actions $a_j^i, a_k^i \in A_i$. It is not hard to show that Eq. 1 can be written in matrix form as

$$\mathbf{p}^T C_{ijk} \mathbf{u}_i \geq 0, \tag{2}$$

where C_{ijk} is an $m \times m$ matrix that has the form

$$(C_{ijk})_{(\mathbf{a}_{\mathrm{row}}, \mathbf{a}_{\mathrm{col}})} = \begin{cases} -1 & \text{if } \mathbf{a}_{\mathrm{row}} = (a_j, \mathbf{a}_{-i}^{\mathrm{col}}) \\ 1 & \text{if } \mathbf{a}_{\mathrm{row}} = (a_k, \mathbf{a}_{-i}^{\mathrm{col}}) \\ 0 & \text{otherwise.} \end{cases}$$

This formulation exposes intriguing symmetry between the equilibrium distribution $\mathbf{p}$ and the utilities $\mathbf{u}_i$. By our earlier definitions, the utilities $\mathbf{u}_i$ in a linear succinct game can be written as $\mathbf{u}_i = O_i \mathbf{v}_i$; this allows us to rewrite Eq. 2 as

$$\mathbf{p}^T C_{ijk} O_i \mathbf{v}_i \geq 0. \tag{3}$$

While C_{ijk} and O_i are exponentially large in n, their product is not, so in Sect. 4 we show that we can compute this product efficiently, without constructing C_{ijk} and O_i explicitly. To do this we let O_i be represented by a small program that for action profile $\mathbf{a}$ and outcome o returns $(O_i)_{\mathbf{a},o}$. These small programs are given in Sect. 4.2.

3.3   Non-Degeneracy Conditions

In general, inferring agent utilities is not a well-defined problem. For instance, Eq. 1 is always satisfied by $\mathbf{v}_i = \mathbf{0}$ and remains invariant under scalar multiplication $\alpha \mathbf{v}_i$. To avoid such trivial solutions, we add an additional non-degeneracy condition on the utilities.

Condition 1 (Non-degeneracy). *A non-degenerate vector* $\mathbf{v} \in \mathbb{R}^d$ *satisfies* $\sum_{k=1}^d v_k = 1$.

3.4   The Inverse Game Theory Problem

We are now ready to formalize two important inverse game theory problems. In the first problem — INVERSE-UTILITY — we observe L games between n players; the structure of every game is known, but can vary. As a motivating example, consider n drivers that play a congestion game each day over a network of roads and on certain days some roads may be closed. Alternatively, consider L scheduling games where different subsets of machines are available on each day. Our goal is to find valuations that rationalize the observed equilibria of all the games at once.

Definition 3 (INVERSE-UTILITY problem). *Given:*

1. *A set of L partially observed succinct n-player games*
   $G_l = [(A_{il})_{i=1}^n, \cdot, (O_{il})_{i=1}^n]$, for $l \in \{1, 2, ..., L\}$.
2. *A set of L correlated equilibria $(\mathbf{p}_l)_{l=1}^L$.*

Determine succinct utilities $(\mathbf{v}_i)_{i=1}^n$ such that $\mathbf{p}_l$ is a valid correlated equilibrium in each G_l, in the sense that Eq. 3 holds for all i and for all $a_j^i, a_k^i \in A_{il}$. Alternatively, report that no such $\mathbf{v}_i$ exist.

Recall that $(O_{il})_{i=1}^n$ have a compact representation (e.g. as small programs, rather than the complete matrix), so that our input is polynomial in the number of players and actions. In the second problem — INVERSE-GAME — the players are again playing in L games, but this time both the utilities and the structure of these games are unknown.

Definition 4 (INVERSE-GAME problem). *Given:*

1. *A set of L partially observed succinct n-player games $G_l = [(A_{il})_{i=1}^n, \cdot, \cdot]$, for $l \in \{1, 2, ..., L\}$.*
2. *A set of L correlated equilibria $(\mathbf{p}_l)_{l=1}^m$.*
3. *Candidate game structures $(\mathcal{S}_l)_{l=1}^L$, one $\mathcal{S}_l$ per game. Each $\mathcal{S}_l = (S_{lh})_{h=1}^p$ contains p candidate structures. A structure $S_{lh} = (O_{lhi})_{i=1}^n$ specifies an outcome matrix O_{lhi} for each player i.*

Determine succinct utilities $(\mathbf{v}_i)_{i=1}^n$ and a structure $S_l^* = (O_{li}^*)_{i=1}^n \in \mathcal{S}_l$ for each game, such that $\mathbf{p}_l$ is a correlated equilibrium in each $[(A_{il})_{i=1}^n, (\mathbf{v}_i)_{i=1}^n, (O_{li}^*)_{i=1}^n]$, in the sense that

$$\mathbf{p}_l^T C_{ijk} O_{il}^* \mathbf{v}_i \geq 0$$

holds for all i, l and for all $a_j^i, a_k^i \in A_{il}$. Alternatively, report that no such $\mathbf{v}_i$ exist.

An example of this problem is when we observe L graphical games among n players and each game has a different and unknown underlying graph chosen among a set of candidates. We wish to infer both the common $\mathbf{v}$ and the graph of each game. Finally, note again that our results also hold for pure Nash equilibria as a special case.

4 Learning Utilities in Succinct Games

In this section, we show how to solve INVERSE-UTILITY in most succinct games. We start by looking at a general linear succinct game, and derive a simple condition under which INVERSE-UTILITY can be solved. Then we consider specific cases of games (e.g. graphical, congestion, network games), and show (1) that they are succinct and linear, and (2) that they satisfy the previous condition.

4.1   General Linear Succinct Games

To solve INVERSE-UTILITY, we need to find valuations $\mathbf{v}_i$ that satisfy the equilibrium condition (3) for every player i and every pair of actions a_j^i, a_k^i. Notice that if we can compute the product $\mathbf{c}_{ijk}^T \triangleq \mathbf{p}^T C_{ijk} O_i$, then Eq. 3 reduces to a simple linear constraint $\mathbf{c}_{ijk}^T \mathbf{v}_i \leq 0$ for $\mathbf{v}_i$. However, the dimensions of C_{ijk} and O_i grow exponentially with n; in order to multiply these objects we must therefore exploit special problem structure. This structure exists in every game for which the following simple condition holds.

Property 1 *Let $A_i(o) = \{\mathbf{a} : (O_i)_{\mathbf{a},o} = 1\}$ be the set of joint-actions that trigger outcome o for player i. The equilibrium summation property holds if*

$$\sum_{\mathbf{a}_{-i}:(a_j^i, \mathbf{a}_{-i}) \in A_i(o)} p(\mathbf{a}_{-i}) \tag{4}$$

can be computed in polynomial time for any outcome o, product distribution $\mathbf{p}$, and action a_j^i.[1]

Informally, Property 1 states that the exact expected utility of players can be computed efficiently. While this is not possible for any game, we establish that this can be done for a large number of linear succinct games in Sect. 4.2.

Lemma 1. *Let G be a linear succinct game and let $\mathbf{p}$ be a PMP correlated equilibrium. Let $\mathbf{c}_{ijk}^T \triangleq \mathbf{p}^T C_{ijk} O_i$ be the constraint on vector $\mathbf{v}_i$ in Eq. 3 for a pair of actions a_k^i, a_j^i. If Property 1 holds, then the components of $\mathbf{c}_{ijkj}^T$ can be computed in polynomial time.*

Proof. For greater clarity, we start with the formulation (1) of constraint (3):

$$\sum_{\mathbf{a}_{-i}} p(a_j^i, \mathbf{a}_{-i}) u(a_j^i, \mathbf{a}_{-i}) \geq \sum_{\mathbf{a}_{-i}} p(a_j^i, \mathbf{a}_{-i}) u(a_k^i, \mathbf{a}_{-i}) \tag{5}$$

We derive from (5) an expression for each component of $\mathbf{c}_{ijk}$.

Recall that we associate the components of $\mathbf{v}_i$ with a set of outcomes $\mathcal{O}_i$. Let $\mathcal{O}_i(\mathbf{a}) = \{o : O_{(\mathbf{a},o)} = 1\}$ denote the set of outcomes that are triggered by $\mathbf{a}$; similarly, let $A(o) = \{\mathbf{a} : (O_i)_{\mathbf{a},o} = 1\}$ be the set of joint-actions that trigger an

---

[1] Property 1 is closely related to the *polynomial expectation property* (PEP) of [5] which states that the expected utility of a player in a succinct game should be efficiently computable for a product distribution. In fact, the arguments we will use to show that this property holds are inspired by arguments for establishing the PEP.

outcome o. The left-hand side of (5) can be rewritten as:

$$\sum_{\mathbf{a}_{-i}} p(a_j^i, \mathbf{a}_{-i}) u_i(a_j^i, \mathbf{a}_{-i}) = \sum_{\mathbf{a}_{-i}} p(a_j^i, \mathbf{a}_{-i}) \sum_{o \in \mathcal{O}_i(a_j^i, \mathbf{a}_{-i})} v_i(o)$$

$$= \sum_{o \in \mathcal{O}_i} \sum_{\substack{\mathbf{a}_{-i}: \\ (a_j^i, \mathbf{a}_{-i}) \in A_i(o)}} p(a_j^i, \mathbf{a}_{-i}) v_i(o)$$

$$= \sum_{o \in \mathcal{O}_i} v_i(o) \sum_{\substack{\mathbf{a}_{-i}: \\ (a_j^i, \mathbf{a}_{-i}) \in A_i(o)}} p(a_j^i, \mathbf{a}_{-i})$$

Similarly, the right-hand side of (5) can be rewritten as

$$\sum_{\mathbf{a}_{-i}} p(a_j^i, \mathbf{a}_{-i}) u_i(a_k^i, \mathbf{a}_{-i}) = \sum_{o \in \mathcal{O}_i} v_i(o) \sum_{\substack{\mathbf{a}_{-i}: \\ (a_k^i, \mathbf{a}_{-i}) \in A_i(o)}} p(a_j^i, \mathbf{a}_{-i}).$$

Substituting these two expressions into (5) and factoring out $p_i(a_j^i)$ (recall that $\mathbf{p}$ is a product distribution) allows us to rewrite (5) as:

$$\sum_{o \in \mathcal{O}_i} v_i(o) \left[\sum_{\substack{\mathbf{a}_{-i}: \\ (a_j^i, \mathbf{a}_{-i}) \in A_i(o)}} p(\mathbf{a}_{-i}) - \sum_{\substack{\mathbf{a}_{-i}: \\ (a_k^i, \mathbf{a}_{-i}) \in A_i(o)}} p(\mathbf{a}_{-i}) \right] \geq 0.$$

Notice that the expression in brackets corresponds to the entries of the vector $\mathbf{c}_{ijk}^T$. If $\mathbf{p}$ is a product distribution, then by Property 1, we can compute these terms in polynomial time. If $\mathbf{p}$ is a correlated equilibrium with a PMP representation $\sum_{k=1}^{K} q_k$, it is easy to see that by linearity of summation we can apply Property 1 K times on each of the terms q_k and sum the results. This establishes the lemma.   □

Lemma 1 suggests solving INVERSE-UTILITY in a game G by means of the following optimization problem.

$$\text{minimize} \quad \sum_{i=1}^{n} f(\mathbf{v}_i) \tag{6}$$

$$\text{subject to} \quad \mathbf{c}_{ijk}^T \mathbf{v}_i \geq 0 \quad \forall i, j, k \tag{7}$$

$$\mathbf{1}^T \mathbf{v}_i = 1 \quad \forall i \tag{8}$$

Constraint (7) ensures that $\mathbf{p}$ is a valid equilibrium; by Lemma 1, we can compute the components of c_{ijk} if Property 1 holds in G. Constaint (8) ensures that the $\mathbf{v}_i$ are non-degenerate. The objective function (6) selects a set of $\mathbf{v}_i$ out of the polytope of all valid utilities. It is possible to incorporate into this program additional prior knowledge on the form of the utilities or on the coupling of valuations across players.

The objective function f may also incorporate prior knowledge, or it can serve as a regularizer. For instance, we may choose $f(\mathbf{v}_i) = ||\mathbf{v}_i||_1$ to encourage sparsity and make the $\mathbf{v}_i$ more interpretable. We may also use f to avoid degenerate $\mathbf{v}_i$'s; for instance, in graphical games, $\mathbf{c}_{ijk}^T \mathbf{1} = \mathbf{0}$ and constant $\mathbf{v}_i$'s are a valid solution. We may avoid this by adding the $\mathbf{v} \geq 0$ constraint (this is w.l.o.g. when $\mathbf{c}_{ijk}^T \mathbf{1} = \mathbf{0}$) and by choosing $f(\mathbf{v}) = \sum_{o \in \mathcal{O}_i} v(o) \log v(o)$ to maximize entropy.

Note that to simply find a valid $\mathbf{v}_i$, we may set $f(\mathbf{v}_i) = 0$ and find a feasible point via linear programming. Moreover, if we observe L games, we simply combine the constraints $\mathbf{c}_{ijk}$ into one program. Formally, this establishes the main lemma of this section:

Lemma 2. *The* INVERSE-GAME *problem can be solved efficiently in any game where Property 1 holds.* □

4.2 Inferring Utilities in Popular Succinct Games

We now turn our attention to specific families of succinct games which represent the majority of succinct games in the literature [5]. We show that these games are linear and satisfy Property 1, so that INVERSE-UTILITY can be solved using the optimization problem (6).

Graphical Games. In graphical games [21], a graph H is defined over the set of players; the utility of a player depends only on their actions and those of their neighbors in the graph.

The outcomes for player i are associated to joint-actions $a_{N(i)}$ by i and its neighbors $N(i)$. A joint-action $\mathbf{a}$ triggers the outcome $a_{N(i)}$ specified by actions of the players in $N(i)$ in $\mathbf{a}$. Formally,

$$(O_i)_{\mathbf{a}, \mathbf{a}_{N(i)}} = \begin{cases} 1 & \text{if } \mathbf{a}, \mathbf{a}_{N(i)} \text{ agree on the actions of } N(i) \\ 0 & \text{otherwise.} \end{cases}$$

It is easy to verify that graphical games possess Property 1. Indeed, for any outcome $o = \mathbf{a}_{N(i)}$ and action a_j^i, and letting $a_{N(i)}^k$ be the action of player k in $\mathbf{a}_{N(i)}$, we have

$$\sum_{\substack{\mathbf{a}_{-i}: \\ (a_j^i, \mathbf{a}_{-i}) \in A_i(o)}} p(\mathbf{a}_{-i}) = \prod_{\substack{k \in N(i) \\ k \neq i}} p_k(a_{N(i)}^k) \prod_{\substack{k \notin N(i) \\ k \neq i}} \sum_{\in A_k} p_k(a^k)$$

$$= \prod_{\substack{k \in N(i) \\ k \neq i}} p_k(a_{N(i)}^k)$$

Polymatrix Games. In a polymatrix game [22], each player plays i in $(n-1)$ simultaneous 2-player games against each of the other players, and utilities are summed across all these games. Formally, each joint-action triggers $n-1$ different outcomes for player i, one for each pair of actions (a^i, a^j) and thus $u_i(\mathbf{a}) = \sum_{j \neq i} v_i(a^i, a^j)$. The associated outcome matrix is

$$(O_i)_{\mathbf{a},(a^i,a^j)} = \begin{cases} 1 & \text{if } a^i_j \text{ and } a^i_i \text{ are played within a} \\ 0 & \text{otherwise.} \end{cases}$$

To establish Property 1, observe that when $o = (a^i, a^j)$ is one of the outcomes affecting the utility of player i, we have

$$\sum_{\substack{\mathbf{a}_{-i}: \\ (a^i_j, \mathbf{a}_{-i}) \in A_i(o)}} p(\mathbf{a}_{-i}) = \sum_{\mathbf{a}_{-i}: a^j \in \mathbf{a}_{-i}} p(\mathbf{a}_{-i}) = p_j(a^j).$$

Hypergraphical Games. Hypergraphical games [5] generalize polymatrix games to the case where the simultaneous games involve potentially more than two players. Each instance of a hypergaphical game is associated with a hypergraph H; the vertices of H correspond to players and a hyperedge e indicates that the players connected by e play together in a subgame; the utility of player i is the sum its utilities in all the subgames in which it participates.

The fact that hypergraphical games are linear and possess Property 1 follows easily from our discussion of polymatrix and graphical games.

Congestion Games. In congestion games [23], players compete for a set of resources E (e.g., roads in a city, modeled by edges in a graph); the players' actions correspond to subsets $a^i \subseteq E$ of the resources. After all actions have been played, each player i incurs a cost that equals the sum $\sum_{e \in a^i} d_e(\ell_e)$ of delays $d_e(\ell_e)$ at each resource e, where $\ell_e(\mathbf{a}) = |\{i : e \in a^i\}|$ denotes the number of players using that resource. In the example involving roads, delays indicate how long it takes to traverse a road based on the congestion.

The outcomes for player i in congestion games are associated with a resource e and the number L of players using that resource; we denote this by $o = (e, L)$. A joint action $\mathbf{a}$ activates the outcomes for the resources in a^i that have $\ell_e(\mathbf{a})$ users. The value $v(o)$ of an outcome $o = (e, L)$ corresponds to the delay experienced on e. Formally, the outcome matrix for a congestion game has the form

$$(O_i)_{\mathbf{a},(e,L)} = \begin{cases} 1 & \text{if } e \in a^i \text{ and } \ell_e(\mathbf{a}) = L \\ 0 & \text{otherwise.} \end{cases}$$

To establish Property 1, we need to show that the expression

$$\sum_{\substack{\mathbf{a}_{-i}: \\ (a^i_j, \mathbf{a}_{-i}) \in A_i(o)}} p(\mathbf{a}_{-i}) = \sum_{\mathbf{a}_{-i}: \ell(\mathbf{a}_{-i}) = L - 1\{e \in a^i_j\}} p(\mathbf{a}_{-i})$$

can be computed for any outcome $o = (e, L)$. Here, $\ell(\mathbf{a}_{-i})$ denotes the number of players other than i using resource e and $1\{e \in a^i_j\}$ equals one if $e \in a^i_j$ and zero otherwise.

The expression $P_L(e) \triangleq \sum_{\mathbf{a}_{-i}: \ell(\mathbf{a}_{-i})=L} p(\mathbf{a}_{-i})$ can be computed via dynamic programming. Indeed, observe that $P_L(e)$ equals $P[\sum_{j \neq i} B_j(p,e) = L]$, where $B_j(p,e)$ is a Bernoulli random variable whose probability of being one corresponds to the probability $P_{j,e} \triangleq \sum_{a^j : e \in a^j} p_j(a^j)$ of player j selecting an action that includes e. The probabilities $P_{j,e}$ are of course easy to compute. From the $P_{j,e}$ it is easy to compute the $P_L(e)$ using dynamic programming via the recursion:

$$P_L(e) = \sum_{j \neq i} P\left[B_j(p,e) = 1 \cap B_k(p,e) = 0 \; \forall k \neq i,j\right] P_{L-1}(e).$$

Facility Location and Network Design Games. In facility location games [24], players choose one of multiple facility locations, each with a certain cost, and the cost of each facility is then divided by all the players who build it. In network design games [25], players choose paths in a graph to connect their terminals, and the cost of each edge is shared among the players that use it.

These two game types are special cases of congestion games with particular delay functions. These can be handled through additional linear constraints. The earlier discussion for congestion games extends easily to this setting to establish Property 1.

Scheduling Games. In a scheduling game [5,26], there are M machines and each player i schedules a job on a machine a^i; the job has a machine-dependent running time $t(m,i)$. The player then incurs a cost $t_i(\mathbf{a}) = \sum_{\{j:a^j=a^i\}} t(a^i, j)$ that equals the sum of the running times of all tasks on its machine.

Player outcomes $o = (m,j)$ are associated with a machine m and the task of a player j. The outcome matrix O_i has the form

$$(O_i)_{\mathbf{a},(m,j)} = \begin{cases} 1 & \text{if } m \in a^i \text{ and } m \in a^j \\ 0 & \text{otherwise.} \end{cases}$$

Property 1 can be established by adapting the dynamic programing argument used for congestion games. Note also that congestion games require adding the constraint $v_i(m,k) = v_j(m,k)$ for all i and j in optimization problem (6). We summarize our results in the following theorem.

Theorem 1. *The* INVERSE-UTILITY *problem can be solved in polynomial time for the classes of succinct games defined above.* □

5   Learning the Structure of Succinct Games

Unlike the INVERSE-UTILITY problem, for which we have sweeping positive results, the INVERSE-GAME problem is generally hard to solve, even for pure-strategy Nash equilibria. We show this under the following non-degeneracy condition on player utilities.

Condition 2 (Non-indifference) *For each player i, there exist $a_j^i, a_k^i, \mathbf{a}_{-i}$ such that $u_i(a_j^i, \mathbf{a}_{-i}) \neq u_i(a_k^i, \mathbf{a}_{-i})$, where $\mathbf{u}_i = O_i \mathbf{v}_i$.*

The interpretation of this condition is that for every action a_j^i, there should be another action a_k^i that gives i a different utility for some $\mathbf{a}_{-i}$. If not, then action j is effectively useless and can be excluded from the model without loss of generality. This in a senses violates the pre-specified model, which selected the action to be there in the first place.

Theorem 2. *Assuming Condition 2, it is NP-Hard to solve* INVERSE-GAME *in the setting of graphical games. However, the corresponding instance of* INVERSE-UTILITY *is easy to solve.*

Proof (Sketch). We reduce from an instance of 3-SAT. There are $n+1$ players in each game j (for $1 \leq j \leq m$) that are indexed by $i = 0, .., n$. Player 0 has only one action: $a^{(0)}$. Every other player $i \geq 1$ has 2 actions: $a_T^{(i)}$ and $a_F^{(i)}$.

Every game j is associated with a clause C_j. Game j has an unknown underlying graph that is chosen in the set of graphs $S_j = \{H_{j1}, H_{j2}, H_{j3}\}$, where H_{jk} is the graph consisting of only a single edge between player 0 and the player associated with the variable that appears as the k-th literal in clause j. In other words, in each game, only one of three possible players is connected to player 0 by an edge.

The utilities v_i of each player $i \geq 1$ are four-dimensional: they specify two values $v_i(a_T^{(i)}), v_i(a_F^{(i)})$ when player i is not connected by an edge to player 0, and two values $v_i(a_T^{(i)}; a^{(0)}), v_i(a_F^{(i)}; a^{(0)})$ when they are.

For every clause C_j, we also define an input equilibrium p_j. Each p_j is a pure strategy Nash equilibrium and decomposes into a product $p_j = \prod_{i=1}^n p_{ji}$. Since player 0 has only one action, p_{j0} is defined trivially. When variable x_i appears in clause C_j, we define the probability of player $i \geq 1$ playing action $a_T^{(i)}$ as

$$p_{ji}\left(a^{(i)} = a_T^{(i)}\right) = \begin{cases} 1 & \text{if } x_i \text{ is positively in clause } C_j \\ 0 & \text{if } x_i \text{ is negated in clause } C_j, \end{cases}$$

and $p_{ji}(a^{(i)} = a_F^{(i)}) = 1 - p_{ji}(a^{(i)} = a_T^{(i)})$.

When variable x_i does not appear in clause C_j, we set the strategy in one such game j (chosen arbitrarily) to be $p_{ji}(a^{(i)} = a_T^{(i)}) = 1$, and in the remaining games we set $p_{ji}(a^{(i)} = a_F^{(i)}) = 1$.

This completes the construction of the game. To complete the proof, it can be shown that finding valid utilities in this game is equivalent to finding a correct assignment; we do so in the full version of the paper. □

References

1. Kalyanaraman, S., Umans, C.: The complexity of rationalizing matchings. In: Hong, S.-H., Nagamochi, H., Fukunaga, T. (eds.) ISAAC 2008. LNCS, vol. 5369, pp. 171–182. Springer, Heidelberg (2008)

2. Kalyanaraman, S., Umans, C.: The complexity of rationalizing network formation. In: 50th Annual IEEE Symposium on Foundations of Computer Science, FOCS 2009, pp. 485–494, October 2009

3. Daskalakis, C., Goldberg, P.W., Papadimitriou, C.H.: The complexity of computing a Nash equilibrium. In: Proceedings of the Thirty-Eighth Annual ACM Symposium on Theory of Computing, STOC 2006, pp. 71–78. ACM, New York (2006)

4. Chen, X., Deng, X., Teng, S.-H.: Settling the complexity of computing two-player Nash equilibria. J. ACM (JACM) 56(3), 14 (2009)

5. Papadimitriou, C.H., Roughgarden, T.: Computing correlated equilibria in multi-player games. J. ACM 55(3), 14:1–14:29 (2008)

6. Foster, D.P., Vohra, R.V.: Calibrated learning and correlated equilibrium. Games Econ. Behav. 21(12), 40–55 (1997)

7. Hart, S., Mas-Colell, A.: A simple adaptive procedure leading to correlated equilibrium. Econometrica 68(5), 1127–1150 (2000)

8. Samuelson, P.A.: Consumption theory in terms of revealed preference. Economica 15(60), 243–253 (1948)

9. Afriat, S.N.: The construction of utility functions from expenditure data. Int. Econ. Rev. 8(1), 67–77 (1967)

10. Varian, H.R.: The nonparametric approach to demand analysis. Econometrica 50(4), 945–973 (1982)

11. Varian, H.R.: Revealed preference. In: Samuelsonian Economics and the Twenty-First Century (2006)

12. Nekipelov, D., Syrgkanis, V., Tardos, E.: Econometrics for learning agents. In: Proceedings of the Sixteenth ACM Conference on Economics and Computation, EC 2015, pp. 1–18. ACM, New York (2015)

13. Bresnahan, T.F., Reiss, P.C.: Empirical models of discrete games. J. Econometrics 48(1), 57–81 (1991)

14. Lise, W.: Estimating a game theoretic model. Comput. Econ. 18(2), 141–157 (2001)

15. Bajari, P., Hong, H., Ryan, S.P.: Identification and estimation of a discrete game of complete information. Econometrica 78(5), 1529–1568 (2010)

16. Ng, A.Y., Russell, S.: Algorithms for inverse reinforcement learning. In: Proceedings of the 17th International Conference on Machine Learning, pp. 663–670. Morgan Kaufmann (2000)

17. Abbeel, P., Ng, A.Y.: Apprenticeship learning via inverse reinforcement learning. In: Proceedings of the Twenty-first International Conference on Machine Learning, ICML 2004, pp. 1–8. ACM, New York (2004)

18. Waugh, K., Ziebart, B.D., Andrew Bagnell, J.: Computational rationalization. The inverse equilibrium problem (2013)

19. Ziebart, B.D., Maas, A., Andrew Bagnell, J., Dey, A.K.: Navigate like a cabbie: probabilistic reasoning from observed context-aware behavior. In: Proceedings of the Ubicomp, pp. 322–331 (2008)

20. Ahuja, R.K., Orlin, J.B.: Inverse optimization. Oper. Res. 49(5), 771–783 (2001)

21. Kearns, M., Littman, M.L., Singh, S.: Graphical models for game theory. In: Proceedings of the Seventeenth Conference on Uncertainty in Artificial Intelligence, UAI 2001, pp. 253–260. Morgan Kaufmann Publishers Inc, San Francisco (2001)

22. Howson Jr., J.T.: Equilibria of polymatrix games. Manage. Sci. 18(5), 312–318 (1972)

23. Rosenthal, R.W.: A class of games possessing pure-strategy Nash equilibria. Int. J. Game Theory 2(1), 65–67 (1973)

24. Chun, B.-G., Chaudhuri, K., Wee, H., Barreno, M., Papadimitriou, C.H., Kubiatowicz, J.: Selfish caching in distributed systems: a game-theoretic analysis. In: Proceedings of the 23rd Annual ACM Symposium on PODC, PODC 2004, pp. 21–30. ACM, New York (2004)

25. Anshelevich, E., Dasgupta, A., Kleinberg, J., Tardos, É., Wexler, T., Roughgarden, T.: The price of stability for network design with fair cost allocation. SIAM J. Comput. **38**(4), 1602–1623 (2008)

26. Fotakis, D.A., Kontogiannis, S.C., Koutsoupias, E., Mavronicolas, M., Spirakis, P.G.: The structure and complexity of nash equilibria for a selfish routing game. In: Widmayer, P., Triguero, F., Morales, R., Hennessy, M., Eidenbenz, S., Conejo, R. (eds.) ICALP 2002. LNCS, vol. 2380, pp. 123–134. Springer, Heidelberg (2002)

Exchange Market Mechanisms without Money

Zeinab Abbassi[1], Nima Haghpanah[2], and Vahab Mirrokni[3]

[1] Columbia University, New York, USA
zeinab@cs.columbia.edu
[2] MIT, Cambridge, USA
nima@csail.mit.edu
[3] Google Research, New York, USA
mirrokni@google.com

Consider a set of agents where each agent has some items to offer, and wishes to receive some items from other agents. A mechanism specifies for each agent a set of items that he gives away, and a set of item that he receives. Each agent would like to receive as many items as possible from the items that he wishes, that is, his utility is equal to the number of items that he receives and wishes. However, he will have a large dis-utility if he gives away more items than what he receives, because he considers such a trade to be unfair. To ensure voluntary participation (also known as individual rationality), we require the mechanism to avoid this. This problem is a generalization of the kidney exchange problem, and is motivated by several barter exchange websites on the Internet.

We show that any individually rational exchange can be viewed as a collection of directed cycles, in which each agent receives an item from the agent before him, and gives an item to the agent after him. In addition to simplifying the statement of the problem, this suggests that we can implement an exchange by separately carrying out one-to-one trades among subsets of agents. In some settings, carrying out cycle-exchanges of large size is undesirable or infeasible. Therefore, we distinguish the restricted problem in which the number of agents in each cycle is bounded above by some given constant $k \geq 2$. The most natural and commonly practiced cycles are of length 2 (i.e., swaps).

For the length-constrained variant of the problem, we rule out the existence of a $1 - o(1)$-approximate truthful mechanism for the length-constrained problem for $k \geq 2$. We show that no truthful deterministic or randomized mechanism can achieve an approximation factor better than $\frac{3k+1}{3k+2}$ or $\frac{3k+1.89}{3k+2}$, respectively. We strengthen the hardness of the problem by proving that even without the truthfulness requirement, the problem is APX-hard for any k. We present a $\frac{1}{8}$-approximately optimal truthful mechanism for the problem with $k = 2$. The mechanism visits pairs of agents in some fixed order, and considers adding a subset of exchanges when visiting a pair. The order is chosen such that an agent can not affect future exchanges involving the agent by misreporting.

The full version of the paper can be downloaded at
http://people.csail.mit.edu/nima/papers/exchanges.pdf.

© Springer-Verlag Berlin Heidelberg 2015
E. Markakis and G. Schäfer (Eds.): WINE 2015, LNCS 9470, pp. 429–430, 2015.
DOI: 10.1007/978-3-662-48995-6

For the unconstrained version, and without the truthfulness constraint, we present a class of polynomial-time algorithms solving the optimal exchange market problem. The algorithms closely resemble algorithms for maximum flow and circulation problems. An algorithm maintains a set of feasible exchanges, and iteratively augments the current solution until the residual graph does not contain any more cycles.

The Storable Good Monopoly Problem
with Indivisible Demand

Gerardo Berbeglia[1], Gautam Rayaprolu[2], and Adrian Vetta[3]

[1] Melbourne Business School, The University of Melbourne, Parkville, Australia
g.berbeglia@mbs.edu
[2] School of Computer Science, McGill University, Montréal, Canada
gautam.rayaprolu@mail.mcgill.ca
[3] Department of Mathematics and Statistics and School of Computer Science,
McGill University, Montréal, Canada
vetta@math.mcgill.ca

Abstract. We study the dynamic pricing problem faced by a monopolist who sells a storable good – a good that can be stored for later consumption. In this framework, the two major pricing mechanisms studied in the theoretic literature are the price-commitment and the threat (no-commitment) mechanisms. We analyse and compare these mechanisms in the setting where the good can be purchased in indivisible atomic quantities and where demand is time-dependent. First, we show that, given linear storage costs, the monopolist can compute an optimal price-commitment strategy in polynomial time. Moreover, under such a strategy, the consumers do not need to store units in order to anticipate price rises. Second we show that, under a threat mechanism rather than a price-commitment mechanism, (i) prices can be lower, (ii) profits can be higher, and (iii) consumer surplus can be higher. This result is surprising, in that these three facts are in complete contrast to the case of a monopolist for divisible storable goods [3]. Third, we quantify exactly how much more profitable a threat mechanism can be with respect to a price-commitment mechanism. Specifically, for a market with N consumers, a threat mechanism can produce a *multiplicative* factor of $\Omega(\log N)$ more profits than a price-commitment mechanism, and this bound is tight. Again, this result is slightly surprising. A special case of this model, is the durable good monopolist model of Bagnoli et al. [1]. For a durable good monopolist, it was recently shown ([2]) that the profits of the price-commitment and the threat mechanisms are always within an *additive constant*. Finally, we consider extensions to the case where inventory storage costs are concave.

This paper is available at http://arxiv.org/abs/1509.07330

References

1. Bagnoli, M., Salant, S., Swierzbinski, J.: Durable-goods monopoly with discrete demand. J. Polit. Econ. **97**, 1459–1478 (1989)
2. Berbeglia, G., Sloan, P., Vetta, A.: Bounds on the profitability of a durable good monopolist. In: Web and Internet Economics, pp. 292–293. Springer (2014)
3. Dudine, P., Hendel, I., Lizzeri, A.: Storable good monopoly: the role of commitment. Am. Econ. Rev. **96**, 1706–1719 (2006)

© Springer-Verlag Berlin Heidelberg 2015
E. Markakis and G. Schäfer (Eds.): WINE 2015, LNCS 9470, p. 431, 2015
DOI: 10.1007/978-3-662-48995-6

Monopoly Pricing in the Presence
of Social Learning
Working Paper

Davide Crapis[1], Bar Ifrach[2], Costis Maglaras[1], and Marco Scarsini[3]

[1] Columbia Business School, Columbia Univeristy, New York, USA
[2] Airbnb, Inc., San Francisco, USA
[3] Dipartimento di Economia e Finanza, LUISS, Rome, Italy

A monopolist offers a product to a market of consumers with heterogeneous quality preferences. Although initially uninformed about the product quality, they learn by observing reviews of other consumers who have previously purchased and experienced the product. Our goal is to analyze the social learning mechanism and its effect on the seller's pricing decision. We postulate a non-Bayesian and fairly intuitive learning mechanism, where consumers assume that all prior decisions were based on the same information, and, under this bounded rationality assumption, consumers pick the *maximum likelihood estimate* (MLE) of the quality level that would best explain the observed sequence of reviews.

First, we characterize the quality estimate resulting from the MLE procedure and show that, under regularity conditions, it converges to the true product quality almost surely. Then, we derive a mean-field asymptotic approximation for the learning dynamics and present the system of differential equations that govern such dynamics. The solution gives a crisp characterization of the dependence of the learning trajectory on the monopolist's price. This approach is flexible and applicable in other settings where the microstructure of the learning process is different.

We then turn to the monopolist's problem of choosing the static price that optimizes her infinite horizon discounted revenues, and characterize the optimal solution, which is unique and lies in the interval of two natural price points: (a) the optimal price assuming that consumers do not learn and always make purchase decisions based on their prior quality estimate; and (b) the optimal price in a setting where consumers knew the true quality all along. Lastly, we give the seller some degree of dynamic pricing capability, namely she can change her price once, at a time of her choosing. We show that, in this case, the monopolist may sacrifice short term revenues in order to influence the social learning process in the desired direction and capitalize on that after changing the price.

A complete version is available at http://ssrn.com/abstract=1957924.

E. Markakis and G. Schäfer (Eds.): WINE 2015, LNCS 9470, p. 432, 2015.
DOI: 10.1007/978-3-662-48995-6

The Stable Matching Linear Program
and an Approximate Rural Hospital Theorem
with Couples

Oliver Hinder

Stanford University, Stanford, CA, USA
ohinder@stanford.edu
http://stanford.edu/~ohinder/stability-and-lp/working-paper.pdf

Abstract. The deferred acceptance algorithm has been the most commonly studied tool for computing stable matchings. An alternate less-studied approach is to use integer programming formulations and linear programming relaxations to compute optimal stable matchings. Papers in this area tend to focus on the simple ordinal preferences of the stable marriage problem. This paper advocates the use of linear programming for computing stable matchings with more general preferences: complements, substitutes and responsiveness, by presenting a series of qualitative and computational results.

First, we show how linear programming relaxations can provide strong qualitative insights by deriving a new approximate rural hospital theorem. The standard rural hospital theorem, which states that every stable outcome matches the same doctors and hospitals, is known to fail in the presence of couples. We show that the total number of doctors and hospitals that change from matched to unmatched, and vice versa, between stable matchings is, at most, twice the number of couples. Next, we move from qualitative to computational insights, by outlining sufficient conditions for when our linear program returns a stable matchings. We show solving the stable matching linear program will yield a stable matching (i) for the doctor-optimal objective (or hospital-optimal), when agent preferences obey substitutes and the law of aggregate demand, and (ii) for any objective, when agent preferences over sets of contracts are responsive. Finally, we demonstrate the computational power of our linear program via synthetic experiments for finding stable matchings in markets with couples. Our linear program more frequently finds stable matchings than a deferred acceptance algorithm that accommodates couples.

Keywords: Matching markets · Linear programming · Optimization · Stable matching

© Springer-Verlag Berlin Heidelberg 2015
E. Markakis and G. Schäfer (Eds.): WINE 2015, LNCS 9470, p. 433, 2015.
DOI: 10.1007/978-3-662-48995-6

Strategic Investment in Protection in Networked Systems

Matt V. Leduc[1,2] and Ruslan Momot[3]

[1] Management Science and Engineering, Stanford University, 475 Via Ortega,
Stanford, CA 94305-4121, USA
[2] IIASA, Schlossplatz 1, 2361 Laxenburg, Austria
mattvleduc@gmail.com
[3] INSEAD, Boulevard de Constance, 77305 Fontainebleau, France
ruslan.momot@insead.edu

We study cascading failures in networks and the incentives that agents have to invest in costly protection. Agents are connected through a network and can fail either intrinsically or as a result of the failure of a subset of their neighbors. Each agent must decide on whether to make a costly investment in protection against cascading failures. This investment can mean vaccination, investing in computer security solutions or airport security equipment, to name a few important examples.

We derive a mean-field equilibrium (MFE), where agents simply consider a mean-field approximation of the cascading process when making their decision of whether to invest in protection. We characterize the equilibrium and derive conditions under which equilibrium strategies are monotone in degree (i.e., in how connected an agent is on the network). We show that different kinds of applications (e.g. vaccination, airport security) lead to very different equilibrium patterns of investments in protection. Indeed, the monotonicity is reversed depending on whether the investment in protection insulates an agent against the failure of his neighbors or just against his own intrinsic failure. The former case defines a game of strategic substitutes in which some agents free-ride on the investment in protection of others, while the latter case defines a game of strategic complements in which agents pool their investments in protection. Risk and welfare implications are discussed.

The mean-field model conveniently allows for comparative statics in terms of the degree distribution and the effect of increasing the level of connectedness on the incentives to invest in protection is discussed. The model also allows us to study global effects (e.g. price feedback, congestion). We can therefore analyze how the presence of both local and global network externalities affects equilibrium behavior. We show that our results are robust to the introduction of such global effects.

Full paper available at http://ssrn.com/abstract=2515968

© Springer-Verlag Berlin Heidelberg 2015
E. Markakis and G. Schäfer (Eds.): WINE 2015, LNCS 9470, p. 434, 2015.
DOI: 10.1007/978-3-662-48995-6

Multilateral Bargaining in Networks:
On the Prevalence of Inefficiencies

Joosung Lee

Business School, University of Edinburgh, 29 Buccluech Place,
Edinburgh EH8 9JS, UK
joosung.lee@ed.ac.uk

Abstract. We introduce a noncooperative multilateral bargaining model
for network-restricted environments. In each period, a randomly selected
proposer makes an offer by choosing 1) a coalition, or bargaining partners, among the neighbors in a given network and 2) monetary transfers
to each member in the coalition. If all the members in the coalition
accept the offer, then the proposer buys out their network connections
and controls the coalition thereafter. Otherwise, the offer dissolves. The
game repeats until the grand-coalition forms, after which the player who
controls the grand-coalition wins the unit surplus. All the players have
a common discount factor.

The main theorem characterizes a condition on network structures
for efficient equilibria. If the underlying network is either *complete* or
circular, an efficient stationary subgame perfect equilibrium exists for
all discount factors: all the players always try to reach an agreement as
soon as practicable and hence no strategic delay occurs. In *any other* network, however, an efficient equilibrium is impossible if a discount factor
is greater than a certain threshold, as some players strategically delay
an agreement. We also provide an example of a *Braess-like paradox*, in
which the more links are available, the less links are actually used. Thus,
network improvements may decrease social welfare.

This paper, at least in two reasons, concentrates on unanimity-game
situations in which only a grand-coalition generates a surplus. First, analyzing unanimity games is enough to show the prevalence of inefficiencies.
If any of proper subcoalitions generates a partial surplus, an efficient
equilibrium is impossible even in complete networks for high discount
factors, as a companion paper[1] shows. Second, in unanimity games we
can investigate the role of network structure on strategic delay controlling network-irrelevant factors.

Keywords: Noncooperative bargaining · Coalition formation · Network
restriction · Buyout · Braess's Paradox

A full version of the paper is available at http://www.research.ed.ac.uk/portal/files/
21746313/NetworkBargaining.pdf
This paper is based on the second chapter of my Ph.D. dissertation submitted to the
Pennsylvania State University. I thank Kalyan Chatterjee, Ed Green, Jim Jordan,
Vijay Krishna, Shih En Lu, Neil Wallace, the three anonymous referees from WINE
2015 for helpful discussions and suggestions.

[1] LEE, J. (2015): Bargaining and Buyout, *working paper*, available at http://www.
research.ed.ac.uk/portal/files/21741650/BargainingBuyout.pdf.

© Springer-Verlag Berlin Heidelberg 2015
E. Markakis and G. Schäfer (Eds.): WINE 2015, LNCS 9470, p. 435, 2015
DOI: 10.1007/978-3-662-48995-6

One-Dimensional Mechanism Design

Hervé Moulin

University of Glasgow, Glasgow, UK
herve.moulin@glasgow.ac.uk

Abstract. In many contexts we cannot design allocation rules that are efficient, fair, and incentive-compatible in the strong sense of strategyproofness. A well known exception is voting over a line of candidates when individual preferences are single-peaked: the median peak defines such a rule. Another instance is the division of a single non disposable commodity (e.g., a workload) when preferences over one's share are single-peaked ([2]). We generalize these two models, and more. We show that the three design goals above are compatible in any problem where individual allocations are one-dimensional, preferences are single-peaked (strictly convex), and the set of feasible allocation profiles is convex.

The general model. The finite set of agents is N. An allocation profile is $x = (x_i)_{i \in N} \in \mathbb{R}^N$; it is feasible only if $x \in X$, a closed and convex subset of $\mathbb{R}^N$. Agent i's preferences over X_i, the i-th projection of X, are single-peaked with peak p_i. A *peak-only* rule f maps a profile of peaks $p \in \Pi_N X_i$ into a feasible allocation $f(p) = x \in X$. We use five axioms.

Efficiency of f means it always selects a Pareto optimal allocation. *Strong-GroupStrategyproofness* (SGSP): when a subset of agents move jointly from reporting their true peaks, either at least one of them is strictly worse off or nobody's welfare changes. We call any permutation $\sigma : N \to N$ leaving X invariant a *symmetry* of X; we call x a *symmetric* allocation of X if x is invariant by all symmetries of X. *Symmetry* (SYM) of f: if σ is a symmetry of X then $f(p^\sigma) = f(p)^\sigma$. *Envy-Freeness* (EF): if permuting i and j is a symmetry of X, then x_i is between p_i and x_j. Fix a feasible allocation $\omega \in X$; ω-*Guarantee* (ω-G): x_i is between p_i and ω_i.

Main Result. *For any symmetric allocation $\omega \in X$, there exists at least one peak-only rule f^ω that is Efficient, SGSP, and meets SYM, EF, and ω-G. This rule is also continuous if X is a polytope or is strictly convex of full dimension.*

The proof is constructive. Our *uniform gains rule* f^ω equalizes benefits from the benchmark allocation ω. For any $p \in \Pi_N X_i$ the rule picks $f^\omega(p) = x$ in $X \cap [\omega, p]$: x is feasible and each x_i is between ω_i and p_i. We choose x so that the profile of individual benefits $|x_i - \omega_i|$, reordered increasingly, is lexicographically maximal in $X \cap [\omega, p]$.

Full length paper: http://www.gla.ac.uk/media/media_409041_en.pdf

© Springer-Verlag Berlin Heidelberg 2015
E. Markakis and G. Schäfer (Eds.): WINE 2015, LNCS 9470, pp. 436–437, 2015.
DOI: 10.1007/978-3-662-48995-6

References

1. Moulin, H.: On strategy-proofness and single-peakedness, Public Choice **35**, 437–455 (1980)
2. Sprumont, Y.: The division problem with single-peaked preferences: a characterization of the uniform allocation rule. Econometrica **59**, 509–519 (1991)

Choosing k from m: Feasible Elimination Procedures Reconsidered

Bezalel Peleg[1] and Hans Peters[2]

[1] The Federmann Center for the Study of Rationality and the Institute of Mathematics, The Hebrew University of Jerusalem, Jerusalem 91904, Israel
pelegba@math.huji.ac.il
[2] Department of Quantitative Economics, Maastricht University, PO Box 616, 6200 MD Maastricht, The Netherlands
h.peters@maastrichtuniversity.nl

This paper considers the classical social choice model with finitely many voters, who have strict preferences over a finite set of $m \geq 2$ alternatives. Fix $1 \leq k \leq m-1$. A *social choice correspondence* assigns to each profile of preferences a set of *committees*, where a committee is an ordered set of k alternatives. The aim of the paper is to find a reasonable (anonymous, (Maskin) monotonic) method which is nonmanipulable in the sense that for a given preference profile no coalition of voters, by not voting truthfully, can ensure a committee preferred by all coalition members to a sincere committee, i.e., a committee resulting from truthful voting.

The main results of the paper imply, that for two natural extensions of preferences from alternatives to committees, this aim is achieved by so-called *feasible elimination procedures*, first introduced in [2]. A feasible elimination procedure depends on (positive integer) weights attached to the alternatives (e.g., by a planner). Given a preference profile one first eliminates an alternative that is at bottom at least as often as its weight, together with as many preferences. By repeating this procedure a ranking of the alternatives is established, and the k last surviving alternatives are chosen.

It is also shown that well-known methods like scoring rules or single-transferable vote do not have this property.

An additional result of the paper is that establishing whether a given committee of size k can result from applying a feasible elimination procedure is equivalent to finding a maximal matching in a specific bipartite graph, which can be done in polynomial time, see [1]. For more background see [3,4]. For the complete paper see [5].

References

1. Hopcroft, J.E., Karp, R.M.: An $n^{5/2}$ algorithm for maximum matchings in bipartite graphs. SIAM J. Comput. **2**, 225–231 (1973)
2. Peleg, B.: Consistent voting systems. Econometrica **46**, 153–161 (1978)
3. Peleg, B.: Game theoretic analysis of voting in committees. Cambridge University Press, Cambridge, Cambridge (1984)

© Springer-Verlag Berlin Heidelberg 2015
E. Markakis and G. Schäfer (Eds.): WINE 2015, LNCS 9470, pp. 438–439, 2015.
DOI: 10.1007/978-3-662-48995-6

4. Peleg, B., Peters, H.: Strategic Social Choice. Springer, Berlin (2010)
5. Peleg, B., Peters, H.: Choosing k from m: feasible elimination procedures reconsidered. See: http://researchers-sbe.unimaas.nl/hanspeters/wp-content/uploads/sites/21/2015/09/fep-km_revised.pdf (2015)

Author Index

Printed in the United States
by Baker & Taylor Publisher Services